Finite Elements for Solids,
Fluids, and Optimization

Finite Elements for Solids, Fluids, and Optimization

G. A. MOHR

*Applied Computational Mechanics
in Engineering, St Kilda,
Victoria*

OXFORD ● NEW YORK ● MELBOURNE
OXFORD UNIVERSITY PRESS
1992

Oxford University Press, Walton Street, Oxford OX2 6DP

Oxford   New York   Toronto
Delhi   Bombay   Calcutta   Madras   Karachi
Kuala Lumpur   Singapore   Hong Kong   Tokyo
Nairobi   Dar es Salaam   Cape Town
Melbourne   Auckland   Madrid

and associated companies in
Berlin   Ibadan

Oxford is a trade mark of Oxford University Press

Published in the United States
by Oxford University Press Inc. New York

A catalogue record for this book is available from the British Library

Library of Congress Cataloging in Publication Data

Mohr. G. A.
Finite elements for solids, fluids, and optimization / G. A. Mohr.
Includes bibliographical references and indexes.
1. Finite element method.   2. Mechanics. Applied—Mathematics.
3. Fluid mechanics—Mathematics.   4. Mathematical optimization.
I. Title.
TA347.F5M64 1992   620.1'.01'.51535—dc20   92-9625
ISBN 0-19-856369-8 (hbk)
ISBN 0-19-856368-X (pbk)

Set by Macmillan India Ltd, Bangalore 25
Printed and bound in Great Britain by
Bookcraft (Bath) Ltd
Midsomer Norton, Avon

Preface

The origins of the *finite element method* (FEM) should be well enough known by now, namely through the pioneering work of Aryris *et al.* at Stuttgart, Clough, Martin *et al.* at Berkeley, and Zienkiewicz, Irons *et al.* at Swansea. Many references to these pioneers and their work will be found in this book, particularly to Argyris' *natural strain* concept and Irons' *isoparametric elements*. Such work laid the foundations for what must surely be the crux of the method, that is, the formulation of the elements themselves.

It was Clough, Martin, Zienkiewicz, and others, however, who did more in the way of application of the method, namely the exploration of applications outside the initial one of plane stress problems, thereby showing that FEM is applicable to any physical problem for which we can first form a governing differential equation or energy principle. The work of these and other researchers brings us to about the mid-1970s, that is, the culmination of some twenty years of research which saw up to 2,000 papers on FEM published in a year. This is, of course, not the end of the story to which this book seeks to contribute and, having done that, summarize.

In the last decade or so there have been many new developments, for example the use of penalty factor and Lagrange multiplier techniques in element formulation, and applications to new problems such as thick plates and shells, compressible and transonic fluid flows, and to the optimization of finite element models. The present text deals with these new developments and contributes others, for example *basis transformation*, which allows existing element formulations to be transformed into new ones, transformation of Lagrange multiplier formulations into equivalent penalty formulations, and application of a perturbation technique to transform functionals into differential equations (or vice versa) or to attack the governing functional or differential equation directly.

Considerable effort is also made to introduce concisely the basic theories of solid and fluid mechanics, numerical methods, and optimization techniques. The text does not include chapter-end problems, preferring to provide fully worked examples as they are needed and to introduce the reader immediately to practical applications of the method using several compact MEGABASIC programs designed specifically for teaching purposes. These can be used on almost any PC available today but for the plate and shell element programs, a fairly fast machine is desirable for larger problems.

MEGABASIC and the FEMSS system used for the larger programs in the text are described in Appendix 1. These are short enough to be typed in directly, itself a useful learning exercise, and are easily translated into BASIC or FORTRAN if required. If not available locally, however, MEGABASIC can be obtained from American Planning Corporation, PO Box 9635, Alexandria, VA22304, USA, at very moderate cost.

The text is full of material not to be found in any other book and therefore represents an independent and fully self-sufficient view of the method under one cover. As the reader may imagine, such a treatise was not born overnight, the first edition (1980) being restricted to structural mechanics applications, the second (1985) incorporating fluid mechanics applications, the third (1989) introducing a range of PC programs, and the present edition extending these to optimization applications.

Usage of the text. Very little can be sensibly learnt about the subject in one dose, however large, and my own practice, in the context of engineering courses, is to introduce material as follows:

Year 1: truss and beam elements (Chapter 1);
Year 2: mechanics of solids, energy methods, interpolation and numerical integration, and a simple plane stress element (Chapters 2–6);
Year 3: isoparametric elements, plate and shell elements, and FEM programming (Chapters 7–10 and 15);
Year 4: differential equations, weighted residual methods and virtual work (perturbation), heat flow and potential flow (Chapters 16–18 and 20);
Elective/postgraduate: more advanced elements, non-linear and time-dependent problems, doubly curved shells, viscous flow, BEM, and optimization (Chapters 11–14 and 20–23).

An alternative approach can be found from the title, namely to break the subject into four modules (elements, solids, fluids, and optimization) which can be taught towards the end of engineering, mathematics, or physics courses or at postgraduate level. In the latter instance a brief introductory course in engineering statics might be required. With either approach, however, a knowledge of mathematics to at least first year university level is desirable though it is my experience that students who are well matriculated in mathematics, that is, well versed in basic algebra, calculus, computing, Newtonian mechanics, and trigonometry, cope well with the material of the present text, perhaps because, as far as possible, it is developed from first principles. In teaching at undergraduate level, however, lecturers should be at pains to direct attention to only the key sections of the more comprehensive chapters. In the case of Chapter 19, for example, I generally concentrate on Sections 19.1, 19.2, 19.5, and 19.8 following these up with the problems and programs of Chapter 20.

Finally, grateful acknowledgement should be given to Prof. Peter Lowe, under whose auspices as Ph.D. supervisor an interest in optimization was inculcated at Cambridge, Prof. Bob Milner who read much of the second edition (also contributing eqns (1.14) and (3.73) and most of Sections 12.4, 12.8, and 13.9), Prof. Ernest Hinton who did much to encourage the present edition and its publication, and Richard Lawrence and the staff of Oxford University Press for a most expeditious job of finally publishing the text.

Indeed I hope the text will be well received by students, researchers, and teachers for a long time to come. Myself, I see the finite element method as a natural computer calculus applicable to an almost unlimited range of problems of mathematical modelling and optimization, and feel that those who share this same faith are, given a little perseverance, bound to find many useful applications for the elements, techniques, and ideas of the present text for many years to come.

St Kilda, Victoria   G.A.M.
February 1992

Contents

Part IV   Finite elements in fluid flow and other field problems

Part I
Introduction to finite elements

1
Skeletal elements

After some preliminary discussion of the finite element method, discrete one-dimensional finite elements of skeletal engineering systems are considered as an introduction to finite element analysis. The matrices for these elements can be derived without recourse to finite element theory, but once these are derived their application to build a complete mathematical model of a system follows the same general lines as with all other finite elements.

1.1 What are finite elements?

In essence the finite element method is a numerical procedure for solving the ordinary and partial differential equations that arise in engineering and mathematical physics. The method is applicable to problems in such diverse fields as frame analysis, stress analysis of solid continua, fluid dynamics, heat flow, and electrical field theory, to name a few. Its development has now reached the stage where it is the most commonly used method of numerical analysis.

A precursor to the finite element method was the development of *matrix structural analysis*, that is, techniques for the analysis of skeletal structures using matrix methods.[1,2] By intuitively lumping physical properties such techniques were also able to be used to analyse skeletal models of plates and shells.[3-7] The matrices for the elements of such models can be derived by direct formation of equilibrium equations for a typical element, for example using the slope–deflection equations to form equations for the flexural action of a beam element.

The first two-dimensional *continuum* elements, however, were developed using energy arguments.[8,9] Although some mathematical foundations already existed for such developments,[10,11] many of the subsequent developments of the finite element method were inspired by intuition,[12] undoubtedly because of the strong physical images which finite elements encourage.

By the early 1970s package programs for finite element analysis, such as SAP IV, became widely available and routine analysis of stresses and displacements in complex plate and shell structures or solid objects was practicable. In recent years the method has been extended to the analysis of non-structural problems and commercial software is now available for solving these.

The following chapters expound the finite element method in an order not greatly dissimilar to the order of its historical development. Before commencing, however, it is worth summarizing some basic concepts of continuum behaviour and of finite elements which prove useful in the following chapters.

1.1.1 Classification of physical problems

The main types of physical problems which we analyse using finite elements may be classified as follows:

(1) Equilibrium problems. These involve systems which are in a steady state, values of stresses, displacements, electrical or magnetic potential, velocities of flow, pressures, temperatures, and so on, remaining constant with time. Such problems are governed by a matrix equation of the form

$$K\{D\} = \{Q\}.$$

Examples of such problems occur in Chapters 1, 6–11, and 18–20.

(2) Diffusion problems. These involve transient systems in which fluxes of matter or energy occur and these change according to a *velocity law*, that is, the time-dependent behaviour involves first derivatives of strain, temperature, potential, and so on, with respect to time. Such problems are governed by a matrix equation of the form

$$C\{\dot{D}\} + K\{D\} = \{Q\}.$$

Examples of such problems occur in Chapters 14, 18, and 19.

(3) Inertial problems. These involve transient systems in which distributions of displacements, potential, and so on, change according to an *acceleration law*, that is, the time-dependent behaviour involves second derivatives with respect to time. Such problems are governed by a matrix equation of the form

$$M\{\ddot{D}\} + C\{\dot{D}\} + K\{D\} = \{Q(t)\}.$$

Examples of such problems occur in Chapters 14 and 19, though in the latter instance the inertia terms are expressed in the form of $u\partial u/\partial x$ where $u = \partial x/\partial t$, which is equivalent to the form $\partial^2 u/\partial x^2$.

(4) Eigenproblems. These are a special type of equilibrium problem in which various equilibrium states are possible, each being characterized by an

eigenvalue and an associated eigenvector. These include problems of incipient instability (metastable equilibrium) or harmonic excitation (dynamic equilibrium). Such problems are governed by a matrix equation of the form

$$(K - \lambda M)\{D\} = \{0\}$$

Examples of such problems occur in Chapters 13 and 19.

Types (2) and (3) may both be described as *propagation problems*[13] but are classified separately here as they involve different forms of finite element equation. Propagation problems are sometimes referred to as *initial value problems* as they generally have all their velocity or acceleration boundary conditions specified at one point in time. Types (1) and (4) may both be described as equilibrium problems and are also sometimes referred to as *boundary value problems* as they generally have their boundary conditions at the extremities or boundary of the domain in which the solution to some physical problems is sought.

1.1.2 Conservation and other requirements in physical systems

Physical systems must satisfy requirements of *conservation*; for example of kinetic, thermal, and potential energy, or of matter; *continuity* of certain problem variables; and *constitutive* relationships between some of these variables. The equations governing mathematical models of physical systems must therefore be based upon these requirements.

1.1.3 Matrix equations for finite element models

A feature of finite element analysis is that, no matter what the physical problem, it can always be expressed as a system of matrix equations. Considering for example a vibration problem, these equations take the form

$$M\{\ddot{D}\} + C\{\dot{D}\} + K\{D\} = \{Q(t)\} \tag{1.1}$$

where M, C, and K are respectively the *mass*, *damping*, and *stiffness* matrices for the complete system and $\{\ddot{D}\}$, $\{\dot{D}\}$, and $\{D\}$ are respectively the acceleration, velocity, and displacement vectors for the system, comprising values at a number of points or *nodes* in the system. Finally $\{Q(t)\}$ is a vector of time-dependent loads applied at these nodes.

In the finite element method we initially develop equations of this form for a small subdomain or finite element and complete details of such elements for a wide variety of problems are given in the following text. These finite elements, which are simple in shape, about one another forming a patchwork model of the complete domain, as illustrated in Fig. 1.1.

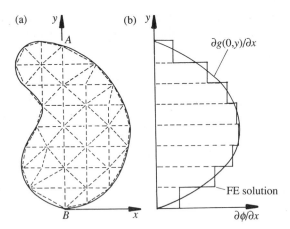

Fig. 1.1 Finite element discretization

A parallel mathematical process involves summing the element matrices m, c, k, and $\{q(t)\}$ in a manner corresponding to the node numbers used by each element to form the matrices M, C, K, and $\{Q(t)\}$ for the complete domain, that is, we perform the operations indicated symbolically by

$$M = \sum m, \qquad C = \sum c, \qquad K = \sum k, \qquad \{Q(t)\} = \sum \{q(t)\}$$

which involves the use of very simple computer coding. Once the matrices for the complete domain have been formed the unknown *nodal freedoms*, in the foregoing example these being the displacements $\{D\}$, are found by standard matrix solution techniques.

The question of how many elements of a given type should be used to obtain reasonable accuracy requires consideration of such factors as truncation and interpolation errors and these are not discussed until later in the text. Simple curve sketching, such as that illustrated in Fig. 1.1(b), however, can be of some help in this regard. Suppose that in Fig. 1.1(a) the distribution $g(x, y)$ of some variable ϕ is to be determined and that the values of $\partial \phi / \partial x$ and $\partial \phi / \partial y$ are also of interest. Then if the general shape of the curve for $\partial \phi / \partial x$ at section AB is as shown in Fig. 1.1(b) we should expect elements which model a constant value of the derivatives of ϕ (and thus approximate $g(x, y)$ linearly in each element) to yield a sufficiently accurate result of the stepped form indicated in Fig. 1.1(b). Hence the number of elements shown in Fig. 1.1(a) would be expected to be adequate whilst if elements of greater accuracy are employed in the analysis, fewer would be needed.

The matrices for the individual finite elements can be derived by direct operation on the governing differential equations or by variational (energy) methods. The former procedure is illustrated for the case of a beam element in the following section and the latter is introduced in Chapter 3.

1.2 Element equations by direct operation on differential equations

In the following section we introduce a direct integration procedure of forming finite element equations. This approach is limited to problems involving ordinary differential equations and it therefore provides a useful introduction to one-dimensional finite elements. Moreover, for the case of the beam element which we shall study, the direct integration procedure clarifies the important concepts of *interelement reactions* and *consistent loads* in the finite element method.

The differential equation governing the shape of the neutral axis of a bent beam (the *elastica*) is

$$\frac{EId^4v}{dx^4} = p(x) \tag{1.2}$$

where v is the transverse deflection of the beam and $p(x)$ is the intensity of distributed loading upon the beam. Here equilibrium considerations require that $p(x) = dV/dx$, where V is the shear force in the beam, and $V = dM_b/dx$, where M_b is the bending couple in the beam. Here we use the subscript b to distinguish this from a moment at a section in the beam and assume that positive M_b causes sagging curvature of the beam.

Consider now, for example, the beam element shown in Fig. 1.2, in which the displacements v and slopes $\phi = dv/dx$ at each node are chosen as *freedoms* which are to become unknowns in the element equations. Assuming that the internal loading is uniformly distributed, that is, $p(x) = p$, a constant, eqn (1.2) can be integrated four times with respect to the coordinate x,

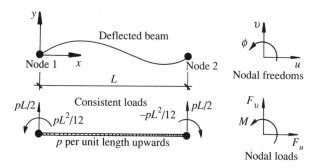

Fig. 1.2 Consistent loads in a beam element

yielding the four equations

$$\frac{EId^3y}{dx^3} = px + c_1 \tag{1.3a}$$

$$\frac{EId^2y}{dx^2} = \frac{px^2}{2} + c_1 x + c_2 \tag{1.3b}$$

$$\frac{EIdv}{dx} = \frac{px^3}{6} + \frac{c_1 x^2}{2} + c_2 x + c_3 \tag{1.3c}$$

$$EIv = \frac{px^4}{24} + \frac{c_1 x^3}{6} + \frac{c_2 x^2}{2} + c_3 x + c_4. \tag{1.3d}$$

Putting $x = 0$ in eqns (1.3c) and (1.3d) we see that $c_3 = EI\phi_1$ and $c_4 = EIv_1$. Also putting $x = L$ we then obtain

$$EI\phi_2 = \frac{pL^3}{6} + \frac{c_1 L^2}{2} + c_2 L + EI\phi_1 \tag{1.4}$$

$$EIv_2 = \frac{pL^4}{24} + \frac{c_1 L^3}{6} + \frac{c_2 L^2}{2} + EIL\phi_1 + EIv_2. \tag{1.5}$$

Solving these for c_1 and c_2 the values of the shear force and moment at the first node are obtained as

$$F_{v1} = \left(\frac{EId^3v}{dx^3}\right)_1 = c_1 = \frac{6EI(\phi_1 + \phi_2)}{L^2} + \frac{12EI(v_1 - v_2)}{L^3} - \frac{pL}{2} \tag{1.6}$$

$$M_1 = -\left(\frac{EId^2v}{dx^2}\right)_1 = -c_2 = \frac{EI(4\phi_1 + 2\phi_2)}{L} + \frac{6EI(v_1 - v_2)}{L^2} - \frac{pL^2}{12}. \tag{1.7}$$

Now using eqns (1.3a) and (1.3b) with $x = L$ one obtains the values of the shear force and moment at the second node as

$$F_{v2} = -\left(\frac{EId^3v}{dx^3}\right)_2 = -pL - F_{v1}$$

$$= -\frac{6EI(\phi_1 + \phi_2)}{L^2} - \frac{12EI(v_1 - v_2)}{L^2} - \frac{pL}{2} \tag{1.8}$$

$$M_2 = \left(\frac{EId^2v}{dx^2}\right)_2 = \frac{pL^2}{2} + c_1 L + c_2$$

$$= \frac{EI(2\phi_1 + 4\phi_2)}{L} + \frac{6EI(v_1 - v_2)}{L^2} + \frac{pL^2}{12}. \tag{1.9}$$

Note that reversed signs have been used to obtain eqns (1.8) and (1.9) because the end forces are now calculated from the beam shear force and moment at the right-hand side of an infinitesimal element adjacent to node 2. Note also in passing that eqns (1.7) and (1.9) are in fact the *slope–deflection* equations for a beam.

Transposing the terms involving p to the left-hand sides of eqns (1.6)–(1.9) and writing the results in matrix form we obtain

$$
\begin{Bmatrix} F_{v1} + pL/2 \\ M_1 + pL^2/12 \\ F_{v2} + pL/2 \\ M_2 - pL^2/12 \end{Bmatrix} = \frac{EI}{L^3} \begin{bmatrix} 12 & 6L & -12 & 6L \\ 6L & 4L^2 & -6L & 2L^2 \\ -12 & -6L & 12 & -6L \\ 6L & 2L^2 & -6L & 4L^2 \end{bmatrix} \begin{Bmatrix} v_1 \\ \phi_1 \\ v_2 \\ \phi_2 \end{Bmatrix}. \tag{1.10}
$$

It is also possible to form two equations for the *extensional* action of the beam by integrating the differential equation $d^2u/dx^2 = 0$ twice, u being the displacement parallel to the axis of the beam. Alternatively, in this simple instance they can be obtained by writing the tensile force in the beam as

$$
T = \sigma A = EA\varepsilon = \frac{EA\delta L}{L} = \frac{EA(u_2 - u_1)}{L}. \tag{1.11}
$$

Then to satisfy equilibrium we require the parallel forces at each end of the beam to be $F_{u1} = T$ and $F_{u2} = -T$. Combining these results with eqn (1.10) one obtains the *element stiffness matrix* for a six-freedom beam/column element as

$$
\begin{bmatrix} F_{u1} \\ F_{v1} \\ M_1 \\ F_{u2} \\ F_{v2} \\ M_2 \end{bmatrix} + \begin{bmatrix} 0 \\ pL/2 \\ pL^2/12 \\ 0 \\ pL/2 \\ -pL^2/12 \end{bmatrix} = \begin{bmatrix} k_1 & 0 & 0 & -k_1 & 0 & 0 \\ 0 & k_2 & k_3 & 0 & -k_2 & k_3 \\ 0 & k_3 & k_4 & 0 & -k_3 & k_5 \\ -k_1 & 0 & 0 & k_1 & 0 & 0 \\ 0 & -k_2 & -k_3 & 0 & k_2 & -k_3 \\ 0 & k_3 & k_5 & 0 & -k_3 & k_4 \end{bmatrix} \begin{bmatrix} u_1 \\ v_1 \\ \phi_1 \\ u_2 \\ v_2 \\ \phi_2 \end{bmatrix}
$$

or

$$
\{q_r\} + \{q_t\} = k\{d\} \tag{1.12}
$$

where $k_1 = EA/L$, $k_2 = 12EI/L^3$, $k_3 = 6EI/L^2$, $k_4 = 4EI/L$, and $k_5 = 2EI/L$. Note that the stiffness matrix is symmetric (and this property ensures reciprocity), terms on the leading diagonal are positive as required by Newton's third law, and the stiffnesses associated with the translations sum to zero, ensuring that rigid body motions do not cause straining of the element.

The vector $\{q_t\}$ in eqn (1.12) contains the *consistent loads* corresponding to the uniformly distributed transverse load of intensity p. (If this is due to the weight of the beam, that is, $p = \rho A$, then we term ρ a *body force* and denote the consistent loads as $\{q_b\}$ but in this case they are obtained by the same formula.) These correspond to the particular integral of the governing differential equation and a general formula for consistent loads in finite elements is derived by an energy argument in Chapter 3. Note too that, the solution for the displacements having been completed, the consistent loads are reversed in sign to calculate the *stress resultants* in the element, as eqns (1.6)–(1.9) show.

We shall illustrate the procedure of assembling the element equations and solving the resulting system equations to determine the displacements in the complete structure in the following section. In the context of the assembly of the element equations, however, it should be noted that the vector $\{q_r\}$ in eqn (1.12) contains the *interelement reactions*. In the element equation assembly process these forces and moments must cancel between elements, that is, in the summation $\{Q\} = \sum \{q_r\} + \sum \{q_l\}$ we have $\sum \{q_r\} = \{0\}$. Thus we have only to sum the consistent loads for the elements, but note that any concentrated loads applied at the nodes must also be added to the structure load vector $\{Q\}$. This concept of cancelling of interelement reactions. is quite general in the finite element method and in fluid flow problems; for example, we require a cancelling of fluid flux terms between elements in order to ensure conservation of mass.

The foregoing direct integration procedure has much in common with the *boundary element method* where again the element equations are formed by integrating the governing differential equations for a beam four times. Indeed such skeletal elements essentially have a boundary element character, the nodes at each end corresponding to boundary elements of the domain in which the differential equation applies.

It is also worth noting that eqn (1.12) has an exact character. This is because eqn (1.3d) is of fourth order and fortuitously this is sufficient to yield the exact solutions for the nodal displacements and forces in a beam carrying a uniformly distributed load. As we shall see in Chapter 3, however, if we follow the usual finite element procedures for a beam element then we model the flexural action by associating a cubic polynomial for the transverse displacement v with the four nodal freedoms. This will not model the shape of a uniformly loaded element between the nodes exactly and the approximate nature of the finite element method is then apparent.

In the present case, however, it is possible to obtain an exact solution for the deflected shape by using the governing differential equation and the boundary conditions $v = v_1$, $dv/dx = \phi_1$ at $x = 0$ and $v = v_2$, and $dv/dx = \phi_2$ at $x = L$ to determine the coefficients of a fourth-order polynomial. The differential equation is $EI d^4 v/dx^4 = p$ and the deflected shape is written as

$$v = c_1 + c_2 x + c_3 x^2 + c_4 x^3 + c_5 x^4. \tag{1.13}$$

Differentiating eqn (1.13) four times we obtain

$$\frac{dv}{dx} = c_2 + 2c_3 x + 3c_4 x^2 + 4c_5 x^3 \tag{1.14a}$$

$$\frac{d^2 v}{dx^2} = 2c_3 + 6c_4 x + 12c_5 x^2 \tag{1.14b}$$

$$\frac{d^3v}{dx^3} = 6c_4 + 24c_5 x^2 \tag{1.14c}$$

$$\frac{d^4v}{dx^4} = 24c_5. \tag{1.14d}$$

Then assuming $EI = 1$ we clearly have $c_5 = p/24$. The constants $c_1, \ldots, c_4$ can be replaced by boundary displacements and rotations, that is, $v = v_1$, $dv/dx = \phi_1$ at $x = 0$ and $v = v_2$, and $dv/dx = \phi_2$ at $x = L$. This leads to

$$c_1 = v_1, \qquad c_3 = \frac{1}{L^2}\left(\frac{-3v_1 - 2\phi_1 L + 3v_2 - \phi_2 L + pL^4}{24}\right)$$

$$c_2 = \phi_1, \qquad c_4 = \frac{1}{L^3}\left(\frac{2v_1 + \phi_1 L - 2v_2 + \phi_2 L - pL^4}{12}\right). \tag{1.14e}$$

Substituting these solutions for the constants into eqns (1.14b) and (1.14c) and writing $F_{v1} = EI(d^3v/dx^3)_1$, $M_1 = -EI(d^2v/dx^2)_1$, $F_{v2} = -EI(d^3v/dx^3)_2$, and $M_2 = EI(d^2v/dx^2)_2$ leads to the expected solution, namely eqn (1.10).

Thus eqn (1.10) does exactly satisfy the boundary conditions for a uniformly loaded beam element and by combining eqns (1.13) and (1.14e) we can also obtain an exact solution for the deflected shape. This is because eqn (1.13) was of the same form as the exact solution to the differential equation. It should be noted, however, that this is an exceptional case. Had the governing differential equation been $d^4v/dx^4 + k^2 d^2v/dx^2 - p = 0$, for example, the exact solution is of the form $v = c_1 \sin kx + c_2 \cos kx + c_3 x + c_4 + px^2/2k^2$ and the usual polynomial bases used in the finite element method can only model such problems approximately.

It is worth noting, however, the fundamental role that polynomials play in numerical methods and in classical methods. An example of this is seen by writing an infinite polynomial for some distribution $g(x)$:

$$g(x) = c_1 + c_2 x + c_3 x^2 + c_4 x^3 + \ldots .$$

Differentiating $g(x)$ several times we can then write

$$g'(x) = c_2 + 2c_3 x + 3c_4 x^2 + \ldots, \qquad \text{that is, } c_2 = g'(0)$$

$$g''(x) = 2c_3 + 6c_4 x + \ldots, \qquad \text{that is, } c_3 = g''(0)$$

$$g'''(x) = 6c_4 + \ldots, \qquad \text{that is, } c_4 = g'''(0)$$

and so on. Thus, the coefficients of the polynomial can be expressed in terms of the various derivatives at the origin. Then substituting these results into the original polynomial for $g(x)$ we obtain by induction

$$g(x) = g(0) + xg'(0) + \frac{x^2 g''(0)}{2} + \frac{x^3 g'''(0)}{6} + \cdots + \frac{x^n g^n(0)}{n!}. \tag{1.15}$$

This is Maclaurin's formula, from which the infinite series for the transcendental functions sin x, cos x, e^x, and so on, follow and Taylor's theorem also follows from this by simply shifting the origin to $x = h$.

If we were able to choose a polynomial long enough then we can capture the solution to any problem to any desired accuracy. Then the objective of the finite element method is to choose elements small enough to limit the length of polynomial required to approximate the solution, that is, if x is kept 'small' then trailing terms in the 'long polynomial' solution become negligible. Then, as we shall see in the following section, assembly of the equations for 'standard' elements is a simple matter and provides us with a powerful automated method of solving problems of mathematical physics.

1.3 Assembly of the element equations

Figure 1.3 shows a beam continuous over three spans divided into three elements. Considering only the rotations at the supports as unknowns each element has only two freedoms. The effect of the distributed load is given by the reversed fixed end moments $\pm pL^2/12$ in each span, as shown. The stiffness matrix for each element is obtained from eqn (1.12), deleting all but the ϕ–ϕ terms,

$$\{q_1\} = \begin{Bmatrix} -24 \\ 24 \end{Bmatrix} = \begin{bmatrix} 4\beta & 2\beta \\ 2\beta & 4\beta \end{bmatrix} \begin{Bmatrix} \phi_1 \\ \phi_2 \end{Bmatrix} = k_1\{d_1\} \tag{1.16a}$$

$$\{q_2\} = \begin{Bmatrix} -16 \\ 16 \end{Bmatrix} = \begin{bmatrix} 4\beta & 2\beta \\ 2\beta & 4\beta \end{bmatrix} \begin{Bmatrix} \phi_2 \\ \phi_3 \end{Bmatrix} = k_2\{d_2\} \tag{1.16b}$$

$$\{q_3\} = \begin{Bmatrix} -16 \\ 16 \end{Bmatrix} = \begin{bmatrix} 6\beta & 3\beta \\ 3\beta & 6\beta \end{bmatrix} \begin{Bmatrix} \phi_3 \\ \phi_4 \end{Bmatrix} = k_3\{d_3\} \tag{1.16c}$$

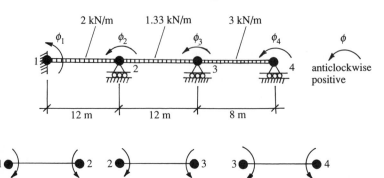

Fig. 1.3 Continuous beam with uniformly loaded spans showing the equivalent nodal loadings

where $\beta = EI/L$ for spans 12 and 23 and coefficients 6 and 3 appear to deal with the stiffer third span. Summing the element load vectors and stiffness matrices the structure load vector and stiffness matrix are obtained, taking care to deploy each element contribution correctly:

$$\{Q\} = \begin{bmatrix} M_1 \\ M_2 \\ M_3 \\ M_4 \end{bmatrix} = \begin{bmatrix} -24 \\ 8 \\ 0 \\ 16 \end{bmatrix} = \beta \begin{bmatrix} 4 & 2 & 0 & 0 \\ 2 & 8 & 2 & 0 \\ 0 & 2 & 10 & 3 \\ 0 & 0 & 3 & 6 \end{bmatrix} \begin{bmatrix} \phi_1 \\ \phi_2 \\ \phi_3 \\ \phi_4 \end{bmatrix} = K\{D\}.$$

Before solving the matrix equation for the structure any boundary conditions must be set (though this type of problem is one in which there may be none). Here the boundary condition is $\phi_1 = 0$ so that the first column of K vanishes and the first row can be omitted as ϕ_1 is known so that this equation is not needed. Hence the reduced load vector and stiffness matrix are obtained as

$$\{Q_R\} = \begin{Bmatrix} 8 \\ 0 \\ 16 \end{Bmatrix} = \beta \begin{bmatrix} 8 & 2 & 0 \\ 2 & 10 & 3 \\ 0 & 3 & 6 \end{bmatrix} \begin{Bmatrix} \phi_2 \\ \phi_3 \\ \phi_4 \end{Bmatrix} = K_R\{D_R\}. \tag{1.17}$$

The solution for the reduced displacement vector (containing only the unsuppressed displacements) is obtained as

$$\{D_R\} = \begin{Bmatrix} \phi_2 \\ \phi_3 \\ \phi_4 \end{Bmatrix} = \frac{1}{384\beta} \begin{bmatrix} 51 & -12 & 6 \\ -12 & 48 & -24 \\ 6 & -24 & 76 \end{bmatrix} \begin{Bmatrix} 8 \\ 0 \\ 16 \end{Bmatrix} = K_R^{-1}\{Q_R\}. \tag{1.18}$$

Though inversion is convenient in the present small example, eqn (1.17) is generally solved iteratively or directly by Gauss reduction or factorization techniques, some of which are briefly described in Section 15.3.

Remembering to include the reactions caused by distributed loading, the element moments are obtained as

$$\begin{Bmatrix} M_1 \\ M_2 \end{Bmatrix} = \begin{Bmatrix} 24 \\ -24 \end{Bmatrix} + k_1 \begin{Bmatrix} \phi_1 \\ \phi_2 \end{Bmatrix} \tag{1.19}$$

for the first element and the same scheme is followed for the remaining elements. Here an exception to the usual finite element procedure is the use of the element stiffness matrices to calculate the element forces. Generally a separate element force matrix must be used for this purpose.

1.4 Coordinate transformation for inclined elements

To deal with inclined elements a coordinate transformation must be used. Referring to Fig. 1.4, consider an element inclined at an angle α to the

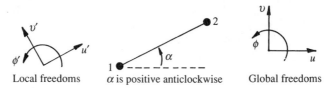

Local freedoms α is positive anticlockwise   Global freedoms

Fig. 1.4 Local and global axes in a beam element

horizontal reference position. Using the full inclined length of the member the standard matrix of eqn (1.12) applies only to the local displacement set shown. At any point on the beam this is related to the global displacement set by the transformation

$$
\begin{bmatrix} u' \\ v' \\ \phi' \end{bmatrix} = \begin{bmatrix} \cos\alpha & \sin\alpha & 0 \\ -\sin\alpha & \cos\alpha & 0 \\ 0 & 0 & 1 \end{bmatrix} \begin{bmatrix} u \\ v \\ \phi \end{bmatrix} \tag{1.20}
$$

Applying this result to the displacements at each node one obtains

$$
\{d'\} = \begin{bmatrix} u'_1 \\ v'_1 \\ \phi'_1 \\ u'_2 \\ v'_2 \\ \phi'_2 \end{bmatrix} \begin{bmatrix} \cos\alpha & \sin\alpha & 0 & 0 & 0 & 0 \\ -\sin\alpha & \cos\alpha & 0 & 0 & 0 & 0 \\ 0 & 0 & 1 & 0 & 0 & 0 \\ 0 & 0 & 0 & \cos\alpha & \sin\alpha & 0 \\ 0 & 0 & 0 & -\sin\alpha & \cos\alpha & 0 \\ 0 & 0 & 0 & 0 & 0 & 1 \end{bmatrix} \begin{bmatrix} u_1 \\ v_1 \\ \phi_1 \\ u_2 \\ v_2 \\ \phi_2 \end{bmatrix} = T\{d\}.
$$

Local $\longleftarrow$ ——————————————————— Global   (1.21)

In eqn 1.21 the arrow emphasizes that the matrix T transforms *from global to local displacements*. The same transformation applies to the local and global element load vectors, $\{q\}$ and $\{q'\}$. If one wishes to reverse the transformation and transform from local to global values one writes

$$
\{d\} = T^{-1}\{d'\} = T^{\mathrm{t}}\{d'\} \tag{1.22}
$$

as the transformation between orthogonal frames of reference is unitary (that is, $T^{\mathrm{t}}T = I$). Then if k' is the local element stiffness matrix, obtained from eqn (1.12), the global matrix k is obtained from

$$
\{q\} = T^{\mathrm{t}}\{q'\} = T^{\mathrm{t}}k'\{d'\} = T^{\mathrm{t}}k'T\{d\} = k\{d\} \tag{1.23}
$$

where

$$k = T^t k' T = \begin{bmatrix} & & -k_3 s & & & -k_3 s \\ k_{uv} & & & -k_{uv} & & \\ & & k_3 c & & & k_3 c \\ -k_3 s & k_3 c & k_4 & k_3 s & -k_3 c & k_5 \\ & & k_3 s & & & k_3 s \\ -k_{uv} & & & k_{uv} & & \\ & & -k_3 c & & & -k_3 c \\ -k_3 s & k_3 c & k_5 & k_3 s & -k_3 c & k_4 \end{bmatrix}$$ (1.24)

and

$$k_{uv} = \begin{bmatrix} k_1 c^2 + k_2 s^2 & k_1 sc - k_2 sc \\ k_1 sc - k_2 sc & k_1 s^2 + k_2 c^2 \end{bmatrix}$$

where $s = \sin \alpha$, $c = \cos \alpha$. Equation (1.24) is referred to as a *congruent transformation*, an important property of which is that if k' is symmetric, so too is k. Once the transformation defined by eqn (1.23) has been carried out for each element, assembly then proceeds as described in Section 1.3.

Following the solution of the system equations for the global displacements, the nodal forces in each element are given by

$$\{q'\} = Tk\{d\} - \{q_t'\}$$ (1.25)

that is, as we saw in eqns (1.6)–(1.9) the consistent loads must be reversed in sign when calculating the element nodal forces. Note that in eqn (1.25), $\{q_t'\}$ are local values. Therefore in the summation of the element consistent loads to form the structure load vector these should first have been transformed to global values, that is, with the usual cancelling of interelement reactions, this summation can be expressed as $\{Q\} = \sum T^t \{q_t'\}$.

1.5 The beam: special cases

The stiffness matrix for a beam with an internal pin joint (Fig. 1.5) is given by writing $M_b = F_{v1} x - M_1 = EI d^2 v / dx^2$ and integrating to give

$$\frac{EI dv}{dx} = \frac{F_{v1} x^2}{2} - M_1 x + c_1$$ (1.26)

$$EI v = \frac{F_{v1} x^3}{6} - \frac{M_1 x^2}{2} + c_1 x + c_2.$$ (1.27)

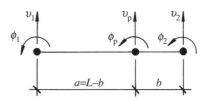

Fig. 1.5 Beam with an internal pin

Denoting the displacements and forces at the pin as v_p, ϕ_p, F_{vp}, M_p and applying eqns 1.26 and 1.27 on both sides of this point, one obtains

$$EIv_p = \frac{F_{v1}a^3}{6} - \frac{M_1 a^2}{2} + EIa\phi_1 + EIv_1 \tag{1.28}$$

$$EIv_2 = \frac{F_{vp}b^3}{6} - \frac{M_p b^2}{2} + EIb\phi_p + EIv_p \tag{1.29}$$

$$EI\phi_2 = \frac{F_{vp}b^2}{2} - M_p b + EI\phi_p. \tag{1.30}$$

Noting that $M_1 = F_{v1}a$, $M_p = 0$, $F_{vp} = -F_{v1}$ and eliminating v_p, ϕ_p one obtains

$$F_{v1} = \frac{3EI(v_1 - v_2 + a\phi_1 + b\phi_2)}{a^3 + b^3} \tag{1.31}$$

and using $M_1 = F_{v1}a$, $F_{v2} = -F_{v1}$, $M_2 = -bF_{v1}$ the element stiffness matrix for flexure is

$$k_b = \frac{3EI}{a^3 + b^3} \begin{bmatrix} 1 & a & -1 & b \\ a & a^2 & -a & ab \\ -1 & -a & 1 & -b \\ b & ab & -b & b^2 \end{bmatrix}. \tag{1.32}$$

In the special cases when the pin is at a node ($a = 0$ or $b = 0$) no stiffness will be associated with ϕ_1 or ϕ_2 and care must be taken to suppress the affected rotation as a boundary condition or use a small arbitrary stiffness as its diagonal entry if no other element contributes stiffness to this freedom. When a limited rotational joint stiffness is specified at each end of the beam the appropriate stiffness matrix can be derived by a similar procedure to that above and the result for this case is given by Livesley.[14]

If the element is moment-free at both ends the result is a spar element with two local freedoms (u_1', u_2') or four global freedoms (u_1, v_1, u_2, v_2). In this case the global element stiffness matrix is

$$k = \begin{bmatrix} k_{uv} & -k_{uv} \\ -k_{uv} & k_{uv} \end{bmatrix} \quad \text{where } k_{uv} = \frac{EA}{L} \begin{bmatrix} c^2 & cs \\ sc & s^2 \end{bmatrix} \tag{1.33}$$

and this result is useful in the analysis of pin-jointed truss structures. However, like the other special cases noted here, it is more usefully incorporated as an option in a general-purpose frame analysis program.

1.6 Duality in skeletal finite elements

An alternative approach to skeletal structures, called the *direct stiffness method*[15, 16] can be demonstrated by the truss structure of Fig. 1.6. Equating

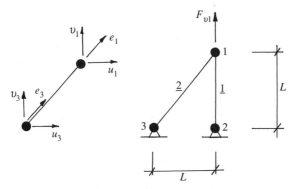

Fig. 1.6 Two-element truss. Element numbers are underscored, node numbers are not

the extensions $\{e\}$ of the members to the global displacements at each end
and writing the result in matrix form one obtains

$$\begin{Bmatrix} e_1 \\ e_2 \end{Bmatrix} = \begin{bmatrix} 0 & 1 & 0 & -1 & 0 & 0 \\ 1/\sqrt{2} & 1/\sqrt{2} & 0 & 0 & -1/\sqrt{2} & -1/\sqrt{2} \end{bmatrix} \{D\}$$

$$= B\{u_1, v_1, u_2, v_2, u_3, v_3\}. \tag{1.34}$$

The element extensions are related to the element thrusts by the equation

$$\{T\} = \begin{Bmatrix} T_1 \\ T_2 \end{Bmatrix} = \begin{bmatrix} EA/L & 0 \\ 0 & EA/\sqrt{2}L \end{bmatrix} \begin{Bmatrix} e_1 \\ e_2 \end{Bmatrix} = D\{e\}. \tag{1.35}$$

By resolving the internal force pairs $\pm T_i$ (assumed inwardly directed or
tensile) the equilibrium equations for each joint are obtained as

$$\{Q\} = \begin{bmatrix} F_{u1} \\ F_{v1} \\ F_{u2} \\ F_{v2} \\ F_{u3} \\ F_{v3} \end{bmatrix} = \begin{bmatrix} 0 & 1/\sqrt{2} \\ 1 & 1/\sqrt{2} \\ 0 & 0 \\ -1 & 0 \\ 0 & -1/\sqrt{2} \\ 0 & -1/\sqrt{2} \end{bmatrix} \begin{Bmatrix} T_1 \\ T_2 \end{Bmatrix} = C\{T\}. \tag{1.36}$$

Then the structure stiffness matrix is given by

$$\{Q\} = C\{T\} = CD\{e\} = CDB\{D\} = K\{D\} \tag{1.37}$$

giving

$$K = \frac{\sqrt{2}EA}{4L} \begin{bmatrix} 1 & 1 & 0 & 0 & -1 & -1 \\ 1 & 2\sqrt{2}+1 & 0 & -2\sqrt{2} & -1 & -1 \\ 0 & 0 & 0 & 0 & 0 & 0 \\ 0 & -2\sqrt{2} & 0 & 2\sqrt{2} & 0 & 0 \\ -1 & -1 & 0 & 0 & 1 & 1 \\ -1 & -1 & 0 & 0 & 1 & 1 \end{bmatrix}.$$

The same result is obtained by assembly of element matrices obtained from eqn (1.33). Comparing eqns (1.34) and (1.36) one observes that

$$B = C^t.$$

This property is called contragredience[17] or *duality*. It arises from reciprocity requirements[18] and it is shown in Section 3.2 that duality applies to any linear system for which a minimum principle exists. As we shall see in Chapter 22 this result is also important in the theory of optimization where it leads to the dual form of the linear programming problem.

Here we have demonstrated duality at structure level. In Section 3.2 we show that it also applies at element level, again using spar elements as an example. Duality also applies, at least in principle, to continuum finite elements but in most of these it is not practically possible to independently derive matrices C and B which are the transpose of each other. Nevertheless duality is a fundamental and useful idea in the finite element method.

1.7 Grillage elements

Figure 1.7 shows a horizontal beam or grillage element, the stiffness matrix for which must include torsion rather than axial thrust. The element is usually used in conjunction with two-dimensional plate elements to analyse beam/slab systems[19] and has three freedoms at each node, as do many plate elements.

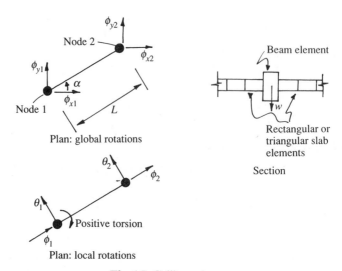

Fig. 1.7 Grillage element

The global rotations are transformed to yield the local flexural and torsional rotations, respectively ϕ and θ, at each node,[†]

$$\phi = \phi_x\cos \alpha + \phi_y\sin \alpha, \qquad \theta = -\phi_x\sin \alpha + \phi_y\cos \alpha. \tag{1.38}$$

The torsional moment is calculated as

$$M_T = \frac{-GJ(\theta_2 - \theta_1)}{L} = \frac{GJ(\theta_1 - \theta_2)}{L} \tag{1.39}$$

and from the slope–deflection equations the flexural end moments are

$$M_1 = \frac{4EI\phi_1}{L} + \frac{2EI\phi_2}{L} + \frac{6EI(w_1 - w_2)}{L^2}$$

$$= \frac{4EI}{L}(\phi_1 + (w_1 - w_2)/L) + \frac{2EI}{L}(\phi_2 + (w_1 - w_2)/L) \tag{1.40}$$

$$M_2 = \frac{2EI}{L}(\phi_1 + (w_1 - w_2)/L) + \frac{4EI}{L}(\phi_2 + (w_1 - w_2)/L) \tag{1.41}$$

where the correction $(w_1 - w_2)/L$ is the rigid body correction to the rotations. Substituting eqns (1.38) in eqns (1.39), (1.40), and (1.41) and writing the resulting equations in matrix form:

$$\{M_1, M_2, M_T\} = DT(w_1, \phi_{x1}, \phi_{y1}, w_2, \phi_{x2}, \phi_{y2}) \tag{1.42}$$

where

$$D = \begin{bmatrix} 4EI/L & 2EI/L & 0 \\ 2EI/L & 4EI/L & 0 \\ 0 & 0 & GJ/L \end{bmatrix}$$

and

$$T = \begin{bmatrix} 1/L & \cos \alpha & \sin \alpha & -1/L & 0 & 0 \\ 1/L & 0 & 0 & -1/L & \cos \alpha & \sin \alpha \\ 0 & -\sin \alpha & \cos \alpha & 0 & \sin \alpha & -\cos \alpha \end{bmatrix}$$

and the element stiffness matrix is obtained as

$$k = T^tDT. \tag{1.43}$$

The matrix T is a coordinate transformation matrix, though here it includes a rigid body transformation in which the effective flexural rotations are

---

[†] Note that in the present text straight arrows in element diagrams show the direction in which $\phi_x = dw/dx$, for example, is calculated.

calculated by subtracting the rigid body rotation $(w_1 - w_2)/L$, that is, the slope of the chord of the deformed beam.

As shown in Fig. 1.7 when combined with plate elements this beam element shares a common neutral axis. Situations in which the beam lies below the slab, as is usual, can be approximately modelled by increasing the moment of inertia specified for the beam (that is, using the parallel axis theorem). A more precise treatment can be obtained by including in-plane displacement freedoms in both the slab and beam elements and assuming that the local rotation freedoms ϕ_1 and ϕ_2 in the beams apply only at the neutral axis of the adjoining slab elements.[20]

Considering the section in a beam element shown in Fig. 1.8 the variation in the horizontal displacement through the depth in the beam is given as

$$u = u_0 + \phi z = u_0 + \frac{dw}{dx} z$$

so that the corresponding strain is given as

$$\varepsilon_x = \frac{du}{dx} = \frac{du_0}{dx} + \frac{d^2w}{dx^2} z$$

and the bending moment in the beam is given by integrating the stress over the entire depth:

$$M_x = Eb \int_{z_1}^{z_2} \left(\frac{du_0}{dx} + \frac{d^2w}{dx^2} z \right) z \, dz \tag{1.44}$$

where b is the breadth of the beam. When $z_1 = -z_2$ this reduces to the usual result

$$M_x = \frac{Ebd^3 \dfrac{d^2w}{dx^2}}{12} = EI\chi_x. \tag{1.45}$$

The stiffness matrix obtained for the general case (eqn (1.44)) is given by Gustafson and Wright.[20]

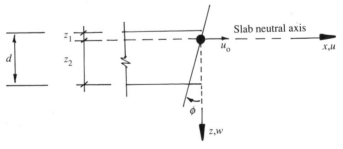

Fig. 1.8 Beam element set below a slab

1.8 Shear in beam elements

The shearing effects in a beam element can be taken into account by defining the slope of the elastica as

$$\frac{dv}{dx} = \phi + \gamma \tag{1.46}$$

where ϕ is the rotation caused by flexure and γ is the additional rotation caused by transverse shearing action. As shown in Section 2.7 this is given by

$$\gamma = \beta V / GA \tag{1.47}$$

where the section parameter β is defined in eqn (2.35) and takes the value 1.2 for a homogeneous rectangular cross-section. Thus, including the shearing rotations in eqns (1.26) and (1.27) one obtains

$$\frac{EIdv}{dx} = \frac{F_{v1}x^2}{2} - M_1 x + c_1 + \frac{F_{v1}EI\beta}{GA} \tag{1.48}$$

$$EIv = \frac{F_{v1}x^3}{6} - \frac{M_1 x^2}{2} + c_1 x + c_2 + \frac{F_{v1}EI\beta x}{GA}. \tag{1.49}$$

Using eqns (1.46) and (1.47) to rewrite (1.48) as

$$EI\phi = \frac{F_{v1}x^2}{2} - M_1 x + c_1 \tag{1.50}$$

and applying the boundary conditions for $x = 0$ and $x = L$ to eqns (1.49) and (1.50) one obtains

$$EI\phi_1 = c_1 \tag{1.51}$$

$$EI\phi_2 = \frac{F_{v1}L^2}{2} - M_1 L + c_1 \tag{1.52}$$

$$EIv_1 = c_2 \tag{1.53}$$

$$EIv_2 = \frac{F_{v1}L^3}{6} - \frac{M_1 L^2}{2} + c_1 L + c_2 + \frac{F_{v1}EI\beta x}{GA}. \tag{1.54}$$

Substituting into eqn (1.54) for c_1, M_1, and c_2 from eqns (1.51), (1.52), and (1.53), respectively, one obtains

$$F_{v1} = \frac{1}{1 + \Omega} \left(\frac{6EI\phi_1}{L^2} + \frac{6EI\phi_2}{L^2} + \frac{12EI(v_1 - v_2)}{L^3} \right) \tag{1.55}$$

where $\Omega = 12EI\beta / GAL^2$. In turn substituting into eqn (1.52) gives

$$M_1 = \frac{1}{1 + \Omega} \left(\frac{EI(4 + \Omega)\phi_1}{L} + \frac{EI(2 - \Omega)\phi_2}{L} + \frac{6EI(v_1 - v_2)}{L^3} \right) \tag{1.56}$$

giving an alternative form of the slope–deflection equations. Note here that

ϕ_1 and ϕ_2 are only the flexural components of the beam slopes. If, in fact, the shear rotations are added to eqns (1.51) and (1.52) the shear effects completely cancel and the usual shallow beam equations are obtained.

Thus the stiffness matrix of eqn (1.12) easily includes the shearing stiffness if the terms k_2, k_3, k_4, and k_5 are redefined as[21]

$$k_4 = \frac{(4 + \Omega)EI}{(1 + \Omega)L} \tag{1.57a}$$

$$k_5 = \frac{(2 - \Omega)EI}{(1 + \Omega)L} \tag{1.57b}$$

$$k_3 = \frac{6EI}{(1 + \Omega)L^2} \tag{1.57c}$$

$$k_2 = \frac{12EI}{(1 + \Omega)L^3}. \tag{1.57d}$$

Using these, one obtains the exact result for the problem cited in Table 11.1 using a single element but the numerical integration formulation of Section 11.1 is more readily generalized to deal with tapered and curved elements, for example.

1.9 Three-dimensional beam elements

Figure 1.9 shows a beam element arbitrarily oriented in three-dimensional space. With the addition of lateral bending and torsion the element now has twelve freedoms. The local element stiffness matrix is obtained as[21]

$$
\begin{bmatrix} F_{u1} \\ F_{v1} \\ F_{w1} \\ M_{x1} \\ M_{y1} \\ M_{z1} \\ F_{u2} \\ F_{v2} \\ F_{w2} \\ M_{x2} \\ M_{y2} \\ M_{z2} \end{bmatrix} =
\begin{bmatrix}
k_1 & 0 & 0 & 0 & 0 & 0 & -k_1 & 0 & 0 & 0 & 0 & 0 \\
0 & k_{2z} & 0 & 0 & 0 & k_{3z} & 0 & -k_{2z} & 0 & 0 & 0 & k_{3z} \\
0 & 0 & k_{2y} & 0 & -k_{3y} & 0 & 0 & 0 & -k_{2y} & 0 & -k_{3y} & 0 \\
0 & 0 & 0 & k_6 & 0 & 0 & 0 & 0 & 0 & -k_6 & 0 & 0 \\
0 & 0 & -k_{3y} & 0 & k_{4y} & 0 & 0 & 0 & k_{3y} & 0 & k_{5y} & 0 \\
0 & k_{3z} & 0 & 0 & 0 & k_{4z} & 0 & k_{3z} & 0 & 0 & 0 & k_{5z} \\
-k_1 & 0 & 0 & 0 & 0 & 0 & k_1 & 0 & 0 & 0 & 0 & 0 \\
0 & -k_{2z} & 0 & 0 & 0 & -k_{3z} & 0 & k_{2z} & 0 & 0 & 0 & -k_{3z} \\
0 & 0 & k_{2y} & 0 & k_{3y} & 0 & 0 & 0 & k_{2y} & 0 & k_{3y} & 0 \\
0 & 0 & 0 & -k_6 & 0 & 0 & 0 & 0 & 0 & k_6 & 0 & 0 \\
0 & 0 & -k_{3y} & 0 & k_{5y} & 0 & 0 & 0 & k_{3y} & 0 & k_{4y} & 0 \\
0 & k_{3z} & 0 & 0 & 0 & k_{5z} & 0 & k_{3z} & 0 & 0 & 0 & k_{4z}
\end{bmatrix}
\begin{bmatrix} u_1 \\ v_1 \\ w_1 \\ \phi_{x1} \\ \phi_{y1} \\ \phi_{z1} \\ u_2 \\ v_2 \\ w_2 \\ \phi_{x2} \\ \phi_{y2} \\ \phi_{z2} \end{bmatrix}.
$$

$$\tag{1.58}$$

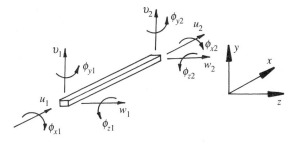

Fig. 1.9 Beam element in three dimensions

Here the terms for thrust (k_1) and vertical bending (with subscript y) correspond to those of eqn (1.12), while the terms for torsion (k_6) and for horizontal bending are obtained from eqn (1.43) by replacing ϕ with ϕ_y and θ with $-\phi_x$ and putting $\alpha = 0$. Then the stiffness parameters of eqn (1.58) are defined by

$$k_1 = \frac{EA}{L}, \qquad k_6 = \frac{GJ}{L}$$

$$k_{2z} = \frac{12EI_z}{(1 + \Omega_z)L^3}$$

$$k_{3z} = \frac{6EI_z}{(1 + \Omega_z)L^2}$$

$$k_{4z} = \frac{(4 + \Omega_z)EI_z}{(1 + \Omega_z)L}$$ (1.59)

$$k_{5z} = \frac{(2 - \Omega_z)EI_z}{(1 + \Omega_z)L}$$

where $\Omega_z = \dfrac{12EI_z\beta_z}{GA_zL^2}$ and k_{2y}, k_{3y}, k_{4y}, and k_{5y} are obtained by replacing the subscript z with y in eqns (1.59).

Note also that A_z denotes an effective shear area in that direction and that the parameters β and Ω may have different values for the y- and z-directions.

For an arbitrarily oriented element the global element stiffness matrix is obtained using eqn (1.23) with

$$T = \begin{bmatrix} t & 0 & 0 & 0 \\ 0 & t & 0 & 0 \\ 0 & 0 & t & 0 \\ 0 & 0 & 0 & t \end{bmatrix}$$ (1.60)

where

$$t = \begin{bmatrix} L_{xy}/L & 0 & Z/L \\ 0 & 1 & 0 \\ -Z/L & 0 & L_{xy}/L \end{bmatrix} \begin{bmatrix} X/L_{xy} & Y/L_{xy} & 0 \\ -Y/L_{xy} & X/L_{xy} & 0 \\ 0 & 0 & 1 \end{bmatrix}$$

and

$$X = x_2 - x_1, \qquad Y = y_2 - y_1$$
$$Z = z_2 - z_1, \qquad L_{xy} = \sqrt{(X^2 + Y^2)}$$
$$L = \sqrt{(X^2 + Y^2 + Z^2)}$$

using two successive transformations for rotation from the x-axis to the plan projection of the element, followed by vertical rotation from the plan projection to the actual spatial position.

1.10 Other skeletal problems in engineering

There are several other engineering systems that take a network form and which can thus be dealt with by matrix techniques.

1.10.1 D.C. circuits

Figure 1.10 shows a d.c. circuit in which the branch currents (assumed positive from low to high node number) must satisfy Ohm's law, which gives the voltage drop in each branch as $\delta V = IR$. Using the reciprocal form $I = G\delta V$, where G is the branch conductance, one can form an element conductance matrix for each branch as

$$G_{ij} \begin{bmatrix} 1 & -1 \\ -1 & 1 \end{bmatrix} \begin{Bmatrix} V_i \\ V_j \end{Bmatrix} = \begin{Bmatrix} I_{ij} \\ -I_{ij} \end{Bmatrix} \tag{1.61}$$

where V_i, V_j are the potentials at each node. Then the assembled matrix for

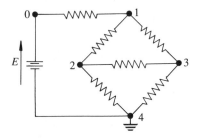

Fig. 1.10 Simple d.c. circuit

the system of Fig. 1.10 is

$$
\begin{bmatrix}
G_{01} & -G_{01} & 0 & 0 & 0 \\
-G_{01} & G_{01} + G_{12} + G_{13} & -G_{12} & -G_{13} & 0 \\
0 & -G_{12} & G_{12} + G_{23} + G_{24} & -G_{23} & -G_{24} \\
0 & -G_{13} & -G_{23} & G_{13} + G_{23} + G_{34} & -G_{34} \\
0 & 0 & -G_{24} & -G_{34} & G_{24} + G_{34}
\end{bmatrix}
$$

$$
\begin{bmatrix}
v_0 \\
v_1 \\
v_2 \\
v_3 \\
v_4
\end{bmatrix} = \{0\} \tag{1.62}
$$

and the right-hand side vanishes to satisfy Kirchhoff's current law, that the algebraic sum of currents at a node is zero. The problem is solved by specifying $V_4 = 0$ (datum voltage) and $V_0 = E$, giving

$$
\begin{bmatrix}
G_{01} + G_{12} + G_{13} & -G_{12} & -G_{13} \\
-G_{12} & G_{12} + G_{23} + G_{24} & -G_{23} \\
-G_{13} & -G_{23} & G_{13} + G_{23} + G_{34}
\end{bmatrix}
\begin{bmatrix}
V_1 \\
V_2 \\
V_3
\end{bmatrix} =
\begin{bmatrix}
G_{01} E \\
0 \\
0
\end{bmatrix} \tag{1.63}
$$

where $G_{01} E$ on the right-hand side is the equivalent 'load' to the specified voltage $V_0 = E$. The form of the problem is comparable to that for truss structures except that coordinate transformation is no longer required. It is also interesting to note that the problem possesses a dual or contragradient form with the loop currents as the variables which is obtained by the application of Kirchhoff's voltage law,[22] but we choose the foregoing formulation here as it is much easier to program and more closely corresponds to the displacement method much used in following chapters.

1.10.2 A.C. circuits

In steady state a.c. circuits one writes for each branch

$$
Z = \frac{\delta V}{I} = \frac{j\omega L + R - j/\omega C}{I} \tag{1.64}
$$

where ω, L, C are respectively the frequency, inductance, and capacitance and $j = \sqrt{-1}$. Here Z is the complex impedance and, as in Section 1.10.1 we prefer to cast the problem in reciprocal form using the complex admittance $A = 1/Z$, which is then

$$
A = \frac{1}{Z} = \frac{R - jB}{R^2 + B^2} \tag{1.65}
$$

where $B = \omega L - 1/\omega C$. The element admittance matrices are given by

$$\frac{R - jB}{R^2 + B^2} \begin{bmatrix} 1 & -1 \\ -1 & 1 \end{bmatrix} \begin{Bmatrix} V_1 \\ V_2 \end{Bmatrix} = \{0\} \tag{1.66}$$

leading to a 4×4 matrix as the problem must be solved for both the real and imaginary voltage components in the complex frequency plane.[13]

Alternating current circuit problems can also be expressed in eigenvalue form[13] or dealt with as time-dependent problems by time stepping techniques (see Chapter 14) and in the latter regard it is useful to recall the mechanical–electrical analogues:

$$L \text{ (inductance)} = m \text{ (mass)}$$
$$R \text{ (resistance)} = \mu \text{ (damping)}$$
$$C \text{ (capacitance)} = k \text{ (stiffness)}$$

much used in analogue computers.

1.10.3 Pipe networks

Steady state pipe network problems must, like all equilibrium problems, satisfy conservation, continuity, and constitutive requirements. In linear skeletal structural frames these are respectively dealt with by ensuring conservation of energy, equilibrium (or compatibility in dual formulations), and using Hooke's law as the constitutive relationship.

In pipe networks, on the other hand, these requirements are met in the same way as in electrical circuits, respectively by:

(1) $\sum q = 0$ at junctions (the *conservation* requirement);

(2) $\sum \delta p = 0$ around closed loops (the *continuity* requirement);

(3) $\delta p = Rq^n$ in each branch (the *constitutive* requirement).

The first two are again Kirchhoff's laws but now the constitutive law (Darcy's law) is non-linear and generally the exponent n is slightly less than 2.[23] Again the problem possesses dual forms, with either the nodal pressure heads[13] or loop flows as freedoms but now a non-linear solution technique is needed and these are discussed in Chapter 12.

1.11 A simple program for truss analysis

The following pin-jointed truss analysis BASIC[†] program is intended as a very simple introduction to finite element programming and can easily be mounted on the smallest of microcomputers. The program uses

[†] The specific form used throughout the text is MEGABASIC (see Appendix 1).

Gauss–Jordan reduction (see Section 15.3) to invert the assembled stiffness matrix, first imposing the zero displacement conditions by making the appropriate row and column in K null, placing a 1 on the diagonal and a 0 in $\{Q\}$, that is, one creates a trivial equation. Rather than using a matrix augmented by an identity matrix, only a single extra variable is used (to store the pivot and later the 'row multiplier') and the original matrix is progressively overwritten by the forming inverse. Applying the same reduction operations to the load vector, this will finally contain the solution. The number of operations is halved if the inverse is not retained by sweeping to the right only (see lines 330 and 380) but retention is needed for the 'initial stiffness method' (see Section 12.3).

Dimensioned arrays are:

CORD(50,2) nodal coordinates, x in column 1, y in column 2;
NOP(100,2) each row stores the node number pair for each element;
ES(4,4)   element stiffness matrices formed according to eqn (1.33);
S(100,100) structure stiffness matrix;
Q(100)   vector of specified loads.

```
10  Dim CORD(50,2),NOP(100,2),ES(4,4),S(100,100),Q(100)
20  Read NP,NE,NB,NL;NN=2*NP           ;Rem NO. NODES,ELEMENTS,B.C.s,LOADED NODES
30  Read E,A                           ;Rem MODULUS AND X-AREA - SAME THROUGHOUT
40  For I=1 to NP
50  Read CORD(I,1),CORD(I,2);Next I ;Rem NODAL COORDINATES
60  For N=1 to NE
70  Read NOP(N,1),NOP(N,2);Next N      ;Rem NODE NUMBERS FOR EACH ELEMENT
80  For N=1 to NE  ;Rem LOOP ON ELEMENTS *************************************
90  NI=NOP(N,1);NJ=NOP(N,2)
100 DX=CORD(NJ,1)-CORD(NI,1);DY=CORD(NJ,2)-CORD(NI,2)
110 EL=sqrt(DX*DX+DY*DY)               ;Rem ELEMENT LENGTH
120 CA=DX/EL;SA=DY/EL;F=E*A/EL         ;Rem DIRECTION COSINES FOR ELEMENT
130 ES(1,1)=F*CA*CA;ES(1,2)=F*SA*CA
140 ES(2,1)=F*SA*CA;ES(2,2)=F*SA*SA;Rem FILL TOP LH CORNER OF ESM
150 For I=1 to 2;For J=1 to 2
160 ES(I,J+2)=-ES(I,J);ES(I+2,J)=-ES(I,J)
170 ES(I+2,J+2)=ES(I,J);Next;Next  ;Rem COPY TOP LH CORNER TO FILL ESM
180 For I=1 to 2;For J=1 to 2
190 For IL=1 to 2;IE=2*(I-1)+IL;NR=2*NOP(N,I)-2+IL
200 For JL=1 to 2;JE=2*(J-1)+JL;NC=2*NOP(N,J)-2+JL
210 S(NR,NC)=S(NR,NC)+ES(IE,JE)        ;Rem DEPLOY ESM IN THE STRUCTURE MATRIX
220 Next;Next;Next;Next
230 Next N       ;Rem END LOOP ON ELEMENTS *************************************
240 For I=1 to NL
250 Read N,Q(2*N-1),Q(2*N);Next        ;Rem READ NODAL LOADS
260 For I=1 to NB                       ;Rem LOOP ON NO. OF BOUNDARY CONDITIONS
270 Read N,NF;NF=2*N-2+NF
280 For J=1 to NN
290 S(NF,J)=0;S(J,NF)=0;Next J          ;Rem ZERO ROW AND COLUMN
300 S(NF,NF)=1;Q(NF)=0;Next I           ;Rem ONE ON DIAGONAL, ZERO ON RHS
310 For I=1 to NN                        ;Rem START GAUSS-JORDAN REDUCTION **********
320 X=S(I,I);Q(I)=Q(I)/X                ;Rem X=PIVOT
330 For J=I+1 to NN
340 S(I,J)=S(I,J)/X;Next J              ;Rem ROW/PIVOT
350 For K=1 to NN
360 If K=I then Goto 400
370 X=S(K,I);Q(K)=Q(K)-X*Q(I)          ;Rem X='ROW MULTIPLIER'
```

```
380 For J=I+1 to NN
390 S(K,J)=S(K,J)-X*S(I,J);Next J   ;Rem ROW SUBTRACTION OPERATION
400 Next K
410 Next I                          ;Rem END GAUSS-JORDAN REDUCTION ************
420 !;! "NODAL DISPLACEMENTS U,V";!
430 For N=1 to NP
440 ! %"I5",N,%"15F6",Q(2*N-1),Q(2*N);Next
450 !;! "ELEMENT TENSIONS"
460 For N=1 to NE                   ;Rem LOOP TO CALCULATE ELEMENT TENSIONS
470 NI=NOP(N,1);NJ=NOP(N,2)
480 DX=CORD(NJ,1)-CORD(NI,1);DY=CORD(NJ,2)-CORD(NI,2)
490 EL=sqrt(DX*DX+DY*DY)
500 CA=DX/EL;SA=DY/EL;F=E*A/EL
510 U2=CA*Q(2*NJ-1)+SA*Q(2*NJ)      ;Rem LOCAL DISPLACEMENTS (I.E. PARALLEL)
520 U1=CA*Q(2*NI-1)+SA*Q(2*NI)      ;Rem AT EACH END OF ELEMENT
530 T=F*(U2-U1)
540 ! %"I5",N,%"15F4",T;Next N
550 End


1000 Data 6,9,3,1                   ;Rem NO. NODES, ELEMENTS, B.C.s & LOADS
1010 Data 100,0.1                   ;Rem E,A (SAME FOR EACH ELEMENT)
1020 Data 0,0                       ;Rem COORDS OF FIRST NODE
1030 Data 0,3
1040 Data 4,0
1050 Data 4,3
1060 Data 8,0
1070 Data 8,3                       ;Rem COORDS OF LAST NODE
1080 Data 1,2                       ;Rem NODE NUMBERS FOR FIRST ELEMENT
1090 Data 1,3
1100 Data 2,3
1110 Data 2,4
1120 Data 3,4
1130 Data 3,5
1140 Data 4,5
1150 Data 4,6
1160 Data 5,6                       ;Rem NODE NUMBERS FOR LAST ELEMENT
1170 Data 5,0,-30                   ;Rem THE ONLY LOAD IN THIS PROBLEM
1180 Data 1,1                       ;Rem U=0 AT NODE 1
1190 Data 2,1                       ;Rem U=0 AT NODE 2
1200 Data 2,2                       ;Rem V=0 AT NODE 2
```

The data of lines 1000–1200 is that for the truss of Fig. 1.11. Here, as for other problems, the data input requirements are

Lines	Data
1	number of nodes (NP); number of elements (NE); number of boundary conditions (NB); number of loaded nodes (NL).
2	E, A (same for each element—this restriction is easily removed);
NP lines	node number (N); x coordinate (CORD(N,1)); y coordinate (CORD(N,2)).
NE lines	element number (N); first node number (NOP(N,1)); second node number (NOP(N,2)).
NL lines	node number (N); x load (Q(2*N − 1)); y load (Q(2*N)).
NB lines	node number (N); degree of freedom suppressed (NF) where NF = 1 for $u = 0$ and NF = 2 for $v = 0$

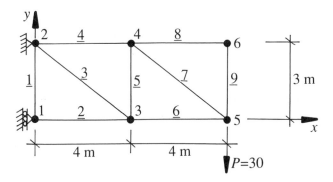

Fig. 1.11 Truss (nodal df u, v), $E = 100$, $A = 0.1$

In the MEGABASIC used for this and subsequent programs in the text note that ! is the abbreviation for PRINT and %" " is used for format specification. The program output will consist of the nodal displacements u, v followed by the member tensions. The reader should have little difficulty obtaining analytical solutions using the method of joints and virtual work to provide comparisons.

1.12 Discussion

Skeletal elements prove particularly useful in engineering practice for the analysis of structural, electrical, and hydraulic networks and compact programs for their implementation in equilibrium (as in the preceding section), eigenvalue,[24] and propagation problems can be written for the smallest of modern computers.

With the use of variational formulations (see Chapter 3) the usefulness of skeletal elements is greatly increased as elements with variable cross-sectional properties, surface tractions, and external elastic restraint, to cite a few examples, are readily dealt with. In this regard it is worth remarking that if the six-freedom beam element of Fig. 1.4 is tapered the element cannot equilibrate the varying thrust thus permitted and this suggests an advantage of the element of Section 11.1 for applications to structural frames.

In connection with skeletal structures some mention of draped cable structures may be worthwhile. These may be analyzed as a series of spar-like segments but generally the geometry is unknown and the situation must be resolved iteratively as a *two-point boundary value problem*,[25] iteration ceasing when the *misclosure* at one end is removed. Numerical methods can also include the cable weight as this is by no means easily done analytically.[26]

At the outset we have noted the early use of skeletal models as an approximation to continua. It is also worth noting that continuum approximations can be used to deal with some skeletal structures, for example the use of a plate analogy to provide approximate solutions for parallel chord space trusses.[27, 28]

It is hoped that the program of Section 1.11 provides an immediate introduction to finite element techniques for those in need of it. The d.c. network problem of Section 1.10 is even easier to program whilst readers may refer to reference 29 for a very useful rigid-jointed frame program.

References

1. Livesley, R. K. (1953). Analysis of rigid frames by an electronic computer. *Engineering*, **176**, 230.
2. Livesley, R. K. (1964). *Matrix Methods of Structural Analysis*. Pergamon, New York.
3. Ebner, H. and Koeller, H. (1937). Zur Berechnung des Krafverlaufes in versteiften Zylindershalen. *Luftfahrforshung*, **14**, no. 12.
4. Hrenikoff, A. (1941). Solution of problems in elasticity by the framework method. *J. Appl. Mech.*, **A8**, 169.
5. McHenry, D. (1943). A lattice analogy for the solution of plane stress problems. *J. Inst. Civ. Eng.*, **21**, 59.
6. Newmark, N. M. (1949). Numerical methods of analysis in bars, plates and elastic bodies. In *Numerical Methods in Analysis in Engineering*, (ed. L. E. Grinter), MacMillan, New York.
7. Argyris, J. H. and Dunne, P. C. (1976). The finite element method applied to fluid mechanics. Proceedings of the Conference on Computer Methods and Problems in Aeronautical Fluid Dynamics, University of Manchester, 1974.
8. Argyris, J. H. (1960). *Energy Theorems and Structural Analysis*, Butterworth, London. (Reprinted from *Aircraft Engrg*, 1954–1955).
9. Turner, M. J., Clough, R. W., Martin, H. C., and Topp, L. J. (1956). Stiffness and deflection analysis in complex structures. *J. Aero. Sci.*, **23**, 805.
10. Courant, R. (1943). Variational methods for the solution of problems of equilibrium and vibration. *Bull. Amer. Math. Soc.*, **49**, 1.
11. Prager, W. and Synge, J. L. (1947). Approximation in elasticity based on the concept of function space. *Quart. Appl. Math.*, **5**, 241.
12. Zienkiewicz, O. C. (1970). The finite element method: from intuition to generality. *Appl. Mech. Rev.*, **23**, 249.
13. Crandall, S. H. (1956). *Engineering Analysis*, McGraw-Hill, New York.
14. Livesley, R. K. (1975). *Matrix Structural Analysis*, (2nd edn). Pergamon, Oxford.
15. Gallagher, R. H. (1975). *Finite Element Analysis Fundamentals*. Prentice-Hall, Englewood Cliffs, NJ.
16. Norris, C. H., Wilbur, J. B. and Utku, S. (1975). *Elementary Structural Analysis*, (3rd edn). Prentice-Hall, Englewood Cliffs, NJ.

17. Asplund, S. O. (1966). *Structural Mechanics: Classical and Matrix Methods.* Prentice-Hall, Englewood Cliffs, NJ.
18. Hall, A. and Kabaila, A. P. (1977). *Basic Concepts of Structural Analysis.* Pitman, London.
19. Mohr, G. A. (1972). Finite element analysis of beam and plate systems. *Civ. Eng. TN*, Monash University (Caulfield) Melbourne.
20. Gustafson, W. C. and Wright, R. N. (1968). Analysis of skewed composite girder bridges. *J. Struct. Div. ASCE*, **94**, 5890.
21. Przemieniecki, J. S. (1968). *Theory of Matrix Structural Analysis*, McGraw-Hill, New York.
22. Jennings, A. (1977). *Matrix Computation for Engineers and Scientists*, Wiley, Chichester.
23. *Manual of British Water Engineering Practice*, (4th edn) (1969). (Ed. W. O. Skeat.) Heffer and Sons, Cambridge.
24. Williams, F. W. and Howson, W. P. (1977). Compact calculation of natural frequencies and buckling loads for plane frames. *Int. J. Num. Meth. Engng*, **11**, 1067.
25. Harrison, H. B. (1973). *Computer Methods in Structural Analysis*. Prentice-Hall, Englewood Cliffs, NJ.
26. O'Brien, W. T. and Francis, A. J. (1964). Cable movement under two-dimensional loads. *Proc. Struct. Div. ASCE*, **90**, no. ST7.
27. Schmidt, L. C. (1982). A simple numerical method for the analysis of rectangular, orthotropic plates. Proceedings of the Fourth Australian International Conference on *Finite Element Methods*, Univ. of Melbourne.
28. Schmidt, L. C. (1983). Behaviour of a single-chorded plane space truss. *Trans I.E. Aust.*, **CE25(3)**.
29. Mohr, G. A. and Milner, H. R. (1986). *A Microcomputer Introduction to the Finite Element Method*. Pitman, Melbourne.

2
Basic relationships
for solid continua

This chapter first introduces some of the basic equations relating the stresses, strains, and displacements in elastic continua and discusses the concept of a continuous domain arbitrarily divided into finite-sized elements. Discussion of energy techniques for the formation of matrix equations for such elements is then taken up in Chapter 3.

2.1 Eulerian and Lagrangian frames of reference

In the mechanics of a continuum, particles at each point are displaced resulting in stresses, strains, and movement of material throughout the continuum. To study such behaviour we must first define a frame of reference, that is, a coordinate system. There are two possible approaches.

2.1.1 Eulerian (fixed) coordinate

In the Eulerian system the coordinates x, y, z are related to a fixed set of axes which is usually, but not necessarily orthogonal. Then if we denote the distribution of some time-dependent property g of a particle in the domain as

$$g = g(x, y, z, t) \tag{2.1}$$

then an increment in this during a time interval δt is given by

$$\delta_g = \frac{\partial g}{\partial t} \delta t + \frac{\partial g}{\partial x} \delta x + \frac{\partial g}{\partial y} \delta y + \frac{\partial g}{\partial z} \delta z \tag{2.2}$$

where δx, δy and δz are the movements of this particle.

Then the *total derivative* or material derivative of g with respect to t is defined by

$$\frac{Dg}{Dt} = \lim_{t \to 0} \frac{\delta g}{\delta t} = \frac{\partial g}{\partial t} + \lim_{t \to 0} \left(\frac{\partial g}{\partial x} \frac{\partial x}{\delta t} + \frac{\partial g}{\partial y} \frac{\delta y}{\delta t} + \frac{\partial g}{\partial z} \frac{\delta z}{\delta t} \right)$$

$$= \frac{\partial g}{\partial t} + \frac{\partial g}{\partial x} V_x + \frac{\partial g}{\partial y} V_y + \frac{\partial g}{\partial z} V_z \tag{2.3}$$

where V_x, V_y and V_z are the components of the particle velocity and the terms involving these are called the *convective inertia* terms.

The Eulerian approach is that usually used in fluid mechanics applications where the inertia terms are obviously of possible significance. In solid mechanics applications it is also usual to employ the Eulerian approach as the movements δx, δy and δz are usually very small so that $\mathrm{D}g/\mathrm{D}t \simeq \partial g/\partial t$ and total derivatives and their associated convective terms therefore need not be introduced. In problems involving large displacements, however, the strain definitions must include additional terms corresponding to the convective terms of the total derivative or, alternatively, employ Lagrangian co-ordinates.

2.1.2 Lagrangian (moving) coordinates

Lagrangian coodinates are attached to a material point and thus convect with the material. The convective terms of eqn (2.3) therefore vanish. Except in specific instances, however, this approach is not generally practical in fluid mechanics problems. In solid mechanics, however, more formal and precise strain definitions can be constructed using the Lagrangian approach. The extensional strain in a small element of a member curved by deformation, for example, is defined by

$$\varepsilon = \frac{(\mathrm{d}s - \mathrm{d}s_0)}{\mathrm{d}s} \qquad (2.4)$$

where s is measured along the curved member and $\mathrm{d}s_0$ is the initial length of the element.

Nevertheless most studies of the mechanics of solids employ the Eulerian approach but as we shall see in Chapter 12 the Lagrangian approach sometimes proves useful in problems involving large displacements.

2.2 Basic problems in solid mechanics

The differential equations governing solid continua provide the basis for most analytical solutions to problems of solid mechanics. In the finite element method, however, solutions to these problems are usually obtained using energy theorems.

The following sections introduce some basic continuum concepts and the fundamental relationships needed to solve the following elasticity problems by energy methods:

(1) one-dimensional problems, that is, spar and beam problems;

(2) plane stress and strain;

(3) three-dimensional and axisymmetric solids;

(4) bending of thin plates and thick plates;

(5) thin shells, thick shells, and axisymmetric shells.

In each case a set of strain–displacement and stress–strain relationships is required and these are used respectively to form a *strain-interpolation matrix* B and a *modulus matrix D*. Then, as we shall show in Chapter 3, an *element stiffness matrix k* is obtained by calculating the strain energy U_e of a finite element as

$$U_e = \tfrac{1}{2} \int \{\varepsilon\}^t \{\sigma\} \, dV = \tfrac{1}{2} \{d\}^t (\int B^t DB \, dV)\{d\} = \tfrac{1}{2} \{d\}^t k \{d\} \tag{2.5}$$

where $\int dV$ denotes integration over the volume of a finite element. Here the matrix B relates the strains to a set of nodal displacements $\{d\}$ for the element, that is, $\{\varepsilon\} = B\{d\}$, and the matrix D relates the stressess to the strains as $\{\sigma\} = D\{\varepsilon\}$.

The energy approach is used in the solid mechanics problems of the following chapters because, except in the simplest cases, this aproach proves the most direct. In thick shell and large displacement analysis, for example, discovery of the governing differential equation is in itself a difficult exercise whereas these problems can be solved with little difficulty by energy methods. In most problems of fluid mechanics, however, energy arguments are less appropriate whereas the differential equations governing these problems are easily obtained. However, we defer discussion of finite element formulations obtained directly from differential equations until we commence a study of fluid mechanics and general field problems later in the text.

2.3 Three-dimensional elasticity

The two direct strains and the shear strain at a point in a two-dimensional continuum are given by the rates of change in the Cartesian displacements u and v:

$$\varepsilon_x = \lim_{x \to 0} \frac{\delta u}{\delta x} = \frac{\partial u}{\partial x} \tag{2.6a}$$

$$\varepsilon_y = \lim_{y \to 0} \frac{\delta v}{\delta y} = \frac{\partial v}{\partial y} \tag{2.6b}$$

$$\varepsilon_{xy} = \lim_{y \to 0} \frac{\delta u}{\delta y} + \lim_{x \to 0} \frac{\delta v}{\delta x} = \frac{\partial u}{\partial y} + \frac{\partial v}{\partial x}. \tag{2.6c}$$

The direct strains are the changes in length per unit length of filaments parallel to the axes and the shear strain is defined by the in-plane rotations $\partial u/\partial y$ and $\partial v/\partial x$.

Generalizing eqns (2.6) the strain–displacement relationships for the three-dimensional case are obtained as

$$\varepsilon_x = \frac{\partial u}{\partial x}, \qquad \varepsilon_y = \frac{\partial v}{\partial y}, \qquad \varepsilon_z = \frac{\partial w}{\partial z}$$

$$\varepsilon_{xy} = \frac{\partial u}{\partial y} + \frac{\partial v}{\partial x}$$

$$\varepsilon_{yz} = \frac{\partial v}{\partial z} + \frac{\partial w}{\partial y} \tag{2.7}$$

$$\varepsilon_{zx} = \frac{\partial w}{\partial x} + \frac{\partial u}{\partial z}.$$

In the finite element method the strains are expressed in terms of an element nodal displacement vector $\{d\}$ by a matrix relationship of the form $\{\varepsilon\} = B\{d\}$. This relationship is obtained by differentiating a relationship of the form $\{\bar{u}\} = F^t\{d\}$ where F^t is an interpolation matrix and $\{\bar{u}\} = \{u, v, w\}$. To clarify the process we firstly restate eqs (2.7) in the form of a differential matrix operator B^*,

$$\{\varepsilon\} = \begin{bmatrix} \partial(\ )/\partial x & 0 & 0 \\ 0 & \partial(\ )/\partial y & 0 \\ 0 & 0 & \partial(\ )/\partial z \\ \partial(\ )/\partial y & \partial(\ )/\partial x & 0 \\ 0 & \partial(\ )/\partial z & \partial(\ )/\partial y \\ \partial(\ )/\partial z & 0 & \partial(\ )/\partial x \end{bmatrix} \begin{Bmatrix} \{u\} \\ \{v\} \\ \{w\} \end{Bmatrix} = B^*\{\bar{u}\}. \tag{2.8}$$

If the displacement u at any point in the continuum is given by an interpolation of a set of point values written as

$$u = \sum f_i u_i = \{f\}^t\{u\} \tag{2.9}$$

and the same interpolation is used for v and w then we can write

$$\{\bar{u}\} = \begin{bmatrix} \{f\}^t & \{0\}^t & \{0\}^t \\ \{0\}^t & \{f\}^t & \{0\}^t \\ \{0\}^t & \{0\}^t & \{f\}^t \end{bmatrix} \begin{Bmatrix} \{u\} \\ \{v\} \\ \{w\} \end{Bmatrix} = F^t\{d\} \tag{2.10}$$

where $\{d\}$ is the complete set of displacements at the n nodes of the element. Note, however, that in practice $\{d\}$ is arranged in the order

$$\{d\} = \{u_1, v_1, w_1, u_2, v_2, w_2, \ldots, u_n, v_n, w_n\} \tag{2.11}$$

and that entries in F must be rearranged accordingly, that is the first row is arranged in the order $f_1, 0, 0, f_2, 0, 0, \cdots, f_n, 0, 0$.

The simplest example of such interpolation is the linear interpolation of a variable u within a length L which can be expressed as

$$u = \left(1 - \frac{x}{L}\right)u_1 + \frac{x}{L}u_2 = f_1 u_1 + f_2 u_2 \tag{2.12}$$

where u_1 is the value at $x = 0$ and u_2 the value at $x = L$.

Once a suitable interpolation has been found then this is differentiated to provide an interpolation for the strains at any point in the element. This operation can be stated symbolically in the form

$$\{\varepsilon\} = B^*F\{d\} = B\{d\} \tag{2.13}$$

where B is the strain–displacement or strain-interpolation matrix.

In an isotropic medium the stress–strain relationships are given by cyclic progression of x, y, z (denoted by $x \rightarrow y, z$) in the equations[1]

$$\varepsilon_x = \frac{\sigma_x}{E} - \frac{v\sigma_y}{E} - \frac{v\sigma_z}{E} \qquad x \rightarrow y, z \tag{2.14a}$$

$$\varepsilon_{xy} = \frac{\sigma_{xy}}{G} \qquad x \rightarrow y, z \tag{2.14b}$$

where E is Young's modulus, v is Poisson's ratio, and G is the shear modulus which can be expressed as $G = E/2(1 + v)$.

Solving eqns (2.14) for the stresses and expressing the result in matrix form we obtain

$$\begin{bmatrix} \sigma_x \\ \sigma_y \\ \sigma_z \\ \sigma_{xy} \\ \sigma_{yz} \\ \sigma_{zx} \end{bmatrix} = \frac{E}{1 - v - 2v^2} \begin{bmatrix} 1 - v & v & v & 0 & 0 & 0 \\ v & 1 - v & v & 0 & 0 & 0 \\ v & v & 1 - v & 0 & 0 & 0 \\ 0 & 0 & 0 & v' & 0 & 0 \\ 0 & 0 & 0 & 0 & v' & 0 \\ 0 & 0 & 0 & 0 & 0 & v' \end{bmatrix} \begin{bmatrix} \varepsilon_x \\ \varepsilon_y \\ \varepsilon_z \\ \varepsilon_{xy} \\ \varepsilon_{yz} \\ \varepsilon_{zx} \end{bmatrix}$$

or

$$\{\sigma\} = D\{\varepsilon\} \tag{2.15}$$

where $v' = (1 - 2v)/2$. Using only eqns (2.5), (2.13) and (2.15) we are able to calculate the strain energy in a subdomain (that is, a finite element) and hence determine the element stiffness matrix.

In the following sections we give the basic relationships needed to form the B and D matrices for a variety of other problems in the mechanics of solid continua.

2.4 Plane stress and strain

In plane stress and plane strain we generally analyse a 'slice' of a three-dimensional solid. The slice is considered to lie in the xy-plane and stresses and strains are assumed not to vary in the z-direction. In both plane stress and strain problems the analysis can be carried out in terms of stresses σ_x, σ_y and σ_{xy} and the corresponding strains. The strain–displacement relationships are

$$\varepsilon_x = \frac{\partial u}{\partial x}, \quad \varepsilon_y = \frac{\partial v}{\partial y} \tag{2.16}$$

$$\varepsilon_{xy} = \frac{\partial u}{\partial y} + \frac{\partial v}{\partial x}$$

that is, one neglects all the terms involving z in eqns (2.7).

For *plane stress* conditions one assumes $\sigma_z = 0$, $\sigma_{yz} = 0$, and $\sigma_{zx} = 0$ and solves eqns (2.14) for the remaining stresses to give

$$\{\sigma\} = \begin{Bmatrix} \sigma_x \\ \sigma_y \\ \sigma_{xy} \end{Bmatrix} = \frac{E}{1 - v^2} \begin{bmatrix} 1 & v & 0 \\ v & 1 & 0 \\ 0 & 0 & 1 - v/2 \end{bmatrix} \begin{Bmatrix} \varepsilon_x \\ \varepsilon_y \\ \varepsilon_{xy} \end{Bmatrix} = D\{\varepsilon\}. \tag{2.17}$$

Note also that a strain $\varepsilon_z = -v(\sigma_x + \sigma_y)/E$ exists but is associated with zero stress and therefore need not be included in an energy-based analysis.

For *plane strain* conditions one assumes $\varepsilon_z = 0$, $\varepsilon_{yz} = 0$ and $\varepsilon_{zx} = 0$ and contracts eqn (2.15) to give

$$\{\sigma\} = \begin{Bmatrix} \sigma_x \\ \sigma_y \\ \sigma_{xy} \end{Bmatrix} = \frac{E}{1 - v - 2v^2} \begin{bmatrix} 1 - v & v & 0 \\ v & 1 - v & 0 \\ 0 & 0 & 1 - 2v/2 \end{bmatrix} \begin{Bmatrix} \varepsilon_x \\ \varepsilon_y \\ \varepsilon_{xy} \end{Bmatrix} = D\{\varepsilon\}. \tag{2.18}$$

A stress $\sigma_z = E(\varepsilon_x + \varepsilon_y)/(1 - v - 2v^2)$ also exists but is associated with zero strain and does not enter into an energy-based analysis.

The strain energy per unit volume, dU, is given by $dU = \frac{1}{2}\{\varepsilon\}^t\{\sigma\}$ and is, for plane stress,

$$dU = \frac{1}{2}(\varepsilon_x\sigma_x + \varepsilon_y\sigma_y + \varepsilon_{xy}\sigma_{xy})$$

$$= \frac{E[\varepsilon_x^2 + \varepsilon_y^2 + 2v\varepsilon_x\varepsilon_y + (1 - v)\varepsilon_{xy}^2/2]}{2(1 - v^2)} \tag{2.19}$$

whilst for plane strain it is

$$dU = \frac{1}{2}(\varepsilon_x\sigma_x + \varepsilon_y\sigma_y + \varepsilon_{xy}\sigma_{xy})$$

$$= \frac{E[(1 - v)(\varepsilon_x^2 + \varepsilon_y^2) + 2v\varepsilon_x\varepsilon_y + (1 - 2v)\varepsilon_{xy}^2/2]}{2(1 - v - 2v^2)} \tag{2.20}$$

and we see that dU becomes infinite when Poisson's ratio is 0.5, corresponding to an incompressible solid. Therefore special variational principles must be used to deal with this situation. The same applies to three-dimensional solids, the strain energy expression for which is a straightforward extension of eqn (2.20).

2.5 Axisymmetric solids

Axisymmetric problems are three-dimensional but possess radial symmetry, that is the solid itself is formed by revolution of a plane shape about an axis. If the loadings applied to it are also axisymmetric then the stresses in the solid have the same variation along any radial coordinate taken at some angle to the axis of revolution. Such problems are most conveniently solved in cylindrical coordinates (r, z, θ) where z defines the axis of revolution, r is a radial coordinate perpendcular to the z-axis, and θ is the angular rotation of the radial arm r.

Figure 2.1 shows an example of an axisymmetric problem. At any point (r, z, θ) three perpendicular axes, the third being tangential to the circle formed by revolution of r, are defined along with corresponding stresses σ_r, σ_z, and σ_θ. A shear stress σ_{rz} also exists but stresses $\sigma_{z\theta}$ and $\sigma_{\theta r}$ must be zero as they would cause a 'rotational bias' and destroy the axisymmetry. The strain–displacement relationships are

$$\varepsilon_r = \frac{\partial u}{\partial r,} \qquad \varepsilon_z = \frac{\partial w}{\partial z}, \qquad \varepsilon_\theta = \frac{u}{r}$$

$$\varepsilon_{rz} = \frac{\partial u}{\partial z} + \frac{\partial w}{\partial r}$$

(2.21)

where u and w are displacements parallel to the r- and z-axes. In eqns (2.21) the *circumferential strain* is given by the fractional change in circumference at

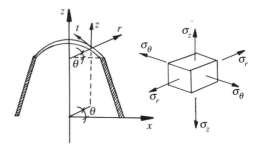

Fig. 2.1 Coordinates and stresses in an axisymmetric solid

radius r caused by a radial displacement u:

$$\varepsilon_\theta = \frac{2\pi(r + u) - 2\pi r}{2\pi r} = \frac{u}{r}. \tag{2.22}$$

The stress–strain relationships are a contraction of eqn (2.15), that is,

$$\begin{Bmatrix} \sigma_r \\ \sigma_z \\ \sigma_\theta \\ \sigma_{rz} \end{Bmatrix} = \frac{E}{1 - v - 2v^2} \begin{bmatrix} 1 - v & v & v & 0 \\ v & 1 - v & v & 0 \\ v & v & 1 - v & 0 \\ 0 & 0 & 0 & 1 - 2v/2 \end{bmatrix} \begin{Bmatrix} \varepsilon_r \\ \varepsilon_z \\ \varepsilon_\theta \\ \varepsilon_{rz} \end{Bmatrix} = D\{\varepsilon\}. \tag{2.23}$$

As we shall see in Chapter 7, analysis of axisymmetric problems is no more difficult than the analysis of plane stress and strain problems.

Note also that such analyses may be extended by the use of Fourier series to deal with non-axisymmetric loadings. If there is a single point load, for example, this is approximated by the half-range series[2]

$$q = \sum_{i=1}^{N} q_i \sin\left(\frac{n\theta}{2\pi}\right) \qquad n = \text{func. } (i) \tag{2.24}$$

where N is large enough to include the significant Fourier amplitudes q_i. Then an axisymmetric analysis is carried out for each of the q_i and each of the numerical results so obtained is combined according to eqn (2.24) to obtain the final result.

2.6 Stress resultants in thin plates

Figure 2.2(a) shows an element of a plate and Fig. 2.2(b) a distorted thin laminar slice of the element at height z above the neutral surface. The direct

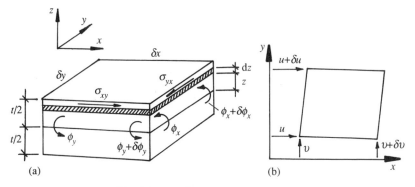

Fig. 2.2 Element of a bent thin plate

curvatures are defined as the rates of change in slope, that is,

$$\chi_x = \frac{\partial \phi_x}{\partial s_x} \simeq \frac{\partial \phi_x}{\partial x}, \qquad \chi_y = \frac{\partial \phi_y}{\partial s_y} \simeq \frac{\partial \phi_y}{\partial y} \qquad (2.25)$$

where s_x and s_y are curvilinear coordinates on the deformed neutral surface for which the approximations $s_x \simeq x$ and $s_y \simeq y$ are made if this deformation is small.

A twist curvature must also be defined and to this end we examine the shear stresses in the laminar slice at height z, the magnitudes of which are given by

$$\frac{\sigma_{xy}}{G} = \frac{\delta \mu}{\delta y} + \frac{\delta v}{\delta x} = -z \frac{\delta \phi_x}{\delta y} - z \frac{\delta \phi_y}{\delta x}. \qquad (2.26)$$

Integrating over the thickness of the element we obtain the twisting moments per unit width m_{xy} and m_{yx} as

$$m_{xy} = -\int_{-t/2}^{t/2} \sigma_{xy} z \, dz = \frac{Gt^3 \chi_{xy}}{12} \qquad \text{and} \qquad m_{yx} = m_{xy} \qquad (2.27)$$

where the twist curvature follows from eqn (2.26) as

$$\chi_{xy} = \frac{\partial \phi_x}{\partial y} + \frac{\partial \phi_y}{\partial x}. \qquad (2.28)$$

As a state of plane stress applies within each lamina such as that shown in Fig. 2.2(b) the plane stress elasticity matrix of eqn (2.17) will appear again but multiplied by $t^3/12$ owing to the integration over the depth operation used in eqn (2.27). The constitutive equations for the *stress resultants* or generalized stresses, which are here the moments per unit width, therefore take the form

$$\left\{ \begin{array}{c} m_x \\ m_y \\ m_{xy} \end{array} \right\} = \frac{Et^3}{12(1-v^2)} \begin{bmatrix} 1 & v & 0 \\ v & 1 & 0 \\ 0 & 0 & 1-v/2 \end{bmatrix} \left\{ \begin{array}{c} \chi_x \\ \chi_y \\ \chi_{xy} \end{array} \right\} = D\{\chi\}. \qquad (2.29)$$

In thin plates it is assumed that $\phi_x = \partial w/\partial x$ and $\phi_y = \partial w/\partial y$ so that the curvatures of eqns (2.25) and (2.28) are usually expressed as

$$\chi_x = \frac{\partial^2 w}{\partial x^2}, \qquad \chi_y = \frac{\partial^2 w}{\partial y^2}, \qquad \chi_{xy} = \frac{2\partial^2 w}{\partial x \partial y} \qquad (2.30)$$

and the last of these implies the compatibility constraint $\partial^2 w/\partial x \partial y = \partial^2 w/\partial y \partial x$.

The strain energy per unit area of the plate can be expressed in terms of the generalized stresses and strains as

$$dU = \tfrac{1}{2} \{\chi\}^t \{m\} = \frac{Et^3 [\chi_x^2 + \chi_y^2 + 2v\chi_x\chi_y + (1-v)\chi_{xy}^2/2]}{24(1-v^2)}. \qquad (2.31)$$

It should be noted that χ_{xy} was defined as $2\partial^2 w/\partial x\partial y$, and m_{xy} calculated as

$$m_{xy} = D_{xy}\chi_{xy} = \frac{Gt^3\chi_{xy}}{12} \tag{2.32}$$

in obtaining eqn (2.31), and this is the usual practice.[3]

If the twisting curvature is calculated as $\partial^2 w/\partial x\partial y$, however, we use $m_{xy} = 2D_{xy}\chi_{xy}$ to calculate the twisting moment but care must be taken in the calculation of element stiffness matrices using $k = \int B^t DB\, dV$ to double those entries relating to the twisting stiffness (otherwise the twisting stiffness is $\frac{1}{2}(2)\frac{1}{2} = \frac{1}{2}$ the correct magnitude).

The same comments apply to the shear strain ε_{xy} if this is defined as $\frac{1}{2}(\partial u/\partial y + \partial v/\partial x)$, that is, as an average rotation, and this is not uncommon practice as it allows one to write eqns (2.7) in the indicial or tensor form

$$\varepsilon_{ij} = \frac{1}{2}\left(\frac{\partial u_i}{\partial x_j} + \frac{\partial u_j}{\partial x_i}\right) \qquad i \to j, k\ (j = i, j, k) \tag{2.33}$$

where $i \to j, k(j = i, j, k)$ denotes progression of i to values j, k with j also cycling for each value of i. Here $u_i = u$, $u_j = v$, $x_i = x$, $x_j = y$ and so on, and ε_{ij} is a *second-order tensor* (a first-order tensor simply being a vector). As they involve the 'double progression' described above, second-order tensors may be expanded to form a two-dimensional matrix and examples of such matrices (in this case for stresses) appear in Section 2.10.

2.7 Stress resultants in thick plates

In thick plates the total slopes of the neutral surface are[4]

$$\frac{\partial w}{\partial x} = \phi_x + \gamma_x, \qquad \frac{\partial w}{\partial y} = \phi_y + \gamma_y \tag{2.34}$$

where ϕ_x and ϕ_y are the rotations caused by flexure and γ_x and γ_y are the additional rotations resulting from transverse shear action. Considering the element shown in Fig. 2.2(a) the strain energy associated with a shear force V_x acting on vertical sections parallel to the x-axis is given by

$$U = \frac{1}{2}\int\left(\frac{\sigma_{xz}^2}{G}\right)\delta y\delta y\, dz = \frac{1}{2}\int\left(\frac{V_x S_y}{\delta y I}\right)^2\left(\frac{\delta y}{G}\right)\delta x\, dz$$

$$= \frac{1}{2}\left(\int\left(\frac{S_y^2 A}{\delta y I^2}\right)dz\right)\left(\frac{V_x^2}{GA}\right)\delta x = \frac{1}{2}\beta\left(\frac{V_x^2}{GA}\right)\delta x \tag{2.35}$$

where $A = \delta yt$ is the area of these vertical sections. S_y is the first moment of that part of the section area outside the level at which the shear stress values are calculated using $\sigma_{xz} = V_x S_y / \delta y I$, an expression which will be familiar to structural engineers. If the plate is homogeneous the parameter β in eqn (2.35) is 1.2 and the shear stress varies parabolically through the depth.

Then at a given section in the plate Castigliano's second theorem can be applied to eqn (2.35) to define an *average shear strain* γ_x as

$$\frac{\partial U}{\partial V_x} = \gamma_x = \frac{\beta V_x}{GA} = \frac{\beta V_x}{G \delta y t} \tag{2.36}$$

neglecting the δx in eqn (2.35) as V_x is the shear force on an area $\delta y t$ of the plate so that in a strip of plate of width δy the shear strain energy associated with V_x is calculated as $\frac{1}{2} \int \gamma_x V_x \, dx$.

Defining Q_x and Q_y as the shear forces per unit width or generalized stresses in each direction in the plate these are then related to their corresponding average shear strains or generalized strains by

$$Q_x = \frac{V_x}{\delta y} = \frac{5Gt\gamma_x}{6}, \qquad Q_y = \frac{V_y}{\delta x} = \frac{5Gt\gamma_y}{6} \tag{2.37}$$

where

$$\gamma_x = \frac{\partial w}{\partial x} - \phi_x, \qquad \gamma_y = \frac{\partial w}{\partial y} - \phi_y. \tag{2.38}$$

It should be noted that these last equations are not complemented by a third equation coupling the actions in both directions, as in preceding sections, and this may contribute to numerical difficulties encountered in their application to finite elements.[5,6] One possible solution is to determine the required constraint from the equation

$$\frac{5Gt}{6}\left(\frac{\partial \gamma_x}{\partial x} + \frac{\partial \gamma_y}{\partial y}\right) + p = 0 \tag{2.39}$$

where p is the transverse load per unit area on the plate. Equation (2.39) follows from the equation $\partial Q_x / \partial x + \partial Q_y / \partial y + q = 0$ which is the condition for vertical equilibrium of an infinitesimal element of the plate.[3] In section 9.1, for example, a rectangular thick plate element in which the constraints $\partial \gamma_x / \partial x = 0$ and $\partial \gamma_y / \partial y = 0$ are imposed is discussed. This is equivalent to splitting the constraint of eqn (2.39) for the case when $p = 0$ and such approximations have been used successfully in the finite element method, for example in the cylindrical shell element of Ashwell and Sabir discussed in Section 11.4.

It is not the usual practice to impose such constraints in thick plate analysis, however, and an alternative remedy to the problem is used in Section 9.3 where three average shear strains are defined parallel to the sides of triangular thick plate elements. To complete the description of thick plate behaviour, therefore, eqns (2.25) and (2.28) are used to define the flexural curvatures, eqn (2.29) gives the associated modulus matrix, and eqns (2.38) define the average shear strains. The modulus matrix is augmented by the stiffnesses $5Gt/6$ which appear in eqn (2.37).

2.8 Equations for thin shells

The generalized stresses in thin shells are, in addition to the moments per unit width m_x, m_y, and m_{xy}, extensional forces per unit width $N_x = \sigma_x t$, $N_y = \sigma_y t$, and $N_{xy} = \sigma_{xy} t$. There are many shell theories giving alternative definitions of the strains (including the curvatures as generalized strains)[7, 8, 9] but for the purposes of approximate finite element analysis it is sufficient to generalize the equations given by Love for cylindrical shells,[9] giving

$$\varepsilon_x = \frac{\partial u}{\partial x} + \frac{w}{R_x}, \qquad \varepsilon_y = \frac{\partial v}{\partial y} + \frac{w}{R_y}$$

$$\varepsilon_{xy} = \frac{\partial u}{\partial y} + \frac{\partial v}{\partial x} + \frac{2w}{R_{xy}}$$

$$\chi_x = -\frac{\partial^2 w}{\partial x^2} + \frac{\partial}{\partial x}\left(\frac{u}{R_x}\right)$$

$$\chi_y = -\frac{\partial^2 w}{\partial y^2} + \frac{\partial}{\partial y}\left(\frac{v}{R_y}\right) \tag{2.40}$$

$$\chi_{xy} = -\frac{2\partial^2 w}{\partial x \partial y} + \frac{\partial}{\partial y}\left(\frac{u}{R_x}\right) + \frac{\partial}{\partial x}\left(\frac{v}{R_y}\right)$$

where R_x and R_y are the radii of curvature and R_{xy} is the twist of the surface.

Equations (2.40) are used as the basis for a variety of special cases in Chapters 10 and 11. The first terms on the right-hand sides are those for plane strain and thin plate bending but additional terms arise owing to the curvature of the surface. The terms of the form w/R are the *radial strain* components, arising in the manner illustrated in eqn (2.22), and the u, v contributions to the curvatures are discussed in Section 11.4. As shown in Section 11.7, however, the latter have a relatively small effect and are often omitted in finite element formulations.[10, 11]

In these equations the coordinates x, y are *curvilinear coordinates* defined by curved lines on the shell surface drawn in the xz- and xy-planes, that is, at any point on the shell surface a unique normal can be defined and axes x and y at this point are perpendicular to this normal and thus lie in the tangent plane to the surface.

The modulus matrix giving the stress resultants corresponding to the strains and curvature changes of eqn (2.40) is obtained by combining eqns (2.17) and (2.29), leading to

$$\{N_x, N_y, N_{xy}, m_x, m_y, m_{xy}\} = \begin{bmatrix} tD_m & 0 \\ 0 & D_b \end{bmatrix} \{\varepsilon\} \tag{2.41}$$

where D_m and D_b are respectively given by eqns (2.17) and (2.29), the former being multiplied by the shell thickness, t, to obtain the generalized extensional stresses.

In eqns (2.40) the curvature $1/R_x$ is given by

$$\frac{1}{R_x} = \frac{\partial^2 z / \partial x^2}{[1 + (\partial z / \partial x)^2]^{3/2}} \tag{2.42}$$

and similarly for $1/R_y$, but the twist curvature $1/R_{xy}$ is more elusive. At any point on the shell surface *principal axes* such that $1/R_{xy} = 0$ can be defined and principal curvatures $1/R_1$ and $1/R_2$ calculated using eqn (2.42). Curvature values for other axes may then be obtained from these using Mohr's circle or, equivalently, eqns (2.51).

The determination of the principal axes and curvatures for the shell surface, however, requires recourse to the classical theory of surfaces which yields solutions only for special cases such as those of cylindrical and spherical shells.[12] An alternative approach which calculates *natural curvatures* on the sides of curved triangular shell elements is described in Sections 11.7 and 11.8. This approach avoids the need to define a twist curvature and can thus be applied to shells of arbitrary shape.

2.9 Anisotropic materials

For anisotropic materials the strains are given, in the case of plane stress, by

$$\varepsilon_x = \frac{\sigma_x}{E_x} - \frac{v\sigma_y}{E_y} + \alpha_x \theta \tag{2.43a}$$

$$\varepsilon_y = \frac{\sigma_y}{E_y} - \frac{v\sigma_x}{E_x} + \alpha_y \theta \tag{2.43b}$$

$$\varepsilon_{xy} = \frac{\sigma_{xy}}{G_{xy}} \tag{2.43c}$$

where α_x and α_y are the coefficients of linear thermal expansion and θ is the temperature rise. Solving eqns (2.43) for the stresses one obtains

$$\begin{Bmatrix} \sigma_x \\ \sigma_y \\ \sigma_{xy} \end{Bmatrix} = \frac{1}{1 - nv^2} \begin{bmatrix} E_x & vE_x & 0 \\ vE_x & E_x/n & 0 \\ 0 & 0 & G_{xy}(1 - nv^2) \end{bmatrix} \begin{Bmatrix} \varepsilon_x - \alpha_x\theta \\ \varepsilon_y - \alpha_y\theta \\ \varepsilon_{xy} \end{Bmatrix} \tag{2.44}$$

where $n = E_x/E_y$.

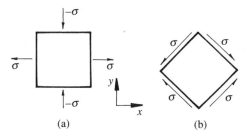

Fig. 2.3 Elements of unit dimension in equivalent stress states

A difficulty arises in estimating G_{xy} when an experimentally measured value is not available. It is possible to estimate a value, however, by considering an element of unit dimensions loaded by equal and opposite stresses, as shown in Fig. 2.3(a). This stress state is equivalent to the stress state shown in Fig. 2.3(b) on an element taken at 45 degrees to the original element, this second stress state being that of pure shear.

As the elements have unit volume, equating the strain energies in each leads to the result

$$\tfrac{1}{2}(\sigma\varepsilon_1 - \sigma\varepsilon_2) = \tfrac{1}{2}\sigma\varepsilon_{xy}$$

or

$$\varepsilon_{xy} = \varepsilon_1 - \varepsilon_2 = \frac{\sigma}{E_x} + \frac{v\sigma}{E_y} - \left(\frac{-\sigma}{E_y} - \frac{v\sigma}{E_x}\right)$$

$$= \frac{\sigma(1 + v)(E_x + E_y)}{E_x E_y} = \frac{\sigma_{xy}}{G_{xy}}$$

or

$$G_{xy} = \frac{E_x E_y}{(1 + v)(E_x + E_y)}. \tag{2.45}$$

An alternative result $G_{xy} = (E_x E_y)^{\frac{1}{2}}/2(1 + v)$ which is suggested by Timoshenko[3] gives similar results but the value given by eqn (2.45) has more formal justification.

For plane strain, appropriate equations for anisotropic materials can be formed in much the same way as eqns (2.44) and (2.45). When the elastic parameters are specified at some angle to the x, y-axes a matrix similar to that in eqn (2.51) must be used to congruently transform (that is, $T^t D T$) the modulus matrix to obtain the stress–strain relatonships for the skew axes.[6]

2.10 Stress transformations

Frequently in stress analysis we require transformations of stress components from one set of Cartesian axes to components in a rotated set. Often this is carried out to obtain principal stresses. We have already encountered axis transformations in Chapter 1 in connection with stiffness matrices, forces, and displacements. We now show how stress components may be transformed when orthogonal axes are rotated.

Suppose global axis stress components shown on the sides of the right-angled wedge of Fig. 2.4 are to be transformed to components corresponding to axes x', y'. If the wedge has unit thickness the forces on the sloping face in the x- and y-directions are

$$X = \sigma_x dy + \sigma_{xy} dx \tag{2.46a}$$

$$Y = \sigma_{yx} dy + \sigma_y dx. \tag{2.46b}$$

Noting that $dx = ds \cos \alpha$ and $dy = ds \sin \alpha$ it follows that

$$(ds) \begin{Bmatrix} X \\ Y \end{Bmatrix} = \begin{bmatrix} \sigma_x & \sigma_{xy} \\ \sigma_{yx} & \sigma_y \end{bmatrix} \begin{Bmatrix} \cos \alpha \\ \sin \alpha \end{Bmatrix}. \tag{2.47}$$

The stress components associated with the forces X and Y and parallel to the axes x', y' are given by resolving X and Y and dividing by ds, this exercise leading to

$$\sigma_x' = \frac{(X \cos \alpha + Y \sin \alpha)}{ds} \tag{2.48a}$$

$$\sigma_{yx}' = \frac{(-X \sin \alpha + Y \cos \alpha)}{ds}. \tag{2.48b}$$

Repeating the procedure for a face perpendicular to the inclined one shown in Fig. 2.4 leads to

$$\begin{bmatrix} \sigma_x' & \sigma_{xy}' \\ \sigma_{yx}' & \sigma_y' \end{bmatrix} = \frac{1}{ds} \begin{bmatrix} \cos \alpha & \sin \alpha \\ -\sin \alpha & \cos \alpha \end{bmatrix} \begin{bmatrix} X & X' \\ Y & Y' \end{bmatrix} \tag{2.49}$$

where X' and Y' are the force components for this perpendicular face. The magnitudes of these components are given by replacing α by $(90 + \alpha)$ in

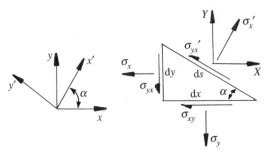

Fig. 2.4 Stress components on a wedge of unit thickness

eqn (2.47). Substituting these results and eqn (2.47) into eqn (2.49) gives

$$\begin{bmatrix} \sigma'_x & \sigma'_{xy} \\ \sigma'_{yx} & \sigma'_y \end{bmatrix} = \begin{bmatrix} \cos\alpha & \sin\alpha \\ -\sin\alpha & \cos\alpha \end{bmatrix} \begin{bmatrix} \sigma_x & \sigma_{xy} \\ \sigma_{yx} & \sigma_y \end{bmatrix} \begin{bmatrix} \cos\alpha & -\sin\alpha \\ \sin\alpha & \cos\alpha \end{bmatrix}$$

or

$$[\sigma'] = T^t[\sigma]\,T. \tag{2.50}$$

This transformation is orthogonal and similar in form to the transformation used for a spar element stiffness matrix in Chapter 1. If the multiplications implied by eqn (2.50) are carried out it will be found that the transformed stresses may be expressed in the alternative form

$$\begin{Bmatrix} \sigma'_x \\ \sigma'_y \\ \sigma'_{xy} \end{Bmatrix} = \begin{bmatrix} \cos^2\alpha & \sin^2\alpha & \sin 2\alpha \\ \sin^2\alpha & \cos^2\alpha & -\sin 2\alpha \\ -\sin 2\alpha & \sin 2\alpha & \cos 2\alpha \end{bmatrix} \begin{Bmatrix} \sigma_x \\ \sigma_y \\ \sigma_{xy} \end{Bmatrix}. \tag{2.51}$$

Equation (2.50) may readily be extended to three dimensions with the T matrix given by

$$T = \begin{bmatrix} \cos(x, x') & \cos(x, y') & \cos(x, z') \\ \cos(y, x') & \cos(y, y') & \cos(y, z') \\ \cos(z, x') & \cos(z, y') & \cos(z, z') \end{bmatrix} \tag{2.52}$$

where $\cos(x, y')$ is the cosine of the angle between the x- and y'-axes.

In eqn (2.51) stress vectors are transformed whereas in (eqn 2.50) stress tensors are transformed. In passing it is worth noting that an interesting property of the three-dimensonal stress tensor

$$\sigma_{ij} = [\sigma] = \begin{bmatrix} \sigma_x & \sigma_{xy} & \sigma_{xz} \\ \sigma_{yx} & \sigma_y & \sigma_{yz} \\ \sigma_{zx} & \sigma_{zy} & \sigma_z \end{bmatrix} \tag{2.53}$$

is that its eigenvalues λ_1, λ_2, and λ_3 are the principal stresses σ_1, σ_2, and σ_3, that is, the roots of the polynomial given by expanding the determinant $|[\sigma] - \lambda I| = 0$ are the principal stresses.

Three important *stress invariants* (that is, stress expressions which do not vary under orthogonal coordinate transformation) follow from σ_{ij}, namely[13]

$$J_1 = \text{Trace}(\sigma_{ij}) = \sigma_x + \sigma_y + \sigma_z \tag{2.54}$$

$$J_2 = \sigma_x\sigma_y + \sigma_y\sigma_z + \sigma_z\sigma_x - \sigma_{xy}^2 - \sigma_{yz}^2 - \sigma_{zx}^2 \tag{2.55}$$

$$J_3 = |\sigma_{ij}| = \sigma_1\sigma_2\sigma_3 \tag{2.56}$$

and these are called the first, second, and third stress invariants. The first two of these prove useful in starting material failure criteria in Chapter 12.

As we shall see in later chapters stress transformations are sometimes of considerable importance. In Chapter 8, for example, a transformation from

Cartesian to *natural stresses* parallel to the sides of triangular elements is introduced and this transformation proves useful in many other chapters of the text.

References

1. Timoshenko, S. P. and Goodier, J. N. (1951). *Theory of Elasticity*, (3rd edn). McGraw-Hill, New York.
2. Argyris, J. H. (1965). Continua and Discontinua. First Conference on Matrix Methods in Structural Mechanics, Wright-Patterson AFB, Ohio.
3. Timoshenko, S. P. and Woinowsky-Krieger, S. (1956). *Theory of Plates and Shells*, (2nd edn). McGraw-Hill, New York.
4. Timoshenko, S. P. (1921). On the correction for shear of the differential equations for transverse vibrations of prismatic bars. *Phil. Mag.* series 6, **41**, 744.
5. Mohr, G. A. (1978). A triangular finite element for thick slabs. *Comput. Struct.* **9**, 595.
6. Brebbia, C. A. and Connor, J. J. (1973). *Fundamentals of Finite Element Techniques for Structural Engineers*. Butterworths, London.
7. Naghdi, P. M. (1957). On the theory of thin elastic shells. *J. Appl. Mech.*, **14**, 369.
8. Novozhilov, V. V. (1959). *Theory of Thin Shells*, (trans. P. G. Lowe). Noordhoff, Gröningen.
9. Love, A. E. H. (1944). *A Treatise on the Mathematical Theory of Elasticity*, (4th edn). Dover, New York.
10. Dawe, D. J. (1976). Some high order elements for arches and shells. In *Finite Elements for Thin Shells*, (ed. D. G. Ashwell and R. H. Gallagher). Wiley, London.
11. Mohr, G. A. (1980). A numerically integrated triangular element for doubly curved thin shells. *Comput. Struct.*, **11**, 565.
12. Dym, C. L. (1974). *Introduction to the Theory of Shells*. Pergamon, Oxford.
13. Southwell, R. V. (1951). *Theory of Elasticity*, (2nd edn). Dover, New York.

3
Energy principles
and applications

The energy basis used in the development of displacement finite elements is demonstrated and the displacement method is then used to establish again the matrices for the spar and beam elements discussed in Chapter 1. The displacement method is then generalized to include initial strains and boundary terms and the chapter closes with a discussion of alternative variational approaches to finite element formulation.

3.1 Theorem of minimum potential energy and virtual work

In Chapter 1 we operated directly on ordinary differential equations to derive stiffness matrices for beam elements. The displacement method described in this chapter is an energy or variational method in which an energy *functional* is minimized with respect to the nodal displacements to obtain the equilibrium equations for an element. In the case of the displacement method the energy functional is the integral of the total potential energy over the element volume, that is the sum of the strain energy and the potential energy of the applied loads.

The total potential energy of a linearly elastic solid in three-dimensional Cartesian coordinates (x, y, z) is given by

$$\Pi = \tfrac{1}{2}\int \{\varepsilon\}^{t}\{\sigma\}\,\mathrm{d}V - \int \{\bar{u}\}^{t}\{T\}\,\mathrm{d}S \qquad (3.1)$$

where

$$\{\varepsilon\} = \{\varepsilon_x, \varepsilon_y, \varepsilon_z, \varepsilon_{xy}, \varepsilon_{yz}, \varepsilon_{zx}\}$$

$$\{\sigma\} = \{\sigma_x, \sigma_y, \sigma_z, \sigma_{xy}, \sigma_{yz}, \sigma_{zx}\}$$

$$\{\bar{u}\} = \{u, v, w\} \quad \text{and} \quad \{T\} = \{t_x, t_y, t_z\}$$

are respectively the strains refered to the x, y, z-axes, the stresses referred to the x, y, z-axes, the displacements parallel to these axes, and the surface tractions (with dimensions of force per unit area) parallel to these axes. In eqn (3.1) $\mathrm{d}S$ and $\mathrm{d}V$ are infinitesimal increments of the surface area and volume of the solid. It is assumed that the stresses and strains are related by

$$\{\sigma\} = D\{\varepsilon\} \tag{3.2}$$

where D is the modulus matrix, numerous examples of which were given in Chapter 2.

In the finite element method U is computed for a number of subdomains or finite elements and the potential energy of the system of elements used to represent the complete domain is given by summation of the element energies, that is, by the summation $U = \Sigma U_e$. In the displacement method, however, we must first express the strains in each element as a function of the element's nodal displacements as these are the variables for which a solution is sought. Within elements, which are thought of as relatively small in relation to the complete domain, it is assumed that the displacements can be represented with sufficient accuracy by the interpolation

$$\{\bar{u}\} = F^t\{d\} \tag{3.3}$$

where $\{d\}$ is the vector of nodal displacements of the element. The form of the matrix F for a three-dimensional finite element was defined in Section 2.3 (in eqn (2.10)) and, as shown there, by differentiating this according to the strain–displacement relationships of eqns (2.7) we obtain the relationship

$$\{\varepsilon\} = B\{d\} \tag{3.4}$$

where B is the strain-interpolation or strain–displacement matrix.

Once the matrices D, B and F have been defined we can determine the stresses, strains, and displacements at any point in an element and hence evaluate its strain energy. To this end we substitute eqns (3.2)–(3.4) into eqn (3.1) to obtain

$$\Pi_e = \tfrac{1}{2}\{d\}^t \int B^t D B \, dV \{d\} - \{d\}^t \int F\{T\} \, ds \tag{3.5}$$

where the nodal displacements $\{d\}$ are placed outside the integrations as they are spatially independent, that is, they are at fixed points in space. Then, minimizing eqn (3.5) with respect to each of the nodal dispacements in turn, which we denote as $\partial \Pi_e / \partial \{d\}$, we obtain

$$\frac{\partial \Pi_e}{\partial \{d\}} = \int B^t D B \, dV \{d\} - \int F\{T\} \, dS = k\{d\} - \{q_t\} = \{0\} \tag{3.6}$$

where k is the *element stiffness matrix* and $\{q_t\}$ is the vector of *consistent loads* corresponding to the surface tractions. These are the nodal loadings which, in the *basis F* of eqn (3.3), contribute the same change in potential energy as the surface tractions when undergoing displacements $\{d\}$.

Then eqns (3.6) are the equilibrium equations for the element as minimization of the total potential energy implies that a state of equilibrium exists.[1] Note, however, that as it stands eqn (3.6) is not correct because it implies that if some elements do not carry surface tractions then these must have zero nodal displacements and this is nonsensical. This anomaly arises because surrounding elements will in fact exert forces on a particular element under consideration. In Section 1.2 we termed these forces the interelement reactions and these first appeared in eqn (1.12) as the vector $\{q_r\}$. Therefore in the finite element method we must include an additional term in eqn (3.5), giving the element potential energy as

$$\Pi_e = \tfrac{1}{2}\{d\}^t \int B^t DB \, dV \{d\} - \{d\}^t \int F(T) \, dS - \{d\}^t \{q_r\}. \qquad (3.7)$$

Minimization of Π_e with respect to $\{d\}$ leads to

$$k\{d\} = \{q_t\} + \{q_r\} = \{q\}. \qquad (3.8)$$

Then summing over all elements and including any loads directly applied at the nodes in a vector $\{Q_n\}$ we obtain

$$\Sigma k\{D\} = \Sigma\{q_t\} + \Sigma\{q_r\} + \{Q_n\}$$

or

$$K\{D\} = \{Q_t\} + \{0\} + \{Q_n\} = \{Q\} \qquad (3.9)$$

where K is the system stiffness matrix and $\{Q\}$ the system load vector. The summation procedure to form K is as described in Section 1.3 and, as noted in Section 1.2, it is assumed that $\Sigma\{q_r\} = \{0\}$, that is, *the interelement reactions cancel between elements.*

In the equations for a six-freedom beam/column element derived in Section 1.2 (eqns (1.12)), for example, we had

$$\{q_r\} = \{F_{u1}, F_{v1}, M_1, F_{u2}, F_{v2}, M_2\}. \qquad (3.10)$$

Considering the moments, for example, in two abutting elements i and j we have moments M_{ni} and M_{nj} at a shared node n. Unless there is a concentrated moment loading applied at this node, and this is rarely the case, the magnitudes of M_{ni} and M_{nj} will equal the magnitude of the bending couple but for such a bending couple M_{ni} and M_{nj} are counterposed moments, that is, $M_{ni} = -M_{nj}$ so that these interelement reactions do indeed cancel. The same argument applies to the shear forces and thrusts between elements whereas in the simple d.c. networks discussed in Section 1.10, to cite another simple example, cancellation of the interelement reactions, in this case being the current outflows at their nodes, is required by Kirchhoff's current law.

In most two- and three-dimensional finite elements, however, the interelement reactions will only cancel approximately, the degree of approximation diminishing as finer meshes of elements are used to model the continuum. Approximation, of course, is fundamental to the philosophy of the finite element method and the questions of interelement continuity and element accuracy are taken up later in the chapter.

It is worth noting that eqn (3.9) can also be arrived at by a *virtual work* argument and in Chapter 17 we shall also see that the virtual work method is equivalent to the Galerkin weighted residual method. The latter is widely used in fluid mechanics and similar problems to form finite element equations by direct operation on the governing differential equation. The virtual work method, therefore, can be used to obtain both energy and weighted residual finite element formulations and is thus of fundamental importance.

To obtain a displacement finite element formulation by the virtual work method we allow an arbitrary variation in the element's nodal displacements $\{\delta d\}$. This will cause variations in the stresses, $\{\delta\sigma\}$, strains $\{\delta\varepsilon\}$, and displacements, $\{\delta\bar{u}\}$, at any point in the element. Then if we equate the resulting change in potential energy of the loads with the change in strain energy in the element (equivalent to putting $\delta\Pi_e = 0$) we obtain

$$\int\{\delta\varepsilon\}^t\{\sigma\}\,dV = \int\{\delta\bar{u}\}^t\{T\}\,dS + \{\delta d\}^t\{q_r\} \tag{3.11}$$

remembering now to include the interelement reactions. From eqns (3.2)–(3.4) it follows that we can substitute $\{\sigma\} = D\{\varepsilon\} = DB\{d\}$, $\{\delta\bar{u}\} = F^t\{\delta d\}$, and $\{\delta\varepsilon\} = B\{\delta d\}$ in eqn (3.11) and hence obtain

$$\{\delta d\}^t \int B^t\, DB\, dV\{d\} = \{\delta d\}^t \int F\{T\}\,dS + \{\delta d\}^t\{q_r\} \tag{3.12}$$

where $\{\delta d\}$ falls outside the integrations because it is spatially independent. Then as the arbitrary variation $\{\delta d\}^t$ is a common factor throughout eqn (3.12) it can be deleted, leading to the same element equations as those obtained from the theorem of minimum potential energy, that is, eqn (3.8). A similar calculation is used in the classical calculus of variations, for example in deriving the Euler–Lagrange equations.[2]

In applications of the virtual work method such as that above it is common practice to imagine that the variations $\{\delta d\}$ are small. In problems of linear elasticity, however, this restriction is not necessary and these variations can indeed be arbitrary in magnitude.[1] In non-linear problems, however, the 'virtual' increment $\{\delta d\}$ is thought of as small otherwise in non-linear material problems, for example, the modulus matrix D alters and results like that of eqn (3.12) would be invalidated.

3.2 Duality in continuum finite elements

In Section 1.6 we discussed the concept of duality in relation to skeletal finite elements, arriving at the duality property $B = C^t$ for a simple truss structure. We can now give a more general proof of this property using the virtual work method, extending the argument to deal with continuum finite elements.

In the case of a continuum element the dual relationship to that of eqn (3.4) is

$$\{q\} = \int C\{\sigma\} \, dV \tag{3.13}$$

that is, integration over the element volume is now needed. This relationship gives the nodal loads which correspond to a given state of stress $\{\sigma\}$ in the element.

Allowing a virtual increment in the nodal displacements $\{\delta d\}$, resulting in stress increments $\{\delta\sigma\}$, strain increments $\{\delta\varepsilon\}$, and displacement increments $\{\delta\bar{u}\}$ throughout the element, the work done by the loads $\{q\}$ in moving through the displacements $\{\delta d\}$ (that is, the change in potential energy of the loads) is the 'external work'

$$W_e = \{q\}^t \{\delta d\}. \tag{3.14}$$

Invoking the Maxwell–Betti reciprocal theorem of linear elasticity we can also write

$$W_e = \{\delta q\}^t \{d\} \tag{3.15}$$

where $\{\delta q\}$ are the loads corresponding to the increment $\{\delta d\}$ and these can be calculated using eqn (3.13).

The strain energy or 'internal work' corresponding to eqn (3.14) is

$$W_i = \int \{\sigma\}^t \{\delta\varepsilon\} \, dV. \tag{3.16}$$

Again applying the Maxwell–Betti reciprocal theorem we can also write

$$W_i = \int (\varepsilon)^t \{\delta\sigma\} \, dV. \tag{3.17}$$

Then equating the results of eqns (3.15) and (3.17) we obtain

$$\{\delta q\}^t \{d\} = \int (\varepsilon)^t \{\delta\sigma\} \, dV. \tag{3.18}$$

Substituting eqns (3.4) and (3.13) into eqn (3.18) leads to

$$(\int C\{\delta\sigma\} \, dV)^t \{d\} = \int (B\{d\})^t \{\delta\sigma\} \, dV$$

or

$$\{d\}^t \int C\{\delta\sigma\} \, dV = \{d\}^t \int B^t \{\delta\sigma\} \, dV \tag{3.19}$$

from which it follows that $C = B^t$ so that eqns (3.4) and (3.13) are indeed dual statements.

As an example consider a four-freedom spar element with freedoms u and v at each end. If the element is at an angle α to the x-axis these displacements are easily resolved to calculate those parallel to the element. Subtracting the results and dividing by the length the strain is given by

$$\varepsilon = [-c/L, -s/L, c/L, s/L] \{u_1, v_1, u_2, v_2\} = B\{d\} \tag{3.20a}$$

where $c = \cos\alpha$ and $s = \sin\alpha$.

If we draw inward-directed tensile forces $T = \sigma A$ at each end of the member we can easily resolve these forces into components parallel to the axes. Reversing the signs of these components the nodal forces required to equilibrate the tension forces in the element are then given by

$$\{q_r\} = \left\{ \begin{matrix} F_{u1} \\ F_{v1} \\ F_{u2} \\ F_{v2} \end{matrix} \right\} = \left\{ \begin{matrix} -c \\ -s \\ c \\ s \end{matrix} \right\} T = \left\{ \begin{matrix} -c/L \\ -s/L \\ c/L \\ s/L \end{matrix} \right\} \sigma(AL) = \int C\sigma \, dV \quad (3.20b)$$

and comparing eqns (3.20a) and (3.20b) we see that indeed $C = B^t$.

Although the matrices for this simple element can be formed by inspection, integration is implicitly involved in eqn (3.20b) so that it does provide an adequate demonstration of the duality of eqns (3.4) and (3.13). As we saw in Section 1.6 this duality property also applies at structure level for simple truss elements. This is because the interelement reactions between these elements cancel exactly. In two- and three-dimensional finite elements, however, there is generally only partial continuity of stresses between elements at most. In principle, duality still holds but in practice it is not generally possible to independently form matrices C and B which are the transpose of each other in most finite elements.

Towards the close of the text we turn our attention to optimization of finite element models and duality is a fundamental concept in the theory of optimization. It appears in Section 22.3, for example, where the primal and dual linear programming problems can be respectively stated as

$$\text{Min}(z = \{c\}^t \{x\}) \quad \text{subject to constraints} \quad A\{x\} + I\{y\} = \{b\}$$

$$\text{Max}(z = \{b\}^t \{y\}) \quad \text{subject to constraints} \quad -A^t\{y\} + I\{x\} = \{c\}.$$

Indeed many other examples of duality appear in modern mathematics. In the truss problem of Section 1.6, for example, the *primal variables* are the nodal displacements and these are the unknowns in the *primal problem*. In the *dual problem* the unknowns are forces or stresses. In the mechanics of solids the primal form of the problem is solved by *stiffness methods* and the dual form is solved by *flexibility methods*; these are discussed at the close of the chapter.

In the d.c. network problem of Section 1.10, on the other hand, the primal variables are the nodal potentials and the dual variables are loop currents around closed loops in the circuit. Likewise, pipe network problems can be solved in terms of either the nodal pressure heads or loop flows around closed circuits. Thus in the mechanics of fluids the primal variables are pressures and the dual variables are flows.

In both the mechanics of solids and fluids *mixed formulations* which use both primal and dual variables as unknowns can also be obtained. These are discussed briefly at the close of the chapter and examples of these will appear later in the text. Finally, if we substitute B^t for C in eqn (3.13) we obtain $\{q\} = \int B^t \{\sigma\} \, dV$ and this relationship can be used to calculate thermal loads in finite elements (see Section 3.6) and *element reactions* in non-linear problems (see Chapter 12).

3.3 Energy formulation for spar and beam finite elements

We have already derived the stiffness matrices for spar and beam finite elements in Chapter 1 by directly integrating their governing ordinary differential equations. It provides a useful demonstration of the displacement method introduced in Section 3.1, however, to use this to reproduce the results obtained in Chapter 1.

First, considering the two-freedom spar element shown in Fig. 3.1, the first step is to derive the interpolation functions f_1 and f_2 for the interpolation

$$u = F^t \{d\} = \{f\}^t \{u_1, u_2\} = f_1 u_1 + f_2 u_2.$$

If the origin is at the centre of the element, as shown in Fig. 3.1, the required functions are easily established by inspection as

$$\{f\} = \{\tfrac{1}{2} - x/L, \tfrac{1}{2} + x/L\}. \tag{3.21}$$

Then the strain in the element is given by differentiating these with respect to the coordinate x:

$$\varepsilon_x = \frac{du}{dx} = \frac{d(\{f\}^t \{d\})}{dx} = [-1/L, 1/L]\{d\} = B\{d\}. \tag{3.22}$$

As the interpolation functions were linear we see that differentiation to form the strain–displacement matrix B results in a matrix with constant entries, that is, the element can be described as a constant strain element.

As there is only a single strain the modulus matrix reduces to a scalar and the element stress is given by $\sigma = E\varepsilon$. Then using eqn (3.6) and replacing

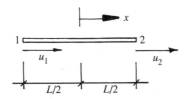

Fig. 3.1 Spar element with two freedoms

matrix D by the scalar E the element stiffness matrix is given by

$$k = \int B^t D B \, dV = \int \left\{ \begin{array}{c} -1/L \\ 1/L \end{array} \right\} \, E[-1/L, 1/L] \, A \, dx = \frac{EA}{L} \left[\begin{array}{cc} 1 & -1 \\ -1 & 1 \end{array} \right]$$

(3.23)

assuming that the element has constant E and constant cross-sectional area, A, so that the volume integral $\int dV$ is replaced by $\int A \, dx = A \int dx = AL$.

The stiffness matrix obtained in eqn (3.23) is identical to that obtained from eqn (1.12) by deleting the flexural stiffnesses and retaining only the extensional stiffnesses, that is, the k_1 terms. The procedure used in eqn (3.23), however, can also be applied to a wide range of two- and three-dimensional finite elements, as we shall see in the following chapters.

As a further example we apply the energy method to the task of forming the element stiffness matrix for the four-freedom beam element shown in Fig. 3.2. This is not a simple constant strain element like the two-freedom spar element of Fig. 3.1 (in fact it allows a linear variation of the generalized strain χ_x, that is, the curvature). Therefore it provides a better example of integration to form the element stiffness matrix than does eqn (3.23).

As the slope ϕ is a function of v (that is $\phi = dv/dx$) again only a single interpolation is needed. This will be for the transverse displacement v and values of the slope follow from this by differentiation. Generally we seek to express such an interpolation in the form $v = \{f\}^t \{d\}$ but the form of the required interpolation functions is no longer obvious, as it was for the simple two-freedom spar, largely because with four-freedoms available a cubic interpolation is permitted and the functions $\{f\}$ for this will be more complicated.

To resolve this difficulty we begin by choosing a simple cubic polynomial

$$v = c_1 + c_2 x + c_3 x^2 + c_4 x^3 = \{M\}^t \{c\}$$

(3.24)

which we term a *displacement function* to distinguish this from the use of interpolation functions $\{f\}$. In eqn (3.24) $\{M\} = \{1, x, x^2, x^3\}$ is the vector of *displacement modes* and $\{c\} = \{c_1, c_2, c_3, c_4\}$ is the vector of *displacement*

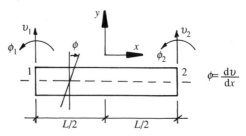

Fig. 3.2 Beam element with four freedoms

amplitudes. Using eqn (3.24) we can form a *kernel stiffness matrix* k' relating to the displacement amplitudes $\{c\}$ and this can be transformed to obtain the stiffness matrix relating to the displacements $\{d\} = \{v_1, \phi_1, v_2, \phi_2\}$. Moreover, as we shall show in Section 3.5, the transformation matrix required for this purpose (an *interpolation matrix* C) can be used to form the interpolation functions $\{f\}$ for the element and, as shown in Section 5.14, the element stiffness matrix can also be directly formed by using these.

First we differentiate eqn (3.24) to obtain a polynomial for the slope:

$$\phi = \frac{dv}{dx} = c_2 + 2c_3 x + 3c_4 x^2. \tag{3.25}$$

Then substituting the nodal values of the displacements on the left-hand sides of eqns (3.24) and (3.25), writing $x = -L/2$ and $x = L/2$ on the right-hand sides, and writing the results in matrix form we obtain

$$\{d\} = \begin{Bmatrix} v_1 \\ \phi_1 \\ v_2 \\ \phi_2 \end{Bmatrix} = \begin{bmatrix} 1 & -L/2 & L^2/4 & -L^3/8 \\ 0 & 1 & -L & 3L^2/4 \\ 1 & L/2 & L^2/4 & L^3/8 \\ 0 & 1 & L & 3L^2/4 \end{bmatrix} \begin{Bmatrix} c_1 \\ c_2 \\ c_3 \\ c_4 \end{Bmatrix} = C^{-1}\{c\}$$

$$\tag{3.26}$$

where we have denoted the connecting matrix as C^{-1} as we are more interested in its inverse C, the interpolation matrix.

Now the generalized strain for the element is the curvature χ_x which, if the displacements are small, is given by $\chi_x = d^2v/dx^2$. Differentiating eqn (3.24) twice we see that the amplitudes c_1 and c_2 do not contribute to χ_x and these are the amplitudes of the *rigid body modes* 1 and x. After inverting the matrix C^{-1} of eqn (3.26), therefore, we partition the result to delineate these modes. Hence we obtain

$$\{c\} = \begin{Bmatrix} c_1 \\ c_2 \\ c_3 \\ c_4 \end{Bmatrix} = \begin{bmatrix} 1/2 & L/8 & 1/2 & -L/8 \\ -3/2L & -1/4 & 3/2L & -1/4 \\ 0 & -1/2L & 0 & 1/2L \\ 2/L^3 & 1/L^2 & -2/L^3 & 1/L^2 \end{bmatrix} \begin{Bmatrix} v_1 \\ \phi_1 \\ v_2 \\ \phi_2 \end{Bmatrix} = \begin{bmatrix} C_{\mathrm{rb}} \\ C_{\mathrm{s}} \end{bmatrix} \{d\}.$$

$$\tag{3.27}$$

Now the slope ϕ shown in Fig. 3.2 for a typical cross-section of the beam causes a horizontal displacement $u = \phi y$ at a distance y from the neutral axis. Hence the stress at this point is given by

$$\sigma_x = E\varepsilon_x = \frac{E\,du}{dx} = \frac{E\,d(\phi y)}{dx} = E\frac{d^2v}{dx^2}y = E\chi_x y. \tag{3.28}$$

Then integrating eqn (3.28) through the depth of the beam we obtain the bending couple M_b as the stress resultant

$$M_b = \int \sigma_x y \, dA = \int (E\chi_x y) y \, dA = EI\chi_x. \tag{3.29}$$

The interpolation for the generalized strain χ_x in the beam is now obtained by differentiating eqn (3.24) twice and substituting for the displacement amplitudes $\{c\}$ using eqn (3.27). As the rigid body modes c_1 and c_2 do not contribute to χ_x we can omit these from subsequent calculations. This is done by using the bottom partition, C_s, of matrix C in eqn (3.27), and the interpolation for χ_x can then be expessed as

$$\chi_x = \frac{d^2 v}{dx^2} = 2c_3 + 6c_4 x = \{2, 6x\}^t C_s \{d\} = S_s C_s \{d\} = B\{d\}. \tag{3.30}$$

As integration over the cross-sectional area is implicit in the use of the stress resultant M_b the element strain energy is given by $U = \frac{1}{2} \int \chi_x M_b \, dx$. Hence the element stiffness matrix is given by

$$k = \int_{-L/2}^{L/2} B^t D B \, dx = C_s^t (\int S_s^t (EI) S_s \, dx) C_s = C_s^t k' C_s \tag{3.31}$$

replacing the modulus matrix D by a scalar (EI), as required by eqn (3.29), and bringing the constant matrix C_s outside the integration. Then if the beam has constant flexural rigidity EI the kernel stiffness matrix k' is obtained as

$$k' = \int S_s^t (EI) S_s \, dx = EI \int_{-L/2}^{L/2} \begin{bmatrix} 4 & 12x \\ 12x & 36x^2 \end{bmatrix} = EI \begin{bmatrix} 4L & 0 \\ 0 & 3L^3 \end{bmatrix}. \tag{3.32}$$

Observing eqn (3.31) the final element stiffness matrix is calculated as $k = C_s^t k' C_s$ and combining this result with eqns (3.27) and (3.32) we obtain the same result as eqn (1.10).

In Section 3.5 we shall show that eqn (3.27) can be used to form interpolation functions $\{f\}$ in terms of a *dimensionless coordinate* $s = \frac{1}{2} + x/L$. Differentiating these twice the matrix B of eqn (3.30) can be formed directly and this is the more usual procedure, eliminating the need for a kernel stiffness matrix. Cartesian polynomial interpolations and the kernel stiffness matrix approach will prove useful on many occasions later in the text, however.

3.4 Inclusion of initial and boundary terms in the displacement method

The discussion of Section 3.1 neglects some other important contributions to the potential energy of a finite element. These include those made by initial stresses and strains, body forces, and strain energy absorbed by elastic boundaries. In the present section we generalize the results of Section 3.1 to include these effects.

First we define a vector of *body forces* $\{X\} = \{X, Y, Z\}$ and a vector of *surface tractions* $\{p\} = \{p_x, p_y, p_z\}$ caused by the reactions on an elastic boundary. Including the potential energy associated with these and with the interelement reactions in eqn (3.1) the total potential energy of an element is given by

$$\Pi_e = \tfrac{1}{2} \int \{\varepsilon\}^t \{\sigma\} \, dV - \{d\}^t \{q_r\} - \int \{\bar{u}\}^t \{T\} \, dS$$
$$- \int \{\bar{u}\}^t \{X\} \, dV + \tfrac{1}{2} \int \{\bar{u}\}^t \{p\} \, dS. \qquad (3.33)$$

For simplicity we assume that the elastic boundary is isotropic, that is it has the same stiffness μ in each direction, so that the additional tractions $\{p\}$ can be expressed as $\{p\} = \mu\{\bar{u}\}$. Using the virtual work method we allow an arbitrary perturbation $\{\delta d\}$ and associated increments $\{\delta\sigma\}$, $\{\delta\varepsilon\}$, and $\{\delta\bar{u}\}$. The corresponding increment in the potential energy is

$$\delta\Pi_e = \int \{\delta\varepsilon\}^t \{\sigma\} \, dV - \{\delta d\}^t \{q_r\} - \int \{\delta\bar{u}\}^t \{T\} \, dS$$
$$- \int \{\delta\bar{u}\}^t \{X\} \, dV + \mu \int \{\delta\bar{u}\}^t \{\bar{u}\} \, dS \qquad (3.34)$$

assuming that the elastic boundary stiffness μ is constant.

To include the initial strains $\{\varepsilon_0\}$ we write

$$\{\sigma\} = D(B\{d\} + \{\varepsilon_0\}) \qquad (3.35)$$

that is, the corresponding initial stresses are simply $\{\sigma_0\} = D\{\varepsilon_0\}$ and the total strain after displacements $\{d\}$ have occurred is $\{\varepsilon\} = B\{d\} + \{\varepsilon_0\}$. Substituting eqn (3.35) and the incremental quantities $\{\delta\bar{u}\} = F^t\{\delta d\}$ and $\{\delta\varepsilon\} = B\{\delta d\}$ into eqn (3.34) we obtain

$$\delta\Pi_e = \{\delta d\}^t \int B^t DB \, dV \{d\} + \{\delta d\}^t \int B^t D\{\varepsilon_0\} \, dV - \{\delta d\}^t \{q_r\}$$
$$- \{\delta d\}^t \int F\{T\} \, dS - \{\delta d\}^t \int F\{X\} \, dV + \{\delta d\}^t \mu \int FF^t \, dS. \quad (3.36)$$

Cancelling the arbitrary variation $\{\delta d\}$ and putting $\delta\Pi_e = 0$ we obtain the equilibrium equations for the element. Placing the element and boundary stiffness terms on the left-hand side these can be written as

$$k\{d\} + k_b\{d\} = \{q_r\} + \{q_t\} + \{q_b\} + \{q_0\} = \{q\} \qquad (3.37)$$

where k is the usual element stiffness matrix and k_b is the *contact stiffness matrix* for the elastic boundary, these being given by

$$k = \int B^t DB \, dV \qquad (3.38a)$$

$$k_b = \mu \int FF^t \, dS. \qquad (3.38b)$$

In eqn (3.37) $\{q_r\}$ are the interelement reactions, $\{q_t\}$ are the surface traction loads, $\{q_b\}$ are the body force loads, and $\{q_0\}$ are the loads corresponding to the initial strains. The interelement reactions are unknown but assumed to cancel upon element summation and the other load terms are given by

$$\{q_t\} = \int F\{T\}\,dS \tag{3.39a}$$

$$\{q_b\} = \int F\{X\}\,dV \tag{3.39b}$$

$$\{q_0\} = -\int B^t D\{\varepsilon_0\}\,dV. \tag{3.39c}$$

Summing the element equations the system equations are obtained as

$$K\{D\} + K_b\{D\} = \{Q_n\} + \{Q_t\} + \{Q_b\} + \{Q_0\} \tag{3.40}$$

where $K = \sum k$, $K_b = \sum k_b$, and so on, and $\sum \{q_r\} = \{0\}$. However, in place of the self-equilibrating interelement reactions a vector containing any concentrated loadings specified at the nodes, $\{Q_n\}$, must be included.

In summary the loading terms on the right-hand side of eqn (3.40) respectively deal with:

(1) loads corresponding to the nodal freedoms applied directly at the nodes;

(2) Nodal loads corresponding to surface traction loads on the element boundaries; these are called *consistent loads* because they are calculated using the same interpolation functions and the same energy argument as is used to calculate the element stiffness matrix;

(3) Consistent loads corresponding to the action of body forces in the elements;

(4) The *element reactions* arising from initial stresses or strains which, for example, may be introduced by thermal expansion or lack of fit between components of a structure.

Examples of the use of consistent loads, the introduction of initial strains into finite element analysis, and the use of the contact stiffness matrix, k_b, are given in the three sections which follow.

In the case of initial stress problems the corresponding loadings are calculated by replacing $D\{\varepsilon_0\}$ by $\{\sigma_0\}$ in eqn (3.39c). Practically, therefore, it is not necessary to distinguish between the two types of problem as only the generalized Hooke's laws used to form D are needed to transform from one type to the other.

3.5 Consistent loads in a beam element

As an example of the calculation of consistent loads consider the four-freedom beam element shown in Fig. 3.3(a). This is subjected to a uniformly distributed transverse surface traction p per unit length acting in the positive

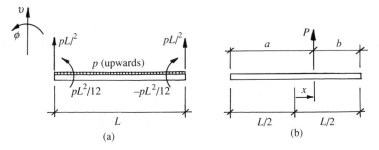

Fig. 3.3 Beam element with distributed and concentrated loadings

y-axis direction, that is, upwards. Alternatively in the case where $p = \rho A$, where ρ is the density of the beam, then the consistent loads to be calculated are those for body forces.

The interpolation functions for the transverse displacement v can be formed by substituting eqn (3.27) into eqn (3.24), yielding

$$v = \{M\}^t\{c\} = \{M\}^t C\{d\} = \{f\}^t\{d\}$$

so that the interpolation functions are given by $\{f\}^t = \{M\}^t C$. Carrying out this multiplication we obtain

$$f_1 = \frac{1}{2} - \frac{3x}{2L} + \frac{2x^3}{L^3}$$

$$f_2 = \frac{L}{8} - \frac{x}{4} - \frac{x^2}{2L} + \frac{x^3}{L^2}$$

$$f_3 = \frac{1}{2} + \frac{3x}{2L} - \frac{2x^3}{L^3} \tag{3.41}$$

$$f_4 = \frac{-L}{8} - \frac{x}{4} + \frac{x^2}{2L} + \frac{x^3}{L^2}.$$

Dividing f_2 and f_4 by L and substituting $x/L = s - \frac{1}{2}$, where s is a *dimensionless coordinate*, we obtain the dimensionless interpolation functions

$$f_1 = 1 - 3s^2 + 2s^3, \qquad f_2 = s - 2s^2 + s^3 \qquad s = 0 \to 1 \tag{3.42}$$

$$f_3 = 3s^2 - 2s^3, \qquad f_4 = s^3 - s^2$$

and the interpolation is written as

$$v = f_1 v_1 + f_2(L\phi_1) + f_3 v_2 + f_4(L\phi_2) = \{f\}^t\{d'\} \tag{3.43}$$

interpolating freedoms $\{d'\}$ of consistent dimensions. As there is only a single interpolation we replace F by $\{f\}$ in eqn (3.39a), yielding

$$\{q_t\} = \int_0^L p\{f\} \, dx = pL \int_0^1 \{f\} \, ds$$

$$= \{pL/2, pL^2/12, pL/2, -pL^2/12\} \qquad (3.44a)$$

where the loads for the second and fourth rotation freedoms have been multiplied by L, as required by eqn (3.43). Generally it is preferable to use these loads rather than the intuitive *lumped loads* $\{q_t\} = pL/2\{1, 0, 1, 0\}$.

For the case of an intermediate concentrated load P (Fig. 3.3(b)) the interpolation functions directly yield the displacement under the load and hence the change in potential energy, so that the consistent loads are obtained simply as

$$\{q_t\} = P\{f_1, f_2 L, f_3, f_4 L\} \qquad (3.44b)$$

where $\{f\}$ is obtained from eqns (3.42) with $s = a/L$.

More detailed discussion of dimensionless interpolations is given in Chapter 5 and many examples of consistent loads are given in later chapters.

3.6 Initial strains

The treatment of thermal stress problems provides a useful example of the introduction of initial strains in finite element analysis. First all nodes are fixed against movement caused by temperature increases, the resulting thermal loads are calculated, and these are added to the load vector for the system. Then the solution for the displacements proceeds in the usual way and finally the element stresses are calculated as the sum of the stresses caused by nodal displacements and the initial restraint stresses required to initially hold the nodes fixed.

Figure 3.4 illustrates this procedure, the *method of strain suppression*,[3] for a two-freedom spar element with a coefficient of linear thermal expansion α and subjected to a rise in temperature θ. For a positive temperature increase expansion of the element is prevented and as shown in Fig. 3.4 the associated restraining stresses are compressive so that, assuming tensile strains to be positive, we have $\varepsilon_0 = -\alpha\theta$. Substituting this result into eqn (3.39c) the

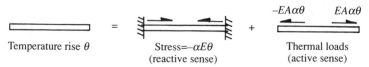

Temperature rise θ		Stress$=-\alpha E\theta$ (reactive sense)		Thermal loads (active sense)

Fig. 3.4 Method of strain suppression

thermal loads are obtained as

$$\{q_0\} = -\int B^t D\{\varepsilon_0\} \, dV = -\int B^t E \varepsilon_0 A \, dx$$

$$= \int_0^L \left\{ \begin{array}{c} -1/L \\ 1/L \end{array} \right\} E A \alpha \theta \, dx = \left\{ \begin{array}{c} -EA\alpha\theta \\ EA\alpha\theta \end{array} \right\} \tag{3.45}$$

and these are added to the system load vector.

In general the thermal loads between elements will not equilibrate at shared nodes and thus the nodal displacements will be affected. Then after the solution for the displacements $\{D\}$ has been obtained the element stresses are calculated from eqn (3.35), that is, $\{\sigma\} = D(B\{d\} + \{\varepsilon_0\})$.

As a simple example consider a single element like that in Fig. 3.4 but with the right-hand node completely free to move. Then the system equations are

$$\frac{EA}{L} \begin{bmatrix} 1 & -1 \\ -1 & 1 \end{bmatrix} \left\{ \begin{array}{c} u_1 \\ u_2 \end{array} \right\} = \left\{ \begin{array}{c} -EA\alpha\theta \\ EA\alpha\theta \end{array} \right\}. \tag{3.46}$$

Imposing the boundary condition $u_1 = 0$ we obtain the solution $u_2 = \alpha\theta L$. Hence the stress in the element is given by

$$\sigma = D(B\{d\} + \{\varepsilon_0\}) = E\left(\frac{u_2}{L} - \alpha\theta\right) = 0$$

giving a zero result as expected because the element is undergoing un-restrained expansion.

An important application of such thermal stress analysis is as a follow-up to heat flow analysis to determine the temperature distribution in a system (see Chapter 18). Thermal stress analysis can then determine the resulting stresses.

3.7 The contact stiffness matrix

The *contact stiffness matrix*[†] given by equation (3.38b) is very useful.[4] As an example, consider a beam simply supported from a rigid foundation at the ends and resting upon an elastic foundation material, as shown in Fig. 3.5.

If each element shown has a width b then eqn (3.38b) gives

$$k_b = \int F(\mu)F^t \, dS = b \int \{f\}\mu\{f\}^t \, dx = \frac{\mu b}{\rho A} m \tag{3.47}$$

where, as there is only a single interpolation, $F = \{f\}$, μ is the foundation stiffness in units of force/area/displacement and k_b is expressed as a scalar

[†] The term 'boundary stiffness matrix' might also seem appropriate but could be confused with the boundary element method.

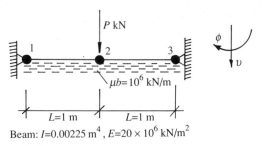

Beam: I=0.00225 m^4, E=20 × 10^6 kN/m^2

Fig. 3.5 Beam on elastic foundation

multiple of the consistent mass matrix, m, of eqn (13.41) as it involves the same vector product $\{f\}\{f\}^t$. Using symmetry and analysing the left half with boundary conditions $v_1 = 0$ and $\phi_2 = 0$, we obtain

$$\left[\frac{\mu b L}{420}\begin{bmatrix} 4L^2 & 13L \\ 13L & 156 \end{bmatrix} + \frac{EI}{L^3}\begin{bmatrix} 4L^2 & -6L \\ -6L & 12 \end{bmatrix}\right]\begin{Bmatrix} \phi_1 \\ v_2 \end{Bmatrix} = \begin{Bmatrix} 0 \\ P/2 \end{Bmatrix} \quad (3.48)$$

and this yields the solution $v_2 = 0.8197P \times 10^{-6}$ m for the deflection under the load whereas without the elastic foundation one obtains $v_2 = P \times 10^{-6}$ m. The exact solution for the problem is[3]

$$v_2 = \frac{P}{2\mu b}\frac{\sinh 2\beta L - \sin 2\beta L}{\cosh 2\beta L + \cos 2\beta L} \quad \text{where } \beta = (\mu b/4EI)^{\frac{1}{4}} \quad (3.49)$$

$$= 0.8366P \times 10^{-6} \text{ m}.$$

If, on the other hand, lumped stiffnesses are used for the boundary (that is, the term 156 in eqn (3.48) becomes 210 and the other terms in k_b vanish) one obtains $v_2 = 0.7874P \times 10^{-6}$ m. Hence the contact stiffness matrix does indeed give better results than the assumption of lumped 'spring' stiffnesses at the nodes.[4]

The foregoing procedure has been frequently applied to plates and shells resting upon elastic foundations[5-9] where the foundation stiffness per unit area may be estimated as $\mu = E_f/H$ where E_f and H are respectively Young's modulus of the foundation and the stressed depth of the foundation. Alternatively, appropriate stiffnesses can be estimated by using the solution to Boussineq's problem of a point-loaded half-space.[5]

Other applications of contact stiffness matrices include:

(1) modelling of adhesion of components to one another;

(2) In dynamical analysis of infinite domains where they allow realistic modelling of energy absorption at the boundaries of the finite element model.[10]

3.8 Continuity in finite element models

As we shall see in the following section the degree of continuity between abutting finite elements affects the rate of convergence with mesh refinement. The simple prismatic spar and beam elements considered thus far ensure thrust and moment continuity so that we consider instead the six- and twelve-freedom plane stress elements of Sections 6.1 and 6.3 as these elements are only C_0 continuous, that is, they provide continuity of the zeroth displacement derivatives and not of the strains.

The six-freedom *constant strain triangle* (CST) has freedoms u and v at each of the three vertices so that appropriate Cartesian interpolation polynomials are

$$u = c_1 + c_2 x + c_3 y \tag{3.50a}$$

$$v = c_4 + c_5 x + c_6 y. \tag{3.50b}$$

The magnitudes of the coefficients $c_1, \ldots, c_6$ can be determined by the matrix inversion procedure introduced for a beam element in Section 3.3. For the present purposes, however, we need only apply the strain–displacement relationships of eqns (2.16) to eqns (3.50), leading to the following expressions for the strains:

$$\varepsilon_x = \frac{\partial u}{\partial x} = c_2, \qquad \varepsilon_y = \frac{\partial v}{\partial y} = c_6$$

$$\varepsilon_{xy} = \frac{\partial v}{\partial x} + \frac{\partial u}{\partial y} = c_5 + c_3. \tag{3.51}$$

Then as the magnitudes of the coefficients c_2, c_3, c_5, and c_6 in eqns (3.51) depend only on the coordinates of the element's three nodes but are constant for a given element, we see that the strains are constant within each element. Hence in a mesh of these constant strain elements the strains will jump from constant values in one element to new values in an abutting element but the displacements will be represented continuously between elements. Hence the constant strain triangle is a C_0 element.

The twelve-freedom *linear strain triangle* (LST) has freedoms u and v at each of six nodes, three of these at the vertices and three at the middle of each side. Hence the appropriate Cartesian interpolation polynomials are

$$u = c_1 + c_2 x + c_3 y + c_4 x^2 + c_5 xy + c_6 y^2 \tag{3.52a}$$

$$v = c_7 + c_8 x + c_9 y + c_{10} x^2 + c_{11} xy + c_{12} y^2. \tag{3.52b}$$

These are termed *complete* quadratic interpolations as they contain all possible quadratic terms and differentiating them according to eqns (3.51) yields the following expressions for the strains:

$$\varepsilon_x = c_2 + 2c_4 x + c_5 y, \qquad \varepsilon_y = c_9 + c_{11} x + 2c_{12} y$$
$$\varepsilon_{xy} = c_8 + 2c_{10} x + c_{11} y + c_3 + c_5 x + 2c_6 y. \tag{3.53}$$

Clearly eqns (3.53) provide a linear variation of strain within each element but we must examine the displacement and strain variation along one side of an element to determine the degree of continuity across that side.

Assuming this side is horizontal and putting $y = 0$ in eqns (3.52) and (3.53) (this is easier than considering a typical side $y = ax + b$ which gives the same results after some rearrangement of terms) one obtains

$$u = c_1 + c_2 x + c_4 x^2, \qquad v = c_7 + c_8 x + c_{10} x^2$$
$$\varepsilon_x = c_2 + 2c_4 x, \qquad \varepsilon_y = c_9 + c_{11} x \tag{3.54}$$
$$\varepsilon_{xy} = c_3 + c_5 x + c_8 + 2c_{10} x.$$

The six coefficients for u, v can be uniquely determined by the six nodal freedoms available on this side but in the strains c_9, c_{11}, c_3, and c_5 are not uniquely determined and the strains will not fully continuous between elements. In the analysis of Fig. 7.7, for example, the stresses found at node 9 are

	Circumferential	Radial
Element 1	45.5044	0.117379
Element 2	45.5564	0.328757
Element 3	45.5564	0.328757
Element 4	45.5044	0.117379
Exact solution	45.4700	0.0

and it is the averaged value that is given in Table 7.1. The discrepancy between elements is not attributable to numerical error, the pairing off of some values arising only by virtue of symmetry, and is evidence that the elements lack C_1 continuity.

Observing eqn (3.6), minimization is with respect to the nodal displacements $\{d\}$ at element level so that the minimum principle (and thus convergence) is put at risk at structure level only if some displacement freedoms are not represented continuously between elements. Thus, though both the CST and LST elements are C_0 continuous, they respectively provide linear and quadratic continuity in u, v and in Section 3.9 we shall see that it is the latter consideration which determines the rates of convergence with these elements.

The question of elements which do not provide full continuity of the nodal freedoms arises with elements which include first derivatives (or slopes) as freedoms. This is seen if one examines the behaviour of a cubic polynomial. Rearranging its terms thus

$$w = c_1 + c_2 x + c_4 x^2 + c_7 x^3 + (c_3 y + c_5 xy + c_6 y^2$$
$$+ c_8 x^2 y + c_9 xy^2 + c_{10} y^3) \tag{3.55}$$

the terms in parentheses vanish on element sides with $y = 0$. Using the same rearrangement for the slopes one obtains

$$\frac{\partial w}{\partial x} = c_2 + 2c_4 x + 3c_7 x^2 + (c_5 y + 2c_8 xy + c_9 y^2) \tag{3.56a}$$

$$\frac{\partial w}{\partial y} = c_3 + c_5 x + c_8 x^2 + (2c_6 y + 2c_9 xy + 3c_{10} y^2). \tag{3.56b}$$

Then cubic elements with freedoms w, $\partial w/\partial x$, $\partial w/\partial y$ at the vertices have six freedoms on each side, sufficient to define c_1, c_2, c_4, c_7, c_3, and c_5 uniquely but not c_8. Thus only linear variation of the normal slope $\partial w/\partial y$ is uniquely defined on this side, which corresponds to quadratic variation of w in this direction. In Section 3.9 we see that this effects the rate of convergence of such *non-conforming* elements.

Such reduced continuity does not preclude convergence, however, so long as the element passes the patch test which is described in Section 3.10. To do so the interpolations assumed must obey certain simple rules and these are discussed in Section 5.2.

3.9 Continuity and errors in finite element solutions

Aside from numerical errors the two main considerations which affect the accuracy of a given finite element in well-behaved problems (that is, not containing singularities, for example) are the energy *truncation error* and *interpolation errors*. The error in the element evaluation of Π_e is the energy truncation error which is estimated by writing an interpolation of order p as

$$v = c_0 + c_1 x + c_2 x^2 + \cdots + c_p x^p + c_{p+1} x^{p+1} \tag{3.57a}$$

and the last term is the largest component of the truncation error for v. If the strains are mth order derivatives one has

$$\varepsilon = m! c_m + (m + 1)! c_{m+1} x + \cdots + \frac{p!}{(p - m)!} c_p x^{p-m}$$

$$+ \frac{(p + 1)!}{(p - m + 1)!} c_{p+1} x^{p-m+1} \tag{3.57b}$$

so that for an element of length h the truncation error for ε is said to be of order h^{p-m+1}, which we denote as $O(h^{p-m+1})$. In the case of two- or three-dimensional elements h is the *characteristic length*. For triangular elements, for example, this is calculated as $h = \sqrt{(2\Delta)}$.

To calculate the energy truncation error $\delta\Pi_e$ for an element suppose that the exact solutions for displacement and stress are $\{\bar{u} + \delta\bar{u}\}$ and $\{\varepsilon + \delta\varepsilon\}$ and that the finite element approximate solution is $\{\bar{u}\}$ and $\{\varepsilon\}$, that is, $\{\delta\bar{u}\}$ and $\{\delta\varepsilon\}$ are the errors in the finite element solution. Then the exact total potential energy in the element subdomain is

$$\Pi_e + \delta\Pi_e = \tfrac{1}{2}\int \{\varepsilon + \delta\varepsilon\}^t D\{\varepsilon + \delta\varepsilon\}\,\mathrm{d}V - \int \{\bar{u} + \delta\bar{u}\}^t\{T\}\,\mathrm{d}S$$

$$= \int (\tfrac{1}{2}\{\varepsilon\}^t D\{\varepsilon\} + \{\delta\varepsilon\}^t D\{\varepsilon\} + \tfrac{1}{2}\{\delta\varepsilon\}^t D\{\delta\varepsilon\})\,\mathrm{d}V$$

$$- \int (\{\bar{u}\}^t\{T\} + \{\delta\bar{u}\}^t\{T\})\,\mathrm{d}S. \tag{3.58a}$$

Assuming that the element is conforming and neglecting the interelement reactions (which cancel on element summation) the total potential energy of the approximate finite element solution in the subdomain occupied by an element is

$$\Pi_e = \tfrac{1}{2}\int \{\varepsilon\}^t D\{\varepsilon\}\,\mathrm{d}V - \int \{\bar{u}\}^t\{T\}\,\mathrm{d}S. \tag{3.58b}$$

Now we can allow an arbitrary perturbation in the strains and displacements of eqn (3.58b) and according to the principle of minimum potential energy the resulting increment in potential energy is zero. This is equivalent to the virtual work argument introduced at the close of Section 3.1.

If we make this perturbation $\{\delta\varepsilon\}$ and $\{\delta\bar{u}\}$, that is, equal to the errors in the finite element solution, we obtain

$$\int \{\delta\varepsilon\}^t D\{\varepsilon\}\,\mathrm{d}V - \int \{\delta\bar{u}\}^t\{T\}\,\mathrm{d}S = 0. \tag{3.58c}$$

Substituting this result in eqn (3.58a) and subtracting eqn (3.58b) from eqn (3.58a) the truncation error in the potential energy of the element is

$$\delta\Pi_e = \tfrac{1}{2}\int \{\delta\varepsilon\}^t D\{\delta\varepsilon\}\,\mathrm{d}V. \tag{3.58d}$$

In eqn (3.58d) we assume that the element volume is computed exactly so that integration does not affect the value of $\delta\Pi_e$. If we assume that k is calculated by numerical integration as $k = \sum B^t DB(\omega_i V)$ then the element volume V is known and indeed does not contribute to the error. Then if $\{\delta\varepsilon\}$ is $O(h^{p-m+1})$, as indicated by eqn (3.57b), it follows from eqn (3.58d) that

$$\delta\Pi_e = (\text{const.})h^{2(p-m+1)} \tag{3.59}$$

that is, the energy truncation error is $O(h^{2(p-m+1)})$ and in constant strain elements, for example, it is therefore $O(h^2)$.

The rate of convergence of the displacement solutions is also governed by eqn (3.59) and for a known truncation error $O(h^N)$ we can construct a very useful extrapolation formula. To this end we write the solutions for a typical displacement as d_i and d_j for two mesh sizes h_i and h_j ($h_j < h_i$) as

$$d_i = d^* - (\text{const.})h_i^N, \qquad d_j = d^* - (\text{const.})h_j^N$$

where d^* is an estimate of the solution in the limit of mesh refinement. Eliminating the constant from these equations we obtain the formula

$$d^* = d_j + \frac{d_j - d_i}{(h_j/h_i)^N - 1} \tag{3.60}$$

which in the case $N = 2$ is known as Richardson's extrapolation formula.[11] Ramstad finds Richardson's extrapolation to yield good results for the twelve-freedom 'ACM' cubic thin plate element (see Section 8.1),[12] despite the fact that for this $p = 3$ and $m = 2$ so that eqn (3.59) predicts h^4 convergence, not h^2 as implicit in Richardson extrapolation. This is because eqn (3.59) applies only when the element is conforming and for elements that are non-conforming or subject to constraints which limit their continuity one should use[13]

$$\delta\Pi_e = (\text{const.})h^{2(s-m+1)} \qquad s < p \tag{3.61}$$

where s is the order of the displacement polynomial defined continuously on each side and this can be determined in the manner described in Section 3.8. Table 3.1 shows predicted and observed rates of convergence for two membrane elements and four thin plate elements. The observed rates are deduced from eqn (3.60) or by using the formula

$$\alpha = \frac{2^N(d_k - d_j) + (\tfrac{1}{2})^N(d_j - d_i)}{d_k - d_i} \tag{3.62}$$

Table 3.1 Predicted and observed rates of convergence for a variety of finite elements.[13]

Element	Rates of convergence				Source of observation
	Displacement		Stresses		
	Predicted	Observed	Predicted[†]	Observed	
CST	h^2	h^2	h/h^2	h^2	Table 6.1
LST	h^4	h^4	h^3/h^2	–	Table 6.1
ACM	h^2	h^2	h/h^2	h^2	Table 8.1
Cubic	h^2	h^2	h/h^2	h^2	Table 8.1
Quartic	h^4	h^4	h^2/h^2	h^2	Table 8.3
Quintic	h^6	$h^{6\,††}$	h^4/h^2	h^2	Table 8.4

[†] from eqns (3.68)
[††] this rate is cited in reference 14

for solutions d_i, d_j, d_k obtained with element sizes successively halved (that is, $h_i = 2h_j = 4h_k$). For the correct rate h^N the value of α will be close to unity (noting that $N = 0$ is the trivial non-convergent case).

For the displacements eqn (3.59) holds for the two membrane elements but for the non-conforming ACM, cubic, and quartic elements and the constrained eighteen-freedom quintic element (eqn 3.61) is verified, these elements respectively providing quadratic, cubic, and quartic continuity in the interpolation. Rates of convergence are more difficult to predict in hybrid elements and in elements in which reduced integration or penalty factors are used and the latter two are discussed in later chapters.

Until now we have considered only the rate of strain energy convergence. The question of the rates of convergence of stress solutions in the displacement method is one in which interpolation errors, rather than the energy truncation error, play a dominant role. To demonstrate the effects of interpolation errors consider the four one-dimensional elements shown in Fig. 3.6 and let these occupy a subspace in which the variation in the displacement v is actually of the form

$$v = a + bx + cx^2 + dx^3 + ex^4 + fx^5. \tag{3.63}$$

Then the approximations a^*, b^*, c^*, d^* to the true amplitudes in eqn (3.63) given by the alternative elements are as follows[13]

(a) $\qquad a^* = \dfrac{v_1 + v_2}{2} - \dfrac{ch^2}{4} - \dfrac{eh^4}{16}$

$\qquad\quad b^* = \dfrac{v_2 - v_1}{h} - \dfrac{dh^2}{4} - \dfrac{fh^4}{16} \tag{3.64}$

(b) $\qquad a^* = v_2, \qquad b^* = \dfrac{v_3 - v_1}{h} - \dfrac{dh^2}{4} - \dfrac{fh^4}{16}$

$\qquad\quad c^* = \dfrac{2(v_3 + v_1 - 2v_2)}{h^2} - \dfrac{eh^4}{4} \tag{3.65}$

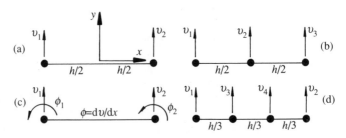

Fig. 3.6 Linear, quadratic, and cubic elements

(c) $a^* = \dfrac{v_1 + v_2}{2} + \dfrac{h(\phi_1 - \phi_2)}{8} + \dfrac{eh^4}{16}$

  $b^* = \dfrac{3(v_2 - v_1)}{2h} - \dfrac{\phi_1 + \phi_2}{4} + \dfrac{fh^4}{16}$

  $c^* = \dfrac{\phi_2 - \phi_1}{2h} - \dfrac{eh^2}{2}$

  $d^* = \dfrac{2(v_1 - v_2)}{h^3} + \dfrac{\phi_1 + \phi_2}{h^2} - \dfrac{fh^2}{2}$ (3.66)

(d) $a^* = \dfrac{9v_3 + 9v_4 - v_1 - v_2}{16} + \dfrac{eh^4}{144}$

  $b^* = \dfrac{27v_4 - 27v_3 + v_1 - v_2}{8h} + \dfrac{fh^4}{144}$

  $c^* = \dfrac{9(v_1 + v_2 - v_3 - v_4)}{4h^2} - \dfrac{5eh^2}{18}$

  $d^* = \dfrac{9(v_2 - v_1 + 3v_3 - 3v_4)}{2h^3} - \dfrac{5fh^2}{18}$ (3.67)

Even if eqn (3.63) were an infinite polynomial, eqns (3.64)–(3.67) capture the largest error terms. In case (a) the truncation error is $O(h^2)$ from eqn (3.61) and the interpolation error consists of two parts. The first arises from the energy truncation error in v_1, v_2 and is $O(h)$ for b^*, and the second is the *truncation error of the interpolation* which is $O(h^2)$. Generally only the first is considered but in the cubic cases the second is larger.

In the remaining cases one sees that the interpolation error is problem-dependent, for instance in the cubic cases with $m = 1$ the interpolation error is $O(h^2)$ if the x^2 displacement mode is dominant in the subspace whereas if the x mode is dominant the error will be $O(h^4)$. To take the pessimistic view the error may be as large $O(h^2)$. Thus in Table 3.1 the two interpolation errors are predicted as

$$O(h^{2(s+1)-3m}), \quad O(h^2) \qquad h < 1. \tag{3.68}$$

For the LST element a rate cannot be predicted as the test problem is one in which d, e, and f in eqn (3.63) are zero. The remaining cases in Table 3.1 show that the $O(h^2)$ truncation error of the interpolation is generally dominant. Observing Tables 8.2–8.4, for example, one sees that higher-order elements do indeed yield more accurate stress solutions with a given number of freedoms so that the stress accuracy with coarse meshes does depend upon the strain energy convergence rate. But thereafter, as the displacement solutions are more closely approximated the energy error cancels by subtraction in eqns 3.64–3.67 and the rate of stress convergence is governed by the $O(h^2)$ truncation error of the interpolation.

Observing eqn (3.65) one observes an advantage of the quadratic element, namely that the centroidal freedom provides better approximation of the rigid body modes, and this same phenomenon has been observed in two-dimensional thick plate and shell elements[15,16] (see also Fig. 9.3).

3.10 The patch test

The patch test was developed by Irons[17,18] to determine whether the effects of incompatibility in non-conforming elements disappear as the element mesh is refined. This test was based on Irons' proposition that elements must exactly model the rigid body and constant strain modes to ensure convergence. These criteria are important considerations in choosing element interpolation functions and this and related questions are taken up in Section 5.2.

Generally the patch test is carried out numerically for a *patch* of elements or a single element by applying a series of loadings which should produce the various rigid body and constant strain modes of behaviour which need to be modelled. If the elements do reproduce these modes the test is passed. Strang and Fix,[19] however, show that in the simple element of Fig. 3.7 the test can be carried out analytically. In what follows, therefore, we use the virtual work arguments developed in preceding sections to apply the test to this element. The interpolaton functions for the in-plane displacements u and v may be written as

$$u = \{f\}^t\{u\}, \qquad v = \{f\}^t\{v\}$$

where
$$f_1 = \tfrac{1}{4}(1 - x)(1 - y), \qquad f_2 = \tfrac{1}{4}(1 + x)(1 - y)$$
$$f_3 = \tfrac{1}{4}(1 + x)(1 + y), \qquad f_4 = \tfrac{1}{4}(1 - x)(1 + y). \tag{3.69}$$

The reader may easily verify the correctness of these functions by substituting the nodal coordinates shown in Fig. 3.7. The particular coordinates chosen simplify subsequent calculations but any other choice of coordinates would not affect our conclusions.

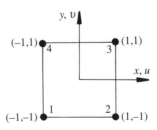

Fig. 3.7 Rectangular finite element for plane stress

We shall test the element by applying the *test functions*

$$u = a + bx + cy, \qquad v = 0 \qquad\qquad (3.70)$$

which clearly produce constant strains. Then considering an arbitrary perturbation of the u displacements, u^*, and assuming $v^* = 0$ to simplify calculations, it follows from eqn (2.19) that the resulting increment in strain energy is

$$\delta U_e = \tfrac{1}{2} \iint \left[\frac{\partial u^*}{\partial x}\frac{\partial u}{\partial x} + \tfrac{1}{2}\frac{\partial u^*}{\partial y}\frac{\partial u}{\partial y} \right] dx\,dy \qquad\qquad (3.71)$$

if Young's modulus is unity, Poisson's ratio is zero, and the element has unit thickness.

Now we integrate eqn (3.71) by parts to introduce the strains at the element boundary into this *energy product calculation*. Then for the two terms we obtain

$$\iint u_x^* u_x\,dx\,dy = \int u^* u_x\,dy - \iint u^* u_{xx}\,dx\,dy$$

$$\iint u_y^* u_y\,dx\,dy = \int u^* u_y\,dx - \iint u^* u_{yy}\,dx\,dy$$

where $u_x = \partial u/\partial x$, $u_{xx} = \partial^2 u/\partial x^2$, and so on, and the single integral terms are evaluated at the element boundary. We substitute these results into eqn (3.71) and equate the resulting expression to the original one, neglecting the second factor of $\tfrac{1}{2}$ which does not affect the final result (the first cancels as a common factor). We then obtain

$$\iint (u_x^* u_x + u_y^* u_y)\,dx\,dy - \int u^* u_x\,dy - \int u^* u_y\,dy + \iint u^*(u_{xx} + u_{yy})\,dx\,dy = 0.$$

Substituting $u^* = \{f\}'\{u^*\}$, $u_x^* = \{f_x\}'\{u^*\}$, and $u_y^* = \{f_y\}'\{u^*\}$ where $\{f_x\} = \{\partial f/\partial x\}$ and $\{f_y\} = \{\partial f/\partial y\}$ and noting that $u_{xx} = u_{yy} = 0$ (see eqn (3.70)) leads to the condition

$$\{u^*\}'[\iint (\{f_x\}u_x + \{f_y\}u_y)\,dx\,dy - \int \{f\}u_x\,dy - \int \{f\}u_y\,dx] = 0. \qquad (3.72)$$

Choosing the perturbation $\{u^*\} = \{u_1^*, 0, 0, 0\}$, substituting $u_x = b$, $u_y = c$, and $f_1 = (1 - x)(1 - y)/4$ and eliminating u_1^* as a common factor eqn (3.72) yields for the element of Fig. 3.7

$$\tfrac{1}{4} \int_{-1}^{1}\int_{-1}^{1} (1 - y)b\,dx\,dy + \tfrac{1}{4}\int_{-1}^{1}\int_{-1}^{1} (1 - x)c\,dx\,dy - \tfrac{1}{4}\int_{-1}^{1} (1 - x - y + xy)c\,dx$$

$$- \tfrac{1}{4}\int_{-1}^{1} (1 - x - y + xy)b\,dy - \tfrac{1}{4}\int_{-1}^{1} (1 - x - y + xy)c\,dx$$

$$- \tfrac{1}{4}\int_{-1}^{1} (1 - x - y + xy)b\,dy$$

$$= b + c - c + 0 + 0 - b = 0$$

noting that $y = -1$, $x = 1$, $y = 1$, and $x = -1$ are used in the last four integrals.

Corresponding to the perturbations u_2^*, u_3^*, and u_4^* the functions f_2, f_3, and f_4 also satisfy eqn (3.72). The use of perturbations v^* leads to a similar equation to (3.72) which is also satisfied by the interpolation functions. Hence the element passes the patch test with any arbitrary combination of displacements u^* and v^* at the nodes.

In the foregoing calculations we assumed Poisson's ratio to be zero and neglected the second factor of $\frac{1}{2}$ in eqn (3.71). Reintroduction of this factor does not affect the final result. This is because consideration of each term arising from eqn (2.19) separately leads to satisfaction of the foregoing test. From this we can conclude that this plane stress element passes the patch test for any value of Poisson's ratio.

The integration by parts procedure leading to eqn (3.72) had the effect of introducing integrals of $\partial u/\partial x$ and $\partial u/\partial y$ at the boundary, that is, strains. If we associate these with corresponding surface tractions $\{T\}$ then we can generalize the patch test by using eqn (3.11) to write[†]

$$\int \{\varepsilon^*\}^t D\{\varepsilon\}\, dV = \int \{\bar{u}^*\}^t \{T\}\, dS \tag{3.73}$$

where $\{\bar{u}^*\}$ are the perturbations and $\{\varepsilon^*\}$ the corresponding strains. Then the surface tractions at the element boundary can be expressed as

$$\{T\} = \begin{Bmatrix} T_x \\ T_y \end{Bmatrix} = G\{\sigma\} = GD\{\varepsilon\} = \begin{bmatrix} c_x & 0 & c_y \\ 0 & c_x & c_y \end{bmatrix} D \begin{Bmatrix} \partial u/\partial x \\ \partial v/\partial y \\ \partial v/\partial x + \partial u/\partial y \end{Bmatrix}$$

where c_x and c_y are the direction cosines of the outward normal at the element boundary. Then eqn (3.73) becomes

$$\{d^*\}^t \int B^t D\{\varepsilon\}\, dV = \{d^*\}^t \int FGD\{\varepsilon\}\, dS \tag{3.74}$$

where the strains $\{\varepsilon\}$ are calculated from the test functions.

Again using the test functions of eqns (3.70), allowing the perturbation $u^* = f_1 u_1^*$, and assuming that D is a unit matrix to simplify calculations the left-hand side of eqn (3.74) is obtained for the element of Fig. 3.7 as

$$u_1^* \int_{-1}^{1} \int_{-1}^{1} \tfrac{1}{4}[(1-y), 0, (1-x)] \begin{Bmatrix} b \\ 0 \\ c \end{Bmatrix} dx\, dy = (b+c)u_1^*$$

[†] Omitting the interelement reactions at the nodes. As usual these are assumed to be self-cancelling, at least in part, after element summation.

and the right-hand side is obtained as

$$u_1^* \int_{-1}^{1} \left[\tfrac{1}{4}(1-x)(1-y), 0\right] \begin{bmatrix} c_x & 0 & c_y \\ 0 & c_x & c_y \end{bmatrix} \begin{Bmatrix} b \\ 0 \\ c \end{Bmatrix} dS = (b+c)u_1^*.$$

The integral vanishes for sides 23 and 34 as these have $x = 1$ and $y = 1$ whilst for side 12 we have $c_x = 0$ and $c_y = -1$ and for side 34 we have $c_x = -1$ and $c_y = 0$. As expected this form of the test is passed.

Strang and Fix[19] establish that the element passes the patch test when distorted into a quadrilateral and also show that the nodeless quadratic interpolants introduced by Wilson[20] (see also eqns (6.44)) in a modified bilinear element also pass the patch test when the element is rectangular but not when distorted to obtain a quadrilateral.

Mitchell and Wait[21] describe another interesting approach to the patch test but both this and the foregoing procedure involve excessive algebra for high-order elements. These tests can be carried out in the computer program by applying loadings which correspond to constant strain fields in single elements (or pairs of triangles)[22] or by using a sequence of refined meshes, as is done on numerous equations in later chapters, and the latter procedure may reveal more about the overall performance of an element in practical situations.

In closing it should be remarked that

(1) we have not considered the question of convergence in the presence of singularities, near which poor approximations may be obtained;[23] such difficulties can be overcome by the use of special elements (see Section 7.7) or by using eqns (3.60) and (3.68) to obtain extrapolated solutions at the singularity (see Section 15.9);[13]

(2) we have considered only the energy truncation error in elliptic (see Section 16.5) equilibrium problems; a discussion of errors in parabolic and hyperbolic problems is given by Chung.[24]

3.11 Other energy methods

The two fundamental approaches first developed for matrix structural analysis were the stiffness and flexibility methods. Originally these were applied to the analysis of skeletal frames and for such structures it is possible to use the direct stiffness method (see Section 1.6) but in the analysis of continua, rather than frames composed of discrete members, the formation of matrices C and B may be difficult or impossible. Thus the standard finite element procedure is to form separate element stiffness or flexibility matrices and then assemble these to obtain the structure matrix in the manner illustrated in Section 1.3.

As we have seen, calculation of element matrices by the displacement method is based upon the principle of minimum potential energy. Alternative variational methods are summarized in Table 3.2 and these are here divided into three categories, not on the basis of the minimum principle employed but on the basis of the nature of the resulting element matrix.

The three major categories are the following.

(1) Stiffness methods. These lead to the formation of an element stiffness matrix and, generally, the nodal freedoms are displacements.

(2) Flexibility methods. These lead to the formation of an element flexibility matrix and the nodal freedoms are forces or stress functions.

(3) Mixed methods. These lead to the formation of an element stiffness matrix which is augmented by additional equilibrium equations and the nodal freedoms are a *mixture* of displacements and stresses or forces.

The displacement method has already been discussed in some detail in preceding sections and is much used in later chapters. The 'method of nested interpolations' is an extension of this in which interpolations on the element boundary are used to redefine the element freedoms prior to interpolation within the element. Such *basis transformation* procedures are of wide applicability in the finite element method and are discussed in Chapter 4 and examples of the method appear in Chapters 8 and 10.

The hybrid elements assume both displacement and stress or force fields, the equilibrium and 'stress-based' elements employ the minimum complementary energy principle whilst mixed elements originally employed the Reissner energy functional.[38] The Lagrange multiplier method is termed the 'generalized potential energy' method by Thomas and Gallagher[36] but is here classified as a mixed method because it employs two types of nodal freedoms and yields a stiffness matrix with a null partition, a characteristic of mixed elements.

It should be noted that the three broad classifications used in Table 3.2 correspond to the *primal, dual,* and *mixed* forms of a problem. In the mechanics of solids the primal form is governed by the constitutive relationship $\{\sigma\} = D\{\varepsilon\}$ and this is inverted in the dual form. The equilibrium method is perhaps the most basic dual method for finite elements and it is based on the *principle of minimum complementary energy.* First we assume an equilibrium stress field in the element

$$\{\sigma\} = C\{q_s\} \tag{3.75}$$

where $\{q_s\}$ are the *redundant forces* at the nodes, that is, the forces over and above those required to prevent rigid body motion of the element.

Table 3.2 Summary of various methods of finite element formulation. *Key*: P.E. potential energy; C.E. complementary energy; T.E. total energy; d element displacement field; F element force field; σ element stress field; ϕ element stress function field; b boundary displacement field; T boundary traction field; $\{d\}$ displacement freedoms at nodes; $\{\sigma\}$ stress freedoms at nodes; $\{\phi\}$ stress function freedoms at nodes; $\{\lambda\}$ Lagrange multiplier freedoms at nodes; $\{F_s\}$ redundant force freedoms at nodes;

	Assumed fields	Minimum principle	Nodal freedoms
Stiffness methods			
Displacement	d	P.E.	$\{d\}$
Nested interpolations[25, 26]	d, b	P.E.	$\{d\}$
Displacement hybrid II[27, 28]	d, b, T	P.E.	$\{d\}$
Stress hybrid[29]	σ, T, b	C.E.	$\{d\}$
Flexibility methods			
Equilibrium[30, 31]	σ, T	C.E.	$\{F_s\}$
Stress functions[32, 33, 34]	ϕ	C.E.	$\{\phi\}$
Displacement hybrid I[29, 35]	d, b, T	P.E.	$\{F_s\}$
Mixed methods			
Lagrange multipliers[36]	d, T	P.E.	$\{d\}, \{\lambda\}$
Reissner energy.[37, 38]	d, F	T.E.–C.E.	$\{d\}, \{\sigma\}$

The element complementary energy is written as

$$\Pi_e^* = \tfrac{1}{2} \int \{\sigma\}^t D^{-1} \{\sigma\} \, dV - \{q_s\}^t \{d_s\} \tag{3.76}$$

where $\{d_s\}$ are the displacement freedoms corresponding to $\{q_s\}$. Substituting eqn (3.75) into eqn (3.76) and minimizing with respect to $\{q_s\}$ we obtain the element equations

$$\int C^t D^{-1} C \, dV \{q_s\} = f_e \{q_s\} = \{d_s\} \tag{3.77}$$

where f_e is the *element flexibility matrix*.

In the four-freedom beam element of Fig. 3.2, for example, choosing the forces at the first node as the redundant or 'straining' forces we write

$$\{\sigma\} = M = [x, -1]\{F_{v1}, M_1\} = C\{q_s\}. \tag{3.78}$$

Substituting eqn (3.78) into eqn (3.77) the element flexibility matrix is obtained as

$$f_e = \frac{L}{6EI} \begin{bmatrix} 2L^2 & -3L \\ -3L & 6 \end{bmatrix} \tag{3.79}$$

and the reader can verify that inversion of this yields the upper left-hand partition of the stiffness matrix of eqn (1.10) for this element. This suggests a straightforward means of obtaining flexibility matrices, that is, one deletes rows and columns corresponding to the redundant forces and inverts the remaining matrix. If this is done for the four 2×2 partitions of the stiffness matrix of eqn (1.10) we obtain the flexibility relationship

$$\frac{L}{6EI} \begin{bmatrix} 2L^2 & -3L & L^2 & 3L \\ -3L & 6 & -3L & -6 \\ L^2 & -3L & 2L^2 & 3L \\ 3L & -6 & 3L & 6 \end{bmatrix} \begin{Bmatrix} F_{v1} \\ M_1 \\ F_{v2} \\ M_2 \end{Bmatrix} = \begin{Bmatrix} v_1 \\ \phi_1 \\ v_2 \\ \phi_2 \end{Bmatrix}. \qquad (3.80)$$

As a simple example of the flexibility method let us solve, using a single element, the problem of a cantilever with a transverse load P at node 2. Then we know $F_{v2} = P$ and $M_2 = 0$ so that we multiply column three of the flexibility matrix by P and transpose to the right side. Then deleting rows and columns three and four and noting that $v_1 = \phi_1 = 0$ we obtain

$$\begin{bmatrix} 2L^2 & -3L \\ -3L & 6 \end{bmatrix} \begin{Bmatrix} F_{v1} \\ M_1 \end{Bmatrix} = \begin{Bmatrix} -PL^2 \\ 3PL \end{Bmatrix} \qquad (3.81)$$

from which we obtain the solutions $F_{v1} = P$ and $M_1 = PL$. These are element forces and should be reversed in sign to obtain the boundary reactions.

To calculate the displacements one notes that F_{v1} and M_1 are equilibrated by the reactions at the boundary. Thus substituting $F_{v1} = M_1 = M_2 = 0$ and $F_{v2} = P$ in eqn (3.80) yields the results $v_2 = PL^3/3EI$ and $\phi_2 = PL^2/2EI$. The reader can easily verify these results by the displacement method by imposing v_1 and $\phi_1 = 0$ in eqn (1.10) and thus deleting the corresponding rows and columns in the stiffness matrix. Specifying the loads as $F_{v2} = P$ and $M_2 = 0$ we obtain

$$\frac{EI}{L^3} \begin{bmatrix} 12 & -6L \\ -6L & 4L^2 \end{bmatrix} \begin{Bmatrix} v_2 \\ \phi_2 \end{Bmatrix} = \begin{Bmatrix} P \\ 0 \end{Bmatrix}.$$

Solving these equations, identical solutions for v_2 and ϕ_2 are obtained.

Then using the first two rows of the original unreduced stiffness matrix the reactions at node 1 are obtained as

$$F_{v1} = \frac{EI(-12v_2 + 6L\phi_2)}{L^3} = -P, \qquad M_1 = \frac{EI(-6Lv_2 + 2L^2\phi_2)}{L^3} = -PL$$

as expected.

As variational formulations in following chapters are based upon the displacement or *nested interpolation* methods the interested reader is referred to the sources noted in Table 3.2 for details of the other methods listed therein and in particular to references 29 and 35 for an overview of them. Like those of basis transformation, Lagrange multiplier techniques, however, are widely applicable in the finite element method and these are discussed further in Chapter 4.

In summary, for the purposes of the present text, displacement and nested interpolation elements are frequently used, mixed-type elements may be obtained by the Lagrange multiplier technique (see Section 4.3), and stress-type elements can be obtained by applying the weighted residual techniques discussed in Chapter 17 directly to the governing differential equation. Indeed in the author's view these lines of approach represent the state of the art.

3.12 Energy bounds of finite element solutions

Compatible stiffness models should provide lower bounds on the strain energy and hence the displacements (see Table 6.1) and to demonstrate this we use an argument equivalent to that leading to eqn (3.58d) but now in terms of displacements rather than strains. The total potential energy of the system equations is

$$\Pi + \delta\Pi = \tfrac{1}{2}\{D + \delta D\}^t K\{D + \delta D\} - \{D + \delta D\}^t\{Q\} \qquad (3.82)$$

where $\{\delta D\}$ are the truncation errors in the approximate solution and these are of the same order as those at element level. The approximate solution is obtained as $\{D\} = K^{-1}\{Q\}$ and substituting this in eqn (3.82) and expanding the result gives, assuming K is symmetric,

$$\Pi + \delta\Pi = -\tfrac{1}{2}\{Q\}^t(K^{-1})^t\{Q\} + \tfrac{1}{2}\{\delta D\}^t K\{\delta D\}$$
or
$$\delta\Pi = \tfrac{1}{2}\{\delta D\}^t K\{\delta D\} \qquad (3.83)$$

so that the error $\delta\Pi$ is positive if K is positive definite and the energy approximation is a lower bound. Thus with 'stress' or equilibrium elements, where one minimizes the complementary energy Π^* (that is, maximizes Π) one obtains upper bounds on Π.[39]

These conclusions do not hold if the elements are incompatible as one is uncertain of the effect of discontinuities. It has been found, however, that some non-conforming displacement elements are similar to those formed by the hybrid or stress methods[40,41,42] and the same is true of elements in which reduced integration is used, as this results in non-conformity (see Section 6.6 for an example).[43] Given this similarity one expects that, in general, non-conforming elements will exhibit upper bound behaviour and this indeed is found to be the case. It has recently been shown that this is because summation of the integrated displacement discontinuities yields a result of larger magnitude than the result of eqn (3.83).[13]

Hybrid and mixed models may also yield either upper bound or lower bound behaviour and the objective is to so construct the model that the degrees of approximation of the interior and boundary fields are comparable in order to accelerate convergence; it is reported that some care in this regard will minimize ill-conditioning problems.[44]

3.13 General conclusions

In the present text variational elements are derived by the minimum potential energy principle and recent work indicates that the displacement method gives results as good as hybrid methods under certain circumstances. Of note in this regard is the work of Argyris *et al.*,[45] Bergan and Hanssen,[46] Razzaque,[41] and the method of nested interpolations,[25,26] where nine-freedom displacement elements approaching in accuracy that of the eighteen-freedom quintic element (see Figs 8.11–8.14) are derived.

Of the other variational methods the assumed stress hybrid elements are of interest as they employ only displacement freedoms and the assumed stress basis might be hoped to yield more accurate stress solutions. Many useful stress hybrid elements have been developed[28,47] though some of these are less accurate than comparable displacement elements[48] whilst displacement elements can be formulated with due consideration given to the stress or strain fields implicitly assumed by the chosen displacement functions. Examples of such elements occur in Sections 6.6 and 11.4. Research on force/flexibility and hybrid methods continues, promising better results in some cases,[49] but the displacement method has now become the generally accepted method.

References

1. Malvern, L. E. (1969). *Introduction to the Mechanics of a Continuous Medium.* Prentice-Hall, Englewood Cliffs NJ.
2. Irving, J. and Mollineaux, N. (1959). *Mathematics in Physics and Engineering.* Academic Press, London.
3. Timoshenko, S. (1956). *Strength of Materials, Part 2*, (3rd edn). Van Nostrand Reinhold, New York.
4. Mohr, G. A. (1980). The finite element contact stiffness matrix for problems involving external elastic restraint. *Comput. Struct.*, **12**, 189.
5. Cheung, Y. K. and Nag, D. K. (1968). Plates and beams on elastic foundations, linear and nonlinear behaviour. *Geotechnique*, **18**, 250.

6. Lee, I. K. (1973). Application of finite element method in geotechnical engineering. In *Finite Element Techniques*, (ed. A. P. Kabaila and A. S. Hall). Clarendon Press, Sydney.

7. Bhattacharya, B. and Ramaswamy, G. S. (1978). A finite element analysis of funicular shells on a two parameter foundation model. *Bull. Int. Assoc. Shell and Spatial Struct.*, **18** (3), 45.

8. Hain, S. J. (1977). A rational analysis for raft and raft-pile foundations. Ph.D. thesis, University of NSW, Sydney.

9. Pircher, H. and Beer, G. (1977). On the treatment of infinite boundaries in the finite element method. *Int. J. Num. Meth. Engng*, **11**, 1194.

10. White, W., Valliappan, S., and Lee, I. K. (1977). A unified boundary for finite dynamic models. *J. Eng. Mech. Div. ASCE*, **103**, 100.

11. Richardson, L. F. (1911). The approximate arithmetical solution by finite differences of physical problems involving differential equations, with an application to the stresses in masonry dams. *Phil. Trans. R. Soc.*, **210A**, 307.

12. Ramstad, H. (1969). Convergence and numerical accuracy with special reference to plate bending. In *Finite Elements in Stress Analysis*, (ed. I. Holand and K. Bell). Tapir, Trondheim.

13. Mohr, G. A. and Medland, I. C. (1983). On convergence of displacement finite elements, with an application to singularity problems. *Engng Fract. Mech.*, **17**, 481.

14. Cowper, G. W., Lindberg, G. M. and Olson, M. D. (1970). A shallow shell finite element of triangular shape. *Int. J. Solids Struct.*, **6**, 1133.

15. Pugh, E. D. L., Hinton, E., and Zienkiewicz, O. C. (1978). A study of quadrilateral plate bending elements with reduced integration. *Int. J. Num. Meth. Engng*, **12**, 1059.

16. Cook, R. D. (1974). *Concepts and Applications of Finite Element Analysis*. Wiley, New York.

17. Irons, B. M. (1966). Numerical integration applied to finite element models. Conference on Digital Computer Engineering, University of Newcastle.

18. Irons, B. M. and Razzaque, A. (1972). Experience with the patch test for convergence of the finite element method. In *Mathematical Foundations of the Finite Element Method*, (ed. A. Aziz). Academic Press, New York.

19. Strang, G. and Fix, G. J. (1973). *An Analysis of the Finite Element Method*. Prentice-Hall, Englewood Cliffs NJ.

20. Wilson, E. L., Taylor, R. L., Doherty, W. P., and Ghaboussi, J. (1971). Incompatible displacement models. Symposium on Numerical Methods, University of Illinois.

21. Mitchell, A. R. and Wait, R. (1977). *The Finite Element Method in Partial Differential Equations*. Wiley, London.

22. Olson, M. D. and Bearden, T. W. (1979). A simple triangular shell element revisited. *Int. J. Num. Meth. Engng*, **14**, 51.

23. Johnson, M. W. and McLay, R. W. (1968). Convergence of the finite element method in the theory of elasticity. *J. Appl. Mech. Div. ASCE*, 274.

24. Chung, T. J. (1978). *Finite Element Analysis in Fluid Dynamics*. McGraw-Hill, New York.

25. Mohr, G. A. (1981). Finite element formulation by nested interpolations: application to cubic elements. *Comput. Struct.* **14**, 211.
26. Mohr, G. A. (1982). Finite element formulation by nested interpolations: application to the drilling freedom problem. *Comput. Struct.*, **15**, 185.
27. McLay, R. W. (1969). A special variational principle for the finite element method. *J. Amer. Inst. Aeron. Astron.*, **7**, 533.
28. Tong, P. (1970). New Displacement hybrid finite element model for solid continua. *Int. J. Num. Meth. Engng*, **2**, 73.
29. Pian, T. H. and Tong, P. (1969). Basis of finite element methods for solid continua. *Int. J. Num. Meth. Engng*, **1**, 3.
30. Fraeijs de Veubeke, B. (1965). Displacement and equilibrium models in the finite element method. In *Stress Analysis*, (ed. G. S. Holister and O. C. Zienkiewicz). Wiley, London.
31. Morley, L. S. D. (1968). The triangular equilibrium element in the solution of plate bending problems. *Aero. Quart.*, **19**, 149.
32. Fraeijs de Veubeke, B. and Zienkiewicz, O. C. (1967). Strain energy bounds in finite element analysis by slab analogy. *J. Strain Anal.*, **2**, 265.
33. Robinson, J. (1973). *Integrated Theory of Finite Element Models*. Wiley, London.
34. Ahmad, S. and Irons, B. M. (1974). An assumed stress approach to refinited isoparametric finite element models in three dimensions. *Proceedings of the First Australian International Conference of Finite Element Methods*, Sydney.
35. Gallagher, R. H. (1975). *Finite Element Analysis Fundamentals*. Prentice-Hall, Englewood Cliffs, NJ.
36. Thomas, G. R. and Gallagher, R. H. (1976). A triangular element based on generalized potential energy concepts. In *Finite Elements for Thin Shells and Curved Members*, (ed. D. G. Ashwell and R. H. Gallagher). Wiley, London.
37. Herrmann, L. R. (1967). Finite element bending analysis for plates. *J. Eng. Mech. Div. ASCE*, **93** (EM 5), 15.
38. Reissner, E. (1950). On a variational theorem in elasticity. *J. Math. Phys.*, **29**, 90.
39. Fraeijs de Veubeke, B. and Sander, G. (1968). An equilibrium model for plate bending. *Int. J. Solids Struct.*, **4**, 447.
40. Morley, L. S. D. (1971). On the constant moment plate bending element. *J. Strain Anal.*, **6**, 20.
41. Razzaque, A. (1973). Program for triangular bending elements with derivative smoothing. *Int. J. Num. Meth. Engng*, **6**, 333.
42. Zienkiewicz, O. C. (1977). *The Finite Element Method*, (3rd edn). McGraw-Hill, London.
43. Kelly, D. W. (1979). Reduced integration to give equilibrium models for assessing the accuracy of finite element analysis. *Proceedings of the Third Australian Conference on Finite Element Methods*, University of NSW, Sydney.
44. Greene, B. E., Jones, R. E., McLay, R. W., and Strome, D. R. (1968). On the application of generalized variational principles in the finite element method. *AIAA/ASME Ninth Structures Dynamics and Materials Conference*, Palm Springs, Ca.
45. Argyris, J. H., Haase, M., and Mlejnek, H. P. (1980). On an unconventional but natural formation of a stiffness matrix. *Comp. Meth. Appl. Mech. Engng*, **22**, 1.

46. Bergan, P. G. and Hanssen, L. (1976). A new approach for deriving good element stiffness matrices. In *The Mathematics of Finite Elements and Applications*, Vol. 2, (ed. J. R. Whiteman). Academic Press, London.
47. Alaylioglu, H. and Ali, R. (1977). A hybrid stress doubly curved shell finite element. *Comput. Struct.*, **7**, 477.
48. Edwards, G. and Webster, J. J. (1976). Hybrid cylindrical shell finite elements. In *Finite Elements for Thin Shells*, (ed. D. G. Ashwell and R. H. Gallagher). Wiley, London.
49. Patnaik, S. N., Berke, L., and Gallagher, R. H. (1991). Integrated force method versus displacement method for finite element analysis. *Comput. Struct.*, **38**, 377.

4
Lagrange multipliers, penalty factors, and basis transformation

In the preceding chapter energy-based finite elements were discussed, paying most attention to the displacement method as this appears to have become dominant in solids mechanics applications. Moreover stress-type elements can be obtained more directly by applying the Galerkin method directly to the governing differential equations (as is done in Section 17.6) whilst the main incentive for the use of hybrid and mixed elements is to increase interelement stress and/or displacement continuity and this can be accomplished more generally, in any physical application, with the use of the techniques to be discussed in this chapter. Indeed the penalty factor and Lagrange multiplier techniques form a linear minimization problem which is a special case of the use of such techniques in non-linear optimization. The latter are further discussed in Chapter 22 where many other useful relationships and applications to the finite element method are to be found.

4.1 Introduction

The solution of any physical problem requires the imposition of boundary conditions. These act as *constraints* and play a large part in determining the nature of the solution obtained. Further, as we shall see throughout the text, many other types of constraint may need to be imposed in the finite element method.

Two techniques of imposing constraints which are now widely used in constrained optimization problems are the penalty factor and Lagrange multiplier methods. In the penalty factor technique the problem of minimizing a function $f(x)$ with several variables x_k, subject to several constraints $c_i(x) \leq 0$ is replaced by the problem of minimizing the augmented function

$$F(x, \beta) = f(x) + \beta \sum |c_i(x)|^2 \qquad (4.1)$$

where the *step function* $|c_i(x)|$ is taken to be zero unless the constraint is violated and β is the penalty factor (a large number). The object is now to minimize $F(x)$ as $\beta \to \infty$ and this implies a minimization of the constraint violations in a least squares sense, that is, as $\beta \to \infty$ the constraint violations must become increasingly small. In most finite element applications of this technique, however, $f(x)$ is a quadratic function, minimization of which leads to a linear problem soluble with a single (large) value of the penalty factor.

In the Lagrange multiplier technique, on the other hand, one seeks stationarity of the augmented function (the *Lagrangian function*)

$$F(x, \lambda) = f(x) + \sum \lambda_i e_i(x) \tag{4.2}$$

with respect to both the variables x_k and the Lagrange multipliers λ_i. Equation (4.2) holds only in the case of equality constraints $e_i(x) = 0$; inequality constraints are discussed in Chapter 22. When $F(x, \lambda)$ is minimized with respect to the variables x_k, however, it transpires that it is a maximum with respect to the Lagrange multipliers λ_i.

There are numerous applications of these techniques in numerical optimization and these are outlined in Chapter 22. The basic statements above, however, are similar to the expressions used in the finite element method where $f(x)$ is replaced by an energy function, leading in most cases to a linear problem.

4.2 Penalty factor techniques

The penalty factor technique, though most frequently used in optimization in recent decades, was first suggested in connection with Rayleigh–Ritz solutions to boundary value problems by Courant.[1] Its application in the finite element method leads to a more rigorous treatment of boundary conditions as well as many new applications.

In matrix form (cf. eqn (4.1)) the penalty factor finite element problem is written as the minimization of the integral

$$I(\{d\}) + \tfrac{1}{2}\beta \int [g\{d\}]^2 \, dV \tag{4.3a}$$

where, in applications to the mechanics of solids, we have

$$I(\{d\}) = \tfrac{1}{2}\int \{\varepsilon\}^t \{\sigma\} \, dV - \{d\}^t \{q\} \tag{4.3b}$$

and g is a constraint matrix which forms the additional constraints we wish to enforce. Equation (4.3a) can be rewritten as

$$I(\{d\}) + \tfrac{1}{2}\beta \{d\}^t (\int g^t g \, dV) \{d\} \tag{4.4a}$$

and after minimization with respect to $\{d\}$ one obtains[†]

$$k\{d\} - \{q\} + \beta (\int g^t g \, dV) \{d\} = \{0\}. \tag{4.4b}$$

[†] Hereafter when, for convenience, we minimize at element level $\{q\}$ is understood to include the interelement reactions which cancel, at least in the limit of mesh refinements, upon summation over the elements.

This differs from the result in eqn (3.8) only by the presence of the last term which may be added to the element stiffness matrix or, when the constraints are strains, the strain–displacement matrix B may include the 'g' rows as is done in the applications of Sections 9.5, 11.1, and 11.6.

Considering the two-element spar problem shown in Fig. 4.1 as an example suppose we wish to enforce the constraints $u_1 = 0$ and $u_2 = u_3$. These are *discrete constraints* (that is, the matrix g has constant coefficients)[†] and in the first element $g\{d\} = [1 \ 0]\{u_1, u_2\}$ whilst in the second element $g\{d\} = [1 \ -1]\{u_2, u_3\}$. The augmented element stiffness matrices are then $(k + \beta g^t g)$ with k given by eqn (3.23). After summation of the augmented element matrices the resulting penalty factor formulation is

$$\begin{bmatrix} c + \beta c & -c & 0 \\ -c & 2c + \beta c & -c - \beta c \\ 0 & -c - \beta c & c + \beta c \end{bmatrix} \begin{Bmatrix} u_1 \\ u_2 \\ u_3 \end{Bmatrix} = \begin{Bmatrix} 0 \\ 0 \\ P \end{Bmatrix} \tag{4.5}$$

where $c = EA/L$ and we replace β by βc so that $\beta \simeq \beta c$ as $\beta \to \infty$.

Solving eqn (4.5) one obtains the solutions

$$u_1 = \frac{P}{\beta c} \to 0 | \beta \to \infty$$

$$u_2 = \frac{P}{c}\left(1 + \frac{1}{\beta}\right) \to \frac{PL}{EA} \bigg| \beta \to \infty \tag{4.6}$$

$$u_3 = \frac{P}{c}\left[\frac{2}{\beta} + 1 - \frac{1}{\beta(1 + \beta)}\right] \to \frac{PL}{EA} \bigg| \beta \to \infty$$

so that the correct solutions are obtained as $\beta \to \infty$. In fact an optimal finite value exists which depends in part upon the precision of computation and this is discussed in Chapter 9.

Also worth noting is that the use of a large penalty number on the diagonal to suppress a displacement, such as u_1 in the example above, is numerically equivalent to the technique used in the program of Section 1.11. This

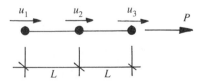

Fig. 4.1 Two-element spar problem

[†] Integral constraints in which g is a function of the coordinates are discussed in Chapter 23.

equivalence is seen if the first row and column in eqn (4.5) is divided by β, division of the column corresponding to a rescaling of the variable u_1. Hence as $\beta \to \infty$ one obtains unity on the diagonal and a null row and column; thus u_1 has been decoupled from the solution.

4.3 Lagrange multiplier techniques

Lagrange multipliers were introduced by Lagrange in connection with minimization of a dynamical energy function[2] and have since found many other applications. In matrix form (cf. eqn (4.2)) the Lagrange multiplier finite element problem is the minimization of

$$I(\{d\}) + \{\lambda\}^t g\{d\} \tag{4.7}$$

where the Lagrange multipliers $\{\lambda\}$ are variables. Once again the constraints are discrete and a Lagrange multiplier must be associated with each constraint (g containing a row for each constraint). The correct values in $\{\lambda\}$ enforce the constraints.

Minimizing eqn (4.7) with respect to $\{d\}$, after transposing the second term so that the remaining variables $\{\lambda\}$ are placed on the right, one obtains

$$k\{d\} - \{q\} + g^t\{\lambda\} = \{0\} \tag{4.8}$$

and minimizing eqn (4.7) with respect to $\{\lambda\}$ gives

$$g\{d\} = \{0\}. \tag{4.9}$$

Combining eqns (4.8) and (4.9) one obtains

$$\begin{bmatrix} \int B^t DB \, dV & g^t \\ g & 0 \end{bmatrix} \begin{Bmatrix} \{d\} \\ \{\lambda\} \end{Bmatrix} = \begin{Bmatrix} \{q\} \\ \{0\} \end{Bmatrix}. \tag{4.10}$$

The original applications of this technique to finite elements were to enforce continuity constraints between elements when, as pointed out by Pian and Tong,[3] the Lagrange multipliers are interpreted as surface tractions so that the freedoms involve a mixture of force and displacement quantities, giving rise to our classification of such elements as mixed in Secton 3.11.

As an example Fig. 4.2 shows two slope-incompatible cubic triangular thin plate bending elements with normal slopes specified in each element at the middle of the common side. (Such incompatibility or non-conformity was discussed in Section 3.8 where an example of how to determine whether thin plate finite elements are slope incompatible was given.) As the elements are incompatible these normal slopes will differ unless constrained to equality and Harvey and Kelsey[4] use Lagrange multipliers to enforce these constraints.

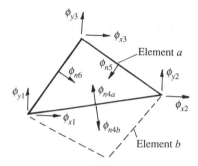

Fig. 4.2 Use of normal slope to develop interelement constraints

Following the procedure described in Section 5.9 the normal slopes may be expressed in terms of the other nine freedoms as

$$\begin{bmatrix} \phi_{n4} \\ \phi_{n5} \\ \phi_{n6} \end{bmatrix} = T\{w_1, \phi_{x1}, \phi_{y1}, w_2, \phi_{x2}, \phi_{y2}, w_3, \phi_{x3}, \phi_{y3}\} \qquad (4.11)$$

and the matrix T is similar in character to that in eqn (8.59). The normal slopes in each element will be inwardly directed, as shown in Fig. 4.2, and, if the global sign corrections of eqn (5.91) are omitted, we automatically obtain, upon assembly of the structure matrix, conditions of the form $\phi_{n4a} + \phi_{n4b} = 0$ yielding the matrix equation

$$G\{D\} = \{\delta\phi_n\} = \{0\} \qquad (4.12)$$

where $G = \sum T$. This corresponds to the bottom rows in eqn (4.10) and thus the complete equations are obtained in the form

$$\begin{bmatrix} K & G^t \\ G & 0 \end{bmatrix} \begin{Bmatrix} \{D\} \\ \{\lambda\} \end{Bmatrix} = \begin{Bmatrix} \{Q\} \\ \{0\} \end{Bmatrix} \qquad (4.13)$$

by summing the element matrices

$$\begin{bmatrix} k & T^t \\ T & 0 \end{bmatrix}. \qquad (4.14)$$

Here the Lagrange multipliers are the generalized stresses (in this instance moments) required to enforce the compatibility conditions at the interelement boundaries.[5]

Observing the form of eqn (4.13), pivoting may be needed during solution of the equations, though in Section 19.5, where a Lagrange multiplier format is obtained for Stokes' flow, it is shown that small numbers added to the null diagonal in eqn (4.13) are sufficient to obviate the need for pivoting. These small numbers associated with each constraint may be regarded as *small slack variables* and this concept is used in Chapter 23 to develop a method of optimization for finite elements.

As another example consider again the problem of Fig. 4.1. Expressing the constraints in Lagrange multiplier format one obtains the structure matrix as

$$
\begin{bmatrix}
c & -c & 0 & 1 & 0 \\
-c & 2c & -c & 0 & 1 \\
0 & -c & c & 0 & -1 \\
1 & 0 & 0 & 0 & 0 \\
0 & 1 & -1 & 0 & 0
\end{bmatrix}
\begin{bmatrix}
u_1 \\
u_2 \\
u_3 \\
\lambda_1 \\
\lambda_2
\end{bmatrix}
=
\begin{bmatrix}
0 \\
0 \\
P \\
0 \\
0
\end{bmatrix}
\tag{4.15}
$$

where again $c = EA/L$. Solving these by Gauss reduction a zero pivot appears in row three and this must be swapped with row four before the solution can proceed, when the final solution is obtained as

$$
u_1 = 0, \qquad u_2 = \frac{PL}{EA}, \qquad u_3 = \frac{PL}{EA}
$$
$$
\lambda_1 = P, \qquad \lambda_2 = -P.
\tag{4.16}
$$

The displacement solutions are correct and one sees that the Lagrange multipliers are the *negative values*[†] of the nodal forces needed to enforce the constraints. Note too that calculation of the tension in the second element in the usual way will yield the correct (zero) result without consideration of the value of λ_2.

The disadvantage of the Lagrange multiplier technique as a means of enforcing element constraints, compared to the penalty factor and basis transformation techniques, is the increased number of solution variables required. However, as shown in Chapter 23, it is possible to convert a Lagrange multiplier formulation to an equivalent penalty factor one. Such condensation is an important advance but the Lagrange multiplier approach nevertheless frequently provides the clearest original statement of the problem and is therefore still useful.

4.4 Basis transformation techniques

As shown in Section 3.9 non-conformity reduces the rates of convergence of finite elements. We can increase interelement continuity by including surface traction or displacement fields along the element boundary and hybrid stiffness, flexibility, and mixed methods which do this were discussed in the preceding chapter. An alternative approach to these is the *method of nested interpolations*[6–10] where immediate attention is focused upon the question of

---

† Shifted to the right side they act as enforcing loads.

interelement continuity by using boundary interpolations to redefine the element freedoms.

This method may be classified as one of *basis transformation*. Before considering examples of basis transformation, however, we shall clarify the term *basis* and some associated concepts, the first of which is that of a *linear space*. A linear space V contains elements which obey the usual addition and multiplication laws of algebra, examples being

(1) the sets of real and complex numbers;

(2) the set of infinite (Maclaurin) series;

(3) a set of unit vectors;

(4) the set of polynomials;

The elements of a bounded linear space also obey the closure axioms:

(1) for every pair x, y there is a unique element $x + y$;

(2) for every element x and real c their is a unique element cx.

A subset S of V which obeys the algebraic and closure axioms is called a *subspace* of V. Such a subset may include linear combinations of the elements of V and subsets comprising alternative linear combinations of the same elements are said to have the same *linear span*. For example, the pairs of subsets

$$\{\mathbf{\mathring{i}}, \mathbf{\mathring{j}}\}, \qquad \{\mathbf{\mathring{i}}, \mathbf{\mathring{j}}, \mathbf{\mathring{i}} + \mathbf{\mathring{j}}, -\mathbf{\mathring{j}}\} \tag{4.17a}$$

and $\qquad \{1, x, x^2\}, \qquad \{1, 1 + x, (1 + x)^2\} \tag{4.17b}$

span the same subspace.

A finite set S is *independent* if all linear combinations $\sum c_i x_i = 0$ only if all $c_i = 0$. For example the polynomial set $x_i = t^i$, $i = 0 \rightarrow n$ is obviously independent as the sum $\sum c_i x_i$ cannot vanish for $t \neq 0$ unless all the coefficients are zero. The set $\{\mathbf{\mathring{i}}, -\mathbf{\mathring{i}}, \mathbf{\mathring{j}}, -\mathbf{\mathring{j}}\}$, on the other hand, is obviously dependent.

A finite set S in a linear space V is called a *finite basis* for V if it is independent and spans V. If such a basis contains n elements then $n = \dim(V)$ is the dimension of S. In the finite element method, for example, we use for the four-freedom beam element a finite cubic basis $\{M\} = \{1, x, x^2, x^3\}$. Thus $\dim(\{M\}) = 4$ and $\{M\}$ is clearly linearly independent. Then when we obtain interpolation functions $\{f\}$ from the original polynomial this is, in fact, an exercise in basis transformation as we shall show.

As a first example of basis transformation consider coordinate transformation in two dimensions. Here the transformation of basis is expressed as

$$\begin{Bmatrix} x' \\ y' \end{Bmatrix} = \begin{bmatrix} c & s \\ -s & c \end{bmatrix} \begin{Bmatrix} x \\ y \end{Bmatrix} = T\{x, y\} \, .$$

If we express the original basis in linear combinatorial form, that is, each element of the basis is expressed as a vector with components x and y, we have

$$\mathbf{e} = \{e_1, e_2\} = \{1x + 0y, 0x + 1y\}.$$

Then the *inner product*, which is a generalization of the vector term dot product used to deal with general linear spaces,[11] is given by

$$(\mathbf{e}_i, \mathbf{e}_j) = 0 \qquad i \neq j \tag{4.18a}$$

$$(\mathbf{e}_i, \mathbf{e}_j) = 1 \qquad i = j. \tag{4.18b}$$

A basis possessing an inner product is called a *Euclidean space*. If such a basis obeys eqn (4.18a) it is said to be an *orthogonal basis* and if it also obeys eqn (4.18b) it is an *orthonormal basis* because its *Euclidean norm* $(\mathbf{e}_i, \mathbf{e}_i)^{1/2}$ is unity.

The reader can easily verify that the transformed basis $\{x', y'\}$ is also orthonormal and consequently the transformation matrix T is said to be orthogonal. As a result it possesses the property $T^{-1} = T^t$.

In the finite element method we use finite polynomial bases as approximating functions in each element. In the cubic beam element, for example, we can use the polynomial basis function

$$v = c_1 + c_2 x + c_3 x^3 + c_4 x^3 = \{c\}^t \{M\} \tag{4.19}$$

or use the interpolation functions derived in Section 3.5:

$$v = \{f\}^t \{v_1, \phi_1, v_2, \phi_2\} = \{f\}^t \{d\} \tag{4.20}$$

where

$$f_1 = 1 - 3d^2 + 2d^3, \qquad f_2 = L(d - 2d^2 + d^3) \qquad d = 0 \to 1.$$
$$f_3 = 3d^2 - 2d^3, \qquad f_4 = L(d^3 - d^2)$$

$\{f\}$ is an alternative basis for the element and these functions are obtained by substituting $d = x/L + \frac{1}{2}$ in eqns (3.41).

Equations (3.41) were obtained by the basis transformation $\{f\} = C^t \{M\}$ and this is equivalent to changing the element variables from coefficients $\{c\}$ to displacements $\{d\}$ using $\{c\} = C\{d\}$. As matrix C has constant entries it falls outside the integration over the element volume so that

$$k = C^t k^* C \tag{4.21}$$

where k^* is derived using the basis $\{M\}$. This was demonstrated in Section 3.3.

As we shall see in the following chapter, however, it is the usual practice to use interpolation functions $\{f\}$ to derive finite element equations and the transformation from a polynomial basis to the basis $\{f\}$ is implicit in the use of the latter.

In the method of nested interpolations *congruent transformation*, that is, eqn (4.21), with a matrix T which itself derives from interpolations is used to carry out some or all of the following operations on finite elements:

(1) alter the number of element freedoms;[12]

(2) change the positions of the element freedoms;[13]

(3) change the nature of the element freedoms, for example from displacements to first derivatives thereof;[6-10]

(4) apply constraints to the boundary of the element with a view to increasing the interelement continuity.[6]

With these changes there is an associated basis transformation in which both the dimension of the basis and the order of its polynomial elements may be changed.

As an example, Fig. 4.3(b) illustrates a four-freedom beam element. We shall transform this to the quadratic 'local' element shown in Fig. 4.3(a) using the method of nested interpolations. Differentiating eqns (4.20) with respect to x and putting $d = 1/2$ to obtain an expression for the slope ϕ_2 at the centre of the beam one obtains the matrix transformation of variables required as

$$\{b\} = \begin{Bmatrix} \phi_1 \\ \phi_2 \\ \phi_3 \end{Bmatrix} = \begin{bmatrix} 0 & 1 & 0 & 0 \\ -3/2L & -1/4 & 3/2L & -1/4 \\ 0 & 0 & 0 & 1 \end{bmatrix} \begin{Bmatrix} v_1 \\ \phi_1 \\ v_2 \\ \phi_2 \end{Bmatrix} = T\{d\}.$$

(4.22)

Using quadratic interpolation for the local freedoms

$$\phi = c_1 + c_2 x + c_3 x^2 = \{M\}^t\{c\}$$ (4.23)

the inverse interpolation matrix C^{-1} is formed as

$$\{b\} = \begin{Bmatrix} \phi_1 \\ \phi_2 \\ \phi_3 \end{Bmatrix} = \begin{bmatrix} 1 & -L/2 & L^2/4 \\ 1 & 0 & 0 \\ 1 & L/2 & L^2/4 \end{bmatrix} \begin{Bmatrix} c_1 \\ c_2 \\ c_3 \end{Bmatrix} = C^{-1}\{c\}$$ (4.24)

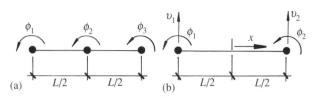

Fig. 4.3 Application of the method of nested interpolations to a beam element: (a) local freedoms; (b) global freedoms

and inverting gives

$$\{c\} = \left\{ \begin{matrix} \{c_{rb}\} \\ \{c_s\} \end{matrix} \right\} = \begin{bmatrix} 0 & 1 & 0 \\ \cdots & \cdots & \cdots \\ -1/L & 0 & 1/L \\ 2/L^2 & -4/L^2 & 2/L^2 \end{bmatrix} \left\{ \begin{matrix} \phi_1 \\ \phi_2 \\ \phi_3 \end{matrix} \right\} = \begin{bmatrix} C_{rb} \\ C_s \end{bmatrix} \{b\}, \quad (4.25)$$

partitioning the rigid body and straining terms. Differentiating eqn (4.23) gives the curvature interpolation matrix S_s,

$$\chi = \frac{d\phi}{dx} = [1, 2x]\{c_s\} = S_s\{c_s\}. \quad (4.26)$$

The kernel stiffness matrix is given as

$$k_m^* = \int_{-L/2}^{L/2} S_s^t(EI)S_s \, dx = \begin{bmatrix} L & 0 \\ 0 & L^3/3 \end{bmatrix} \quad (4.27)$$

and the local element stiffness matrix is

$$k^* = C_s^t k_m^* C_s = \frac{EI}{3L} \begin{bmatrix} 7 & -8 & 1 \\ -8 & 16 & -8 \\ 1 & -8 & 7 \end{bmatrix}. \quad (4.28)$$

The global element stiffness matrix is given by $k = T^t k^* T$ which yields the correct result (eqn (1.10)). The method is seen to advantage in Sections 8.10 and 10.6 where complete quadratic interpolations can be used in triangular elements in which complete interpolations·do not otherwise exist. Compared to the Lagrange multiplier technique it has the advantage that no additional freedoms are required whilst a more convenient interpolation can be obtained.

The 'nested interpolation' method of basis transformation has also been applied to quadrilateral elements[9] and many further applications, for example to the calculation of geometric stiffness matrices,[9] appear possible. Moreover it can be used in conjunction with any variational or weighted residual formulation and also in conjunction with penalty factor techniques, the latter being used to enforce constraints in the interior of an element at appropriate integration points.

To illustrate changing the number of freedoms via basis transformation the quadratic triangle of Fig. 4.4 provides a very simple example. Here we may wish to suppress some midside nodes to allow the element to merge with other elements which do not possess midside nodes. Suppose the element possesses a single degree of freedom ϕ at each node. To remove ϕ_4, for example, one simply introduces the linear constraint $\phi_4 = (\phi_1 + \phi_2)/2$ and if

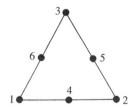

Fig. 4.4 Quadratic triangular element

all three midside-node freedoms are to be suppressed this can be accomplished by the use of the matrix transformation

$$\{\phi^*\} = \begin{bmatrix} \phi_1 \\ \phi_2 \\ \phi_3 \\ \phi_4 \\ \phi_5 \\ \phi_6 \end{bmatrix} = \begin{bmatrix} 1 & 0 & 0 & 0 & 0 & 0 \\ 0 & 1 & 0 & 0 & 0 & 0 \\ 0 & 0 & 1 & 0 & 0 & 0 \\ \frac{1}{2} & \frac{1}{2} & 0 & 0 & 0 & 0 \\ 0 & \frac{1}{2} & \frac{1}{2} & 0 & 0 & 0 \\ \frac{1}{2} & 0 & \frac{1}{2} & 0 & 0 & 0 \end{bmatrix} \begin{bmatrix} \phi_1 \\ \phi_2 \\ \phi_3 \\ \phi_4 \\ \phi_5 \\ \phi_6 \end{bmatrix} = C\{\phi\} \quad (4.29)$$

where $\{\phi\}$ are the global freedoms and $\{\phi^*\}$ are the 'local freedoms' to which the original quadratic element matrix k^* still applies. Then the modified global element stiffness matrix is given by eqn (4.21) and with the use of the complete transformation of eqn (4.29) one obtains the element stiffness matrix for a three node linear element. One should also place a small artificial stiffness on the diagonal for each freedom suppressed or eliminate the freedom from the solution as a boundary condition to prevent overflow during solution of the assembled equations.

Another interesting example of basis transformation also occurs in the element of Fig. 4.4 where it is found that shifting the midside nodes on two sides to the quarter points provides the square root displacement variation on these sides required to model the effect of cracks. This application is further discussed in Section 7.7.[13]

Further simple examples of basis transformation occur in Chapter 8 where linear constraints are applied to normal slopes or curvatures in triangular finite elements for the analysis of thin plates to eliminate inconvenient normal slope freedoms at the midsides.

4.5 Demonstration program for constraint techniques

In the present section it is instructive to use a BASIC program to demonstrate the application of the penalty factor, Lagrange multiplier, and basis transformation techniques to the simple problem of Fig. 4.1. The program loads the complete stiffness and load matrices as data, putting $c = EA/L = 1$ and $P = 1$. For the penalty problem the penalty factor β is set at 10^5, an appropriate choice for the eight-digit computation employed by MEGA-BASIC. Then the data matrices are given by substituting these parameters in eqn (4.5).

```
10 Rem SIMPLE PROGRAM TO ILLUSTRATE PENALTY, LAGRANGE MULTIPLIER AND
20 Rem BASIS TRANSFORMATION TECHNIQUES
22 Restore 400
23 Rem 300=PENALTY DATA,400=LAGRANGE DATA,500=BASIS TRANSF. DATA
30 Read N;Rem N = MATRIX SIZE
40 Dim S(N,N),Q(N);Rem S( , ) = STIFFNESS MATRIX,Q( ) = LOAD MATRIX
50 For I=1 to N;For J=1 to N;Read S(I,J);Next;Next
60 For I=1 to N;Read Q(I);Next
80 Rem GAUSS JORDAN REDUCTION AS USED IN THE TRUSS PROGRAM OF CHAPTER 1
100 For I=1 to N
110 X=S(I,I);Q(I)=Q(I)/X
120 For J=I+1 to N
130 S(I,J)=S(I,J)/X;Next
140 For K=1 to N
150 If K=I then Goto 190
160 X=S(K,I);Q(K)=Q(K)-X*Q(I)
170 For J=I+1 to N
180 S(K,J)=S(K,J)-X*S(I,J);Next
190 Next K
195 Next I
200 For I=1 to N;! Q(I);Next
290 Rem PENALTY DATA AS PER EQN 4.5
300 Data 3
310 Data 1+1E5,-1,0
320 Data -1,2+1E5,-1-1E5
330 Data 0,-1-1E5,1+1E5
340 Data 0,0,1
390 Rem LAGRANGE MULTIPLIER DATA AS PER EQN 4.15 - BUT HERE U1=0 ENFORCED
391 Rem USING A PENALTY - THUS ROW & COL 4 OF EQN 4.15 ARE OMITTED
400 Data 4
410 Data 1+1E5,-1,0,0
420 Data -1,2,-1,1
430 Data 0,-1,1,-1
440 Data 0,1,-1,0
450 Data 0,0,1,0
490 Rem BASIS TRANSFORMATION DATA FOR PROBLEM OF FIG. 4.1
491 Rem AS U2 IS ELIMINATED A SMALL NUMBER IS PLACED ON THE DIAGONAL FOR
492 Rem THIS (AS PER NOTE FOLLOWING EQN 4.29)
500 Data 3
510 Data 1E5,0,-1
520 Data 0,1E-5,0
530 Data -1,0,1
540 Data 0,0,1
```

For the Lagrange multiplier problem the boundary condition $u_1 = 0$ is enforced using a penalty, by adding $\beta = 10^5$ to the first diagonal position in the stiffness matrix. Hence row and column four of eqn (4.15), the Lagrange multiplier specification of this boundary condition, are omitted. Penalty enforcement of $u_1 = 0$ is used to prevent singularity of the stiffness matrix. Note that line 22 of the program is used to run it with the Lagrange multiplier data.

For the basis transformation problem the appropriate transformation matrix to enforce the constraint $u_2 = u_3$ is simply

$$C = \begin{bmatrix} 1 & 0 & 0 \\ 0 & 1 & 0 \\ 0 & 1 & 0 \end{bmatrix}. \tag{4.30}$$

The stiffness matrix is that of eqn (4.5) with $\beta = 0$. Applying eqn (4.21) with these matrices the constrained stiffness matrix is obtained as

$$k = \begin{bmatrix} c & 0 & -c \\ 0 & 0 & 0 \\ -c & 0 & c \end{bmatrix}. \tag{4.31}$$

As noted after eqn (4.29) a small artificial stiffness, which is here chosen as 10^{-5}, must then be placed in the second diagonal location in k to prevent singularity. Consequently the (correct) solution for u_2 will be lost but, of course, the purpose of using the matrix of eqn (4.30) is to eliminate this freedom from the solution.

Finally, as in the penalty and Lagrange multiplier data, $u_1 = 0$ is enforced by placing the penalty $\beta = 10^5$ in the first diagonal location. Thus, simply by altering line 22, the program can be used to demonstrate all three constraint techniques discussed in the present chapter.

4.6 Concluding remarks

The penalty factor, Lagrange multiplier, and basis transformation techniques will reappear frequently in the following chapters. As Reddy remarks[14] of the penalty method, such techniques represent the state of the finite element art. For this reason Chapter 22 discusses numerical optimization techniques as these provide further insight into penalty factors and Lagrange multipliers. Chapter 23 is then able to extend our applications of these techniques, in particular to problems involving *integral* constraints (the constraints dealt with in the foregoing chapter we should term *discrete* constraints), in conjunction with weighted residual methods and to problems involving inequality constraints.[15]

References

1. Courant, R. (1943). Variational methods for the solution of problems of equilibrium and vibrations. *Bull. Amer. Math. Soc.*, **49**, 1.
2. Lagrange, J. L. (1788) *Mécanique Analytique*. Paris. (Cited in Pars, L. A. (1965) *A Treatise on Analytical Dynamics*, Heinemann, London.)
3. Pian, T. H. and Tong, P. (1969). Basis of finite element methods for solid continua. *Int. J. Num. Meth. Engng*, **1**, 3.
4. Harvey, J. W. and Kelsey, S. (1971). Triangular plate bending elements with enforced compatibility. *J. Amer Inst. Aeron. Astron.*, **9**, 1023.
5. Thomas, G.R. and Gallagher, R. H., (1976). A triangular element based on generalised potential energy concepts. In *Finite Elements for Thin Shells and Curved Members*, (ed. D. G. Ashwell and R. H. Gallagher). Wiley, London.

6. Mohr, G. A. (1981). Finite element formulation by nested interpolations: application to cubic elements. *Comput. Struct.* **14**, 211.
7. Mohr, G. A. (1982). Finite element formulation by nested interpolations: application to the drilling freedom problem. *Comput. Struct.*, **15**, 185.
8. Mohr, G. A. (1982). On displacement finite elements for the analysis of shells. Proceedings of the Fourth Australian International Conference on Finite Element Methods, University of Melbourne.
9. Mohr, G. A. (1983). Finite element formulation by nested interpolations: application to a quadrilateral thin plate bending element. *Civil Engng Trans I.E. Aust.*, **CE25** (3), 211.
10. Mohr, G. A. and Mohr, R. S. (1986). A new thin plate bending element by basis transformation. *Comput. Struct.*, **22**, 239.
11. Apostol, T. (1972). *Calculus*, Vol. 2, (2nd edn). Wiley, New York.
12. Mohr, G. A. and Milner, H. R. (1986). *A Microcomputer Introduction to the Finite Element Method*. Pitman, Melbourne.
13. Okabe, M., Yamada, Y., and Nishiguohi, I. (1980). Basis transformation of trial function space in Lagrangian interpolation. *Comp. Meth. Appl. Mech. Engng*, **23**, 85.
14. Reddy, J. N. (1982). On penalty function methods in the finite element analysis of flow problems. *Int. J. Num. Meth. Fluids*, **2**, 151.
15. Mohr, G. A. and Caffin, D. A. (1985). Penalty factors, Lagrange multipliers and basis transformation in the finite element method. *Civ. Engng Trans I.E. Aust.*, **CE27**(2), 174.

5
Interpolation functions
and numerical integration

In this chapter interpolation functions for two and three dimensions are discussed. First Cartesian displacement functions are considered and examples of these are used to demonstrate some of the constraints which should be considered in choosing interpolations for a finite element formulation. Particular emphasis is then given to the definition of suitable dimensionless coordinates for rectangular and triangular regions and the derivation of interpolation functions for such regions. As we saw in Chapter 3 finite element stiffness matrices are then obtained by differentiating these interpolation functions to obtain a strain-interpolation matrix B and then integrating the product $B^t DB$ over the element volume. In simple cases analytical integration can be used to obtain explicit matrices but in general numerical integration is used. The chapter closes with a discussion of numerical integration techniques for finite elements.

5.1 Introduction

Perhaps the vital step in the finite element method is the selection of element interpolations. For an element with freedoms ϕ_i at n nodes we use n linearly independent interpolation functions f_i, writing the interpolation for ϕ as

$$\phi = \{f\}^t\{\phi\} = \sum_{i=1}^{n} f_i \phi_i. \tag{5.1}$$

Many examples of the formation of the f_i are given in following sections. In all but the simplest cases this is achieved by formation of an interpolation matrix C or A by inversion (C denoting the use of Cartesian coordinates and A denoting the use of dimensionless coordinates). When such inversion is not possible (that is, C^{-1} or A^{-1} is singular) the element is said *not to exist* and a simple degree of freedom count, matching the length of the polynomial interpolation to the number of freedoms, is not sufficient to ensure existence of an element.[1] This is the reason for the use of normal slope freedoms (rather than more convenient parallel slopes) in the triangular elements of Sections 5.9 and 5.10.

For an assemblage of elements the interpolation in the complete domain is

$$\phi = \{f\}_k^t \{\phi\}_j \delta_{kj}$$

where

$$\delta_{kj} = \begin{cases} 1 & k = j \\ 0 & k \neq j \end{cases}.$$

the subscripts k, j are element numbers and δ_{kj} is the Kronecker delta function. If ϕ is uniquely defined at the interface between two elements they are said to posess C_0 *continuity*. Interpolations for the first derivatives of ϕ are obtained by differentiating eqn (5.1)

$$\frac{\partial \phi}{\partial x} = \frac{\partial \{f\}^t}{\partial x} \{\phi\} = \sum_{i=1}^{n} \frac{\partial f_i}{\partial x} \phi_i,$$

repeating the process for higher derivatives. If $\partial^\alpha \phi / \partial(\ )^\alpha$ is uniquely defined at the interface between elements they posses C_α *continuity*.

Differentiating $\{f\}$ in this manner the strain matrix B is formed and the element stiffness matrix is given by $k = \int B^t DB \, dV$. In practice numerical integration techniques are used for this purpose and the chapter closes with a discussion of such techniques.

5.2. Constraints in the choice of interpolation functions

The question of constraints in the choice of interpolation functions is a complicated one. A good example of this difficulty is provided by the cubic thin plate bending elements shown in Fig. 5.1. The complete cubic polynomial for two dimensions is

$$w = c_1 + c_2 x + c_3 y + c_4 x^2 + c_5 xy + c_6 y^2 + c_7 x^3$$
$$+ c_8 x^2 y + c_9 xy^2 + c_{10} y^3 \tag{5.2}$$

and in the twelve-freedom rectangular Adini–Clough–Melosh (ACM) element[2] the interpolation is obtained by adding the terms $c_{11} x^3 y$, $c_{12} xy^3$. The element passes the patch test but its rate of convergence is only h^2 (see Section 3.9) which could be achieved with a quadratic element of only six freedoms.

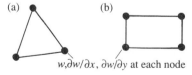

(a) (b)

$w, \partial w/\partial x, \partial w/\partial y$ at each node

Fig. 5.1 Simple bending elements

For the nine-freedom triangle one central node is required to allow the use of eqn (5.2). This is inconvenient but removal of this node by static condensation (see Section 6.5) does not yield satisfactory results. One possibility is to combine the c_8 and c_9 modes[1] as $c_8(x^2 y + xy^2)$ but this yields a singular C^{-1} matrix for some element geometries.[2] Another alternative is to omit either the c_8 or c_9 modes but the resulting element performs poorly. Bazeley *et al.*[3] obtain a more satisfactory result by adjusting the areal coordinate interpolations judiciously (see Section 8.4) though this element fails the patch test (but not seriously) for some mesh geometries. Such behaviour emphasizes the desirability of complete polynomials in triangular elements and no difficulty occurs if these are used.

In plane stress and three-dimensional elements complete interpolations are generally used in *simplexes* (elements having the minimum number of sides or faces for a given number of dimensions), examples being the constant strain triangle (linear), the linear strain triangle (quadratic), and the four- and ten-node tetrahedra (respectively linear and quadratic). An exception is the cubic triangle for which removal of the central node by static condensation proves successful[4] but this parallels the situation in many thin plate elements which are based upon complete polynomials and then subjected to constraints which remove some freedoms.

The *multivariate* quadrilateral and brick elements, on the other hand, involve incomplete polynomials, the simplest example being the bilinear plane stress quadrilateral for which the displacement functions are

$$u = c_1 + c_2 x + c_3 y + c_4 xy, \quad v = c_5 + c_6 x + c_7 y + c_8 xy. \tag{5.3}$$

However, it is with such elements that performance is improved by *reduced integration*, in which the incomplete modes (that is, the xy modes in eqn (5.3)) are not precisely integrated, or by omitting the contributions of these terms to the shearing energy.[5] Thus the incomplete terms serve only to match the nodal freedoms in number. The additional terms are useful in some instances, the bilinear quadrilateral yielding good results in cantilever problems (see, for example Fig. 6.3) because it allows $\varepsilon_x = f(y)$ which is desirable in this case. However in problems such as that of Fig. 7.4 it fares much less well than the (complete) linear strain triangle. Such observations, and Wilson's bilinear and trilinear elements in which the quadratic is completed by the addition of nodeless variables (see eqns (6.44)), again provide evidence of the desirability of complete polynomials.

The bicubic plate element (eqn (5.46)), though, gives excellent results and yet contains incomplete fourth-, fifth-, and sixth-order modes. But considering that some nine-freedom cubic plate elements, in which C_1 continuity is assured, also exhibit excellent convergence (see Figs 8.11–8.14) and that the bicubic element gives similar results when constrained to twelve freedoms (eqn (8.11)) it can be concluded that incompleteness, though causing no harm in this instance, is not necessarily a virtue.

Thus completeness is a sufficient but not a necessary condition for convergence. However, completeness up to some minimum degree is a necessary condition because to ensure satisfaction of the governing equations satisfactory interpolations must *in the limit of mesh refinement* yield[3, 6]

(1) the exact rigid body modes;

(2) the exact constant strain modes;

(3) interelement displacement continuity;

(4) interelement equilibrium;

for without (2) the energy cannot be correctly evaluated and (2) cannot be ensured without satisfaction of (1). In triangular plate elements the exact rigid body modes are given by the first three terms of eqn (5.2) and the use of exact expressions for the rigid body modes proves advantageous in rectangular shell elements.[7] In this last instance periodic function modes are used and these have also been included in rectangular plate elements,[8, 9] demonstrating that polynomials are convenient but not obligatory in rectangular finite elements.

Satisfaction of the governing differential equations will also require conditions (3) and (4) above. In constructing conforming and hybrid elements one is at pains to hasten convergence towards these ends but in meshes of practical fineness continuity and equilibrium need only be satisfied *in the mean sense* at element boundaries, so that many non-conforming elements are able to yield acceptable results.

However, there must be some minimum continuity requirement. Consider for example a Hermitian pentagon with freedoms w, $\partial w/\partial x$, $\partial w/\partial y$ at each vertex. A complete quartic polynomial can be achieved but eqn (3.61) predicts non-convergence. Therefore one concludes that interpolations should:

(1) be complete to order p, $p \geq m$, where mth order derivatives are required to calculate the strains;

(2) ensure interelement continuity of this 'constant strain part' of the interpolation;

(3) allow convergent mean approximations to the stresses and any non-conforming displacements at interelement boundaries.

If the interpolations are non-polynomial they should approximate to polynomials which meet these requirements.

In relation to (1) the desirability of complete polynomials (aside from the imposition of later constraints) in simplexes must be noted. The second requirement implies that elements with first-order strains (such as for plane stress) must ensure continuity of the linear part of the interpolation and that elements with second-order strains (such as for thin plate bending) must ensure continuity of the quadratic part of the interpolation. The third requirement is the inverse of the well-known one for *admissible* solutions to differential equations, the true solution of which involves *jumps*; the trial solution must approximate these jumps in the mean sense.[10] In the present context any jumps in the trial finite element solutions must approximate the true solution in the mean sense.

5.3 Dimensionless interpolations for line elements

5.3.1 Lagrangian interpolation

The interpolation functions for the Lagrangian interpolation $u = \sum_{i=1}^{N} f_i u_i$ can be obtained from the formula

$$f_i = \frac{\prod\limits_{j=1}^{N}(x - x_j)}{\prod\limits_{j=1}^{N}(x_i - x_j)} \quad i \neq j \tag{5.4}$$

which, for the linear element shown in Fig. 5.2(a), with dimensionless coordinates $d = x/L$ or $s = 2d - 1$, yields

$$u = \{f\}^{t}\{u\} \text{ where } f_1 = 1 - d,\ f_2 = d,\ d = 0 \to 1 \tag{5.5a}$$

and

$$u = \{f\}^{t}\{u\} \text{ where } f_1 = (1 - s)/2,\ f_2 = (1 + s)/2,\ s = -1 \to 1. \tag{5.5b}$$

The quadratic interpolation requires the third freedom shown in Fig. 5.2(b) and in this and the following examples we begin by writing an interpolation polynomial or *displacement function,*[†]

$$u = a_1 + a_2 s + a_3 s^2 = \{M\}^{t}\{a\}\ s = -1 \to 1. \tag{5.6}$$

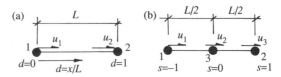

Fig. 5.2 Linear and quadratic interpolation

[†] The resulting matrix inversion procedure is then able to be used for more general interpolations than Lagrangian.

Substituting nodal values on both sides,

$$\{u\} = \begin{Bmatrix} u_1 \\ u_2 \\ u_3 \end{Bmatrix} = \begin{bmatrix} 1 & -1 & 1 \\ 1 & 0 & 0 \\ 1 & 1 & 1 \end{bmatrix} \begin{Bmatrix} a_1 \\ a_2 \\ a_3 \end{Bmatrix} = A^{-1}\{a\} \quad (5.7)$$

and the interpolation is given by inverting to obtain the interpolation matrix A, giving[†]

$$u = f_1 u_1 + f_2 u_2 + f_3 u_3 = \{f\}^t \{u\} \quad (5.8)$$

where

$$\{f\}^t = [f_1 \ f_2 \ f_3] = \{M\}^t A$$

$$= \tfrac{1}{2}[1 \ s \ s^2] \begin{bmatrix} 0 & 2 & 0 \\ -1 & 0 & 1 \\ 1 & -2 & 1 \end{bmatrix}. \quad (5.9)$$

Finally the required interpolation functions are

$$f_1 = \tfrac{1}{2}(s^2 - s)$$
$$f_2 = 1 - s^2 \quad (5.10)$$
$$f_3 = \tfrac{1}{2}(s^2 + s) \quad (s = -1 \to 1)$$

and the interpolation functions for $d = 0 \to 1$ are obtained with the substitution $s = 2d - 1$, giving

$$f_1 = 1 - 3d + 2d^2$$
$$f_2 = 4d(1 - d) \quad (5.11)$$
$$f_3 = 2d^2 - d \quad (d = 0 \to 1).$$

An alternative approach is to use the two coordinates L_1 and L_2 shown in Fig. 5.3. The interpolation polynomial is now written in the form

$$u = a_1 L_1^2 + a_2 L_2^2 + a_3 L_1 L_2 = \{M\}^t \{a\} \quad (5.12)$$

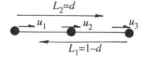

Fig. 5.3 Quadratic interpolation

[†] Note that when Cartesian coordinates are employed we denote the interpolation matrix as C and the modal amplitudes as $\{c\}$. When dimensionless coordinates are used we emphasize this by using the symbols A and $\{a\}$.

and, substituting nodal values, we find

$$\{u\} = \begin{Bmatrix} u_1 \\ u_2 \\ u_3 \end{Bmatrix} = \begin{bmatrix} 1 & 0 & 0 \\ \frac{1}{4} & \frac{1}{4} & \frac{1}{4} \\ 0 & 0 & 1 \end{bmatrix} \begin{Bmatrix} a_1 \\ a_2 \\ a_3 \end{Bmatrix} = A^{-1}\{a\} \qquad (5.13)$$

Comparing eqns (5.7) and (5.13) there is little difference in complexity but the L_1, L_2 approach leads to simpler calculation of the quartic and quintic interpolations discussed at the close of this section.

Now the interpolation functions are given by

$$\{f\}^t = \{M\}^t A \qquad (5.14)$$

$$= [L_1^2 L_2^2 L_1 L_2] \begin{bmatrix} 1 & 0 & 0 \\ 0 & 0 & 1 \\ -1 & 4 & -1 \end{bmatrix} \qquad (5.15)$$

giving
$$f_1 = L_1^2 - L_1 L_2$$
$$f_2 = 4L_1 L_2 \qquad (5.16)$$
$$f_3 = L_2^2 - L_1 L_2 \qquad (L_1 = 1 \to 0,\ L_2 = 0 \to 1).$$

It is readily verified that these are equivalent to eqns (5.10) and (5.11) by substituting $L_1 = 1 - d$ and $L_2 = d$ or $L_1 = (1 - s)/2$ and $L_2 = (s + 1)/2$.

Each of the three alternative forms (eqns (5.10), (5.11), or (5.16)) may prove useful. Equations (5.11), for example, are sometimes preferred for use in developing the stiffness properties of line elements, whereas eqns (5.10) prove useful in establishing the interpolations for rectangular and quadrilateral domains (see Section 5.4). Equations 5.16, on the other hand, are useful as an introduction to the use of natural freedoms in triangular elements (for example, compare eqns (5.26) and (5.27) to (5.68)–(5.70) and (5.94)–(5.96)).

The displacement function assumed in eqn (5.12) is at variance with those assumed previously in that it is a *homogeneous* polynomial of the form

$$u = \sum a_i L_1^a L_2^b \qquad \text{where } a + b = p.$$

Alternatively one could assume the displacement function as an ascending polynomial in L_1 and L_2:

$$u = a_1 L_1 + a_2 L_2 + a_3 L_1 L_2 \qquad (5.17)$$

and, following this procedure, the interpolation functions are now obtained as
$$f_1 = L_1 - 2L_1 L_2$$
$$f_2 = 4L_1 L_2 \qquad (5.18)$$
$$f_3 = L_2 - 2L_1 L_2.$$

This alternative form to eqns. (5.11) gives identical results when applied to the development of element stiffness properties. This can be verified by substituting $L_1 = 1 - d$ and $L_2 = d$ in eqns (5.18) when once again eqns (5.11) are obtained.

5.3.2 Hermitian interpolation

In contrast to Lagrangian interpolations, Hermitian interpolations involve both a function and its derivatives as nodal freedoms. Hence Hermitian interpolation functions must be such that their derivatives vanish at one end of the region of interpolation and are equal to unity at the other. The cubic interpolation for the four-freedom flexural element of Fig. 5.4(a) has already been obtained (see eqns (3.41)) by beginning with a Cartesian polynomial but we shall begin with polynomials in a *dimensionless coordinate* to simplify calculations. (In the second of these ϕ^*, as distinct from ϕ, is introduced in order that the resulting interpolation matrix is dimensionless.) Then

$$v = a_1 + a_2 s + a_3 s^2 + a_4 s^3 = \{M\}^t\{a\}$$

$$\phi^* = \frac{dv}{ds} = a_2 + 2a_3 s + 3a_4 s^2 \qquad s = -1 \to 1$$

(5.19)

and substituting the nodal values $v_1, \phi_1^*, v_2, \phi_2^*$ on the left-hand sides and the nodal coordinates $s = -1$ and $s = 1$ on the right-hand sides one obtains four simultaneous equations. Writing these in matrix form gives

$$\{d^*\} = \begin{bmatrix} v_1 \\ \phi_1^* \\ v_2 \\ \phi_2^* \end{bmatrix} = \begin{bmatrix} 1 & -1 & 1 & -1 \\ 0 & 1 & -2 & 3 \\ 1 & 1 & 1 & 1 \\ 0 & 1 & 2 & 3 \end{bmatrix} \begin{bmatrix} a_1 \\ a_2 \\ a_3 \\ a_4 \end{bmatrix} = A^{-1}\{a\}.$$

Noting that $v = \{f\}^t\{d^*\} = \{M\}^t\{a\} = \{M\}^t A\{d^*\}$ we have

$$\{f\}^t = \{M\}^t A = \begin{bmatrix} 1 \\ s \\ s^2 \\ s^3 \end{bmatrix}^t \begin{bmatrix} 1/2 & 1/4 & 1/2 & -1/4 \\ -3/4 & -1/4 & 3/4 & -1/4 \\ 0 & -1/4 & 0 & 1/4 \\ 1/4 & 1/4 & -1/4 & 1/4 \end{bmatrix}$$

(5.20)

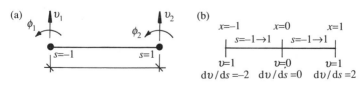

Fig. 5.4 (a) Cubic element; (b) formation of cubic splines

and the interpolation is

$$v = f_1 v_1 + f_2 \left(\frac{dv}{ds}\right)_1 + f_3 v_2 + f_4 \left(\frac{dv}{ds}\right)_2.$$

Now noting that the actual slope freedom ϕ is given by

$$\phi = \frac{dv}{dx} = \frac{dv}{ds}\frac{ds}{dx} = \frac{2}{L}\frac{dv}{ds} = \frac{2}{L}\phi^* \qquad (5.21)$$

that is, $\phi^* = dv/ds = L\phi/2$, the final interpolation can be written as

$$v = \{f\}^t\{v_1, L\phi_1, v_2, L\phi_2\} = \{f\}^t\{d^*\}$$

where

$$
\begin{aligned}
f_1 &= \tfrac{1}{4}(2 - 3s + s^3), & f_2 &= \tfrac{1}{8}(1 - s - s^2 + s^3)\\
f_3 &= \tfrac{1}{4}(2 + 3s - s^3), & f_4 &= \tfrac{1}{8}(-1 - s + s^2 + s^3)
\end{aligned}
\qquad (5.22)
$$

and the interpolation functions are dimensionless and the freedoms $\{d^*\}$ are dimensionally consistent; these are desirable features from the numerical viewpoint. Then the interpolation for the slope ϕ is obtained by applying eqn (5.21) to eqns (5.22) and, repeating this exercise, one can obtain the interpolation for the 'curvature' d^2v/dx^2. (See eqn (5.137) where this result is used.)

The result of eqns (5.22) has useful application in piecewise curve fitting with cubic *splines*. Suppose we wish to obtain two cubic splines to approximate $v = x^4$ in the two regions shown in Fig. 5.4(b). Noting that $dv/ds = \tfrac{1}{2}dv/dx$ one obtains the required amplitudes of the left-hand spline as

$$\{a\} = A\left\{v_1, \left(\frac{dv}{ds}\right)_1, v_2, \left(\frac{dv}{ds}\right)_2\right\} = A\{1, -2, 0, 0\} \qquad (5.23)$$

where matrix A is given in eqn (5.20). Then the required spline can be written

$$v = \{M\}^t\{a\} = [1, s, s^2, s^3]\{a\} = -\frac{s}{4} + \frac{s^2}{2} - \frac{s^3}{4}.$$

Substituting $s = 2x + 1$ to transform this result to the original x-coordinate one obtains the spline as $v = -x^2 - 2x^3$ and the reader can verify that in like fashion the right-hand spline is obtained as $v = -x^2 + 2x^3$.

Alternative forms of the Hermitian cubic interpolation functions can be obtained as

$$
\begin{aligned}
f_1 &= 1 - 3d^2 + 2d^3 & &\text{(for } v_1)\\
f_2 &= d - 2d^2 + d^3 & &\text{(for } L\phi_1)\\
f_3 &= 3d^2 - 2d^3 & &\text{(for } v_2)\\
f_4 &= d^3 - d^2 & &\text{(for } L\phi_2) \qquad d = 0 \to 1
\end{aligned}
\qquad (5.24)
$$

by substituting $s = 2d - 1$ in eqns (5.22) or as

$$f_1 = L_1 + L_1^2 L_2 - L_1 L_2^2, \qquad f_2 = L_1^2 L_2$$
$$f_3 = L_2 - L_1^2 L_2 + L_1 L_2^2, \qquad f_4 = -L_1 L_2^2$$

(5.25)

by using the approach used for eqns (5.17) and (5.18). The reader can verify eqns (5.25) by substituting $L_1 = 1 - d$ and $L_2 = d$ into them, when eqns (5.24) are obtained. The L_1, L_2 approach to one-dimensional interpolations yields simpler interpolation matrices A and is comparable to the use of three area coordinates in *two-dimensional simplexes* (triangles) and four volume coordinates in *three-dimensional simplexes* (tetrahedra). The latter are discussed in following sections.

When eqns (5.25) are used to interpolate a variable the interpolation for its first derivative is obtained by using the chain rule of differentiation to write

$$\phi = \frac{\partial v}{\partial x} = \frac{\partial v}{\partial L_2} \frac{\partial L_2}{\partial x} + \frac{\partial v}{\partial L_1} \frac{\partial L_1}{\partial x}$$

$$= \frac{1}{L} \left(\frac{\partial v}{\partial L_2} - \frac{\partial v}{\partial L_1} \right) = \frac{\phi^*}{L}$$

(5.26)

to establish the matrix A^{-1} and hence obtain the interpolation matrix A. Second derivatives, required for the calculation of the curvature in flexural elements, are given by

$$\frac{\partial^2 v}{\partial x^2} = \frac{1}{L^2} \left(\frac{\partial^2 v}{\partial L_1^2} + \frac{\partial^2 v}{\partial L_2^2} - \frac{2 \partial^2 v}{\partial L_1 \partial L_2} \right).$$

(5.27)

The quartic interpolation for a beam element requires the freedoms shown in Fig. 5.5(a) and is obtained by writing

$$v = a_1 L_1^4 + a_2 L_2^4 + a_3 L_1^3 L_2 + a_4 L_1 L_2^3 + a_5 L_1^2 L_2^2$$

(5.28)

or

$$v = a_1 L_1 + a_2 L_2 + a_3 L_1^3 L_2 + a_4 L_1 L_2^3 + a_5 L_1^2 L_2^2$$

(5.29)

$$= \{M\}^t \{a\}$$

where eqn (5.29) explicitly includes the rigid body modes L_1 and L_2, that is, the terms which give zero curvature in eqn (5.27). The interpolation functions

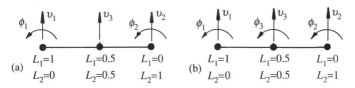

Fig. 5.5 (a) Quartic element; (b) quintic element

are obtained in the usual way by substituting the nodal values of the dimensionless coordinates and using eqn (5.26) for the slopes:

$$\phi^* = \left(\frac{\partial v}{\partial L_2} - \frac{\partial v}{\partial L_1} \right)$$

$$= [-1, 1, L_1^3 - 3L_1^2 L_2, 3L_1 L_2^2 - L_2^3, 2L_1^2 L_2 - 2L_1 L_2^2]\{a\}. \quad (5.30)$$

Substituting the nodal coordinates in eqns (5.29) and (5.30) to form A^{-1}, inverting, and calculating $\{f\}^t = \{M\}^t A$ the interpolation functions are obtained as

$$f_1 = L_1 + L_1^3 L_2 - L_2^3 L_1 - 8L_1^2 L_2^2$$

$$f_2 = L_2 - L_1^3 L_3 + L_2^3 L_1 - 8L_1^2 L_2^2$$

$$f_3 = L_1^3 L_2 - L_1^2 L_2^2, \quad f_4 = -L_1 L_2^3 + L_1^2 L_2^2 \quad (5.31)$$

$$f_5 = 16L_1^2 L_2^2$$

and using the substitutions $L_2 = d$ and $L_1 = 1 - d$ one obtains the alternative form

$$f_1 = 1 - 11d^2 + 18d^3 - 8d^4, \quad f_2 = -5d^2 + 14d^3 - 8d^4$$

$$f_3 = d - 4d^2 + 5d^3 - 2d^4, \quad f_4 = d^2 - 3d^3 + 2d^4 \quad (5.32)$$

$$f_5 = 16d^2 - 32d^3 + 16d^4 \quad d = 0 \rightarrow 1.$$

As an exercise the reader may verify by following the same procedure as above that the interpolation for the quintic element of Fig. 5.5(b) is given by

$$v = f_1 v_1 + f_2 v_2 + f_3 (L\phi_1) + f_4 (L\phi_2) + f_5 v_3 + f_6 (L\phi_3) \quad (5.33)$$

where

$$f_1 = 1 - 23d^2 + 66d^3 - 68d^4 + 28d^5$$

$$f_2 = 7d^2 - 34d^3 + 52d^4 - 24d^5$$

$$f_3 = d - 6d^2 + 13d^3 - 12d^4 + 4d^5$$

$$f_4 = -d^2 + 5d^3 + 8d^4 + 4d^5$$

$$f_5 = 16d^2 - 32d^3 + 16d^4$$

$$f_6 = -8d^2 + 32d^3 - 40d^4 + 16d^5 \quad d = 0 \rightarrow 1.$$

This interpolation is of potential use in arch elements and in geometry calculations in shells where, to calculate the curvatures of the shell, higher-order interpolations are needed for good accuracy. For example, the above interpolation yields the curvature of one quadrant of a circle to about 0.1% per cent (G.W. Sinclair, private communication).

5.4 Lagrangian interpolation in rectangular and 'brick' elements

The interpolation functions for the *bilinear* and *biquadratic* elements of Fig. 5.6 are easily obtained by defining the dimensionless coordinates $a = 2x/L_a$ and $b = 2y/L_b$ referred to the element centroid. The Cartesian interpolation required for the element of Fig. 5.6(a) is

$$u = c_1 + c_2 x + c_3 y + c_4 xy \tag{5.34}$$

which is the shape of a hyperbolic paraboloid, and this can be obtained by multiplying together two linear functions in x and y:

$$u = (c_1 + c_2 x)(c_3 + c_4 y)$$
$$= c_1 c_3 + c_2 c_3 x + c_1 c_4 y + c_2 c_4 xy \tag{5.35}$$

Thus the interpolation in the dimensionless coordinates a and b can be obtained by substituting these in place of s in eqn (5.5b) and multiplying together the resulting functions. Then we obtain the *bilinear* interpolation

$$u = f_1 u_1 + f_2 u_2 = f_3 u_3 + f_4 u_4 = \{f\}^t \{u\}$$

where

$$f_1 = \tfrac{1}{4}(1 - a)(1 - b), \qquad f_2 = \tfrac{1}{4}(1 + a)(1 - b)$$
$$f_2 = \tfrac{1}{4}(1 + a)(1 + b), \qquad f_4 = \tfrac{1}{4}(1 - a)(1 + b) \tag{5.36}$$

or

$$f_i = \tfrac{1}{4}(1 + a_i a)(1 + b_i b) \tag{5.37}$$

where (a_i, b_i) are the nodal values of (a, b) shown in Fig. 5.6(a). Thus we see that $(1 - a)/2$ is applied to edge 14 and $(1 - b)/2$ to edge 12 and these are multiplied to obtain the interpolation function for the node at the intersection of these edges.

The same interpolation, of course, is used for the freedom v. The advantage of the use of dimensionless coordinates a, b is that the same interpolation is applicable to quadrilateral elements, given that a suitable transformation rule can be found to calculate the Cartesian strains ($\partial u/\partial x$, $\partial v/\partial y$, and so on) from

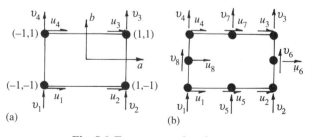

Fig. 5.6 Two rectangular elements

the derivatives $\partial u/\partial a$, $\partial u/\partial b$, $\partial v/\partial a$, and $\partial v/\partial b$ available from the interpolation functions. Such *Jacobian* transformations lead to the development of isoparametric elements and these are discussed in Chapter 7.

The interpolation functions of eqn (5.37) can now be extended to deal with three-dimensional brick elements with dimensionless coordinates a, b, and c. The resulting *trilinear* functions are

$$f_i = \tfrac{1}{8}(1 + a_i a)(1 + b_i b)(1 + c_i c) \tag{5.38}$$

for a brick element with eight nodes at the corners, that is, twenty-four freedoms.

For the eight-noded rectangle of Fig. 5.6(b) multiplication of two quadratic polynomials in x and y gives

$$u = (c_1 + c_2 x + c_3 x^2)(c_4 + c_5 y + c_6 y^2) \tag{5.39}$$

$$= c_1 c_4 + c_2 c_4 x + c_1 c_5 y + c_3 c_4 x^2 + c_2 c_5 xy + c_1 c_6 y^2$$

$$+ c_3 c_5 x^2 y + c_2 c_6 xy^2 + c_3 c_6 x^2 y^2$$

in which the last term (the ninth) cannot be determined and must be eliminated (unless an internal node is allowed). The required interpolation is obtained by combining quadratic and linear interpolations as

$$u = (c_1 + c_2 x + c_3 x^2)(c_4 + c_5 y) + (c_1 + c_2 x)(c_4 + c_5 y + c_6 y^2)$$

$$- (c_1 + c_2 x)(c_4 + c_5 y) \tag{5.40}$$

which eliminates the unwanted term $c_3 c_6 x^2 y^2$. The interpolation functions for the corner nodes are obtained by combining linear and quadratic interpolations in a manner consistent with eqn (5.40), giving

$$4f_i = a_i a(1 + a_i a)(1 + b_i b) + b_i b(1 + b_i b)(1 + a_i a) - (1 + a_i a)(1 + b_i b)$$

$$= (1 + a_i a)(1 + b_i b)(a_i a + b_i b - 1), \qquad i = 1 \rightarrow 4. \tag{5.41}$$

For the midside nodes, only the first or second term of eqn (5.40) is required so that we obtain

$$\begin{aligned} 2f_5 &= (1 - a^2)(1 - b), & 2f_6 &= (1 - b^2)(1 + a) \\ 2f_7 &= (1 - a^2)(1 + b), & 2f_8 &= (1 - b^2)(1 - a). \end{aligned} \tag{5.42}$$

These functions can be extended for the twenty-node brick element shown in Fig. 5.7, yielding

$$f_i = \tfrac{1}{8}(1 + a_i a)(1 + b_i b)(1 + c_i c)(a_i a + b_i b + c_i c - 2) \qquad i = 1 \rightarrow 8$$

$$f_i = \tfrac{1}{4}(1 - a^2)(1 + b_i b)(1 + c_i c) \qquad\qquad i = 9, 11, 17, 19$$

$$f_i = \tfrac{1}{4}(1 - b^2)(1 + c_i c)(1 + a_i a) \qquad\qquad i = 10, 12, 18, 20$$

$$f_i = \tfrac{1}{4}(1 - c^2)(1 + a_i a)(1 + b_i b) \qquad\qquad i = 13, 14, 15, 16.$$

$$\tag{5.43}$$

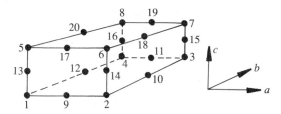

Fig. 5.7. Brick element with twenty nodes

5.5 Hermitian interpolation in rectangular elements

For the rectangular element of Fig. 5.8 the Cartesian displacement function is found by adding $c_{11}x^3y + c_{12}xy^3$ to eqn (5.2). Using dimensionless coordinates a and b the interpolation functions may be obtained by the matrix inversion procedure introduced in Section 5.3 and the results are given in Section 8.1. The resulting element is incompatible, that is, the displacement or interpolation function does not guarantee continuity of the slopes $\partial w/\partial x$ and $\partial w/\partial y$, only continuity of w itself is guaranteed (see Section 3.8). Such elements are said to possess C_0 continuity, whereas compatible elements possess C_1 continuity (continuity of the first derivatives).

A direct means of obtaining a compatible rectangular element is by combining one-dimensional Hermitian interpolations for each side. Writing eqns (5.22), for interpolation over a region of length L, in the form

$$f_1(s) = \tfrac{1}{4}(2 - 3s + s^3), \qquad f_2(s) = \tfrac{1}{4}(2 + 3s - s^3),$$

$$g_1(s) = \tfrac{1}{8}L(1 - s - s^2 + s^3), \quad g_2(s) = \tfrac{1}{8}L(-1 - s + s^2 + s^3)$$

$$s = -1 \rightarrow 1 \tag{5.44}$$

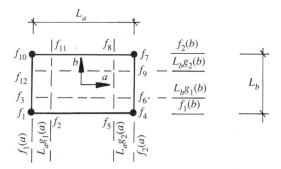

Fig. 5.8 Hermitian interpolation scheme for twelve-freedom rectangle

these functions are applied to each direction a, b following the scheme shown in Fig. 5.8. This is a generalization of the scheme used for the bilinear interpolation in Section 5.4. Here the terms g_1 and g_2 are considered to apply to freedoms $L\phi_1$ and $L\phi_2$ at fictitious nodes adjacent to the corner nodes, as shown. Then the interpolation function for the freedom $(\partial w/\partial y)_2$, for example, is obtained by multiplying the interpolation functions applying to the lines intersecting at the corresponding (fictitious) node, yielding

$$f_6 = L_b f_2(a) g_1(b)$$

so that the twelve functions required are given as

$$
\begin{aligned}
f_1 &= f_1(a)f_1(b) & f_2 &= L_a g_1(a) f_1(b) & f_3 &= L_b f_1(a) g_1(b) \\
f_4 &= f_2(a)f_1(b) & f_5 &= L_a g_2(a) f_1(b) & f_6 &= L_b f_2(a) g_1(b) \\
f_7 &= f_2(a)f_2(b) & f_8 &= L_a g_2(a) f_2(b) & f_9 &= L_b f_2(a) g_2(b) \\
f_{10} &= f_1(a)f_2(b) & f_{11} &= L_a g_1(a) f_2(b) & f_{12} &= L_b f_1(a) g_2(b).
\end{aligned}
\tag{5.45}
$$

This is equivalent to the multiplication

$$
\begin{aligned}
w &= (c_1 + c_2 x + c_3 x^2 + c_4 x^3)(a_1 + a_2 y + a_3 y^2 + a_4 y^3) \\
&= c_1 a_1 + c_2 a_1 x + c_1 a_2 y + c_3 a_1 x^2 + c_2 a_2 xy + c_1 a_3 y^2 \\
&\quad + c_4 a_1 x^3 + c_3 a_2 x^2 y + c_2 a_3 xy^2 + c_1 a_4 y^3 + c_4 a_2 x^3 y \\
&\quad + c_2 a_4 xy^3 + c_3 a_3 x^2 y^2 + c_4 a_3 x^3 y^2 + c_3 a_4 x^2 y^3 + c_4 a_4 x^3 y^3
\end{aligned}
\tag{5.46}
$$

if the last four terms can be removed. In applications to thin plates, however, the element does not provide a constant twist mode[2] so that convergence to the correct solutions is not guaranteed. The problem is most easily overcome by adding an additional twisting curvature freedom (that is $\partial^2 w/\partial x \partial y$) at each node and again following the scheme of Fig. 5.8 the four additional interpolation functions are given as

$$
\begin{aligned}
f_{13} &= L_a L_b g_1(a) g_1(b) & f_{14} &= L_a L_b g_2(a) g_1(b) \\
f_{15} &= L_a L_b g_2(a) g_2(b) & f_{16} &= L_a L_b g_1(a) g_2(b)
\end{aligned}
\tag{5.47}
$$

In eqns (5.45) and (5.47) the interpolation functions contain the side lengths L_a and L_b so that the global slopes and twist do not need to be first converted to 'natural' values, that is, $\phi_{Nx} = L_a \phi_x$, and so on (see Section 5.8). This *bicubic* element was first derived by Bogner, Fox, and Schmidt[11] and gives very rapid convergence. The additional cross derivative freedoms do not complicate use of the element significantly but they do make it more difficult to generalize to a quadrilateral shape.

5.6 Coordinate systems for triangles

The constant strain triangle (CST)[†] shown in Fig. 5.9 is one of the few triangular elements for which explicit solution for the interpolation matrix C (for Cartesian displacement functions) is practical and hence serves as a useful basis for the development of dimensionless coordinate systems for triangular domains. Using the linear interpolation $u = c_1 + c_2 x + c_3 y$ and substituting nodal values one obtains

$$\begin{Bmatrix} u_1 \\ u_2 \\ u_3 \end{Bmatrix} = \begin{bmatrix} 1 & x_1 & y_1 \\ 1 & x_2 & y_2 \\ 1 & x_3 & y_3 \end{bmatrix} \begin{Bmatrix} c_1 \\ c_2 \\ c_3 \end{Bmatrix} = C^{-1}\{c\}. \tag{5.48}$$

Inverting, the matrix C is obtained as

$$C = \frac{1}{|C|} \begin{bmatrix} x_2 y_3 - x_3 y_2 & x_3 y_1 - x_1 y_3 & x_1 y_2 - x_2 y_1 \\ -(y_3 - y_2) & -(y_1 - y_3) & -(y_2 - y_1) \\ x_3 - x_2 & x_1 - x_3 & x_2 - x_1 \end{bmatrix} \tag{5.49}$$

where

$$|C| = (x_2 y_3 - x_3 y_2) + (x_3 y_1 - x_1 y_3) + (x_1 y_2 - x_2 y_1)$$
$$= a_1 + a_2 + a_3. \tag{5.50}$$

After a little algebraic manipulation is easy to show that $a_1 = a_2 = a_3 = 2\Delta/3 = a$ where Δ is the area of the triangle. Then the interpolation functions are given by

$$\{f\} = C^t\{M\} = \frac{1}{3a} \begin{bmatrix} a & -y_{32} & x_{32} \\ a & -y_{13} & x_{13} \\ a & -y_{21} & x_{21} \end{bmatrix} \begin{Bmatrix} 1 \\ x \\ y \end{Bmatrix} \tag{5.51}$$

where $x_{32} = x_3 - x_2, y_{32} = y_3 - y_2$, and so on.

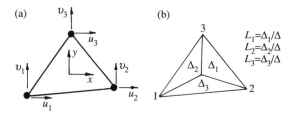

Fig. 5.9 Constant strain triangle

[†] It derives this name from the original applications to plane stress but of course its interpolation can be used as a basis for other applications.

For the six-node triangle formed by adding three midside nodes in Fig. 5.11(a), however, the Cartesian form for the the matrix C is much more complicated, whereas for still higher-order elements it becomes virtually intractable. This difficulty is overcome by using eqn (5.51) to define *area coordinates* in triangular elements. To this end we define three dimensionless coordinates L_1, L_2, and L_3 such that the linear interpolation in a triangle may be written in the simple form

$$u = L_1 u_1 + L_2 u_2 + L_3 u_3.$$

Comparison of this result with eqn (5.51) shows that we must define the area coordinates as

$$L_1 = \frac{a - y_{32}x + x_{32}y}{2\Delta}$$

$$L_2 = \frac{a - y_{13}x + x_{13}y}{2\Delta} \qquad (5.52)$$

$$L_3 = \frac{a - y_{21}x + x_{21}y}{2\Delta}$$

where

$$y_{32} = y_3 - y_2, \qquad y_{13} = y_1 - y_3, \qquad y_{21} = y_2 - y_1$$

$$x_{32} = x_3 - x_2, \qquad x_{13} = x_1 - x_3, \qquad x_{21} = x_2 - x_1$$

and the area Δ is usually calculated from the formula $2\Delta = x_{21}y_{32} - x_{32}y_{21}$ or one of its two permutations.

As $y_1 + y_2 + y_3 = 0$ and $x_1 + x_2 + x_3 = 0$ it is clear that the area coordinates possess the property

$$L_1 + L_2 + L_3 = 1 \qquad (5.53)$$

so that only two of these coordinates are independent. It can be shown that L_1, L_2, L_3 have the simple physical interpretation shown in Fig. 5.9(b). Given the area coordinates of a point within a triangle the Cartesian coordinates are obtained by interpolating from the corner coordinates:

$$x = L_1 x_1 + L_2 x_2 + L_3 x_3$$
$$y = L_1 y_1 + L_2 y_2 + L_3 y_3. \qquad (5.54)$$

An alternative oblique coordinate system for triangles is obtained by taking a dimensionless coordinate along the first side of the triangle (preferably the longest one) and measuring a second coordinate perpendicular to this side, as shown in Fig. 5.10.

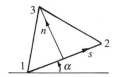

Fig. 5.10 Oblique coordinates

The transformation to the oblique coordinate s, n is

$$\begin{Bmatrix} s \\ n \end{Bmatrix} = \begin{bmatrix} \cos\alpha & \sin\alpha \\ -\sin\alpha & \cos\alpha \end{bmatrix} \begin{Bmatrix} x - x_1 \\ y - y_1 \end{Bmatrix} \quad (5.55)$$

or, reversing this,

$$\begin{Bmatrix} x \\ y \end{Bmatrix} = \begin{bmatrix} \cos\alpha & -\sin\alpha \\ \sin\alpha & \cos\alpha \end{bmatrix} \begin{Bmatrix} s \\ n \end{Bmatrix} - \begin{Bmatrix} x_1 \\ y_1 \end{Bmatrix} \quad (5.56)$$

where $\cos\alpha = x_{21}/L_{21}$, $\sin\alpha = y_{21}/L_{21}$, and $L_{21} = \sqrt{(x_{21}^2 + y_{21}^2)}$.

Area coordinates remain the most popular choice because they are *invariant* with rotation of axes and do not require the arbitrary choice of one particular side of the triangle upon which to base oblique coordinates.

5.7 Lagrangian interpolation in triangles

We have already dealt with the first member of the Lagrangian triangle family in Section 5.6, using it to define area coordinates for triangles. For the six-node triangle of Fig. 5.11(a) we can use the interpolation[†]

$$u = a_1 L_1^2 + a_2 L_2^2 + a_3 L_3^2 + a_4 L_1 L_2 + a_5 L_2 L_3 + a_6 L_3 L_1 = \{M\}^t\{a\}. \quad (5.57)$$

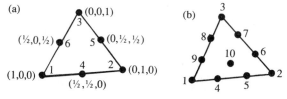

Fig. 5.11 Quadratic and cubic Lagrangian interpolations for triangular elements

---

[†] This element is sometimes referred to as the linear strain triangle (LST) owing to its original applications to plane stress but its interpolation functions are widely used as a basis in other applications.

Writing the nodal values for u on the left side of eqn (5.57) with the corresponding nodal coordinates substituted on the right-hand side, six simultaneous equations are obtained. Writing these in matrix form we obtain

$$
\begin{bmatrix} u_1 \\ u_2 \\ u_3 \\ u_4 \\ u_5 \\ u_6 \end{bmatrix} = \begin{bmatrix} 1 & 0 & 0 & 0 & 0 & 0 \\ 0 & 1 & 0 & 0 & 0 & 0 \\ 0 & 0 & 1 & 0 & 0 & 0 \\ \frac{1}{4} & \frac{1}{4} & 0 & \frac{1}{4} & 0 & 0 \\ 0 & \frac{1}{4} & \frac{1}{4} & 0 & \frac{1}{4} & 0 \\ \frac{1}{4} & 0 & \frac{1}{4} & 0 & 0 & \frac{1}{4} \end{bmatrix} \begin{bmatrix} a_1 \\ a_2 \\ a_3 \\ a_4 \\ a_5 \\ a_6 \end{bmatrix} = A^{-1}\{a\} \qquad (5.58)
$$

and inverting, the interpolation functions are obtained as

$$
\{f\}^t = \{M\}^t A = \{M\}^t \begin{bmatrix} 1 & 0 & 0 & 0 & 0 & 0 \\ 0 & 1 & 0 & 0 & 0 & 0 \\ 0 & 0 & 1 & 0 & 0 & 0 \\ -1 & -1 & 0 & 4 & 0 & 0 \\ 0 & -1 & -1 & 0 & 4 & 0 \\ -1 & 0 & -1 & 0 & 0 & 4 \end{bmatrix} \qquad (5.59)
$$

or

$$
\begin{aligned}
f_1 &= L_1^2 - L_1 L_2 - L_3 L_1 = L_1^2 - L_1(L_2 + L_3) \\
&= L_1^2 - L_1(1 - L_1) = L_1(2L_1 - 1) \\
f_2 &= L_2(2L_2 - 1), \qquad f_3 = L_3(2L_3 - 1) \\
f_4 &= 4L_1 L_2, \qquad f_5 = 4L_2 L_3, \qquad f_6 = 4L_3 L_1.
\end{aligned} \qquad (5.60)
$$

These functions can also be obtained from the formula[12]

$$
f_{abc}(L_1, L_2, L_3) = f_a(L_1) f_b(L_2) f_c(L_3) \qquad (5.61)
$$

where

$$
f_a(L_1) = \begin{cases} \displaystyle\prod_{i=1}^{a} \frac{nL_1 - i + 1}{i}, & a \geq 1 \\[2ex] 1, & a = 0. \end{cases}
$$

$a = nL_1$, $b = nL_2$, and $c = nL_3$ are given by the node's coordinates, and n is the order of the interpolation, that is, $n = 1$ for linear interpolation, $n = 2$ for quadratic interpolation, and so on. For f_1, for example, one has

$$
(a, b, c) = (2, 0, 0)
$$

giving $f_1 = \{(2L_1 - 1 + 1)/1\}\{(2L_1 - 2 + 1)/2\} = L_1(2L_1 - 1)$.

For the ten-node cubic Lagrangian triangle shown in Fig. 5.11(b) the area coordinate modes are

$$\{M\} = \{L_1^3, L_2^3, L_3^3, L_1^2 L_2, L_1^2 L_3, L_2^2 L_3, L_2^2 L_1, L_3^2 L_1, L_3^2 L_2, L_1 L_2 L_3\}.$$

$$(5.62)$$

(Note that with the modes in this order pivoting is needed to obtain A. The order $L_1^3, L_2^3, L_3^3, L_1^2 L_2, L_1 L_2^2, L_2^2 L_3, L_2 L_3^2, L_3^2 L_1, L_3 L_1^2, L_1 L_2 L_3$ can be used to avoid pivoting.) Following the matrix formation and inversion procedure of eqns (5.58) and (5.59) one obtains the interpolation matrix as

$$A = \tfrac{1}{2} \begin{bmatrix} 2 & 0 & 0 & 0 & 0 & 0 & 0 & 0 & 0 & 0 \\ 0 & 2 & 0 & 0 & 0 & 0 & 0 & 0 & 0 & 0 \\ 0 & 0 & 2 & 0 & 0 & 0 & 0 & 0 & 0 & 0 \\ -5 & 2 & 0 & 18 & 0 & 0 & -9 & 0 & 0 & 0 \\ -5 & 0 & 2 & 0 & 18 & 0 & 0 & -9 & 0 & 0 \\ 0 & -5 & 2 & 0 & 0 & 18 & 0 & 0 & -9 & 0 \\ 2 & -5 & 0 & -9 & 0 & 0 & 18 & 0 & 0 & 0 \\ 2 & 0 & -5 & 0 & -9 & 0 & 0 & 18 & 0 & 0 \\ 0 & 2 & -5 & 0 & 0 & -9 & 0 & 0 & 18 & 0 \\ 2 & 2 & 2 & -9 & -9 & -9 & -9 & -9 & -9 & 54 \end{bmatrix}.$$

The interpolation functions are given by $\{f\}^t = \{M\}^t A$, yielding

$$f_i = \tfrac{1}{2}L_i(3L_i - 1)(3L_i - 2) \qquad i = 1, 2, 3$$

$$f_n = 4.5 L_i L_j (3L_i - 1) \qquad n = 4, 6, 8, \quad i = 1, 2, 3, \quad j = 2, 3, 1$$

$$(5.63)$$

$$f_n = 4.5 L_j L_i (3L_j - 1) \qquad n = 5, 7, 9, \quad i = 1, 2, 3, \quad j = 2, 3, 1$$

$$f_{10} = 27 L_1 L_2 L_3.$$

In eqn (5.57) the modes for the interpolation have been written in the homogeneous form

$$L_1^a L_2^b L_3^c \qquad \text{where } a + b + c = 2.$$

As shown in eqn (5.17) one can equally well follow the practice for Cartesian displacement functions and use an ascending polynomial, that is,

$$\{M\} = \{L_1, L_2, L_3, L_1 L_2, L_2 L_3, L_3 L_1\}$$

and this approach leads to exactly the same interpolation functions for both the quadratic and cubic cases. It is the usual practice with area coordinates to use the homogeneous form, though in the case of thin plate bending elements it is useful to choose the first three modes as L_1, L_2, and L_3 as this alters only the three interpolation functions corresponding to the rigid body freedoms w_1, w_2, and w_3. This has the advantage that the rigid body modes $(L_1, L_2,$ and $L_3)$ are represented explicitly in the displacement function. These do not contribute to strain energy and may be omitted in the calculation of the element stiffness matrix, shortening calculations substantially.

5.8 Hermitian interpolation in triangles

Figure 5.12(a) shows a ten-freedom cubic Hermitian triangle. Derivation of the interpolation functions for this element is much simplified if, in place of Cartesian slope freedoms, we use *natural slopes*[13] parallel to the element sides, as shown in Fig. 5.12(b). We use the term 'slopes' with thin plate elements in mind but these could be any first-derivative freedoms.

The natural slopes are defined as *dimensionless derivatives* with respect to coordinates s_a, s_b, and s_c measured along each side of the element. Then parallel to side 12 we define the natural slope

$$\phi_a = L_{21} \frac{\partial w}{\partial s_a} = L_{21}\left(\frac{\partial w}{\partial x}\frac{\partial x}{\partial s_a} + \frac{\partial w}{\partial y}\frac{\partial y}{\partial s_a}\right)$$

$$= L_{21}(c_{ax}\phi_x + c_{ay}\phi_y) = x_{21}\phi_x + y_{21}\phi_y \qquad (5.64)$$

where c_{ax}, c_{ay} are the direction cosines of side 12 and are given by

$$c_{ax} = \frac{x_2 - x_1}{L_{21}} = \frac{x_{21}}{L_{21}}$$

$$\qquad\qquad (5.65)$$

$$c_{ay} = \frac{y_2 - y_1}{L_{21}} = \frac{y_{21}}{L_{21}}.$$

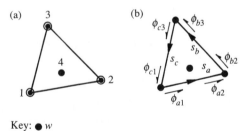

Key: ● w

◉ $w, \partial w/\partial x, \partial w/\partial y$ (global)

Fig. 5.12 (a) Global and (b) local freedoms for a cubic Hermitian triangle

Similarly

$$\phi_b = x_{32}\phi_x + y_{32}\phi_y$$

$$\phi_c = x_{13}\phi_x + y_{13}\phi_y. \tag{5.66}$$

Expressing eqns (5.64) and (5.66) in matrix form we obtain a transformation matrix T relating the local freedoms (including the natural slopes) to the global freedoms:

$$
\{d_N\} =
\begin{bmatrix}
w_1 \\
w_2 \\
w_3 \\
\phi_{a1} \\
\phi_{c1} \\
\phi_{b2} \\
\phi_{a2} \\
\phi_{c3} \\
\phi_{b3} \\
w_4
\end{bmatrix}
=
\begin{bmatrix}
1 & 0 & 0 & 0 & 0 & 0 & 0 & 0 & 0 & 0 \\
0 & 0 & 0 & 1 & 0 & 0 & 0 & 0 & 0 & 0 \\
0 & 0 & 0 & 0 & 0 & 0 & 1 & 0 & 0 & 0 \\
0 & x_{21} & y_{21} & 0 & 0 & 0 & 0 & 0 & 0 & 0 \\
0 & x_{13} & y_{13} & 0 & 0 & 0 & 0 & 0 & 0 & 0 \\
0 & 0 & 0 & 0 & x_{32} & y_{32} & 0 & 0 & 0 & 0 \\
0 & 0 & 0 & 0 & x_{21} & y_{21} & 0 & 0 & 0 & 0 \\
0 & 0 & 0 & 0 & 0 & 0 & 0 & x_{13} & y_{13} & 0 \\
0 & 0 & 0 & 0 & 0 & 0 & 0 & x_{32} & y_{32} & 0 \\
0 & 0 & 0 & 0 & 0 & 0 & 0 & 0 & 0 & 1
\end{bmatrix}
\begin{bmatrix}
w_1 \\
\phi_{x1} \\
\phi_{y1} \\
w_2 \\
\phi_{x2} \\
\phi_{y2} \\
w_3 \\
\phi_{x3} \\
\phi_{y3} \\
w_4
\end{bmatrix}
= T\{d\} \tag{5.67}
$$

and $\{d\}$ and $\{d_N\}$ are respectively regarded as global and local freedoms.

The element matrices are determined in terms of the local freedoms and later transformed to the global system using matrix T. To establish the interpolation functions the natural slopes may be expressed as dimensionless derivatives of the (cubic) displacement function,

$$\phi_a = L_{21}\frac{\partial w}{\partial s_a} = L_{21}\left(\frac{\partial w}{\partial L_1}\frac{\partial L_1}{\partial s_a} + \frac{\partial w}{\partial L_2}\frac{\partial L_2}{\partial s_a}\right)$$

or

$$\phi_a = \frac{\partial w}{\partial L_2} - \frac{\partial w}{\partial L_1} \tag{5.68}$$

$$\phi_b = \frac{\partial w}{\partial L_3} - \frac{\partial w}{\partial L_2} \tag{5.69}$$

$$\phi_c = \frac{\partial w}{\partial L_1} - \frac{\partial w}{\partial L_3}. \tag{5.70}$$

Applying these results to the cubic modes (eqns (5.62)) and substituting the nodal area coordinates $(1, 0, 0)$, $(0, 1, 0)$ $(0, 0, 1)$, and $(1/3, 1/3, 1/3)$ we obtain

$$
\{d_N\} =
\begin{bmatrix}
w_1 \\ w_2 \\ w_3 \\ \phi_{a1} \\ \phi_{c1} \\ \phi_{b2} \\ \phi_{a2} \\ \phi_{c3} \\ \phi_{b3} \\ w_4
\end{bmatrix}
\begin{bmatrix}
1 & 0 & 0 & 0 & 0 & 0 & 0 & 0 & 0 & 0 \\
0 & 1 & 0 & 0 & 0 & 0 & 0 & 0 & 0 & 0 \\
0 & 0 & 1 & 0 & 0 & 0 & 0 & 0 & 0 & 0 \\
-3 & 0 & 0 & 1 & 0 & 0 & 0 & 0 & 0 & 0 \\
3 & 0 & 0 & 0 & -1 & 0 & 0 & 0 & 0 & 0 \\
0 & -3 & 0 & 0 & 0 & 1 & 0 & 0 & 0 & 0 \\
0 & 3 & 0 & 0 & 0 & 0 & -1 & 0 & 0 & 0 \\
0 & 0 & -3 & 0 & 0 & 0 & 0 & 1 & 0 & 0 \\
0 & 0 & 3 & 0 & 0 & 0 & 0 & 0 & -1 & 0 \\
\frac{1}{27} & \frac{1}{27} & \frac{1}{27} & \frac{1}{27} & \frac{1}{27} & \frac{1}{27} & \frac{1}{27} & \frac{1}{27} & \frac{1}{27} & \frac{1}{27}
\end{bmatrix}
\{a\} = A^{-1}\{a\
$$

(5.71)

giving on inversion

$$
\{f\}^t =
\begin{bmatrix}
L_1^3 \\ L_2^3 \\ L_3^3 \\ L_1^2 L_2 \\ L_1^2 L_3 \\ L_2^2 L_3 \\ L_2^2 L_1 \\ L_3^2 L_1 \\ L_3^2 L_2 \\ L_1 L_2 L_3
\end{bmatrix}^t
\begin{bmatrix}
1 & 0 & 0 & 0 & 0 & 0 & 0 & 0 & 0 & 0 \\
0 & 1 & 0 & 0 & 0 & 0 & 0 & 0 & 0 & 0 \\
0 & 0 & 1 & 0 & 0 & 0 & 0 & 0 & 0 & 0 \\
3 & 0 & 0 & 1 & 0 & 0 & 0 & 0 & 0 & 0 \\
3 & 0 & 0 & 0 & -1 & 0 & 0 & 0 & 0 & 0 \\
0 & 3 & 0 & 0 & 0 & 1 & 0 & 0 & 0 & 0 \\
0 & 3 & 0 & 0 & 0 & 0 & -1 & 0 & 0 & 0 \\
0 & 0 & 3 & 0 & 0 & 0 & 0 & 1 & 0 & 0 \\
0 & 0 & 3 & 0 & 0 & 0 & 0 & 0 & -1 & 0 \\
-7 & -7 & -7 & -1 & 1 & -1 & 1 & -1 & 1 & 27
\end{bmatrix}
= \{M\}^t A
$$

(5.72)

whence the interpolation functions are

$$f_1 = L_1^3 + 3L_1^2 L_2 + 3L_1^2 L_3 - 7L_1 L_2 L_3 \quad \text{(for } w_1)$$

$$f_2 = L_2^3 + 3L_2^2 L_3 + 3L_2^2 L_1 - 7L_1 L_2 L_3 \quad \text{(for } w_2)$$

$$f_3 = L_3^3 + 3L_3^2 L_1 + 3L_3^2 L_2 - 7L_1 L_2 L_3 \quad \text{(for } w_3)$$

$$f_4 = L_1^2 L_2 - L_1 L_2 L_3, \qquad f_5 = -L_1^2 L_3 + L_1 L_2 L_3 \quad \text{(for } \phi_{a1}, \phi_{c1}) \quad (5.73)$$

$$f_6 = L_2^2 L_3 - L_1 L_2 L_3, \qquad f_7 = -L_2^2 L_1 + L_1 L_2 L_3 \quad \text{(for } \phi_{b2}, \phi_{a2})$$

$$f_8 = L_3^2 L_1 - L_1 L_2 L_3, \qquad f_9 = -L_3^2 L_2 + L_1 L_2 L_3 \quad \text{(for } \phi_{c3}, \phi_{b3})$$

$$f_{10} = 27 L_1 L_2 L_3 \quad \text{(for } w_4).$$

Note that if the first three modes are taken as L_1, L_2, and L_3 (rather than as L_1^3, L_2^3, and L_3^3 as above) the first three interpolation functions alter, the first becoming

$$f_1 = L_1 + L_1^2 L_2 + L_1^2 L_3 - L_2^2 L_1 - L_3^2 L_1 - 9L_1 L_2 L_3$$

and f_2, f_3 follow by subscript progression, that is these forms explicitly include the thin plate rigid body modes L_1, L_2, and L_3.

Equations (5.73) apply with natural slopes and for Cartesian slopes, observing eqn (5.67), one can write $x_{21} f(\phi_{a1}) + x_{13} f(\phi_{c1}) = x_{21} f_4 + x_{13} f_5$ as the function for ϕ_{x1} and similarly for $\phi_{y1}, \phi_{x2}, \phi_{y2}, \phi_{x3}$, and ϕ_{y3}.

5.9 Interpolations including normal slopes in triangles

The triangular element of Fig. 5.13 is, in terms of the number of freedoms required, the simplest possible plate bending element. Although it gives very slow convergence in thin plate applications and is rarely used in practice, it provides a useful introduction to the manner in which we introduce normal slope freedoms. As shown, it requires the definition of a normal slope at the middle of each side and this considerably complicates the interpolation functions.

Defining ψ_{n4} as the global normal slope at node 4 ($\partial w / \partial y'$ in Fig. 5.13) it is also necessary to define ϕ_{n4} as the local dimensionless derivative. Correspondingly, one must define ψ_a, ψ_b, and ψ_c as the natural slopes $\partial w / \partial s_a$, $\partial w / \partial s_b$, and $\partial w / \partial s_c$, in contrast to the natural slopes ϕ_a, ϕ_b, and ϕ_c defined in eqns (5.64) and (5.66). Observing Fig. 5.13 one can write[13]

$$\psi_{c4} = -\psi_{n4} \sin \beta - \psi_{a4} \cos \beta, \qquad \psi_{b4} = \psi_{n4} \sin \gamma - \psi_{a4} \cos \gamma \qquad (5.74)$$

where $\psi_{c4} = \phi_{c4}/L_{13}$, $\psi_{b4} = \phi_{b4}/L_{32}$, $\psi_{a4} = \phi_{a4}/L_{21}$
or

$$\frac{\phi_{c4}}{L_{13}} = -\psi_{n4} \sin \beta - \frac{\phi_{a4} \cos \beta}{L_{21}}$$

$$\frac{\phi_{b4}}{L_{32}} = \psi_{n4} \sin \gamma - \frac{\phi_{a4} \cos \gamma}{L_{21}} \qquad (5.75)$$

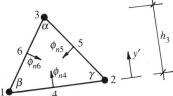

Fig. 5.13 Six-freedom constant moment thin plate element

so that

$$\psi_{n4} L_{13} \sin \beta = -\phi_{c4} - \frac{\phi_{a4} L_{13} \cos \beta}{L_{21}}$$

$$\psi_{n4} L_{32} \sin \gamma = \phi_{b4} + \frac{\phi_{a4} L_{32} \cos \gamma}{L_{13}}$$
(5.76)

where $L_{13} \sin \beta = L_{32} \sin \gamma = h_3 = 2\Delta/L_{21}$. Therefore, averaging the two eqns (5.76) one obtains

$$\psi_{n4} = \frac{\phi_{b4} - \phi_{c4} + \phi_{a4}(L_{32} \cos \gamma - L_{13} \cos \beta)/L_{21}}{2h_3}$$

$$= \frac{L_{21}(\phi_{b4} - \phi_{c4} + e_a \phi_{a4})}{4\Delta}$$
(5.77)

or

$$\psi_{n4} = \frac{\phi_{n4} L_{21}}{4\Delta}.$$
(5.78)

Here e_a may be regarded as the 'eccentricity' of node 4 with respect to vertex 3. The expressions for ϕ_{n5} and ϕ_{n6} follow using cyclic progression:

$$\phi_{n5} = \phi_{c5} - \phi_{a5} + e_b \phi_{b5}$$

$$\phi_{n6} = \phi_{a6} - \phi_{b6} + e_c \phi_{c6}$$
(5.79)

where

$$e_b = \frac{L_{13} \cos \alpha - L_{21} \cos \gamma}{L_{32}}, \qquad e_c = \frac{L_{21} \cos \beta - L_{32} \cos \alpha}{L_{13}}$$
(5.80)

$$\psi_{n5} = \frac{\phi_{n5} L_{32}}{4\Delta}, \qquad \psi_{n6} = \frac{\phi_{n6} L_{13}}{4\Delta}.$$
(5.81)

The angles α, β, γ required for e_a, e_b, e_c are readily calculated from the cosine or sine rules,

$$\cos \alpha = \frac{L_{32}^2 + L_{13}^2 - L_{21}^2}{2L_{32} L_{13}}$$
(5.82)

and

$$\sin \beta = \frac{L_{32} \sin \alpha}{L_{21}}, \qquad \sin \gamma = \frac{L_{13} \sin \alpha}{L_{21}}$$
(5.83)

and Δ is calculated using the 'semi-perimeter' formula (see eqn (11.15)).

To establish the interpolation functions, eqns (5.77) and (5.79) are applied to the displacement modes (eqn (5.57)). For ϕ_{n4} one writes

$$\phi_{n4} = \left(\frac{\partial w}{\partial L_3} - \frac{\partial w}{\partial L_2}\right) - \left(\frac{\partial w}{\partial L_1} - \frac{\partial w}{\partial L_3}\right) + e_a\left(\frac{\partial w}{\partial L_2} - \frac{\partial w}{\partial L_1}\right)$$

$$= 2\frac{\partial w}{\partial L_3} - \frac{\partial w}{\partial L_1} - \frac{\partial w}{\partial L_2} + e_a\left(\frac{\partial w}{\partial L_2} - \frac{\partial w}{\partial L_1}\right) \tag{5.84}$$

and for the first mode L_1^2 one obtains

$$\phi_{n4} = -2L_1 - 2e_aL_1 = -1 - e_a. \tag{5.85}$$

In like fashion the displacement modes for both w and the normal slopes are evaluated at each node, giving

$$\{d_N\} = \begin{vmatrix} w_1 \\ w_2 \\ w_3 \\ \phi_{n4} \\ \phi_{n5} \\ \phi_{n6} \end{vmatrix} \begin{vmatrix} 1 & 0 & 0 & 0 & 0 & 0 \\ 0 & 1 & 0 & 0 & 0 & 0 \\ 0 & 0 & 1 & 0 & 0 & 0 \\ -1-e_a & -1+e_a & 0 & -1 & 1 & 1 \\ 0 & -1-e_b & -1+e_b & 1 & -1 & 1 \\ -1+e_c & 0 & -1-e_c & 1 & 1 & -1 \end{vmatrix} \begin{vmatrix} a_1 \\ a_2 \\ a_3 \\ a_4 \\ a_5 \\ a_6 \end{vmatrix} = A^{-1}\{a\}$$

Using partitioning the required inverse can, in this instance, be obtained manually, giving the interpolation functions as

$$\{f\}^t = \{M\}^t A = [L_1^2, L_2^2, L_3^2, L_1L_2, L_2L_3, L_3L_1]A \tag{5.87}$$

where

$$A = \tfrac{1}{2}\begin{bmatrix} 2 & 0 & 0 & 0 & 0 & 0 \\ 0 & 2 & 0 & 0 & 0 & 0 \\ 0 & 0 & 2 & 0 & 0 & 0 \\ 1-e_c & 1+e_b & 2+e_c-e_b & 0 & 1 & 1 \\ 2+e_a-e_c & 1-e_a & 1+e_c & 1 & 0 & 1 \\ 1+e_a & 2+e_b-e_a & 1-e_b & 1 & 1 & 0 \end{bmatrix} \tag{5.88}$$

giving

$$f_1 = L_1^2 + \tfrac{1}{2}L_1L_2 + L_2L_3 + \tfrac{1}{2}L_3L_1 + \tfrac{1}{2}e_a(L_2L_3 + L_3L_1)$$
$$\quad - \tfrac{1}{2}e_c(L_1L_2 + L_2L_3)$$
$$f_2 = L_2^2 + \tfrac{1}{2}L_1L_2 + \tfrac{1}{2}L_2L_3 + L_3L_1 + \tfrac{1}{2}e_b(L_3L_1 + L_1L_2)$$
$$\quad - \tfrac{1}{2}e_a(L_2L_3 + L_3L_1)$$
$$f_3 = L_3^3 + L_1L_2 + \tfrac{1}{2}L_2L_3 + \tfrac{1}{2}L_3L_1 + \tfrac{1}{2}e_c(L_1L_2 + L_2L_3) \qquad (5.89)$$
$$\quad - \tfrac{1}{2}e_b(L_3L_1 + L_1L_2)$$
$$f_4 = \tfrac{1}{2}(L_2L_3 + L_3L_1)$$
$$f_5 = \tfrac{1}{2}(L_1L_2 + L_3L_1)$$
$$f_6 = \tfrac{1}{2}(L_1L_2 + L_2L_3).$$

For this element the interpolation functions are relatively simple and the curvatures for the element are obtained by applying the appropriate curvature–displacement relations directly to the interpolation functions. Once the B matrix is established eqns (5.78) and (5.81) are used to modify the terms corresponding to the normal rotations, giving the dimensionless slopes from the global values. These adjustements must be carefully signed to ensure a consistent global direction for the normal slopes as the local normal slopes are always directed inwards in each element (see eqn (5.91)).

Figure 5.14 shows a quartic bending element[†] which again uses a normal slope freedom at each midside node. Before interpolation is undertaken the global rotations are transformed to obtain the dimensionless local slopes using eqns (5.64), (5.66), (5.78) and (5.81). The global and local freedoms are, respectively,

$$\{d\} = \{w_1, \phi_{x1}, \phi_{y1}, w_2, \phi_{x2}, \phi_{y2}, w_3, \phi_{x3}, \phi_{y3}, w_4, \psi_{n4}, w_5, \psi_{n5}, w_6, \psi_{n6}\}$$
$$\{d_N\} = \{w_1, w_2, w_3, \phi_{a1}, \phi_{c1}, \phi_{b2}, \phi_{a2}, \phi_{c3}, \phi_{b3}, w_4, w_5, w_6, \phi_{n4}, \phi_{n5}, \phi_{n6}\}$$

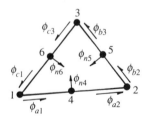

Fig. 5.14 Quartic triangular bending element

[†] Quartic elements such as this have been rarely used to date, and then only for plates and shells. This element, however, provides an introduction to the quintic element interpolation developed in Section 5.10. The latter is widely used.

so that the transformation is given by

$$
\{d_N\} = \begin{bmatrix} T_1\,(9\times9) & & & 0\,(9\times6) & & \\ & 1 & 0 & 0 & 0 & 0 & 0 \\ & 0 & 0 & 1 & 0 & 0 & 0 \\ 0\,(6\times9) & 0 & 0 & 0 & 0 & 1 & 0 \\ & 0 & m_a & 0 & 0 & 0 & 0 \\ & 0 & 0 & 0 & m_b & 0 & 0 \\ & 0 & 0 & 0 & 0 & 0 & m_c \end{bmatrix} \{d\} \tag{5.90}
$$

where T_1 is as given by eqn (5.67) and $m_a = 4\Delta/L_{21}$ or

$$
m_a = \begin{cases} -\,4\Delta/L_{21} & \text{if } x_{21} < 0 \\ -\,4\Delta/L_{21} & \text{if } x_{21} = 0 \text{ and } y_{21} > 0. \end{cases} \tag{5.91}
$$

m_b and m_c are signed in like fashion to m_a. Now representing the rigid body modes explicitly as L_1, L_2, L_3 in the displacement modes, the interpolation matrix is obtained as

$$
A_s = \begin{bmatrix}
1 & -1 & 0 & 1 & 0 & 0 & 0 & 0 & 0 & 0 & 0 & 0 & 0 & 0 & 0 \\
1 & 0 & -1 & 0 & -1 & 0 & 0 & 0 & 0 & 0 & 0 & 0 & 0 & 0 & 0 \\
0 & 1 & -1 & 0 & 0 & 1 & 0 & 0 & 0 & 0 & 0 & 0 & 0 & 0 & 0 \\
-1 & 1 & 0 & 0 & 0 & 0 & -1 & 0 & 0 & 0 & 0 & 0 & 0 & 0 & 0 \\
-1 & 0 & 1 & 0 & 0 & 0 & 0 & 1 & 0 & 0 & 0 & 0 & 0 & 0 & 0 \\
0 & -1 & 1 & 0 & 0 & 0 & 0 & 0 & -1 & 0 & 0 & 0 & 0 & 0 & 0 \\
-8 & -8 & 0 & -1 & 0 & 0 & 1 & 0 & 0 & 16 & 0 & 0 & 0 & 0 & 0 \\
0 & -8 & -8 & 0 & 0 & -1 & 0 & 0 & 1 & 0 & 16 & 0 & 0 & 0 & 0 \\
-8 & 0 & -8 & 0 & 1 & 0 & 0 & -1 & 0 & 0 & 0 & 16 & 0 & 0 & 0 \\
& & & & & & & & & 16 & -16 & 16 & 2 & -2 & 2 \\
& A_1(3\times3) & & & A_2(3\times6) & & & & & 16 & 16 & -16 & 2 & 2 & -2 \\
& & & & & & & & & -16 & 16 & 16 & -2 & 2 & 2
\end{bmatrix}
$$

where

$$
A_1 = \begin{bmatrix}
-10+3e_a-3e_c & -3-3e_a-3e_b & -3+3e_b+3e_c \\
-3+3e_a+3e_c & -10+3e_b-3e_a & -3-3e_b-3e_c \\
-3-3e_a-3e_c & -3+3e_a+3e_b & -10+3e_c-3e_b
\end{bmatrix}
$$

$$
A_2 = \frac{1}{2}\begin{bmatrix}
e_a-1 & e_c+1 & -e_b-1 & e_a-1 & e_c+1 & -e_b+1 \\
e_a+1 & -e_c+1 & e_b-1 & e_a+1 & -e_c-1 & e_b-1 \\
-e_a-1 & e_c-1 & e_b+1 & -e_a+1 & e_c-1 & e_b+1
\end{bmatrix}. \tag{5.92}
$$

Here the interpolation matrix is truncated, that is, the first three rows corresponding to the rigid body modes are omitted as they are not required in the formation of the element stiffness matrix. In establishing matrix A the quartic modal vector is

$$\{M\} = \{L_1, L_2, L_3, L_1^3 L_2, L_1^3 L_3, L_2^3 L_3, L_2^3 L_1, L_3^3 L_1, L_3^3 L_2,$$

$$L_1^2 L_2^2, L_2^2 L_3^2 L_3^2 L_1^2, L_1^2 L_2 L_3, L_1 L_2^2 L_3, L_1 L_2 L_3^2\}.$$

In the inversion to obtain matrix A the terms involving e_a, e_b, and e_c are obtained by first omitting all three and inverting A^{-1}. Then the terms involving e_a are placed in A^{-1} assuming $e_a = 1$ and inversion again carried out to yield the coefficients of e_a required in A. Similarly, with two further inversions, the terms for e_b and e_c are established.

Clearly the interpolation functions are now too complex to write explicitly and curvature and other calculations are carried out upon the mode vector $\{M\}$ to form a kernel stiffness matrix k^* (an example of this procedure is given for the beam in Section 3.3). The final element stiffness matrix is then given by $k = T^t A_s^t k^* A_s T$.

5.10 Interpolation including curvatures in triangles[†]

The quintic element of Fig. 5.15 includes curvatures as freedoms at each vertex, as well as the midside normal slopes. Constraints may be used to suppress the latter, yielding a much more convenient and pratical element. The global curvature freedoms are $\chi_x = \partial^2 w / \partial x^2$, $\chi_y = \partial^2 w / \partial y^2$, and

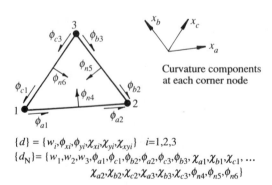

Curvature components at each corner node

$$\{d\} = \{w_i, \phi_{xi}, \phi_{yi}, \chi_{xi}, \chi_{yi}, \chi_{xyi}\} \quad i=1,2,3$$
$$\{d_N\} = \{w_1, w_2, w_3, \phi_{a1}, \phi_{c1}, \phi_{b2}, \phi_{a2}, \phi_{c3}, \phi_{b3}, \chi_{a1}, \chi_{b1}, \chi_{c1}, \cdots$$
$$\chi_{a2}, \chi_{b2}, \chi_{c2}, \chi_{a3}, \chi_{b3}, \chi_{c3}, \phi_{n4}, \phi_{n5}, \phi_{n6}\}$$

Fig. 5.15 Quintic triangular bending element

[†] We use the term curvatures, again with thin plates in mind, but the same quintic basis has also been used in fluid flow analysis.

$\chi_{xy} = \partial^2 w / \partial x \partial y$ and these are transformed to the natural curvatures χ_a, χ_b, and χ_c parallel to the sides of the triangle, that is, $\chi_a = (L_a)^2 \partial^2 w / \partial s_a^2$ $a \rightarrow b, c$ where the factor $(L_a = L_{21})^2$ yields a dimensionless derivative which still has the dimensions of w and $a \rightarrow b, c$ denotes cyclic progression.

The curvature freedom χ_a can be written in terms of the global Cartesian curvatures using

$$\frac{\chi_a}{L_{21}^2} = \chi_x \cos^2 \alpha + \chi_y \sin^2 \alpha + 2\chi_{xy} \cos \alpha \sin \alpha$$

$$= \frac{x_{21}^2 \chi_x + y_{21}^2 \chi_y + 2x_{21} y_{21} \chi_{xy}}{L_{21}^2} \tag{5.93}$$

where $\sin \alpha = y_{21}/L_{21}$ and $\cos \alpha = x_{21}/L_{21}$. Similar expressions apply to the curvatures χ_b and χ_c. Note that χ_a, χ_b and χ_c are dimensionless derivatives, that is, in the calculation of the derivative the dimension of the lateral displacement w is retained. To establish the interpolation matrix these may be calculated directly using eqns (5.68), (5.69), and (5.70),[13]

$$\chi_a = L_{21} \frac{\partial \phi_a}{\partial s_a} = \frac{\partial \phi_a}{\partial L_2} - \frac{\partial \phi_a}{\partial L_1}$$

$$= \frac{\partial^2 w}{\partial L_1^2} + \frac{\partial^2 w}{\partial L_2^2} - \frac{2\partial^2 w}{\partial L_1 \partial L_2}. \tag{5.94}$$

The expressions for χ_b and χ_c follow by cyclic progression

$$\chi_b = \frac{\partial^2 w}{\partial L_2^2} + \frac{\partial^2 w}{\partial L_3^2} - \frac{2\partial^2 w}{\partial L_2 \partial L_3} \tag{5.95}$$

$$\chi_c = \frac{\partial^2 w}{\partial L_3^2} + \frac{\partial^2 w}{\partial L_1^2} - \frac{2\partial^2 w}{\partial L_3 \partial L_1}. \tag{5.96}$$

Suppression of the normal slopes is included as part of the transformation from global to local freedoms. Taking side 12 in Fig. 5.16, for example, the normal slope ϕ_{n4} (the dimensionless derivative) is written as a cubic interpo-

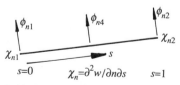

Fig. 5.16 Cubic interpolation to suppress normal slope

lation of the four corner freedoms shown. Hence one obtains, using eqns (5.24),

$$\phi_n = (1 - 3s^2 + 2s^3)\phi_{n1} + (3s^2 - 2s^3)\phi_{n2}$$
$$+ (s - 2s^2 + s^3)\chi_{n1} + (s^3 - s^2)\chi_{n2} \tag{5.97}$$

and putting $s = 0.5$, this gives

$$\phi_{n4} = \tfrac{1}{2}(\phi_{n1} + \phi_{n2}) + \tfrac{1}{8}(\chi_{n1} - \chi_{n2}). \tag{5.98}$$

Now using eqn (5.77) one writes

$$\phi_{n1} = \phi_{b1} - \phi_{c1} + e_a\phi_{a1} \tag{5.99}$$

$$\phi_{n2} = \phi_{b2} - \phi_{c2} + e_a\phi_{a2} \tag{5.100}$$

and as, from eqns (5.68), (5.69), and (5.70) one finds $\phi_a + \phi_b + \phi_c = 1$, eqns (5.99) and (5.100) can be written

$$\phi_{n1} = 1 - 2\phi_{c1} + \phi_{a1}(e_a - 1) \tag{5.101}$$

$$\phi_{n2} = 2\phi_{b2} - 1 + \phi_{a2}(e_a + 1). \tag{5.102}$$

Now from eqn (5.77),

$$\chi_n = \frac{\partial\phi_n}{\partial s} = \frac{\partial\phi_b}{\partial s} - \frac{\partial\phi_c}{\partial s} + e_a\left(\frac{\partial\phi_a}{\partial s}\right) \tag{5.103}$$

where, using eqns (5.69), (5.70), (5.95), and (5.96) one obtains

$$\frac{\partial\phi_b}{\partial s} - \frac{\partial\phi_c}{\partial s} = \frac{\partial^2 w}{\partial L_1^2} - \frac{\partial^2 w}{\partial L_2^2} = \chi_c - \chi_b. \tag{5.104}$$

Hence eqn (5.103) becomes

$$\chi_{n4} = \chi_c - \chi_b + e_a\chi_a \tag{5.105}$$

and substituting eqns (5.105), (5.101), and (5.102) into eqn (5.98) one obtains

$$\phi_{n4} = \tfrac{1}{2}(e_a - 1)\phi_{a1} - \phi_{c1} + \phi_{b2} + \tfrac{1}{2}(e_a + 1)\phi_{a2}$$
$$+ \tfrac{1}{8}(e_a\chi_{a1} - \chi_{b1} + \chi_{c1} - e_a\chi_{a2} + \chi_{b2} - \chi_{c2}). \tag{5.106}$$

The expressions for ϕ_{n5} and ϕ_{n6} follow by cyclic progression. Hence once the T matrix is formed for the eighteen unsuppressed freedoms at the corner nodes the normal slopes are obtained by multiplying the three expressions of the form of eqn (5.106) into this matrix. Thus the global to local transformation matrix T is given as

$$T = \begin{bmatrix} T_1 \ (18 \times 18) \\ T_3 \ (3 \times 18) \end{bmatrix} \quad \text{where } T_3 = [T_4 \ T_5] \ T_1 \tag{5.107}$$

$$T_1 = \begin{bmatrix}
1 & & x_{21} & y_{21} & & & & & & & & & & & & & & & & & \\
 & & x_{13} & y_{13} & & & & & & T_2\,(3\times6) & & & & & & & & & & & \\
 & 1 & & & x_{32} & y_{32} & & & & & & & & & T_2 & & & & & & \\
 & & & & x_{21} & y_{21} & & & & & & & & & & & & & & & \\
 & & 1 & & & & & & & & & & & & & & & & & & \\
 & & & & & & x_{13} & y_{13} & & & & & & & & & & & & & \\
 & & & & & & x_{32} & y_{32} & & & & & & & & T_2 & & & & &
\end{bmatrix}$$

$$T_2 = \begin{bmatrix} 0 & 0 & 0 & x_{21}^2 & y_{21}^2 & 2x_{21}y_{21} \\ 0 & 0 & 0 & x_{32}^2 & y_{32}^2 & 2x_{32}y_{32} \\ 0 & 0 & 0 & x_{13}^2 & y_{13}^2 & 2x_{13}y_{13} \end{bmatrix}$$

$$T_4 = \tfrac{1}{2}\begin{bmatrix} 0 & 0 & 0 & e_a-1 & -2 & 2 & e_a+1 & 0 & 0 \\ 0 & 0 & 0 & 0 & 0 & e_b-1 & -2 & 2 & e_b+1 \\ 0 & 0 & 0 & 2 & e_c+1 & 0 & 0 & e_c-1 & -2 \end{bmatrix}$$

$$T_5 = \tfrac{1}{8}\begin{bmatrix} e_a & -1 & 1 & -e_a & 1 & -1 & 0 & 0 & 0 \\ 0 & 0 & 0 & 1 & e_b & -1 & -1 & -e_b & 1 \\ 1 & -1 & -e_c & 0 & 0 & 0 & -1 & 1 & e_c \end{bmatrix}.$$

Now the interpolation for the local freedoms $\{d_N\}$ is obtained from the quintic modal vector (given below) using eqns (5.68), (5.69), and (5.70) for the slopes, eqns (5.94), (5.95), and (5.96) for the curvatures, and the three equations of the form of eqn (5.84) for the normal slopes. Finally one obtains

$$f_i - L_i = \{M\}^t A_i \qquad i = 1 \to 3 \ (A_i = \text{column } i \text{ of } A_s)$$

$$f_i = \{M\}^t A_s \qquad i = 4 \to 21 \tag{5.108}$$

where

$$\{M\} = \{L_1^4 L_2,\ L_1^4 L_3,\ L_2^4 L_3,\ L_2^4 L_1,\ L_3^4 L_1,\ L_3^4 L_2, ..$$
$$L_1^3 L_2^2,\ L_1^3 L_3^2,\ L_2^3 L_3^2,\ L_2^3 L_1^2,\ L_3^3 L_1^2,\ L_3^3 L_2^2, \ldots ,$$
$$L_1^3 L_2 L_3,\ L_1 L_2^3 L_3,\ L_1 L_2 L_3^3,\ L_1^2 L_2^2 L_3,\ L_1 L_2^2 L_3^2,\ L_1^2 L_2 L_3^2\}$$

and

$$A_s = \begin{bmatrix}
1 & -1 & 0 & 1 & 0 & 0 & 0 & 0 & 0 & 0 & 0 & 0 & 0 & 0 & 0 & 0 & 0 & 0 & 0 & 0 & 0 \\
1 & 0 & -1 & 0 & -1 & 0 & 0 & 0 & 0 & 0 & 0 & 0 & 0 & 0 & 0 & 0 & 0 & 0 & 0 & 0 & 0 \\
0 & 1 & -1 & 0 & 0 & 1 & 0 & 0 & 0 & 0 & 0 & 0 & 0 & 0 & 0 & 0 & 0 & 0 & 0 & 0 & 0 \\
-1 & 1 & 0 & 0 & 0 & 0 & -1 & 0 & 0 & 0 & 0 & 0 & 0 & 0 & 0 & 0 & 0 & 0 & 0 & 0 & 0 \\
1 & 0 & 1 & 0 & 0 & 0 & 0 & 1 & 0 & 0 & 0 & 0 & 0 & 0 & 0 & 0 & 0 & 0 & 0 & 0 & 0 \\
0 & -1 & 1 & 0 & 0 & 0 & 0 & 0 & -1 & 0 & 0 & 0 & 0 & 0 & 0 & 0 & 0 & 0 & 0 & 0 & 0 \\
4 & -4 & 0 & 4 & 0 & 0 & 0 & 0 & 0 & \tfrac12 & 0 & 0 & 0 & 0 & 0 & 0 & 0 & 0 & 0 & 0 & 0 \\
4 & 0 & -4 & 0 & -4 & 0 & 0 & 0 & 0 & \tfrac12 & 0 & 0 & 0 & 0 & 0 & 0 & 0 & 0 & 0 & 0 & 0 \\
0 & 4 & -4 & 0 & 0 & 4 & 0 & 0 & 0 & 0 & \tfrac12 & 0 & 0 & 0 & 0 & 0 & 0 & 0 & 0 & 0 & 0 \\
-4 & 0 & 4 & 0 & 0 & 0 & -4 & 0 & 0 & 0 & \tfrac12 & 0 & 0 & 0 & 0 & 0 & 0 & 0 & 0 & 0 & 0 \\
-4 & 0 & 4 & 0 & 0 & 0 & 0 & 4 & 0 & 0 & 0 & 0 & \tfrac12 & 0 & 0 & 0 & 0 & 0 & 0 & 0 & 0 \\
0 & -4 & 4 & 0 & 0 & 0 & 0 & 0 & -4 & 0 & 0 & 0 & \tfrac12 & 0 & 0 & 0 & 0 & 0 & 0 & 0 & 0 \\
8 & -4 & -4 & 4 & -4 & 0 & 0 & 0 & 0 & \tfrac12 & -\tfrac12 & \tfrac12 & 0 & 0 & 0 & 0 & 0 & 0 & 0 & 0 & 0 \\
-4 & 8 & -4 & 0 & 0 & 4 & -4 & 0 & 0 & 0 & 0 & \tfrac12 & \tfrac12 & -\tfrac12 & 0 & 0 & 0 & 0 & 0 & 0 & 0 \\
-4 & -4 & 8 & 0 & 0 & 0 & 0 & 4 & -4 & 0 & 0 & 0 & 0 & -\tfrac12 & \tfrac12 & \tfrac12 & 0 & 0 & 0 & 0 & 0 \\
& & & & & & & & & & & & & & & & & & 8 & 0 & 0 \\
& & & & & & & & & & & & & & & & & & 0 & 8 & 0 \\
& & & & & & & & & & & & & & & & & & 0 & 0 & 8
\end{bmatrix}$$

(left block A_1 (3×9); middle block $A_2/4$ (3×9))

Here the first three (rigid body) rows of A are removed to obtain the truncated matrix A_s, and the submatrices A_1 and A_2 are given by

$$
1_1^t =
\begin{bmatrix}
3 + 15e_a & -6 & 3 - 15e_c \\
3 - 15e_a & 3 + 15e_b & -6 \\
-6 & 3 - 15e_b & 3 + 15e_c \\
3.5e_a + 8.5 & 0 & -5 \\
5 & 0 & 3.5e_c - 8.5 \\
-5 & 3.5e_b + 8.5 & 0 \\
3.5e_a - 8.5 & 5 & 0 \\
0 & -5 & 3.5e_c + 8.5 \\
0 & 3.5e_b - 8.5 & 5
\end{bmatrix}
\quad
A_2^t =
\begin{bmatrix}
3 + e_a & 0 & -2 \\
2 & 0 & 2 \\
-2 & 0 & 3 - e_c \\
3 - e_a & -2 & 0 \\
-2 & 3 + e_b & 0 \\
2 & 2 & 0 \\
0 & 2 & 2 \\
0 & 3 - e_b & -2 \\
0 & -2 & 3 + e_c
\end{bmatrix}.
$$

Although the interpolation matrix A_s is relatively complicated it is, except for a small part, constant and can be fed to the computer program from a permanent data file. Matrix T, on the other hand, is relatively sparse and is easily formed numerically. These matrices are used to calculate the element stiffness matrix as $k = T^t A_s^t k^* A_s T$ where k^* is a kernel stiffness matrix obtained by direct operation upon $\{M\}$.

5.11 Interpolation functions for tetrahedra and pentahedra

Interpolation functions for pentahedra of the shape shown in Fig. 5.17 are simply obtained by linearly interpolating between the area coordinate interpolations on each of the triangular faces. Hence for the eighteen-freedom pentahedral element shown

$$u = \tfrac{1}{2}(1 - h)(L_1 u_1 + L_2 u_2 + L_3 u_3)$$
$$+ \tfrac{1}{2}(1 + h)(L_1 u_4 + L_2 u_5 + L_3 u_6) \tag{5.109}$$

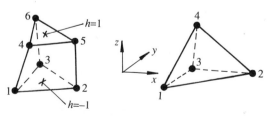

Fig. 5.17 Pentahedron and tetrahedron

and similarly for v, w, that is, one uses the constant strain triangle interpolation on each face and linearly interpolates between these faces. Thus the element can be formulated explicitly when restricted in shape to a right prism.[14] Similarly the interpolation for a triangular prism with midside nodes is obtained by combining eqns (5.10) and (5.60).[15] Formulation for the skew shape shown in Fig. 5.17 can be accomplished numerically with the use of a 3×3 *Jacobian transformation* obtained by substituring L_1, L_2, h for a, b, c in eqns (7.26) and (7.27).

For the four-node tetrahedron shown in Fig. 5.17 the Cartesian interpolation polynomial is

$$u = c_1 + c_2 x + c_3 y + c_4 z$$

applied to the u, v, and w freedoms, so that the element yields constant strains. Alternatively the area coordinate approach used for plane triangles can be extended, yielding four dimensionless *volume coordinates* L_1, L_2, L_3, and L_4. The interpolation for the four-node tetrahedron can then be written as

$$u = L_1 u_1 + L_2 u_2 + L_3 u_3 + L_4 u_4. \tag{5.110}$$

Explicit definitions of the volume coordinates in terms of the Cartesian coordinates can be obtained by following the procedure used in eqns (5.48)–(5.52) but this leads to a good deal of tedious algebra in formulating the element stiffness matrix.

A more concise approach to tetrahedral elements is achieved by extending the isoparametric mapping technique used for the six-node plane triangle in Section 7.3. Now the Jacobian matrix will be 3×3, L_4 being eliminated with the use of the identity $L_1 + L_2 + L_3 + L_4 = 1$. This approach leads to much more concise coding and, in the case of the four-node tetrahedron, only a one-point integration (at the centroid, that is, $L_1 = L_2 = L_3 = L_4 = 1/4$) is required.

The next element of the tetrahedral family has midside nodes and thus a total of thirty freedoms, giving complete quadratic interpolations for u, v, and w. The volume coordinate interpolation for this and higher-order tetrahedral elements is obtained by extending eqn (5.61) to include the fourth coordinate L_4.[16] In the case of the ten-node quadratic element the resulting interpolation functions are

$$f_i = L_i(2L_1 - 1) \qquad i = 1 \to 4$$

for the nodes at the vertices and

$$f_{kj} = 4L_k L_j$$

for a typical midside node on side kj of the tetrahedron (that side joining nodes k and j). Using these functions the element can be very concisely coded using the isoparametric approach described above.

An alternative approach to tetrahedral elements is to calculate six natural strains parallel to each side. These completely define the state of strain in the element and are related to the Cartesian strains by an extension of the transformation given in eqn (8.23) to three dimensions.[17] In three dimensions, however, the numerical isoparametric approach is much more straightforward, particularly for higher-order elements.[18]

5.12 Numerical integration using Gaussian quadrature

A wide variety of numerical integration schemes exist, including the mid-ordinate rule, the trapezoidal rule, and Simpson's rule. Gaussian quadrature is more accurate than these for polynomials (and hence for finite elements). This method chooses integration points symmetrically positioned so that only one extra integration point need be added for every second term added to the polynomial (namely those with even powers in x). Figure 5.18 illustrates the two-point rule, the required integral being given by

$$\int_{-1}^{1} g(x)\,dx = \sum_{i=1}^{n} g_i\Omega_i = g_1\Omega_1 + g_2\Omega_2 \tag{5.111}$$

where x is a dimensionless coordinate and Ω_1 and Ω_2 are the weighting factors, here both unity. The largest error in the integration is given by

$$e = O\left(\frac{d^{2n}g(x)}{dx^{2n}}\right) \tag{5.112}$$

for an integration with n sampling points, so that the two-point integration will exactly integrate a cubic function $g(x)$, and in general an n-point integration will exactly integrate polynomials of order $2n - 1$.

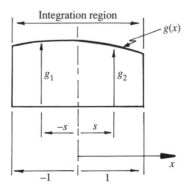

Integration region

$g(x)$

g_1 g_2

$-s$ s

-1 1

x

Fig. 5.18 Gaussian quadrature for $n = 2$

If one chooses to integrate over an interval $0 \to 1$ (rather than $-1 \to 1$) the limits of integration are changed by the transformation

$$\int_0^1 g(u)\,du = g(u_1)\omega_1 + g(u_2)\omega_2 \tag{5.113}$$

where

$$u_i = \tfrac{1}{2}(1 - x_i), \qquad \omega_1 = \tfrac{1}{2}\Omega_1, \qquad \omega_2 = \tfrac{1}{2}\Omega_2. \tag{5.114}$$

The first of eqns (5.114) corresponds to the transformation used in eqns (5.10) to change the limits of an interpolation (yielding eqns (5.11)). To derive the two-point Gauss formula the function $g(x)$ is written as a cubic polynomial

$$g(x) = a + bx + cx^2 + dx^3. \tag{5.115}$$

Substituting $x = \pm s$ into eqn (5.115) we obtain

$$g_1 = a - bs + cs^2 - ds^3, \qquad g_2 = a + bs + cs^2 + ds^3. \tag{5.116}$$

Applying eqn (5.111) with equal weights ω the numerical integral is

$$\sum_{i=1}^{2} g_i\omega = \omega(g_1 + g_2) = 2\omega(a + cs^2). \tag{5.117}$$

The true integral may be evaluated from first principles as

$$\int_{-1}^1 g(x)\,dx = \left(ax + \frac{bx^2}{2} + \frac{cx^3}{3} + \frac{dx^4}{4} \right)\Bigg|_{-1}^1 = 2\left(a + \frac{c}{3} \right). \tag{5.118}$$

Comparing eqns (5.117) and (5.118) term by term gives

$$s^2 = 1/3 \text{ or } s = \pm 1/\sqrt{3} \text{ and } \omega = 1. \tag{5.119}$$

The formulas up to $N = 5$ can also be expressed in surd form and are given in Table 5.1 but the higher-order formulas must be obtained by numerical solution of the Gauss–Legendre polynomials.[19] Note that the limits of integration are $-1, 1$ and the weights are made to sum to unity, so that integration over a length L with a dimensioned coordinate X is given by

$$\int_0^L g(X)\,dX = \int_{-1}^1 g(x)\left(\frac{dX}{dx} \right)dx = \frac{L}{2}\int_{-1}^1 g(x)\,dx = L\Sigma g_i\omega_i. \tag{5.120}$$

To obtain an $n \times n$ integration rule for rectangles from a one-dimensional n-point rule one follows the scheme shown in Fig. 5.19

$$\int_0^{L_a}\int_0^{L_b} f(x, y)\,dxdy = \int_{-1}^1 \int_{-1}^1 Af(a, b)\,dadb = \sum_{j=1}^n \sum_{i=1}^n A\omega_i\omega_j f(a_i b_j) \tag{5.121}$$

where $A = L_a L_b$, the element area. Gauss formulas for triangular domains

Table 5.1 Gauss data for $\int_{-1}^{1} f(x)\,dx = \sum_{i=1}^{n} \omega_i f(x_i)$

n	p	i	x_i	ω_i
1	1	1	0.0	1.0
2	3	1, 2	$\pm 1/\sqrt{3}$	0.5
3	5	1	0.0	8/18
		2, 3	$\pm(\sqrt{15})/5$	5/18
4	7	1, 2	$\pm\sqrt{((15 + 2\sqrt{30})/35)}$	$0.25 - 5/12\sqrt{30}$
		3, 4	$\pm\sqrt{(3/(15 + 2\sqrt{30}))}$	$0.25 + 5/12\sqrt{30}$
5	9	1	0.0	64/225
		2, 3	$\pm\sqrt{(35 + 2\sqrt{70})/63)}$	$(322 - 13\sqrt{70})/1800$
		4, 5	$\pm\sqrt{(15/(35 + 2\sqrt{70}))}$	$(322 + 13\sqrt{70})/1800$

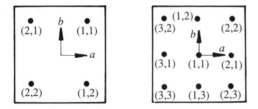

Fig. 5.19 Gaussian quadrature for rectangular domains

may be developed from the one-dimensional integrations

$$\int_{-1}^{1} f(x)\,dx \tag{5.122}$$

$$\int_{-1}^{-1} (1 + x)^\beta f(x)\,dx. \tag{5.123}$$

The first of these is the usual one-dimensional integration and, referring to Fig. 5.20, the second is used with $\beta = 1$ to obtain integrations for triangles. For example, to derive a four-point triangle formula the integration data for v_i is first taken from Table 5.1. Transforming the limits to 0, 1, the points and weights are given as

$$v_1 = \frac{1 - 0.5773503}{2} = 0.2113249, \qquad B_1 = 0.5$$

$$v_2 = \frac{1 + 0.5773503}{2} = 0.7886751, \qquad B_2 = 0.5. \tag{5.124}$$

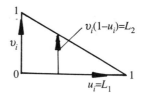

Fig. 5.20 Development of Gaussian quadrature for a triangular domain

Table 5.2 Gauss data for $\int_{-1}^{1} f(x)(1 + x)\,dx = \sum_{i=1}^{n} \omega_i f(x_i)$

n	p	i	x_i			ω_i		
1	1	1	0.0			1.0		
2	3	1	− 0.28989	79485	56636	0.36391	72365	12046
		2	0.68989	79485	56636	0.63608	27634	87954
3	5	1	− 0.57531	89235	21694	0.13965	39598	02908
		2	0.18106	62711	18531	0.45848	22127	19172
		3	0.82282	40809	74592	0.40186	38274	77920
4	7	1	− 0.72048	02713	12439	0.06236	19419	00016
		2	− 0.16718	08647	37834	0.25969	50952	16465
		3	0.44631	39727	23752	0.40692	91360	20543
		4	0.88579	16077	70965	0.27101	38268	62976
5	9	1	− 0.80292	98284	02347	0.03149	58290	43385
		2	− 0.39092	85467	07272	0.14781	77401	45233
		3	0.12405	03795	05228	0.29277	39741	69340
		4	0.60397	31642	52784	0.33434	92761	88739
		5	0.92038	02858	97063	0.19356	31804	53301

The integration for u_i is given by eqn (5.123) and, taking the data from Table 5.2 transforming the limits to 0, 1, the points and weights are

$$u_1 = \frac{1 - 0.6898979}{2} = 0.1550510, \qquad A_1 = 0.6360828$$

$$(5.125)$$

$$u_2 = \frac{1 + 0.2898979}{2} = 0.6449490, \qquad A_2 = 0.3639172$$

Then the integration formula for the triangle is given by the *conical product formula*[20]

$$L_1 = u_1, \qquad L_2 = v_j(1 - u_i), \qquad \omega_i = A_i B_i, \qquad i = 1, 2, \qquad j = 1, 2 \qquad (5.126)$$

giving, with $L_3 = 1 - L_1 - L_2$, the four-point formula

$$
\begin{aligned}
(L_1, L_2)_1 &= (0.1550510, 0.1785587), & \omega_1 &= 0.3180413 \\
(L_1, L_2)_2 &= (0.1550510, 0.6663902), & \omega_2 &= 0.3180413 \\
(L_1, L_2)_3 &= (0.6449490, 0.0750311), & \omega_3 &= 0.1819586 \\
(L_1, L_2)_4 &= (0.6449490, 0.2800198), & \omega_4 &= 0.1819586.
\end{aligned}
$$

In general the work is best carried out with a computer program using eqns (5.124)–(5.126) and the data of Tables 5.1 and 5.2. Similarly formulae for tetrahedra may be established using $\beta = 2$ in eqn (5.123).[20]

5.13 Triangular integration

For triangular domains some special integrations can be derived directly in areal coordinates[21, 22] rather than through a conical product formula, involving the weighted integral $\int f(x)(1 + x)\,dx$. The direct *triangular integrations* are more convenient, require fewer points, and there is now some evidence that they give better results (see Section 9.4).

To derive a quadratic triangular integration one assumes combinations of two coordinates α, β to yield three integration points,

$$(L_1, L_2, L_3) = (\alpha, \beta, \beta), \quad (\beta, \alpha, \beta), \quad (\beta, \beta, \alpha) \tag{5.127}$$

and the integration must exactly integrate the linear and quadratic functions

$$L_1, L_2, L_3 \tag{5.128}$$

$$L_1^2, L_2^2, L_3^2, L_1 L_2, L_2 L_3, L_3 L_1 \tag{5.129}$$

over the triangle area. The exact results for these integrations are given by the formula[23]

$$\iint L_1^a L_2^b L_3^c \, dx\, dy = \frac{a!\,b!\,c!}{(a + b + c + 2)!} (2\Delta) \tag{5.130}$$

so that numerical integration of (5.128) and (5.129) must yield

$$\sum_i \omega_i(\alpha + \beta + \beta) = \tfrac{1}{3} \tag{5.131}$$

$$\sum_i \omega_i(\alpha^2 + \beta^2 + \beta^2) = \tfrac{1}{6} \tag{5.132}$$

$$\sum_i \omega_i(\alpha\beta + \beta\alpha + \beta^2) = \tfrac{1}{12}, \quad i = 1, 2, 3. \tag{5.133}$$

Assuming $\omega_1 = \omega_2 = \omega_3 = 1/3$, so that eqn (5.127) yields $\alpha = 1 - 2\beta$, and eliminating α from eqns (5.132) and (5.133) gives a quadratic equation in β^2:

$$144\beta^4 - 40\beta^2 + 1 = 0 \qquad (5.134)$$

giving the roots $\beta = 1/2,\ 1/6$. Applying eqn (5.131) one obtains the two alternative integrations

$$(L_1, L_2, L_3) = (\tfrac{1}{2}, \tfrac{1}{2}, 0),\quad (0, \tfrac{1}{2}, \tfrac{1}{2}),\quad (\tfrac{1}{2}, 0, \tfrac{1}{2}) \qquad (5.135)$$

$$(L_1, L_2, L_3) = (\tfrac{4}{6}, \tfrac{1}{6}, \tfrac{1}{6}),\quad (\tfrac{1}{6}, \tfrac{4}{6}, \tfrac{1}{6}),\quad (\tfrac{1}{6}, \tfrac{1}{6}, \tfrac{4}{6}) \qquad (5.136)$$

the first of which is the more widely used.[24] However, the second, like Gauss quadrature, uses internal points and gives a better approximation for integrands higher than quadratic (For example, for L_i^4 eqn (5.135) gives $\Delta/24$ and eqn (5.136) gives $\Delta/15.07$, whereas the exact result is $\Delta/15$). It thus performs better in finite element applications to coupled equations (see Sections 9.4 and 11.6).

Table 5.3 Triangular integrations

n	p	L_1	L_2	L_3	ω_i
3	2	1/2	1/2	0	1/3
		0	1/2	1/2	1/3
		1/2	0	1/2	1/3
3	2	4/6	1/6	1/6	1/3
		1/6	4/6	1/6	1/3
		1/6	1/6	4/6	1/3
4	3	3/5	1/5	1/5	25/48
		1/5	3/5	1/5	25/48
		1/5	1/5	3/5	25/48
		1/3	1/3	1/3	$-27/48$
6	4	$1-2\alpha$	α	α	δ
		α	$1-2\alpha$	α	δ
		α	α	$1-2\alpha$	δ
		$1-2\beta$	β	β	$1/3-\delta$
		β	$1-2\beta$	β	$1/3-\delta$
		β	β	$1-2\beta$	$1/3-\delta$

$\alpha = 0.09157\ 62135\ 09771,\ \beta = 0.44594\ 84909\ 15965,\ \delta = 0.10995\ 17436\ 55322$

n	p	L_1	L_2	L_3	ω_i
7	5	α	δ	$1-\alpha-\delta$	ω
		δ	α	$1-\delta-\alpha$	ω
		δ	δ	$1-2\delta$	ω
		β	γ	$1-\beta-\gamma$	Ω
		γ	β	$1-\gamma-\beta$	Ω
		γ	γ	$1-2\gamma$	Ω
		1/3	1/3	1/3	270/1200

$\gamma = (6 + \sqrt{15})/21,\ \delta = (6 - \sqrt{15})/21,\ \Omega = (155 + \sqrt{15})/1200$
$\alpha = (9 + 2\sqrt{15})/21,\ \beta = (9 - 2\sqrt{15})/21,\ \omega = (155 - \sqrt{15})/1200$

Table 5.3 gives some triangular integrations, and others up to septimal accuracy are given by Cowper.[22] Note that the negative weight in the four-point formula should not cause difficulty when one seeks an 'exact' integration[21] but as a reduced integration for L_i^4, for example, it gives $\Delta/16.51$, less accurate than eqn (5.136).

5.14 Numerical integration for a beam element

As an example of numerical integration consider the four-freedom beam element. The curvature interpolation is given, using eqns (5.22), as

$$\frac{\partial^2 v}{\partial x^2} = \frac{\partial^2 v}{\partial s^2}\left(\frac{\partial s}{\partial x}\right)^2 + \frac{\partial v}{\partial s}\frac{\partial^2 s}{\partial x^2} = \frac{4}{L^2}\frac{\partial^2 v}{\partial s^2}$$

$$= \frac{4}{L^2}\frac{\partial^2\{f\}}{\partial s^2\{d\}} = [6s,\ 3s-1,\ -6s,\ 3s+1]\frac{1}{L^2\{d\}} = B\{d\} \quad (5.137)$$

as $\partial^2 s/\partial x^2$ vanishes in a straight element. The element stiffness matrix is given by

$$k = \sum B^t(EI)B(\omega_i L). \quad (5.138)$$

Substitution of eqn (5.137) into eqn (5.138) will give terms in s^2 so that a two-point Gauss rule is needed. At the first point, $s = +1/\sqrt{3}$, $\omega_1 = 1/2$, giving

$$B = \frac{[2\sqrt{3},\ \sqrt{3}-1,\ -2\sqrt{3},\ \sqrt{3}+1]}{L^2} \quad (5.139)$$

and for the second point $s = -1/\sqrt{3}$, $\omega_2 = 1/2$, giving

$$B = \frac{[-2\sqrt{3},\ -\sqrt{3}-1,\ 2\sqrt{3},\ -\sqrt{3}+1]}{L^2}. \quad (5.140)$$

Substituting eqns (5.139) and (5.140) into eqn (5.138) the element stiffness matrix is accumulated at the two integration points in the two parts,

$$\frac{EI}{2L^3}\begin{bmatrix} 12 & 6-2\sqrt{3} & -12 & 6+2\sqrt{3} \\ 6-2\sqrt{3} & 4-2\sqrt{3} & -6+2\sqrt{3} & 2 \\ -12 & -6+2\sqrt{3} & 12 & -6-2\sqrt{3} \\ 6+2\sqrt{3} & 2 & -6-2\sqrt{3} & 4+2\sqrt{3} \end{bmatrix}$$

$$\frac{EI}{2L^3}\begin{bmatrix} 12 & 6+2\sqrt{3} & -12 & 6-2\sqrt{3} \\ 6+2\sqrt{3} & 4+2\sqrt{3} & -6-2\sqrt{3} & 2 \\ -12 & -6-2\sqrt{3} & 12 & -6+2\sqrt{3} \\ 6-2\sqrt{3} & 2 & -6+2\sqrt{3} & 4-2\sqrt{3} \end{bmatrix}.$$

Summing these and multiplying all ϕ–ϕ terms by L^2 and all v–ϕ and ϕ–v terms by L (see eqn (5.21) one obtains the exact solution of eqn (1.10).

The example serves well to demonstrate that numerical integration should lose only the last digit of computation compared to an explicit formulation. In this case Simpson's rule could have been used (the integration points being conveniently positioned for stress sampling) but the Gauss rule is more economical and gives better results in coupled problems as a reduced integration (see Section 11.1).

Note also that in the foregoing example one-point integration will give nonsensical results. The questions of minimum orders of integration and reduced integration are taken up at the close of Chapter 15.

5.15 Discussion

A wide variety of elements are possible and only a few of the most common members of some element families have been presented in the foregoing. These families include:

(1) *The Lagrange family.* Elements with Lagrangian interpolation of boundary nodes, for example, Figs 5.2, 5.9, and 5.11(a)

(2) The Hermitian family. Elements with Hermitian interpolation of boundary nodes, for example, Fig. 5.1.

(3) Multivariate elements. Elements formed by multiplying one-dimensional interpolations for example, Figs 5.6, 5.7, and 5.8. Multivariate elements in which internal nodes are omitted are sometimes termed *Serendipity elements*,[18] for example, Fig. 9.3 cases A, B, and D.

(4) The TRIB family.[13] Elements with displacement and slope freedoms at the vertices and with slope freedoms distributed along the sides, for example, Fig. 5.14.

(5) The TUBA family.[13] Elements which include three curvature freedoms at some of the nodes, for example, Fig. 5.15.

(6) The nested interpolation family. Elements obtained by the method of 'nested interpolations', for example, Figs 8.10 and 10.8.

Families (1), (2), and (3) include many isoparametric elements some of which are discussed in Chapter 7. Moreover, in any discussion of element families the method of nested interpolations is of special interest as, for example, Hermitian elements with only vertex freedoms can be transformed to become Lagrangian elements with vertex and midside freedoms.

References

1. Irons, B. M. (1968) Comment on 'Complete polynomial displacement fields for finite element methods' by P. C Dunne. *Trans R. Aero. Soc.*, **72**, 709.
2. Przemieniecki, J. S. (1968). *Theory of Matrix Structural Analysis.* McGraw-Hill, New York.
3. Bazeley, G. P., Cheung, Y. K., Irons, B. M., and Zienkiewicz, O. C. (1965). Triangular elements in plate bending–conforming and nonconforming solutions. Proceedings of the First Conference of Matrix Methods in Structural Mechanics, Wright-Patterson AFB, Ohio.
4. Cowper, G. R., Kosko, E., Lindberg, G. M., and Olson, M. D. (1968). Formulation of a new triangular plate bending element. *Trans. Canad. Aero-Space Inst.*, **1**, 86.
5. Wilson, E. L., Taylor, R. L., Doherty, W. P., and Ghaboussi, J. (1971). Incompatible displacement models. Proceedings of the ONR Conference on Numerical Methods, McGraw-Hill, Illinois.
6. Oden, J. T. (1972). *Finite Elements of Nonlinear Continua.* McGraw-Hill, New York.
7. Cantin, G. and Clough, R. W. (1968). A curved cylindrical shell finite element. *J. Amer. Inst. Aeron. Astron.*, **6**, 1057.
8. Golley, B. W. (1979). A rectangular variable degree of freedom plate bending element. Proceedings of the Third Australian Conference on Finite Element Methods, University of NSW, Sydney.
9. Cheung Y. K. (1976). *Finite Strip Method in Structural Analysis.* Pergamon, Oxford.
10. Rainville, E. D. and Beafait, W. P. (1978). *Elementary Differential Equations*, (5th edn). Macmillan, New York.
11. Bogner, F. K., Fox, R. L., and Schmidt, L. A. (1965). The generation of inter-element compatible stiffness and mass matrices by the use of interpolation formulas. Proceedings of the First Conference of Matrix Methods in Structural Mechanics, Wright-Patterson AFB, Ohio.
12. Silvester, P. (1971). Higher-order polynomial triangular finite elements for potential problems. *Int. J. Eng. Sci.*, **7**, 849.
13. Argyris, J. H., Fried, I., and Scharpf, D. W. (1968). The TUBA family of plate elements for the matrix displacement method. *J. Aero. Sci.*, **72**, 701.
14. Mohr, G. A. (1976). *Analysis and Design of Plate and Shell Structures using Finite Elements.* Ph.D. thesis, University of Cambridge.
15. Zienkiewicz, O. C., Irons, B. M., Ergatoudis, J., and Scott, F. C. (1969). Isoparametric and associated element families for two and three dimensional analysis. In *Finite Elements in Stress Analysis*, (ed. I. Holand and K. Bell). Tapir, Trondheim.
16. Silvester, P. (1972). Tetrahedral polynomial finite elements for the Helmholtz equation. *Int. J. Num. Meth. Engng*, **4**, 405.
17. Argyris, J. H. (1965). Three-dimensional anisotropic and inhomogeneous media: matrix analysis for small and large displacements. *Ingenieur Archiv.*, **31**, 35.
18. Ergatoudis, J. G., Irons, B. M., and Zienkiewicz, O. C. (1968). Curved isoparametric quadrilateral elements for finite element analysis. *Int. J. Solids Struct.*, **4**, 31.

19. Irons, B. M. and Ahmad, S. (1980). *Techniques of Finite Elements*. Ellis-Horwood, Chichester.
20. Stroud, A. H. and Secrest, D. (1966). *Gaussian Quadrature Formulas*. Prentice-Hall, Englewood Cliffs, NJ.
21. Lannoy, F. G. (1977). Triangular finite elements and numerical integration. *Comput. Struct.*, **7**, 613.
22. Cowper, G. R. (1973). Gaussian quadrature formulas for triangles. *Int. J. Num. Meth. Engng*, **7**, 405.
23. Morley, L. S. D. (1963). *Skew Plates and Structures*. Pergamon, New York.
24. Zienkiewicz, O. C. and Taylor, R. L. (1984). *The Finite Element Method*, (4th edn). McGraw-Hill, London.

Part II
Applications to the statics of solids

6
Plane stress and plane strain

The basic constitutive relationships of Chapter 2, the energy-based displacement method of Chapter 3, and some of the interpolation functions of Chapter 5 are used to develop several finite elements for the analysis of plane stress and plane strain. In the first two sections explicit formulations are obtained using Cartesian polynomial interpolations. These provide a useful introduction to two-dimensional finite elements but for higher-order elements interpolation functions expressed in terms of dimensionless coordinates and numerical integration are used in the remainder of the text.

6.1 The constant strain triangle

The constant strain triangle was the first two-dimensional element and is sometimes referred to as the Turner triangle.[1] The element has six freedoms, namely the u and v displacements at each corner node. The nodes are usually numbered anticlockwise so that the formula shown in Fig. 6.1(a) is positive. (Alternatively one can simply take the absolute value). The interpolation polynomials are

$$u = c_1 + c_2 x + c_3 y = \{M\}^t\{c_1, c_2, c_3\} \tag{6.1}$$

$$v = c_4 + c_5 x + c_6 y = \{M\}^t\{c_4, c_5, c_6\}. \tag{6.2}$$

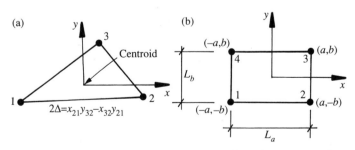

Fig. 6.1 (a) Constant strain triangle, and (b) bilinear rectangle

The interpolation matrix for c_1, c_2, c_3 is given by eqn (5.51) that is,

$$\{c_u\} = \begin{Bmatrix} c_1 \\ c_2 \\ c_3 \end{Bmatrix} = \frac{1}{3a} \begin{bmatrix} a & a & a \\ -y_{32} & -y_{13} & -y_{21} \\ x_{32} & x_{13} & x_{21} \end{bmatrix} \begin{Bmatrix} u_1 \\ u_2 \\ u_3 \end{Bmatrix} = c_u\{u\} \quad (6.3)$$

where $a = 2\Delta/3$, and the interpolation for u may be written

$$\{u\} = \{M\}^t C_u\{u\} = \{f\}^t\{u\}. \quad (6.4)$$

Using the same interpolation for v the solution for all six displacement function coefficients may be written as

$$\{c\} = \begin{bmatrix} c_1 \\ c_2 \\ c_3 \\ c_4 \\ c_5 \\ c_6 \end{bmatrix} = \frac{1}{3a} \begin{bmatrix} a & 0 & a & 0 & a & 0 \\ -y_{32} & 0 & -y_{13} & 0 & -y_{21} & 0 \\ x_{32} & 0 & x_{13} & 0 & x_{21} & 0 \\ 0 & a & 0 & a & 0 & a \\ 0 & -y_{32} & 0 & -y_{13} & 0 & -y_{21} \\ 0 & x_{32} & 0 & x_{13} & 0 & x_{21} \end{bmatrix} \begin{bmatrix} u_1 \\ v_1 \\ u_2 \\ v_2 \\ u_3 \\ v_3 \end{bmatrix}$$

$$= C\{d\}. \quad (6.5)$$

Expressing the strains as derivatives of the displacements

$$\begin{Bmatrix} \partial u/\partial x \\ \partial v/\partial y \\ \partial u/\partial y + \partial v/\partial x \end{Bmatrix} = \begin{bmatrix} 0 & 1 & 0 & 0 & 0 & 0 \\ 0 & 0 & 0 & 0 & 0 & 1 \\ 0 & 0 & 1 & 0 & 1 & 0 \end{bmatrix} \{c\} = S\{c\}. \quad (6.6)$$

Defining an element stress matrix N such that $\{\sigma\} = N\{d\}$ at any point, N and the element stiffness matrix are given by

$$N = DB = DSC \quad (6.7)$$

$$k = \int B^t DB\,dV = C^t k_m C \quad (6.8)$$

where

$$k_m = t \int\int S^t DS\,dx\,dy. \quad (6.9)$$

As the matrix S has constant coefficients the integration in eqn (6.9) reduces to scalar multiplication by the element volume Δt. In the case of plane stress,

D is given by eqn (2.17) and the kernel stiffness matrix k_m is then

$$k_m = \frac{E\Delta t}{1 - v^2} \begin{bmatrix} 0 & 0 & 0 & 0 & 0 & 0 \\ & 1 & 0 & 0 & 0 & v \\ & & v' & 0 & v' & 0 \\ & \text{symmetric} & & 0 & 0 & 0 \\ & & & & v' & 0 \\ & & & & & 1 \end{bmatrix} \tag{6.10}$$

where $v' = (1 - v)/2$. The element stresses may be assigned to the element centroid or used to calculate the average nodal values.

The consistent loads for a constant body force $X = \rho$ acting in the x-direction in an element of constant thickness are given by

$$\{q_b\} = \int F\{X\}\,dV = \int F\{\rho, 0\}\,dV = t \iint \{\rho f_1, 0, \rho f_2, 0, \rho f_3, 0\}\,dx\,dy$$

$$= (\rho t/3a)\iint \{a - y_{32}x + x_{32}y, 0, a - y_{13}x + x_{13}y, 0, a$$

$$- y_{21}x + x_{21}y, 0\}\,dx\,dy = \rho t\Delta\{1/3, 0, 1/3, 0, 1/3, 0\} \tag{6.11}$$

as, using coordinates x, y measured from the element centroid, the terms in x and y vanish upon integration. This is the intuitively expected result, one third of the total load upon the element being distributed to each node.

The thermal loads are $\{q_0\} = \int B^t D\{\varepsilon_0\}\,dV$ where for plane stress

$$D\{\varepsilon_0\} = D\{\alpha\theta, \alpha\theta, 0\} \tag{6.12}$$

giving

$$\{q_0\} = (E\alpha\theta t/2(1 - v^2))\{-y_{32}, x_{32}, -y_{13}, x_{13}, -y_{21}, x_{21}\} \tag{6.13}$$

and $\{\varepsilon_0\}$ for plane strain is given in Section 7.3.

6.2 The bilinear rectangle

The appropriate Cartesian displacement functions for this element, shown in Fig. 6.1(b) and sometimes referred to as the Turner rectangle,[1] are

$$u = c_1 + c_2 x + c_3 xy + c_4 y$$
$$v = c_5 + c_6 x + c_7 xy + c_8 y. \tag{6.14}$$

As with the triangular element, the matrix C is made up of identical parts for u and v, giving

$$
\begin{bmatrix} c_1 \\ c_2 \\ c_3 \\ c_4 \\ c_5 \\ c_6 \\ c_7 \\ c_8 \end{bmatrix} = \frac{1}{L_a L_b} \begin{bmatrix} ab & 0 & ab & 0 & ab & 0 & ab & 0 \\ -b & 0 & b & 0 & b & 0 & -b & 0 \\ 1 & 0 & -1 & 0 & 1 & 0 & -1 & 0 \\ -a & 0 & -a & 0 & a & 0 & a & 0 \\ 0 & ab & 0 & ab & 0 & ab & 0 & ab \\ 0 & -b & 0 & b & 0 & b & 0 & -b \\ 0 & 1 & 0 & -1 & 0 & 1 & 0 & -1 \\ 0 & -a & 0 & -a & 0 & a & 0 & a \end{bmatrix} \begin{bmatrix} u_1 \\ v_1 \\ u_2 \\ v_2 \\ u_3 \\ v_3 \\ u_4 \\ v_4 \end{bmatrix}
$$

$$(6.15)$$

For plane stress the matrix product DS is obtained as

$$
DS = \frac{E}{1-v^2} \begin{bmatrix} 0 & 1 & y & 0 & 0 & 0 & vx & v \\ 0 & v & vy & 0 & 0 & 0 & x & 1 \\ 0 & 0 & v'x & v' & 0 & v' & v'y & 0 \end{bmatrix} \tag{6.16}
$$

and stresses may be calculated at any point in the element using $\{\sigma\} = N\{d\}$ where $N = DSC$. The element stiffness matrix is given by

$$
k = C^t k_m C = C^t \left(\iint S^t (tD) S \, dx \, dy \right) C \tag{6.17}
$$

and using the integration formula

$$
\int_{-a}^{a} \int_{-b}^{b} x^n y^m \, dx \, dy = \frac{\{1 - (-1)^{m+n+1}\} L_a^{n+1} L_b^{m+1}}{2^{n+m+1}(n+1)(m+1)} \tag{6.18}
$$

the kernel stiffness matrix is obtained as

$$
k_m = \frac{E L_a L_b t}{1-v^2} \begin{bmatrix} 0 & 0 & 0 & 0 & 0 & 0 & 0 & 0 \\ & 1 & 0 & 0 & 0 & 0 & 0 & v \\ & & k_{33} & 0 & 0 & 0 & 0 & 0 \\ & & & v' & 0 & v' & 0 & 0 \\ & & \text{symmetric} & & 0 & 0 & 0 & 0 \\ & & & & & v' & 0 & 0 \\ & & & & & & k_{77} & 0 \\ & & & & & & & 1 \end{bmatrix} \tag{6.19}
$$

where $k_{33} = (L_b^2 + v' L_a^2)/12$, $k_{77} = (L_a^2 + v' L_b^2)/12$.

Rectangular elements have rather limited application but such elements can be generalized using dimensionless coordinates and isoparametric mapping to allow a quadrilateral shape, and this approach is described in Section 7.2.

Owing to its explicit formulation, however, the present element provides a useful opportunity to examine the effects of *reduced integration* (in this case *selective reduced integration* of the shearing energy) and this question is taken up in Section 6.6, having first seen in Section 6.4 that such reduced integration significantly accelerates the covergence of the element.

6.3 The linear strain triangle

The twelve-freedom linear strain triangle (LST), shown in Fig. 6.2, was first introduced by Argyris[2] and Fraeijs de Veubeke[3] and, because it naturally yields a complete polynomial interpolation, is a particularly useful element. The stiffness matrix can be derived explicitly with the use of the following integration formulae

$$\iint x^2 \, dx \, dy = \frac{A\Sigma(x_i - \Sigma x_i/3)^2}{12} \tag{6.20}$$

$$\iint y^2 \, dx \, dy = \frac{A\Sigma(y_i - \Sigma y_i/3)^2}{12} \tag{6.21}$$

$$\iint xy \, dx \, dy = \frac{A\Sigma(x_i - \Sigma x_i/3)(y_i - \Sigma y_i/3)}{12} \tag{6.22}$$

or it can be derived from that for the constant strain triangle.[4] Numerical integration, however, provides a more versatile solution because variations in element thickness, for example, are more easily included. Using area co-ordinates the interpolation functions are eqns (5.60), that is,

$$f_1 = L_1(2L_1 - 1), \quad f_2 = L_2(2L_2 - 1), \quad f_3 = L_3(2L_3 - 1)$$
$$f_4 = 4L_1L_2, \quad f_5 = 4L_2L_3, \quad f_6 = 4L_3L_1. \tag{6.23}$$

Using the chain rule of differentiation the strain ε_x, for example, can be calculated as

$$\varepsilon_x = \frac{\partial u}{\partial x} = \frac{\partial u}{\partial L_1}\frac{\partial L_1}{\partial x} + \frac{\partial u}{\partial L_2}\frac{\partial L_2}{\partial x} \tag{6.24}$$

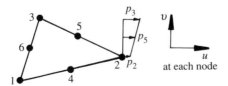

Fig. 6.2 Linear strain triangle

and substituting the interpolation $u = \{f\}^t\{u\}$ we obtain

$$\varepsilon_x = \left[\left\{ \frac{\partial f}{\partial L_1} \right\}^t \frac{\partial L_1}{\partial x} + \left\{ \frac{\partial f}{\partial L_2} \right\}^t \frac{\partial L_2}{\partial x} \right] \{u\} \tag{6.25}$$

where $\{\partial f/\partial L_1\}$ denotes $\{\partial f_1/\partial L_1, \partial f_2/\partial L_1, \ldots, \partial f_6/\partial L_1\}$.

The required local derivatives of the interpolation functions are easily obtained from eqns (6.23) as a matrix S, replacing L_3 by $1 - L_1 - L_2$ throughout so that two independent coordinates only are used, as in the Cartesian system.[†]

Then the interpolation matrix, S, for the local derivatives with respect to L_1 and L_2 is obtained as

$$S = \left[\begin{array}{c} \{\partial f/\partial L_1\}^t \\ \{\partial f/\partial L_2\}^t \end{array} \right]$$

$$= \left[\begin{array}{cccccc} 4L_1 - 1 & 0 & 4L_1 + 4L_2 - 3 & 4L_2 & -4L_2 & 4 - 8L_1 - 4L_2 \\ 0 & 4L_2 - 1 & 4L_1 + 4L_2 - 3 & 4L_1 & 4 - 4L_1 - 8L_2 & -4L_1 \end{array} \right]$$
$$\tag{6.26}$$

Writing the necessary transformation to Cartesian derivatives with respect to both x and y, again using the chain rule, in matrix form:

$$\left[\begin{array}{c} \{\partial f/\partial x\}^t \\ \{\partial f/\partial y\}^t \end{array} \right] = \left[\begin{array}{cc} \partial L_1/\partial x & \partial L_2/\partial x \\ \partial L_1/\partial y & \partial L_2/\partial y \end{array} \right] \left[\begin{array}{c} \{\partial f/\partial L_1\}^t \\ \{\partial f/\partial L_2\}^t \end{array} \right] = J^{-1}S. \tag{6.27}$$

From the area coordinate definitions of eqns (5.52) the matrix J^{-1} (the inverse Jacobian matrix) in eqn (6.27) is given as

$$J^{-1} = \frac{1}{2\Delta} \left[\begin{array}{cc} -y_{32} & -y_{13} \\ x_{32} & x_{13} \end{array} \right]. \tag{6.28}$$

Hence we obtain an interpolation matrix for the Cartesian derivatives as

$$G = J^{-1}S = \left[\begin{array}{cccccc} g_{11} & g_{12} & g_{13} & g_{14} & g_{15} & g_{16} \\ g_{21} & g_{22} & g_{23} & g_{24} & g_{25} & g_{26} \end{array} \right] \tag{6.29}$$

which can be evaluated numerically at any point (L_1, L_2) in the element. Redeploying this matrix, the strain displacement matrix is given as

$$\left\{ \begin{array}{c} \partial u/\partial x \\ \partial v/\partial y \\ \partial u/\partial y + \partial v/\partial x \end{array} \right\}$$

$$= \left[\begin{array}{cccccc} g_{11} \, 0 & g_{12} \, 0 & g_{13} \, 0 & g_{14} \, 0 & g_{15} \, 0 & g_{16} \, 0 \\ 0 \, g_{21} & 0 \, g_{22} & 0 \, g_{23} & 0 \, g_{24} & 0 \, g_{25} & 0 \, g_{26} \\ g_{21} \, g_{11} & g_{22} \, g_{12} & g_{23} \, g_{13} & g_{24} \, g_{14} & g_{25} \, g_{15} & g_{26} \, g_{16} \end{array} \right] \{d\}$$

[†] It is possible to add the term $(\partial u/\partial L_3)(\partial L_3/\partial x)$ to eqn (6.24) and correspondingly a third row $\{\partial f/\partial L_3\}^t$ is added to matrix S and this approach is used in the program of Section 20.7.

or

$$\{\varepsilon\} = B\{u_1, v_1, u_2, v_2, u_3, v_3, u_4, v_4, u_5, v_5, u_6, v_6\} = B\{d\}. \quad (6.30)$$

As the stiffness matrix calculation involves terms of the form L_1^2 and $L_1 L_2$ three-point integration (at the midside nodes) exactly integrates the element stiffness matrix,

$$k = \frac{\Sigma B^t D B(\Delta t)}{3}. \quad (6.31)$$

Numerical integration is recommended here as the algebraic manipulations required for an explicit solution are tedious. Further, the procedure described above can be extended to deal with elements having curved sides and this exercise is undertaken in Section 7.3.

The consistent loads for a constant body force ρ in a given direction are given by $\rho \int \{f\} \, dV$. Integrating the interpolation functions using eqn (5.130):

$$\iint L_1^a L_2^b L_3^c \, dx \, dy = \frac{a! \, b! \, c!}{(a + b + c + 2)!} \, 2\Delta \quad (6.32)$$

gives

$$\{q_b\} = \frac{\rho t \Delta}{3} \{0, 0, 0, 1, 1, 1\} \quad (6.33)$$

if the element has constant thickness t. Thus one third of the total load applied in a given direction is assigned to each of the midside nodes in that direction. If ρ or t varies, however, numerical integration is needed to calculate the consistent loads.

Considering also the case of a varying surface traction applied to side 23 of the element, as shown in Fig. 6.2, the consistent nodal loads are given as[5]

$$\{q_t\} = \int \{f_s\} \{f_s\}^t \{p_2, p_5, p_3\} \, ds \quad (6.34)$$

where $\{f_s\}$ contains the interpolation functions for nodes on side 23:

$$\{f_s\} = \{L_2(2L_2 - 1), 4L_2 L_3, L_3(2L_3 - 1)\}$$
$$= \{L_2(2L_2 - 1), 4L_2(1 - L_2), (1 - L_2)(1 - 2L_2)\}. \quad (6.35)$$

The second appearance of $\{f_s\}$ in eqn (6.34) is to interpolate the nodal pressure intensities quadratically. Finally, replacing $\int ds$ by $L_{23} \int_0^1 dL_2$, where L_{23} is the length of the side joining nodes 2 and 3,

$$\begin{Bmatrix} q_{u2} \\ q_{u5} \\ q_{u3} \end{Bmatrix} = \frac{t L_{23}}{30} \begin{bmatrix} 4 & 2 & -1 \\ 2 & 16 & 2 \\ -1 & 2 & 4 \end{bmatrix} \begin{Bmatrix} p_2 \\ p_5 \\ p_3 \end{Bmatrix}. \quad (6.36)$$

For the case of a linear pressure distribution, substituting $p_2 = p_{\max}$ $p_3 = p_{\min}$, and $p_5 = \bar{p} = (p_{\min} + p_{\max})/2$, we obtain

$$\{q_{u2}, q_{u5}, q_{u3}\} = \left\{ \frac{p_{\max} t L_{23}}{6}, \frac{2\bar{p} t L_{23}}{3}, \frac{p_{\min} t L_{23}}{6} \right\} \tag{6.37}$$

and if the pressure is constant, $\bar{p} = p_{\max} = p_{\min}$ so that the middle node in this case receives four times the loading of the two vertex nodes.

6.4 Accuracy of some C_0 plane stress elements

Figure 6.3 shows one half of a simply supported beam discretized into rectangular and triangular elements. The triangular elements are either the constant strain triangle or the linear strain triangle and the rectangular elements are the eight-freedom rectangle and a twelve-freedom element in which a rotational freedom is included at each corner.[6]

The problem is, of course, a standard test problem except that the load at the end is not parabolically distributed as classical theory requires.[7] Stress results are taken at points A and B, as far away as possible from the ends where deviations from simple beam behaviour occur. Using increasingly refined meshes the displacement and stress solutions (apart from two exceptions the stresses are average nodal values) are given in Table 6.1. Table 6.1 includes results obtained by extrapolating the last pair of solutions for each element using eqn (3.60) with $N = 2$ (except for the 12 df triangle for which $N = 4$ is used for the displacements whilst the first and third shear stress solutions are extrapolated, the reason for this being discussed below).

With the 6 df triangle and the centroidal shear terms in the 12 df rectangle, however, a Southwell plot is used.[6] This is done by replacing h_i by d_i/n_i in eqn

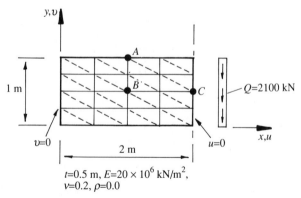

$t=0.5$ m, $E=20 \times 10^6$ kN/m^2, $v=0.2$, $\rho=0.0$

Fig. 6.3 Simply supported beam represented by triangular and rectangular elements

Table 6.1 Numerical results with several membrane elements; centroidal values are shown in parentheses

Element	Freedoms n_i	$- v_C$ mm	$(\sigma_x)_A$ MPa	$(\tau_{xy})_B$ MPa
6 df triangle	12	3.99	7.5	3.70
	24	4.77	12.7	4.06
	40	5.70	16.4	5.00
	60	6.29	18.7	5.16
	84	6.69	20.2	5.58
8 df rectangle	12	5.43	17.9	0.67
	24	6.55	21.5	2.70
	40	7.08	23.0	3.40
	60	7.37	23.8	4.89
	84	7.54	24.4	5.48
12 df rectangle	18	5.71	18.1	0.85
	32	6.72	21.6	(5.48) 2.97
	60	7.18	23.1	4.86
	90	7.42	23.8	(5.94) 5.11
	126	7.56	24.2	5.64
12 df triangle	12	6.78	25.2	3.08
	40	7.77	25.2	7.33
	84	7.86	25.2	5.75
12 df rectangle,	18	7.31	25.2	0.70
constant shear.	32	7.60	25.5	2.83
	60	7.73	25.4	5.13
	90	7.79	25.4	5.28
	126	7.82	25.4	5.79
Analytical solution		7.93	25.2	6.30
Extrapolated solutions:				
6 df triangle		8.06	25.4	6.25
8 df rectangle		7.93	25.6	6.82
12 df rectangle		7.88	25.1	(6.29) 6.84
12 df triangle		7.88	25.2	6.10
12 df rectangle (constant τ)		7.89	25.4	6.95

(3.60), and doing likewise for h_j, using $N = 1$. This form of the Southwell plot is derived in Section 13.10. It gives better results for constant strain elements (but not for higher-order elements) because it more clearly reflects the series nature of the truncation error. This is because use of the Southwell plot is equivalent to hyperbolic rather than h^2 extrapolation (and the corresponding series follows the binomial expansion).

The displacement solutions for the twelve-freedom rectangle, as it models the rigid body behaviour more accurately, show a slight improvement over those for the eight-freedom element in coarse meshes but the stress results are much the same.

The linear strain triangle shows the best overall performance but note that here the first and third shear stress results occur at midside nodes whereas the second is at a vertex node. This results in non-monotonic convergence so that the first and third values are used for this extrapolation.[†]

Both the rectangular elements give poor extrapolated shear stresses but Table 6.1 gives the centroidal shear stresses obtained in the 12 df element (for two meshes in which an element centroid concides with point B in Fig. 6.3) and these converge satisfactorily. It is for this reason that Gauss point stress sampling is frequently advocated with such elements and, as they are closer to the centroid, such results are generally satisfactory. Both the rectangular elements unduly weight the shearing stiffness (a phenomenon called *parasitic shear*) and converge more rapidly when limited to constant shear.[8] In the 8 df element this involves deletion of the diagonal entries k_{33} and k_{77} in eqn (6.19) and in the 12 df element selective reduced integration (one point, the centroid) is used to compute the shear strain contribution to the stiffness matrix. The last series of results in Table 6.1 are for the 12 df element when thus constrained. The displacement and direct stress solutions are much improved but the nodal shears (still calculated with the linear shear terms) are still poor, whereas the centroidal shears are much the same as those without reduced integration. The same behaviour is observed when the 8 df element is restricted to constant shear but no improvement occurs in the 12 df triangle as it has a complete quadratic interpolation.

Finally it is worth noting that a 2×1 nodal layout was used in Fig. 6.3 to match the dimensions of the rectangular domain. In square domains, as might be expected, a 1×1 layout will yield the best rate of convergence (S. Valliappan, private communication, 1978).

6.5 Other elements for plane stress and strain

The twenty-freedom cubic plane triangle may take either of the forms shown in Fig. 6.4 for which the interpolation functions for u and v are eqns (5.63) and (5.73) respectively. In either case it is convenient to remove the inconvenient central node by *static condensation*.

To undertake static condensation we partition the element stiffness equations into the form

$$\left\{ \begin{array}{c} q \\ q_c \end{array} \right\} = \left[\begin{array}{cc} k & k_c \\ k_c^t & k_{cc} \end{array} \right] \left\{ \begin{array}{c} d \\ d_c \end{array} \right\} \tag{6.38}$$

[†] This phenomenon is related to the 'chequerboard syndrome' encountered in Section 19.5.

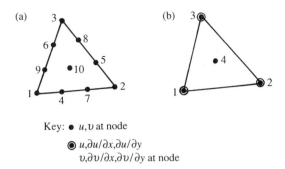

Key: • u,v at node

⦿ $u, \partial u/\partial x, \partial u/\partial y$

$v, \partial v/\partial x, \partial v/\partial y$ at node

Fig. 6.4 Cubic plane elements

where subscript c refers to the freedoms u and v at the central node. Expanding these equations one obtains

$$\{q\} = k\{d\} + k_c\{d_c\}, \quad \{q_c\} = k_c^t\{d\} + k_{cc}\{d_c\} \tag{6.39}$$

and rearranging the second of these

$$\{d_c\} = k_{cc}^{-1}\{\{q_c\} - k_c^t\{d\}\}. \tag{6.40}$$

Substituting this result into eqns (6.39) gives

$$\{q\} = k\{d\} + k_c k_{cc}^{-1}\{\{q_c\} - k_c^t\{d\}\}$$

or

$$\{q_s\} = k_s\{d\}$$

where

$$\{q_s\} = \{q\} - k_c k_{cc}^{-1}\{q_c\} \tag{6.41}$$

$$k_s = k - k_c k_{cc}^{-1} k_c^t \tag{6.42}$$

are the condensed element load and stiffness matrices. As we have seen, this result is an exercise in matrix partitioning and a similar process is used in the *substructure* method of analysis.[9]

For the Lagrangian case (Fig. 6.4(a)) an alternative and more accurate condensation is achieved by suppressing the centroidal freedom using

$$u_{10} = \tfrac{1}{4}(u_4 + u_5 + u_6 + u_7 + u_8 + u_9) - \tfrac{1}{6}(u_1 + u_2 + u_3) \tag{6.43}$$

and the derivation of this is given in Mohr and Mohr.[10]

Another refined plane element can be obtained by combining four LST elements and condensing out the internal freedoms,[11] giving a quadrilateral with sixteen freedoms but this element is derived more directly by biquadratic interpolation (see Section 7.2). The sixteen-freedom quadrilateral, however, is only marginally more accurate than the LST element.[8] To obtain greater accuracy a C_1 quintic formulation (Section 5.10) can be used but this requires at least thirty-six freedoms and more economical results are obtained by improving existing elements using reduced integration (see Section 6.6) or basis transformation (see Section 10.6).

The excessive shear stiffness in bilinear plane stress elements was seen in Section 6.4 (the 12 df rectangular element using bilinear interpolation).[6] While restriction to constant shear improved the displacements and direct stresses it still left the problem of poor shear stress modelling.

A comparable improvement in convergence is obtained by Wilson[12, 13] by adding two 'nodeless' quadratic interpolation functions to the original four. These correspond to fictitious freedoms and are

$$f_9 = (1 - a^2), \qquad f_{10} = (1 - b^2) \tag{6.44}$$

where a, b are defined in Fig. 5.6 and f_9, f_{10} are from eqns (5.10). The two freedoms corresponding to these functions are suppressed by static condensation but the resulting element more equitably weights the direct and shear stiffnesses.

Wilson's element does not even possess C_0 continuity but it passes the patch test[14] and Lesaint[15] shows that it gives h^2 and h^1 convergence respectively for displacements and stresses (the h^1 result does not take into account the truncation error of the interpolation discussed in Section 3.9, that is, the second term in eqn (3.68).

Wilson's approach has also been used in trivariate brick elements which also suffer parasitic shear[16] but a simpler remedy is had by reduced integration,[13] a worked example of which is given in the following section. Many other examples of reduced integration occur in later chapters and these are summarized in Section 15.11.

6.6 An example of reduced integration

As an example of reduced integration we explicitly evaluate the plane stress stiffness matrix for the bilinear rectangle (using eqns (6.15), (6.17), and (6.19), giving

$$k = \frac{Et}{12(1 - v^2)} \begin{bmatrix} k_{\mathrm{I}} & k_{\mathrm{II}} \\ k_{\mathrm{II}}^{\mathrm{t}} & k_{\mathrm{I}} \end{bmatrix} \tag{6.45}$$

where

$$k_{\mathrm{I}} = \begin{bmatrix} k_1 & k_3 & k_5 & -k_4 \\ & k_2 & k_4 & -k_{10} \\ \text{symmetric} & k_1 & -k_3 \\ & & & k_2 \end{bmatrix} \quad k_{\mathrm{II}} = \begin{bmatrix} -k_7 & -k_3 & -k_9 & k_4 \\ & -k_8 & -k_4 & k_6 \\ \text{symmetric} & -k_7 & k_3 \\ & & & -k_8 \end{bmatrix}$$

$$\tag{6.46}$$

and

$$k_1 = 4\lambda + 2(1 - v)/\lambda, \qquad k_2 = 4/\lambda + 2(1 - v)\lambda$$

$$k_3 = 3v + 3(1 - v)/2, \qquad k_4 = -3v + 3(1 - v)/2$$

$$k_5 = -4\lambda + (1 - v)/\lambda, \qquad k_6 = -4/\lambda + (1 - v)\lambda \qquad (6.47)$$

$$k_7 = 2\lambda + (1 - v)/\lambda, \qquad k_8 = 2/\lambda + (1 - v)\lambda$$

$$k_9 = -2\lambda + 2(1 - v)/\lambda, \qquad k_{10} = -2/\lambda + 2(1 - v)\lambda$$

where $\lambda = L_b/L_a$ is the aspect ratio. For $k_1, \ldots, k_{10}$ the second terms are those contributed by shearing and if this is restricted to constancy by retaining only the constant terms in the third row of B, one obtains[17]

$$k_1 = 4\lambda + 3(1 - v)/2\lambda, \qquad k_2 = 4/\lambda + 3(1 - v)\lambda/2$$

$$k_3 = 3v + 3(1 - v)/2, \qquad k_4 = -3v + 3(1 - v)/2$$

$$k_5 = -4\lambda + 3(1 - v)/2\lambda, \quad k_6 = -4/\lambda + 3(1 - v)\lambda/2 \qquad (6.48)$$

$$k_7 = 2\lambda + 3(1 - v)/2\lambda, \qquad k_8 = 2/\lambda + 3(1 - v)\lambda/2$$

$$k_9 = -2\lambda + 3(1 - v)/2\lambda, \, k_{10} = -2/\lambda + 3(1 - v)\lambda/2.$$

Equations (6.48) correspond to the use of selective reduced integration of the shearing energy, that is, one uses only one-point integration for this. As an example, analysing the problem of Fig. 6.3 with a single element for the left half of the beam, one obtains after imposing v_1, u_2, u_3, and $v_4 = 0$ the reduced problem as

$$\frac{Et}{12(1 - v^2)} \begin{bmatrix} k_1 & -k_4 & -k_3 & -k_9 \\ -k_4 & k_2 & k_6 & k_3 \\ -k_3 & k_6 & k_2 & k_4 \\ k_9 & k_3 & k_4 & k_1 \end{bmatrix} \begin{bmatrix} u_1 \\ v_2 \\ v_3 \\ u_4 \end{bmatrix} = \begin{bmatrix} 0 \\ -1050 \\ -1050 \\ 0 \end{bmatrix} \qquad (6.49)$$

which, using eqns (6.47) with $\lambda = 1/2$ and putting $u_1 = -u_4$, gives $v_3 = -2.87$ mm, whereas eqns (6.48) with constant shear strain) give $v_3 = -5.85$ mm, a much improved result as expected with this element. The improved result with reduced integration occurs because k_1 and k_2, and thus trace (k), is significantly reduced, making the element more flexible. Such phenomena are summarized in Section 15.11.

In passing it is also worth noting that the bilinear element can also be formulated by assuming the stress distributions

$$\sigma_x = a_1 + a_2 y, \qquad \sigma_y = a_3 + a_4 x, \qquad \tau_{xy} = a_5 \qquad (6.50)$$

the three constants a_1, a_3, and a_5 being the constant strain modes. Integrating eqns (6.50) according to the strain definitions and constitutive relationships for plane stress, Przemieniecki[9] obtains the interpolations

$$u = c_1 x + c_2 y - c_3(vx^2 + y^2) + 2c_4 xy + c_5$$
$$v = c_6 x + c_7 y - c_4(x^2 + vy^2) + 2c_3 xy + c_8. \tag{6.51}$$

These lead to the same results as eqns (6.48) except that one substitutes $(4 - v^2)$ for 4 and $(2 + v^2)$ for 2 in the first terms of $k_1, \ldots, k_{10}$. In the preceding example where $v = 0.2$ almost identical results will be obtained but as $v \to 0.5$ the difference will become significant. However, when v^2 is relatively small one finds here an interesting example of the comparatively similar behaviour of stress-based and reduced integration displacement elements cited by numerous authors.[17-19]

The three additional constants in eqns (6.51) are introduced as integration constants and correspond to the rigid body modes. Evidently an examination of stress interpolations of the form of eqns (6.50), employing in these the total number of freedoms less the number of rigid body modes, may indicate the degree to which integration can be reduced. Moreover the reader can verify that eqns (6.51) give $\varepsilon_{xy} = c_2 + c_6$ and that these coefficients do not appear in the direct strains. Hence the coupling effect of the shear strain $\varepsilon_{xy} = \partial u/\partial y + \partial v/\partial x$ on the freedoms u and v is reduced. Likewise reduction to constant shear in the usual bilinear element (based on eqns (6.14)) reduces this shear coupling. As we shall see in Section 15.11 this is mathematically equivalent to applying a constraint to the element. In later chapters we shall see that such constraints are frequently needed to obtain satisfactory finite elements.

6.7 Concluding remarks

The first continuum finite elements were the constant strain triangle and the bilinear rectangle for plane stress and it is perhaps appropriate to have commenced our discussion of the applications to the mechanics of solids with these. At first sight the finite element method appears straightforward. We assume polynomial interpolations, differentiate these to form a strain interpolation matrix, B, and calculate the stiffness matrix $k = \int B^t DB \, dV$. In Table 6.1, however, we see that complications can occur with the assumption of simple independent displacement fields. Here the shear stresses converge slowly and, generally, incorrectly. This is because of the coupling effect of the shear strains upon the displacement fields for u and v. Mathematically this coupling is equivalent to the *continuity constraint* $\partial u/\partial x + \partial v/\partial y = 0$ in incompressible fluids and in plane stress the shear strains cause difficulty by *overconstraining* each element in a manner inconsistent with adjoining elements. In the case of plane stress, results are improved by selective reduced integration of the shearing energy, equivalent to applying a constraint to remove inconsistencies between elements. Then, as Table 6.1 shows, the shear stress modelling will be no worse but the general rate of convergence is much improved.

Similar difficulties occur in the finite difference method, for example, and in applications of finite elements to multivariate problems in the mechanics of fluids. Resolution of such difficulties requires the introduction of constraints and with the introduction of these the finite element method is brought to its fullest potential.

References

1. Clough, R. W., Martin, H. C., Topp, L. J., and Turner, M. J. (1956). Stiffness and deflection analysis of complex structures. *J. Aero. Sci.*, **23**, 803.
2. Argyris, J. H. (1965). Continua and discontinua. Proceedings of the First Conference on Matrix Methods in Structural Mechanics, Wright-Patterson AFB, Ohio.
3. Fraeijs de Veubeke, B. (1965). Displacement and equilibrium models in the finite element method. In *Stress Analysis*, (ed. G. S. Holister and O. C. Zienkiewicz). Wiley, London.
4. Pedersen, P. (1973). Some properties of linear strain triangles and optimum finite element models. *Int. J. Num. Meth. Engng*, **7**, 415.
5. Meek, J. L. (1971) *Matrix Structural Analysis*. McGraw-Hill, New York.
6. Mohr, G. A. (1981). A simple rectangular membrane element including the drilling freedom. *Comput. Struct.*, **13**, 483.
7. Timoshenko, S. P. and Goodier, J. N. (1951). *Theory of Elasticity*, (3rd edn). McGraw-Hill, New York.
8. Brebbia, C. A. and Connor, J. J. (1973). *Fundamentals of Finite Element Techniques for Structural Engineers*. Butterworths, London.
9. Przemieniecki, J. S. (1968). *Theory of Matrix Structural Analysis*. McGraw-Hill, New York.
10. Mohr, G. A. and Mohr, R. S. (1986). A new thin plate element by basis transformation. *Comput. Struct.*, **22**, 239.
11. Pulmano, V. A. (1973). Conforming quadrilateral plane stress element of varying thickness. In *Finite Element Techniques*, (ed. A. P. Kabaila and A. S. Hall). Clarendon Press, Sydney.
12. Wilson, E. L., Taylor, R. L., Doherty, W. P., and Ghaboussi, J. (1971). Incompatible displacement models. Symposium on Numerical Methods, University of Illinois.
13. Gallagher, R. H. (1975). *Finite Element Analysis Fundamentals*. Prentice-Hall, Englewood Cliffs, NJ.
14. Strang, G. and Fix, G. J. (1973). *An Analysis of the Finite Element Method*. Prentice-Hall, Englewood Cliffs, NJ.
15. Lesaint, P. (1976). On the convergence of Wilson's nonconforming element for solving the elastic problems. *Comp. Methd. Appl. Mech. Engng.*, **7**, 1.

16. Bretland, J. L. and Cook, R. D. (1979). A new eight node solid element. *Int. J. Num. Meth. Engng*, **14**, 593.
17. Mohr, G. A. and Cook, P. L. (1985). On near equivalence of assumed stress and reduced integration formulations of the bilinear plane stress finite element *Comput. Struct.* **21**, 475.
18. Zienkiewicz, O. C. (1977). *The Finite Element Method*, (3rd edn). McGraw-Hill, London.
19. Kelly, D. W. (1979). Reduced integration to give equilibrium models for assessing the accuracy of finite element analysis. Proceedings of the Third Australian International Conference on Finite Element Methods, University of NSW, Sydney.

7
Isoparametric mapping
and its applications

Isoparametric elements are defined as elements for which the equations required to define the element shape are identical to those used to define the displacement fields within the element. Comparing eqns (6.1) and (5.54) the constant strain triangle is thus an isoparametric element as the number of parameters needed to define the element shape (six: $x_1, y_1, x_2, y_2, x_3, y_3$) equals the number of parameters used to define the displacement field within the element (again, six: $u_1, v_1, u_2, v_2, u_3, v_3$). Further, the parameters correspond in type so that, for example, the six-freedom constant moment element described in Section 8.7 is not isoparametric.

The advantage of isoparametric elements is that the displacement interpolations can be used directly to provide a mapping of elements of simple basic shape (such as rectangles and triangles) into more general shapes (such as quadrilaterals and triangles with curved sides) and thus a whole new series of versatile elements is produced.

7.1 Isoparametric mapping

The term *isoparametric mapping* is now much used in the finite element method but originally relates to the classical calculus of surface integrals. The reader will be familiar with the parametric equations $x = a \cos u$ and $y = b \sin u$ for an ellipse. These map the line $u = 0 \rightarrow 2\pi$ into an ellipse. For a sphere, on the other hand, the parametric equations are[1]

$$x = a \cos u \cos v, \quad y = a \sin u \cos v, \quad z = a \sin v$$

and these map points in the rectangle $u = 0 \rightarrow 2\pi$, $v = -\pi/2 \rightarrow \pi/2$ onto the surface of the sphere.

The position vector $\mathbf{r}(u, v)$ of a point on a sphere is given by

$$\mathbf{r}(u, v) = X(u, v)\hat{\mathbf{i}} + Y(u, v)\hat{\mathbf{j}} + Z(u, v)\hat{\mathbf{k}}$$

where $X(u, v) = a \cos u \cos v$, and so on, and this is called the *vector equation* for the surface. Then if X, Y, and Z are differentiable we define the *fundamental vector product* of the surface as

$$\frac{\partial \mathbf{r}}{\partial u} \times \frac{\partial \mathbf{r}}{\partial v} = \begin{vmatrix} \hat{\mathbf{i}} & \hat{\mathbf{j}} & \hat{\mathbf{k}} \\ \partial X/\partial u & \partial Y/\partial u & \partial Z/\partial u \\ \partial X/\partial v & \partial Y/\partial v & \partial Z/\partial v \end{vmatrix}$$

that is, the magnitudes of the components are expressed as *Jacobian determinants* the first of which, for example, may be written in the abbreviated form $\partial(Y, Z)/\partial(u, v)$.

In fact it is easy to show that $\partial \mathbf{r}/\partial u$ and $\partial \mathbf{r}/\partial v$ define the *tangent plane* to the surface (that is, their directions define the *curvilinear coordinates* on the surface) and their cross product thus defines the normal to the surface, a result which will prove useful in Section 11.8. Then the parametric equations for a surface S map a rectangle of area $\delta u \delta v$ into a *curvilinear parallelogram* on the surface whose area is given by $|\partial \mathbf{r}/\partial u \times \partial \mathbf{r}/\partial v| \delta u \delta v$ and the area of the surface S can be found by integrating this result.

Similar mapping processes allow us to map finite elements of general shape into simple reference shapes in terms of which calculations are much simplified. Indeed, it is possible to use the interpolation functions derived for the simple 'base shape' to carry out such a mapping, that is, we write the parametric equations for a two-dimensional element as

$$x = \Sigma f_i x_i, \quad y = \Sigma f_i y_i \tag{7.1}$$

where x_i, y_i are the nodal coordinates of the original element and x, y are the coordinates of a point in the simple 'base element.'

In triangular elements in which area coordinates are used, for example, we have $f_i = f_i(L_1, L_2, L_3)$ and eqns (7.1) map from a tetrahedron $L_1 = 0 \to 1$, $L_2 = 0 \to 1$, $L_3 = 0 \to 1$ to a triangle in x, y.

When the interpolations used for the geometric mapping are the same as those used to form the element matrices we term such elements *isoparametric* and numerous examples of such elements are given in following sections.

When the element shape is defined by lower-order interpolations than those used to calculate its stiffness matrix, however, the element is termed *subparametric*. When higher-order interpolations are used to define the shape, on the other hand, the element is termed *superparametric*. In some shell elements, for example, coordinates must be specified at the nodes and at additional points in order to calculate the curvatures of the shell. This question is discussed in Section 11.8.

7.2 Bilinear and biquadratic quadrilaterals

The first isoparametric finite element developed was the eight-freedom quadrilateral element for plane stress.[2] As shown in Fig. 7.1 the element is mapped from a square in the local coordinates a, b, for which simple interpolations exist, into a quadrilateral in the Cartesian coordinates. The orientation of the local axes a, b relative to the Cartesian axes will vary throughout the element. Near node 1, for example, they are nearly parallel to the sides meeting at that node, as shown in Fig. 7.1. The orientation of these axes can be calculated by differentiating the·interpolation functions for the square element.

Applying the bilinear interpolation of eqn (5.37) to both the coordinates and the displacements we write

$$x = \Sigma f_i x_i, \quad y = \Sigma f_i y_i \quad i = 1, 2, 3, 4 \tag{7.2a}$$

$$u = \Sigma f_i u_i, \quad v = \Sigma f_i v_i \quad i = 1, 2, 3, 4 \tag{7.2b}$$

where

$$f_i = \tfrac{1}{4}(1 + a_i a)(1 + b_i b) \tag{7.2c}$$

are the interpolation functions, and a_i, b_i are the local nodal coordinates shown in Fig. 7.1. Applied to an elevation z these functions describe the shape of a hyperbolic paraboloid, the generators for which are perpendicular straight lines of varying slope. Thus when applied to x and y they produce mapping into a quadrilateral.

Interpolations of the form of eqn (7.2c) that is, expressed as a product of one-dimensional interpolations, allow very simple and compact coding of the element routine.

Using the chain rule the two direct strains can be expressed as

$$\frac{\partial u}{\partial x} = \left(\left\{ \frac{\partial f}{\partial a} \right\}^t \frac{\partial a}{\partial x} + \left\{ \frac{\partial f}{\partial b} \right\}^t \frac{\partial b}{\partial x} \right) \{u\} \tag{7.3}$$

$$\frac{\partial v}{\partial y} = \left(\left\{ \frac{\partial f}{\partial a} \right\}^t \frac{\partial a}{\partial y} + \left\{ \frac{\partial f}{\partial b} \right\}^t \frac{\partial b}{\partial y} \right) \{v\}. \tag{7.4}$$

The vectors $\{\partial f/\partial a\}$ and $\{\partial f/\partial b\}$ in eqns (7.3) and (7.4) are obtained by differentiating eqns (7.2c), yielding the interpolation matrix

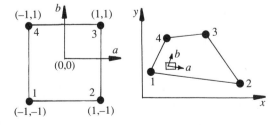

Fig. 7.1 Mapping for a quadrilateral element

$$S = \begin{bmatrix} \{\partial f/\partial a\}^t \\ \{\partial f/\partial b\}^t \end{bmatrix} = \begin{bmatrix} \partial f_1/\partial a & \partial f_2/\partial a & \partial f_3/\partial a & \partial f_4/\partial a \\ \partial f_1/\partial b & \partial f_2/\partial b & \partial f_3/\partial b & \partial f_4/\partial b \end{bmatrix}$$

$$= \tfrac{1}{4} \begin{bmatrix} -(1-b) & (1-b) & (1+b) & -(1+b) \\ -(1-a) & -(1+a) & (1+a) & (1-a) \end{bmatrix}$$

(7.5)

which is easily evaluated at any point in the element. The transformation to Cartesian derivatives is obtained by first writing for the derivative of any variable (taking *u* for example)

$$\begin{Bmatrix} \partial u/\partial a \\ \partial u/\partial b \end{Bmatrix} = \begin{bmatrix} \partial x/\partial a & \partial y/\partial a \\ \partial x/\partial b & \partial y/\partial b \end{bmatrix} \begin{Bmatrix} \partial u/\partial x \\ \partial u/\partial y \end{Bmatrix}.$$

(7.6)

Here the connecting matrix is the 2×2 Jacobian matrix, J, which is evaluated directly from the nodal coordinates by differentiating the Cartesian coordinate interpolations for the element. Thus, differentiating the first row of the following equation with respect to *a* and the second with respect to *b*

$$\begin{bmatrix} x & y \\ x & v \end{bmatrix} = \begin{bmatrix} f_1 & f_2 & f_3 & f_4 \\ f_1 & f_2 & f_3 & f_4 \end{bmatrix} \begin{bmatrix} x_1 & y_1 \\ x_2 & y_2 \\ x_3 & y_3 \\ x_4 & y_4 \end{bmatrix}$$

(7.7)

gives the Jacobian matrix as

$$J = \begin{bmatrix} \partial x/\partial a & \partial y/\partial a \\ \partial x/\partial b & \partial y/\partial b \end{bmatrix} = S[\{x\} \quad \{y\}]$$

(7.8)

where the matrix S is as defined in eqn (7.5). Hence the interpolation for the Cartesian derivatives is given by

$$\begin{Bmatrix} \partial(\ )/\partial x \\ \partial(\ )/\partial y \end{Bmatrix} = J^{-1} \begin{Bmatrix} \partial(\ )/\partial a \\ \partial(\ )/\partial b \end{Bmatrix} = J^{-1}S = G.$$

(7.9)

Then at any point the matrix G can be evaluated numerically to give

$$G = \begin{bmatrix} g_{11} & g_{12} & g_{13} & g_{14} \\ g_{21} & g_{22} & g_{23} & g_{24} \end{bmatrix}$$

(7.10)

so that the strain–displacement matrix B is evaluated at any point as

$$\{\varepsilon\} = \begin{Bmatrix} \partial u/\partial x \\ \partial v/\partial y \\ \partial u/\partial y + \partial v/\partial x \end{Bmatrix} = \begin{bmatrix} g_{11} & 0 & g_{12} & 0 & g_{13} & 0 & g_{14} & 0 \\ 0 & g_{21} & 0 & g_{22} & 0 & g_{23} & 0 & g_{24} \\ g_{21} & g_{11} & g_{22} & g_{12} & g_{23} & g_{13} & g_{24} & g_{14} \end{bmatrix} \{d\}$$

$$= B\{d\}.$$

(7.11)

As the stiffness calculation will involve terms of the form a^2, ab, and b^2 a four-point Gaussian quadrature is required, giving the element stiffness matrix for an element of constant thickness as

$$k = t \iint B^t DB \, dx \, dy = t \iint B^t DB |J|_{abs} \, da \, db$$

$$= \Sigma B^t DB(\omega_i t |J|_{abs}) \tag{7.12}$$

where ω_i are the weighting factors, all equal to 1 as $\iint da \, db = 4$ so that $\Sigma |J|_{abs}$ gives the total area of the element.

Equation (7.12) assumes that $dx \, dy = |J|_{abs} \, da \, db$. To prove this we denote unit vectors parallel to three Cartesian axes as $\hat{\mathbf{i}}, \hat{\mathbf{j}}, \hat{\mathbf{k}}$ and unit vectors parallel to the two non-orthogonal local axes as $\hat{\mathbf{a}}, \hat{\mathbf{b}}$. These unit vectors are related by the equations

$$\hat{\mathbf{a}} = \frac{\partial x}{\partial a}\hat{\mathbf{i}} + \frac{\partial y}{\partial a}\hat{\mathbf{j}}, \quad \hat{\mathbf{b}} = \frac{\partial x}{\partial b}\hat{\mathbf{i}} + \frac{\partial y}{\partial b}\hat{\mathbf{j}}.$$

The differential area $dx \, dy$ can be expressed as the vector cross product

$$\hat{\mathbf{i}} dx \times \hat{\mathbf{j}} dy = \hat{\mathbf{k}} dx dy = \hat{\mathbf{a}} da \times \hat{\mathbf{b}} db = \hat{\mathbf{k}} |J| \, da \, db$$

giving the required result. In practice $|J|_{abs}$ is used because with some nodal orderings a negative result may be obtained.

The element should give the same accuracy as that described in Section 6.2 and, as shown in Section 6.6, converges more rapidly if restricted to constant shear strain by using selective reduced integration.

In summary the following steps are required to develop isoparametric elements.

1. The interpolation functions are differentiated to provide interpolations giving the local derivatives (that is, the derivatives with respect to the local coordinates) at any point in the element.

2. The Jacobian matrix is calculated using the local derivative interpolations and the nodal coordinates and inverted to provide the chain rule transformation required to obtain Cartesian derivative interpolations from the local derivative interpolations.

3. The interpolations for the Cartesian derivatives are then used to evaluate the strain interpolation matrix B numerically at a number of integration points within the element.

4. The element stiffness matrix is accumulated by numerical integration,

$$k = \Sigma B^t DB(\omega_i |J|_{abs} t)$$

where

$$J = \begin{bmatrix} \partial\{f\}^{t}/\partial a \\ \partial\{f\}^{t}/\partial b \end{bmatrix} [\{x\} \quad \{y\}] = S[\{x\} \quad \{y\}]$$

$$B = \begin{bmatrix} \overbrace{\partial\{f\}^{t}/\partial x}^{u \text{ columns}} & \overbrace{\{0\}^{t}}^{v \text{ columns}} \\ \{0\}^{t} & \partial\{f\}^{t}/\partial y \\ \partial\{f\}^{t}/\partial y & \partial\{f\}^{t}/\partial x \end{bmatrix} \quad \text{using} \begin{bmatrix} \partial\{f\}^{t}/\partial x \\ \partial\{f\}^{t}/\partial y \end{bmatrix} = J^{-1}S.$$

A schematic representation is used for B, that is, the entries actually follow the order u_1, v_1, u_2, v_2, and so on.

The sixteen-freedom biquadratic quadrilateral is shown in Fig. 7.2 and, like the linear strain triangle, this element may be allowed to have curved sides. The interpolation functions are eqns (5.41) and (5.42):

$$f_i = \tfrac{1}{4}(1 + a_i a)(1 + b_i b)(a_i a + b_i b - 1), \quad i = 1 \to 4$$

$$f_5 = \tfrac{1}{2}(1 - a^2)(1 - b), \quad f_6 = \tfrac{1}{2}(1 - b^2)(1 + a) \qquad (7.13)$$

$$f_7 = \tfrac{1}{2}(1 - a^2)(1 + b), \quad f_8 = \tfrac{1}{2}(1 - b^2)(1 - a).$$

Realizing that for the corner nodes $a_i^2 = b_i^2 = 1$, the strain interpolation matrix in local curvilinear coordinates is obtained as

$$S^{t} = [\{\partial f/\partial a\}, \quad \{\partial f/\partial b\}] = \begin{bmatrix} \alpha_1 & \beta_1 \\ \alpha_2 & \beta_2 \\ \alpha_3 & \beta_3 \\ \alpha_4 & \beta_4 \\ ab - a & (a^2 - 1)/2 \\ (1 - b^2)/2 & -ab - b \\ -ab - a & (1 - a^2)/2 \\ (b^2 - 1)/2 & ab - b \end{bmatrix}$$

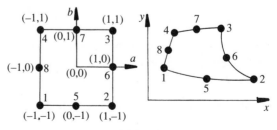

Fig. 7.2 Mapping the eight-node quadrilateral element

where

$$\alpha_i = \tfrac{1}{4}(2a + a_i b_i b + 2b_i ab + a_i b^2)$$

$$\beta_i = \tfrac{1}{4}(2b + a_i b_i a + 2a_i ab + b_i a^2) \quad i = 1 \to 4$$

and, except that some of the matrices are larger, the rest of the formulation is exactly as for the bilinear element. The complete displacement expansion includes all quadratic terms (see eqn (5.39)) so that a minimum four-point Gaussian quadrature should be used. Reduction to constant shear, which proved effective with the bilinear element, is no longer advantageous.

The LST element, which uses complete quadratic interpolations and thus avoids reduced integration considerations, is of comparable accuracy to the biquadratic element yet is no more complicated. This versatile element was introduced in Section 6.3 and in the following section we generalize its formulation using isoparametric mapping, resulting in curved sided elements.

7.3 The linear strain triangle

Introduction of isoparametric mapping into the linear strain triangle formulation of Section 6.3 maps it from the triangular shape in area coordinates to the curved shape in Cartesian coordinates shown in Fig. 7.3. If node 5, for example, is on the perpendicular offset from the midpoint of the straight line joining nodes 2 and 3 then it is evident that quadratic interpolation fits a parabola to points 3, 5, and 2. Applying the interpolation functions of eqns (5.60) to both displacements and Cartesian coordinates,

$$u = \Sigma f_i u_i, \quad v = \Sigma f_i v_i \quad i = 1 \to 6$$

$$x = \Sigma f_i x_i, \quad y = \Sigma f_i y_i \quad i = 1 \to 6$$

where

$$f_1 = L_1(2L_1 - 1), \quad f_2 = L_2(2L_2 - 1), \quad f_3 = L_3(2L_3 - 1)$$

$$f_4 = 4L_1 L_2, \quad f_5 = 4L_2 L_3, \quad f_6 = 4L_3 L_1. \tag{7.14}$$

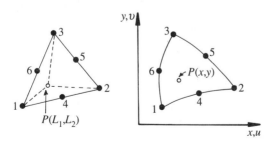

Fig. 7.3 Isoparametric triangular element

Eliminating L_3 from eqns (7.14) using the identity $L_3 = 1 - L_1 - L_2$ (otherwise we would obtain a 2×3 Jacobian matrix which could not be inverted[3]) and differentiating with respect to L_1 and L_2 we obtain

$$S = \begin{bmatrix} \{\partial f/\partial L_1\}^t \\ \{\partial f/\partial L_2\}^t \end{bmatrix} =$$

$$\begin{bmatrix} 4L_1 - 1 & 0 & 4L_1 + 4L_2 - 3 & 4L_2 & -4L_2 & 4 - 8L_1 - 4L_2 \\ 0 & 4L_2 - 1 & 4L_1 + 4L_2 - 3 & 4L_1 & 4 - 4L_1 - 8L_2 & -4L_1 \end{bmatrix}$$

(7.15)

If the element has straight sides the Jacobian matrix is known explicitly (eqn (6.28)) but if the sides are curved it is calculated numerically at each integration point as

$$J = \begin{bmatrix} \partial x/\partial L_1 & \partial y/\partial L_1 \\ \partial x/\partial L_2 & \partial y/\partial L_2 \end{bmatrix} = S[\{x\} \quad \{y\}]$$

(7.16)

giving the Cartesian derivatives as

$$\begin{Bmatrix} \partial u/\partial x \\ \partial x/\partial y \end{Bmatrix} = J^{-1} S\{u\} = G\{u\}$$

(7.17)

at each integration point. The elements of the matrix G are redeployed as shown in eqn (6.30) to give the numerical strain–displacement matrix B and finally the element stiffness matrix is given by three-point integration (eqn (5.135) or eqn (5.136)) as

$$k = \Sigma B^t DB\left(\frac{t|J|_{\text{abs}}}{6}\right)$$

(7.18)

noting that $\Sigma|J|_{\text{abs}}$ gives twice the element area when area coordinates are used. The stresses at any point in the element are given by

$$\{\sigma\} = N\{d\} = DB\{d\}$$

and the element gives good results when average nodal stresses are used. The consistent loads for body forces and the thermal loads caused by a uniform temperature rise are also given by numerical integration as

$$\{q_b\} = \Sigma F\{X\}(\tfrac{1}{2}|J|_{\text{abs}} t\omega_i)$$

(7.19)

$$\{q_0\} = \Sigma B^t D\{\varepsilon_0\}(\tfrac{1}{2}|J|_{\text{abs}} t\omega_i)$$

(7.20)

where $\{\varepsilon_0\}$ is for the case of plane stress as given in eqn (6.12) and for plane strain $\{\varepsilon_0\} = (1 + v)\alpha\theta\{1, 1, 0\}$.

In any of the calculations for the element stiffnesses or loadings one may interpolate the element thickness from the vector of thicknesses at the nodes as

$$t = \{f\}^t\{t_i\}$$

and this result is used in eqns (7.18), (7.19), and (7.20). Note, however, that this will increase the order of integration required to compute the element stiffness matrix exactly, and that temperature and stiffness variations can also be included in the numerical integration process without difficulty.

If Lagrange multipliers are also included as nodal variables these can be used to enforce constraints, for example, suppression of transverse strains across a crack within the element, that is, $\varepsilon_n = c_x(\partial u/\partial x) + c_y(\partial v/\partial y) = 0$ where c_x, c_y are the direction cosines of the normal to the crack. With curved sided isoparametric elements, of course, the curvature of the sides must not be excessive. It the sides are curved inwards, for example, it is generally sufficient to ensure that they do not pass over the centroid of the straight sided triangle formed by the corner nodes, though slightly greater curvature may be permissible if only one side is curved.[4, 5]

Figure 7.4 shows an example of the application of the isoparametric six-node triangle to a thermal stress problem in which a heated pipe expands within an infinite medium. The infinite medium is dealt with by providing appropriate elastic boundary conditions at the finite element boundary (see Section 21.7). The radial displacements obtained in the pipe and surrounding medium are compared to an analytical solution in Fig. 7.5, demonstrating a considerable improvement in the numerical results with the use of an elastic

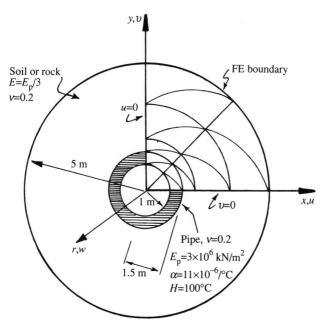

Fig. 7.4 Analysis of heated pipe buried in an infinite medium

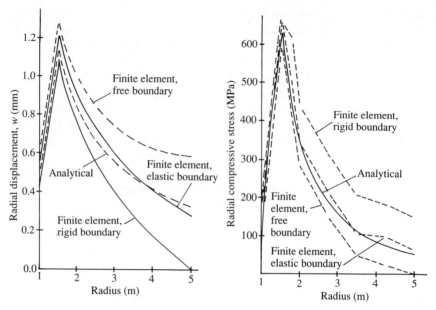

Fig. 7.5 Radial displacements and stresses produced in an expanding buried pipe[7]

boundary rather than a rigid or free boundary. A similar improvement was observed in the computed stresses.

Here parabolic elements provide a convenient approximation for the boundary and it has been shown that interpolation of the boundary conditions in these curved elements will not spoil the h^4 strain energy convergence expected from eqn (3.61).[6]

7.4 Axisymmetric elements

Many three-dimensional problems in the theory of elasticity are axisymmetric, that is, both the structure and the loadings are radially symmetric about an axis which is taken here to be the z-axis. Advantage can be taken of such symmetry to reduce the problem to an equivalent two-dimensional one.

The appropriate strain–displacement equations were derived in Section 2.5 and are

$$\varepsilon_r = \frac{\partial u}{\partial r}, \qquad \varepsilon_z = \frac{\partial w}{\partial z}$$

$$\varepsilon_{zr} = \frac{\partial u}{\partial z} + \frac{\partial w}{\partial r}, \qquad \varepsilon_\phi = \frac{u}{r}. \tag{7.21}$$

These are a contraction of those for three dimensions, shear strains on planes perpendicular to the zr-plane being omitted (otherwise rotational displacements about the axis of revolution would occur). ε_ϕ is the *circumferential strain*. This arises from radial expansion of the annulus-shaped finite elements, that is,

$$\varepsilon_\phi = \frac{[2\pi(r + \delta r) - 2\pi r]}{2\pi r}$$

$$= \frac{\delta r}{r} = \frac{u}{r}.$$

Omitting two shear moduli from the elasticity matrix for three dimensions, the stress–strain equations are

$$\{\sigma\} = \begin{bmatrix} \sigma_r \\ \sigma_z \\ \tau_{zr} \\ \sigma_\phi \end{bmatrix} = \frac{E}{1 - v - 2v^2} \begin{bmatrix} 1 - v & v & 0 & v \\ v & 1 - v & 0 & v \\ 0 & 0 & (1 - 2v)/2 & 0 \\ v & v & 0 & 1 - v \end{bmatrix} = D\{\varepsilon\}.$$

(7.22)

In the case of the three-node triangle, the Cartesian interpolation matrix is given by eqn (6.5), interchanging r with x and z with y. The strain matrix, S, is obtained by applying eqns (7.21) to the interpolation functions (eqns (6.1) and (6.2) with $x = r$, $y = z$) but here the circumferential strain varies with the radial position in the element. If one calculates only the value at the element centroid, the element stiffness matrix is easily obtained from eqn (6.8). As observed in Table 6.1, though, the 6 df three-node triangle converges very slowly and with the additional approximation implied by a constant radius throughout the element, the rate of convergence to the correct solution wll be further impaired. In addition, for very small elements with one side coinciding with the z-axis, the term $1/r$ in eqns (7.21) approaches infinity and this may cause numerical overflow during solution of the assembled equations.

The twelve-freedom six-node triangle gives a much more satisfactory solution if it is formulated using numerical integration. From eqn (7.18),

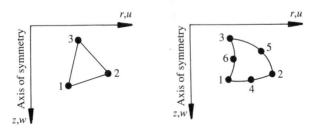

Fig. 7.6 Axisymmetric triangular elements

substituting $2\pi r$ for the thickness (t) in the element volume calculation, one obtains

$$k = \Sigma B^{\text{t}} DB(\tfrac{1}{2}\omega_i 2\pi r |J|_{\text{abs}}) \qquad (7.23)$$

noting that $|J|_{\text{abs}}$ is divided by 2 as $\Sigma\omega_i|J|_{\text{abs}}$ gives twice the cross-sectional area of the annular elements.

The derivatives required to assemble the strain interpolation matrix (B) are again obtained from eqn (7.17) whilst the circumferential strain ε_ϕ is evaluated at each integration point as

$$\varepsilon_\phi = \frac{u}{r} = \frac{\{f\}^{\text{t}}\{u\}}{\{f\}^{\text{t}}\{r\}} \qquad (7.24)$$

where $\{r\}$ is the vector of nodal radii of the element. Note that care should be taken to choose a numerical integration scheme which does not use the nodes as sampling points when some of these lie on the axis of symmetry (that is, at $r = 0$).

Figure 7.7 shows a useful application of this element using three-point quadratic integration in a segment of a pressurized thick cylinder.[7] The consistent loads used for the problem are shown at each node, these being obtained from eqn (6.37). The sum of the nodal loadings shown is 12 kN, so that the equivalent internal pressure is

$$p = \frac{12}{2\pi Rh} = \frac{12}{0.8\pi} = 4.775 \text{ kN/m}^2.$$

The solutions obtained are compared to the exact classical solutions[8] in Table 7.1, indicating excellent agreement, considering that only four elements are used. Note that unless the boundary conditions $w = 0$ are set for the three nodes arrowed in Fig. 7.7 the system is metastable and, particularly with

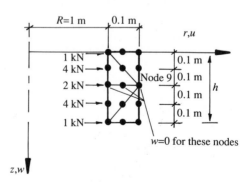

Fig. 7.7 Segment of internally pressurized thick cylinder with the consistent loads shown and $E = 10^6$ kN/m^2 and $v = 0.2$

Table 7.1 Displacements and stresses in thick cylinder

	Radial displacement $(m \times 10^{-4})$	Stresses (kN/m²) Radial	Circumferential
$r = 1.00$			
FEM, $z = 0$	0.5120	− 4.543	50.30
FEM, $z = 0.2$	0.5120	− 4.768	50.25
Exact	0.5120	− 4.775	50.25
$r = 1.05$			
FEM, $z = 0$	0.5054	− 2.278	47.68
FEM, $z = 0.2$	0.5054	− 2.277	47.67
Exact	0.5054	− 2.217	47.69
$r = 1.10$			
FEM, $z = 0$	0.5002	− 0.005	45.47
FEM, $z = 0.2$	0.5002	0.223	45.53
Exact	0.5002	0.0	45.47

numerically integrated solutions, round-off error will lead to unpredictable results in the unrestrained direction or in some cases the complete solution may be lost.

Figure 7.8 shows a model for the Boussinesq problem, again using curved triangular elements, but now a four-point Gaussian integration is used. This gives marginally better results than three points,[†] as in this problem the

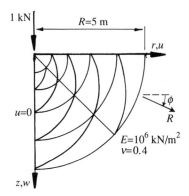

Fig. 7.8 Finite element grid for the Boussinesq problem

[†] That is, eqn (5.135). Alternatively the three internal points of eqn (5.136) would give almost identical results to four-point Gaussian integration.

domain extends to the z-axis where the circumferential strain is indeterminate. The numerical results are compared to the exact solution in Fig. 7.9, the use of appropriate elastic boundary conditions again enhancing the accuracy of the finite element model.[7]

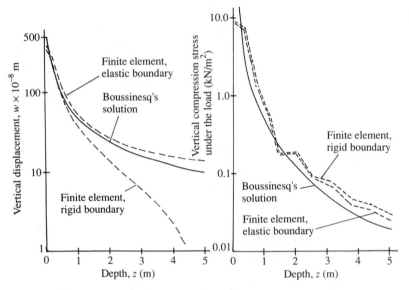

Fig. 7.9 Finite element solution to Boussinesq's problem

7.5 Isoparametric brick elements

Figure 7.10 shows a hexahedral or brick-shaped element, in this case in the form of a parallelepiped. The required interpolation is, from eqns (5.38),

$$u = \Sigma f_i u_i$$

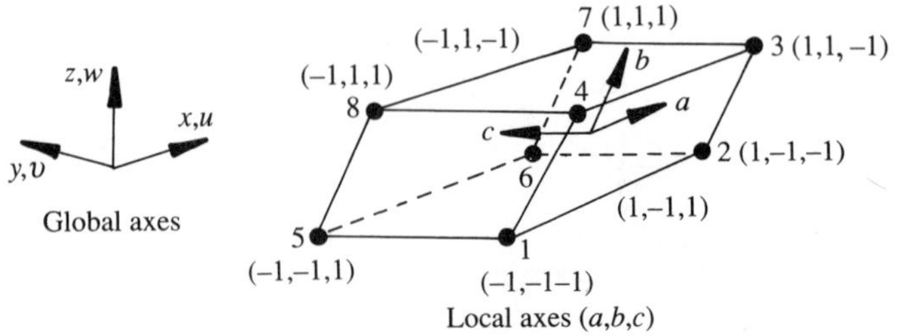

Fig. 7.10 Isoparametric brick element

where

$$f_i = (1 + a_i a)(1 + b_i b)(1 + c_i c)/8, \quad i = 1 \to 8. \tag{7.25}$$

Now the chain rule gives, for any variable (taking u for example),

$$\begin{Bmatrix} \partial u/\partial a \\ \partial u/\partial b \\ \partial u/\partial c \end{Bmatrix} = \begin{bmatrix} \partial x/\partial a & \partial y/\partial a & \partial z/\partial a \\ \partial x/\partial b & \partial y/\partial b & \partial z/\partial b \\ \partial x/\partial c & \partial y/\partial c & \partial z/\partial c \end{bmatrix} \begin{Bmatrix} \partial u/\partial x \\ \partial u/\partial y \\ \partial u/\partial z \end{Bmatrix} \tag{7.26}$$

where the connecting matrix is the Jacobian and this is evaluated as

$$J = S[\{x\} \ \{y\} \ \{z\}] = \begin{bmatrix} \partial f_1/\partial a & \partial f_2/\partial a & \dots & \partial f_8/\partial a \\ \partial f_1/\partial b & \partial f_2/\partial b & \dots & \partial f_8/\partial b \\ \partial f_1/\partial c & \partial f_2/\partial c & \dots & \partial f_8/\partial c \end{bmatrix} [\{x\} \ \{y\} \ \{z\}]. \tag{7.27}$$

Thus the Cartesian derivatives are given by

$$\{\partial u/\partial x, \partial u/\partial y, \partial u/\partial z\} = J^{-1} S\{u\} = G\{x\}. \tag{7.28}$$

Now the 6×24 strain–displacement matrix B at each integration point is given by redeployment of the matrix G (similar to that of eqn (7.11)). This result can be written in the form (not literal, in practice the columns of B must be rearranged in the manner used in eqn (7.11))

$$\{\varepsilon\} = \begin{bmatrix} \partial f/\partial x & 0 & 0 \\ 0 & \partial f/\partial y & 0 \\ 0 & 0 & \partial f/\partial z \\ \partial f/\partial y & \partial f/\partial x & 0 \\ 0 & \partial f/\partial z & \partial f/\partial y \\ \partial f/\partial z & 0 & \partial f/\partial x \end{bmatrix} \begin{Bmatrix} \{u\} \\ \{v\} \\ \{w\} \end{Bmatrix} = B\{d\} \tag{7.29}$$

where $\partial f/\partial x = \{\partial f_1/\partial x, \partial f_2/\partial x, \dots, \partial f_8/\partial x\}^t$, and so on.

Although the element has a total of twenty-four freedoms, its coding is particularly simple and the derivatives required to form the matrix S in eqn (7.27) are given by[9]

$$\frac{\partial f_i}{\partial a} = \frac{a_i(1 + b_i b)(1 + c_i c)}{8}$$

$$\frac{\partial f_i}{\partial b} = \frac{b_i(1 + a_i a)(1 + c_i c)}{8} \tag{7.30}$$

$$\frac{\partial f_i}{\partial c} = \frac{c_i(1 + a_i a)(1 + b_i b)}{8}.$$

Finally the element stiffness matrix is given by

$$k = \Sigma B^t DB(\omega_i |J|_{\text{abs}}) \qquad (7.31)$$

where the modulus matrix is given in Section 2.3 and the integration points and weights are

$$(a_i, b_i, c_i) = (\pm 1/\sqrt{3}, \; \pm 1/\sqrt{3}, \; \pm 1/\sqrt{3})$$

$$\omega_i = \tfrac{1}{64}, \quad i = 1 \to 8$$

as $|J|_{\text{abs}}$ gives eight times the element volume with these coordinates. The great attraction of the isoparametric brick elements is the brevity with which they can be coded.[2] Against this, three-dimensional elements generally require large bandwidths to represent practical problems and Irons' Frontal solution system was developed to overcome this problem.[10]

A difficulty with the eight-node brick is that, like its two-dimensional counterpart, the bilinear rectangle, it experiences difficulty in representing flexural behaviour unless several layers of elements are used. Thus the twenty-node brick (Fig. 5.7) is frequently preferred.[11] For both these brick elements the questions of parasitic shear and reduced integration are of importance and it is now established practice to use reduced integration with these elements[11,12] or to modify them to improve their performance. In the case of the trilinear brick, Wilson's quadratic functions (eqns (6.44)) can be added as nodeless variables:

$$f_{25} = (1 - a^2), \quad f_{26} = (1 - b^2), \quad f_{27} = (1 - c^2) \qquad (7.32)$$

which are removed by static condensation. The resulting element is non-conforming but is better able to model flexural behaviour.[13]

Finally, the use of constrained three-dimensional elements for the analysis of thick plates and shells is worth remarking upon. In the case of the trilinear brick, for example, this is achieved with constraints of the form

$$u_{\text{brick}} = \pm \tfrac{1}{2}t(\phi_x)_{\text{plate}}, \quad v_{\text{brick}} = \pm \tfrac{1}{2}t(\phi_y)_{\text{plate}}, \quad w_{\text{brick}} = w_{\text{plate}}$$

but the thick plate elements discussed in Chapter 9 are more economical.

7.6 An isoparametric joint element

Joint elements are used in geomechanics applications to model the behaviour of fissured rock and Fig. 7.11 shows a modification of the joint element of Goodman *et al.*[14] in which shear strains are also included. The new element is isoparametric in the sense that no interpolation is used in the transverse direction (which has nearly zero thickness) and linear interpolation is used in

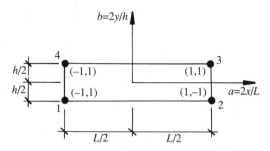

Fig. 7.11 Joint element with eight freedoms

the longitudinal direction. Noting that the inverse Jacobian is

$$J^{-1} = \begin{bmatrix} \partial a/\partial x & \partial b/\partial x \\ \partial a/\partial y & \partial b/\partial y \end{bmatrix} = \begin{bmatrix} 2/L & 0 \\ 0 & 2/h \end{bmatrix} \tag{7.33}$$

and combining eqns (7.5) and (7.9) and suppressing the lateral interpolation by putting $b = 0$ the strain interpolation matrix is given as

$$= \frac{1}{2h} \begin{bmatrix} -h/L & 0 & h/L & 0 & h/L & 0 & -h/L & 0 \\ 0 & a-1 & 0 & -a-1 & 0 & a+1 & 0 & 1-a \\ a-1 & -h/L & -a-1 & h/L & a+1 & h/L & 1-a & -h/L \end{bmatrix}$$

$$\tag{7.34}$$

The element stiffness matrix is obtained by two-point Gaussian quadrature (that is, with $a = \pm 1/\sqrt{3}$) as

$$k = (Lth/2)\Sigma B^t DB. \tag{7.35}$$

Generally the elastic parameters will be anisotropic and the shear modulus will be independent of Young's modulus. If the element is not horizontal, coordinate transformation must be used in the same manner as for the 6 df beam element of Chapter 1. If 'no tension' behaviour is to be taken into account this must be done iteratively.

More refined joint elements in which relative displacements δu, δv are used at nodes 3 and 4 in Fig. 7.11 have also been developed.[15,16] These are designed to reduce numerical accuracy problems. One disadvantage of this approach is that the relative displacement freedoms on one side of the joint are inconsistent with those of the adjacent plane strain elements so that some alteration of the assembly and solution routine is required to allow for this.

7.7 Application of isoparametric elements to model singularities

Special techniques are required to model singularities caused by cracks, holes, or other discontinuities in a continuum or at its boundary. As an example of such a singularity, Fig. 7.12 shows the variation in moment found along the edge of a square hole centrally located in a square slab with simply supported edges carrying a blanket load.[17] The problem is analysed using a sequence of meshes with increasing refinement around the hole and one finds that the moment at the corner is divergent. Confining discussion to crack problems for the moment (We return to the problem of Fig. 7.12 in Section 15.9) several procedures have been developed to deal with these:

(1) special mesh grading techniques;

(2) 'analytic' elements that are formulated by integrating the exact equations for the displacement fields around the crack tip;

(3) isoparametric elements with shifted nodes;

(4) extrapolation techniques.

In the first category, substructuring techniques[18] can be used for a detailed reanalysis of the crack region. By a *chain substructuring* technique, in fact, Walsh[19] obtains element size reductions of the order of 10^{10} without difficulty and the procedure is found to give economical predictions of stress concentration factors.

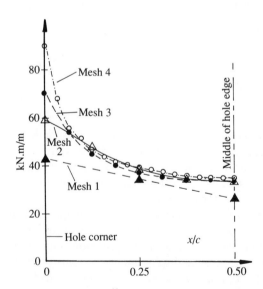

Fig. 7.12 Variation of major principal moment along the edge of an opening in a simply supported slab with blanket load

In the second category the analytical equations for the displacement distribution near the crack tip (which contain $\sqrt{r}$, implying a $r^{-1/2}$ strain singularity) have been incorporated into the formulation of hybrid crack elements by Walsh[20] and Tong and Pian.[21] Such procedures bear some resemblance to the cylinder element of Section 11.4, but now $\sqrt{r}$ modes are included rather than circular function modes and the finite element solution will reveal the stress concentration factors directly as the amplitudes of these modes.

The third approach bears some resemblance to the use of analytic elements, in that correct displacement and strain variation adjacent to the crack tip are approximated using isoparametric elements with shifted nodes. It has the attraction that it can be implemented without any alteration in the computer program; only the nodal coordinate data at the crack tip is changed. This approach does not reveal the stress concentration factors directly but these can be deduced by an extrapolation technique such as that described in Section 15.9.

An example is shown in Fig. 7.13 where six-node crack tip elements are generated by mapping elements with quarter-point nodes adjacent to the crack tip in the xy-system into the standard form (with midpoint nodes) in the $L_1 L_2 L_3$-system. In the xy-plane we have

$$x_4 = \frac{3x_1 + x_2}{4}, \quad y_4 = \frac{3y_1 + y_2}{4}$$

$$x_5 = \frac{x_2 + x_3}{2}, \quad y_5 = \frac{y_2 + y_3}{2} \tag{7.36}$$

$$x_6 = \frac{3x_1 + x_3}{4}, \quad y_6 = \frac{3y_1 + y_3}{4}$$

(a)   (b)

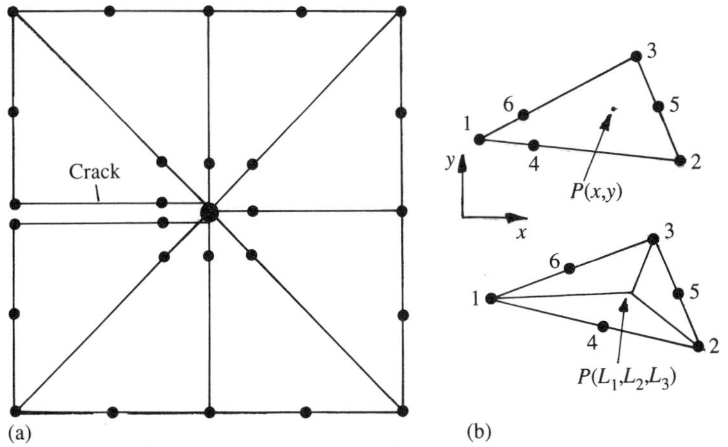

Fig. 7.13 Quadratic crack tip elements

that is, the midside nodes on the sides meeting at the crack tip are shifted towards it. We shall see that this leads to improved modelling of $r^{-1/2}$ stress singularities at crack tips.

In the $L_1 L_2 L_3$-plane the coordinates of these nodes are respectively (0.5, 0.5, 0), (0, 0.5, 0.5), and (0.5, 0, 0.5) as usual, so that the area coordinate interpolations are eqns (5.60). Hence the mapping is

$$x = \Sigma f_i x_i, \quad y = \Sigma f_i y_i \quad i = 1 \to 6 \tag{7.37}$$

where the f_i are given by eqns (5.60). Substituting eqns (7.36) into eqns (7.37) the mapping can be expressed, after some rearrangement with the use of the identity $L_1 + L_2 + L_3 = 1$, as

$$x - x_1 = \{(x_2 - x_1)L_2 + (x_3 - x_1)L_3\}(L_2 + L_3)$$

$$y - y_1 = \{(y_2 - y_1)L_2 + (y_3 - y_1)L_3\}(L_2 + L_3). \tag{7.38}$$

Along side 12, where $L_3 = 0$, one can express the distance from the singularity at node 1 as

$$r^2 = (x - x_1)^2 + (y - y_1)^2 = \{(x_2 - x_1)^2 + (y_2 - y_1)^2\}L_2^4 = L_{21}^2 L_2^4 \tag{7.39}$$

giving $L_2 = (r/L_{21})^{1/2}$. Hence using eqns (5.11) for the interpolation along this side with $d = L_2$ one obtains the interpolation for u as

$$u = (1 - 3L_2 + 2L_2^2)u_1 + 4L_2(1 - L_2)u_2 + (4L_2^2 - L_2)u_4 \tag{7.40}$$

so that terms in $\sqrt{r}$ are obtained for the displacement interpolation along this side as required near the crack. Now the direct strains parallel to this side may be calculated as

$$\frac{\partial u}{\partial L_2}\frac{\partial L_2}{\partial r} = \left(\frac{1}{2L_{21}r}\right)^{1/2}\frac{\partial u}{\partial L_2} \tag{7.41}$$

so that the strains vary according to $r^{-1/2}$ as required to model the crack. In the two-dimensional element the Cartesian strains will be calculated using the inverse Jacobian transformation. From eqns (7.38) the Jacobian is obtained as

$$J = \begin{bmatrix} \partial x/\partial L_2 & \partial y/\partial L_2 \\ \partial x/\partial L_3 & \partial y/\partial L_3 \end{bmatrix} = \begin{bmatrix} x_{21}(2L_2 + L_3) & y_{21}(2L_2 + L_3) \\ x_{31}(2L_3 + L_2) & y_{31}(2L_3 + L_2) \end{bmatrix}. \tag{7.42}$$

Noting that $x_{21}x_{31} - y_{21}x_{31} = 2\Delta$, it follows that

$$|J| = 2\Delta(2L_2 + L_3)(L_3 + L_2) \tag{7.43}$$

so that the Jacobian and hence the strains are singular at node 1 where $L_2 = L_3 = 0$. This is exactly what we require to model crack tips using only a single element rather than highly refined meshes which yield singularity only in the limit of mesh refinement. These quarter-point elements, on the other hand, can be introduced into a mesh without any modification of the LST formulation given in Section 7.3. Further, as the actual nature of the interpolation is changed, such nodal shifts may be regarded as a special instance of the basis transformation procedures discussed in Section 4.4.

An unobserved difficulty to this point, though, is that a higher-order integration is required to integrate accurately elements with nodal singularities.[22] This difficulty may be overcome by the use of special integrations and Dunham[23] shows that, using triangular polar coordinates

$$\alpha = 1 - L_1, \quad \beta = \frac{L_3}{L_1 + L_2} \tag{7.44}$$

one obtains the mapping shown in Fig. 7.14.

Expressing the interpolation in these coordinates the $\sqrt{r}$ behaviour of the displacements will be exactly integrated by applying the standard one-dimensional two-point Gauss formula to the β coordinate, and for the α coordinate applying

$$\alpha_1 = \tfrac{1}{4} \quad \omega_1 = \tfrac{2}{3}$$

$$\alpha_2 = 1 \quad \omega_2 = \tfrac{1}{3}. \tag{7.45}$$

The procedure for combining the two integrations is similar to that described in relation to the conical product formulae in Section 5.12 and results in a four-point formula.

Another instance of the foregoing approach is the use of cubic isoparametric elements with shifted nodes to represent the $r^{2/3}$ displacement behaviour found at the corners of rectangular holes in plates.[5]

7.8 Closing remarks

Until now we have considered only a Jacobian transformation for first derivatives. To transform second derivatives a *Hessian* matrix (for example eqn (22.1)) must be inverted and indeed this operation appears in Section 8.4, although the element shape is not mapped. Complete generalization of isoparametric mapping to deal with thin plate elements directly is a difficult

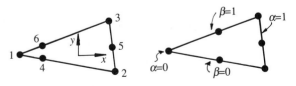

Fig. 7.14 Polar coordinates in a crack tip element

exercise which can be circumvented by beginning with a thick plate formulation which requires only first derivatives (see eqns (9.2)) and using penalty factors to suppress the transverse shear strains, as is done in Sections 9.5, 11.1, and 11.6.

References

1. Apostol, T. (1972). *Calculus*, Vol. II, (2nd edn). Wiley, New York.
2. Irons, B. M. (1966). Engineering applications of numerical integration in stiffness methods. *J. Amer. Inst. Aeron. Astron.*, **4**, 2037.
3. Ergatoudis, J., Irons, B. M., and Zienkiewicz, O. C. (1968). Curved isoparametric elements for finite element analysis. *Int. J. Solids Struct.*, **7**, 31.
4. Mitchell, A. R. (1977). An introduction to the mathematics of the finite element method. In *The Mathematics of Finite Elements and Applications*, Vol. 1, (ed. J. R. Whiteman). Academic Press, London.
5. Mitchell, A. R. and Wait, R. (1977). *The Finite Element Method in Partial Differential Equations*. Wiley, London.
6. Strang, G. (1972). Variational crimes in the finite element method. In *The Mathematical Foundations of the Finite Element Method*, (ed. A. K. Aziz). Academic Press, New York.
7. Mohr, G. A. and Power, A. S. (1978). Elastic boundary conditions for finite elements of infinite and semi-infinite media. *Proc. Inst. Civ. Eng.*, Part 2, **65**, 675.
8. Timoshenko, S. P. and Goodier, J. N. (1951). *Theory of Elasticity*, (3rd edn). McGraw-Hill, New York.
9. Abel, J. F. and Desai, C. S. (1972) *Introduction to the Finite Element Method*. Van Nostrand Reinhold, New York.
10. Irons, B. M. (1970). A frontal solution program for finite element analysis. *Int. J. Num. Meth. Engng*, **2**, 5.
11. Hellen, T. K. (1976). Numerical integration considerations in two and three dimensional isoparametric elements. In *The Mathematics of Finite Elements and Applications*, Vol. 2, (ed. J. R. Whiteman). Academic Press, London.
12. Gallagher, R. H. (1975). *Finite Element Analysis Fundamentals*. Prentice-Hall, Englewood Cliffs NJ.
13. Wilson, E. L. and Taylor, R. L. (1971). Incompatible displacement models. Symposium on Numerical and Computer Methods in Structural Engineering, University of Illinois.
14. Goodman, R. E., Taylor, R. L., and Brekke, T. L. (1968). A model for the mechanics of jointed rock. *J. Soil Found. Mech. Div. ASCE*, **94**, 637.
15. Wilson, E. L. (1977). Finite elements for foundations, joints and fluids. In *Finite Elements for Geomechanics*, (ed. G. Güdehus). Wiley, London.
16. Ghaboussi, J., Wilson, E. L., and Isenberg, G. J. (1973). Finite elements for rock joints and interfaces. *J. Soil Found. Mech. Div. ASCE*, **99**, 833.
17. Mohr, G. A. (1979). Displacement of reinforcement in perforated slabs. *Trans I. E. Aust.*, **CE21** (1), 16.

18. Przemieniecki, J. S. (1968). *Theory of Matrix Structural Analysis.* McGraw-Hill, New York.
19. Walsh, P. F. (1978). Intensive finite element grading for stress concentrations. *Engng. Fract. Mech.* **10**, 211.
20. Walsh, P. F. (1971). The computation of stress intensity factors by a finite element technique. *Int. J. Solids Struct.,* **7**, 1333.
21. Tong, P. and Pian, T. H. H. (1973). A hybrid element approach to crack problems in plane elasticity. *Int. J. Num. Meth. Engng,* **7**, 297.
22. Hibbit, H. D. (1977). Some properties of singular isoparametric elements. *Int. J. Num. Meth. Engng,* **11**, 180.
23. Dunham, R. S. (1979). A quadrature rule for conforming quadratic crack tip elements. *Int. J. Num. Meth. Engng,* **14**, 287.

8
Thin plate bending elements

Finite thin plate bending elements generally differ from plane stress or three-dimensional elements in that, rather than separate displacement functions for displacements parallel to orthogonal axes, it is customary to choose a single function for the lateral displacement. As the nodal freedoms it is frequent practice to allow lateral displacement and its first derivatives (or slopes) at each node, giving three degrees of freedom per node. Considering elements with corner nodes this leads to triangular and rectangular elements with nine and twelve freedoms, respectively. These and many other thin plate elements are discussed in this chapter.

8.1 The ACM rectangle

Figure 8.1 shows the earliest thin plate bending element, a rectangle with twelve freedoms. The element was first derived by Adini, Clough, and Melosh and is sometimes referred to as the ACM rectangle.[1] It is most simply developed using the dimensionless coordinates a and b shown. The displacement function is chosen as a symmetric polynomial,

$$w = c_1 + c_2 a + c_3 b + c_4 a^2 + c_5 ab + c_6 b^2 + c_7 a^3 + c_8 a^2 b$$
$$+ c_9 ab^2 + c_{10} b^3 + c_{11} a^3 b + c_{12} ab^3. \qquad (8.1)$$

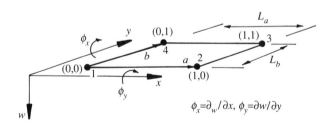

Fig. 8.1 Rectangular bending element

Defining dimensionless coordinates $a = x/L_a$ and $b = y/L_b$, using the relationships

$$\frac{\partial w}{\partial a} = L_a\left(\frac{\partial w}{\partial x}\right), \qquad \frac{\partial w}{\partial b} = L_b\left(\frac{\partial w}{\partial y}\right), \qquad (8.2)$$

and substituting the nodal coordinates (a_i, b_i) in eqn (8.1), the interpolation functions are obtained by the inversion procedure described in Section 5.3. The resulting interpolation functions are:

$$\left.\begin{array}{l} f_1 = (a-1)(b-1)(1+a+b-2a^2-2b^2) \\[2mm] f_2 = L_a(b-1)(a-1)^2 a, \qquad f_3 = L_b(b-1)^2(a-1)b \end{array}\right\} \text{node 1}$$

$$\left.\begin{array}{l} f_4 = (b-1)a(2a^2+2b^2-3a-b) \\[2mm] f_5 = L_a(b-1)a^2(a-1), \qquad f_6 = L_b(b-1)^2 ab \end{array}\right\} \text{node 2} \qquad (8.3)$$

$$\left.\begin{array}{l} f_7 = ab(-2a^2-2b^2-1+3a+3b) \\[2mm] f_8 = L_a a^2 b(a-1), \qquad f_9 = L_b ab^2(b-1) \end{array}\right\} \text{node 3}$$

$$\left.\begin{array}{l} f_{10} = b(a-1)(2a^2+2b^2-a-3b) \\[2mm] f_{11} = L_a ab(a-1)^2, \qquad f_{12} = L_b b^2(a-1)(b-1). \end{array}\right\} \text{node 4}$$

The curvatures are defined, assuming small displacements, as

$$\chi_x = \frac{\partial^2 w}{\partial x^2}, \qquad \chi_y = \frac{\partial^2 w}{\partial y^2}, \qquad \chi_{xy} = \frac{2\partial^2 w}{\partial x \partial y} \qquad (8.4)$$

so that the 3×12 curvature–displacement matrix B is given as

$$B = \begin{bmatrix} 1/L_a^2 & 0 & 0 \\ 0 & 1/L_b^2 & 0 \\ 0 & 0 & \dfrac{1}{L_a L_b} \end{bmatrix} \begin{bmatrix} \partial^2 f_1/\partial a^2 & \partial^2 f_2/\partial a^2 & \cdots & \partial^2 f_{12}/\partial a^2 \\ \partial^2 f_1/\partial b^2 & \partial^2 f_2/\partial b^2 & \cdots & \partial^2 f_{12}/\partial b^2 \\ \partial^2 f_1/\partial a \partial b & \partial^2 f_2/\partial a \partial b & \cdots & \partial^2 f_{12}/\partial a \partial b \end{bmatrix}.$$

$$(8.5)$$

where, using 'the chain rule of differentiation,

$$\frac{\partial^2 f_i}{\partial x^2} = \frac{\partial^2 f_i}{\partial a^2}\left\{\frac{\partial a}{\partial x}\right\}^2 + \frac{\partial f_i}{\partial a}\frac{\partial^2 a}{\partial x^2} = \frac{1}{L_a^2}\frac{\partial^2 f_i}{\partial a^2} \qquad (8.6)$$

as $\partial a/\partial x = 1/L_a$ and $\partial^2 a/\partial x^2 = 0$. Note that $\partial^2 a/\partial x \partial b \neq 0$ for quadrilateral elements. The other entries in the first matrix of eqn (8.5) are obtained in similar fashion. The element stiffness matrix is given by four-point Gaussian quadrature as

$$k = \Sigma B^{t}(\omega_i L_a L_b) D B \tag{8.7}$$

with

$$\frac{(a_i, b_i) = (1 \pm 1/\sqrt{3})}{2}, \qquad \frac{(1 \pm 1/\sqrt{3})}{2}, \qquad \omega_i = \frac{1}{4}, \qquad i = 1 \rightarrow 4. \tag{8.8}$$

The consistent loads for an element with a transverse surface traction p are calculated as $\{q_t\} = \iint F\{T\}\,dx\,dy = p L_a L_b \int_0^1 \int_0^1 \{f\}\,da\,db$, yielding the following results

$$(F_w)_i = \frac{p L_a L_b}{4}$$

$$(M_x)_i = \frac{(1 - 2a_i)p L_a^2 L_b}{24} \tag{8.9}$$

$$(M_y)_i = \frac{(1 - 2b_i)p L_a L_b^2}{24}, \qquad i = 1 \rightarrow 4.$$

The element gives satisfactory results and, though non-conforming, passes the patch test.[2] If nodal moments are calculated these show considerable discontinuities between elements. Although these diminish with mesh refinement[3], nodal averaging of the moments is recommended.

8.2 Conforming rectangular elements

To obtain a conforming rectangular element, Deak and Pian[4] combined four triangular elements to obtain an element with twelve freedoms. The element converges more rapidly than the ACM element but this approach does not lead to interpolation functions applicable to the whole element and is therefore less convenient and more costly, particularly when numerical integration is used.

A more economical approach was proposed by Bogner *et al.*[5] applying the Hermitian interpolations for a beam element (eqns (5.22)) in both the x- and y-directions. For the 12 df rectangle the *crossed beam*[6] or *bicubic* interpolation functions thus obtained are given in eqns (5.45). Unfortunately these functions do not provide a constant twisting curvature mode[7] and hence convergence is not guaranteed. This difficulty is overcome if an additional twisting curvature freedom is included at each node, so that the element then has sixteen freedoms. The appropriate interpolation functions are eqns (5.45) and (5.47). Using eqns (8.5) these functions lead to a simple curvature–displacement matrix and the element stiffness matrix follows from eqn (8.7). Observing eqn (5.46), a 4×4 Gauss rule is needed to integrate the term in $x^3 y^3$ exactly, but nine-point integration integrates all terms except this one exactly and will give satisfactory results.

Inclusion of the twisting freedom $\partial^2 w / \partial x \partial y$ causes little inconvenience and Fig. 8.2 shows the boundary conditions required for the analysis of a quadrant of a simply supported plate. Along side AB, for example, we have $\phi_x = 0$, and along side BC ϕ_x is also zero. As these sides are parallel to the axes, at point B we have

$$\frac{\partial \phi_x}{\partial x} = 0, \qquad \frac{\partial \phi_x}{\partial y} = 0,$$

and the second condition requires $\chi_{xy} = 0$. The boundary conditions for χ_{xy} are non-essential (that is, convergence with mesh refinement is obtained without them) but their inclusion (particularly $\chi_{xy} = 0$ at point C) leads to more accurate results in coarse meshes.

The twisting freedoms can be replaced by finite difference approximations at each corner.[8] At node 1, observing Fig. 8.1, one takes

$$L_a L_b \left(\frac{\partial^2 w}{\partial x \partial y} \right)_1 = \tfrac{1}{2} \left(L_a \left(\frac{\partial w}{\partial x} \right)_4 - L_a \left(\frac{\partial w}{\partial x} \right)_1 + L_b \left(\frac{\partial w}{\partial y} \right)_2 - L_b \left(\frac{\partial w}{\partial y} \right)_1 \right)$$

(8.10)

and expressing the four such transformations in matrix form one obtains

$$\begin{bmatrix} (\partial^2 w / \partial x \partial y)_1 \\ (\partial^2 w / \partial x \partial y)_2 \\ (\partial^2 w / \partial x \partial y)_3 \\ (\partial^2 w / \partial x \partial y)_4 \end{bmatrix}$$

$$= \frac{1}{2 L_a L_b} \begin{bmatrix} -L_a & -L_b & 0 & L_b & 0 & 0 & L_a & 0 \\ 0 & -L_b & -L_a & L_b & L_a & 0 & 0 & 0 \\ 0 & 0 & -L_a & 0 & L_a & L_b & 0 & -L_b \\ -L_a & 0 & 0 & 0 & 0 & L_b & L_a & -L_b \end{bmatrix} \begin{bmatrix} \phi_{x1} \\ \phi_{y1} \\ \phi_{x2} \\ \phi_{y2} \\ \phi_{x3} \\ \phi_{y3} \\ \phi_{x4} \\ \phi_{y4} \end{bmatrix}$$

(8.11)

Applying eqn (8.11) to the $\partial^2 w / \partial x \partial y$ columns of the 3×16 curvature–displacement matrix for the sixteen-freedom element, the reduced 3×12 matrix for a twelve-freedom element is obtained. This new element, though

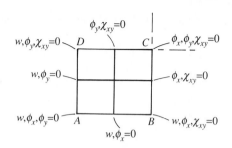

Fig. 8.2 Boundary conditions for a quadrant of a simply supported plate with rectangular bicubic elements

non-conforming, gives results very nearly as good as those with the sixteen-freedom element[9] and can be very concisely coded.

8.3 Quadrilateral elements

The non-conforming ACM element described in Section 8.1 can be generalized into quadrilateral shape using appropriate Hessian transformations for the curvatures (that is, the diagonal matrix of eqn (8.5) becomes fully populated)[10] but unless restricted to parallelogram shape it may violate the constant curvature requirement.[11]

The sixteen-freedom bicubic rectangle has been generalized to parallelogram shape[12] and, in principle, so can the twelve-freedom version given by eqn (8.11). However, generalization of these bicubic elements to quadrilateral is a difficult exercise.

An alternative means of obtaining quadrilateral elements is by combining triangular elements, and Sander[13] accomplished this in a novel manner by prescribing a full cubic displacement field in a quadrilateral and adding two additional linear functions in each of two triangular subdomains of the element. Thus an element with sixteen freedoms is obtained with w, $\partial w/\partial x$, $\partial w/\partial y$ at the vertices, and normal slopes at the midpoints of the sides.

Fraeijs de Veubeke[14] simplifies this element by combining four triangular elements and, through the use of skew coordinates along the element diagonals, obtains explicit interpolation functions for each triangular subdomain.

Observing the nodal freedoms and skew coordinates (ξ, η) shown in Fig. 8.3 the cubic displacement function in each triangular subdomain is given by eqn (5.2) and for region 012, for example, de Veubeke obtains the amplitudes of this function as

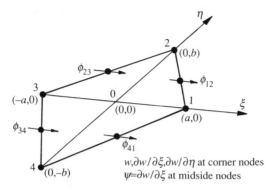

Fig. 8.3 Subdomain formulation of a quadrilateral bending element

$$c_1 = w_0, \quad c_2 = \phi_{\xi0}, \quad c_3 = \phi_{\eta0}, \quad a^2 c_4 = -3w_0 - 2a\phi_{\xi0} + 3w_1 - a\phi_{\xi1}$$

$$2abc_5 = -6w_0 - 2a\phi_{\xi1} - 2b\phi_{\xi0} + 6w_1 - a\phi_{\xi1} + 2b\phi_{\eta1} + a\phi_{\xi2} - 4a\psi_{12}$$

$$b^2 c_6 = -3w_0 - 2b\phi_{\eta0} + 3w_2 - b\phi_{\eta2},$$

$$4a^3 c_7 = 2w_0 + a\phi_{\xi0} - 2w_1 + a\phi_{\xi1} \tag{8.12}$$

$$4a^2 bc_8 = 6w_0 + 2a\phi_{\xi0} + b\phi_{\eta0} - 6w_1 + a\phi_{\xi1} - b\phi_{\eta1} - a\phi_{\xi2} + 4a\psi_{12}$$

$$4ab^2 c_9 = 6w_0 + a\phi_{\xi0} + 2b\phi_{\eta0} - 6w_1 + a\phi_{\xi1} - 2b\phi_{\eta1} + 4a\psi_{12}$$

$$4b^3 c_{10} = 2w_0 + b\phi_{\eta0} - 2w_2 + b\phi_{\eta2}.$$

From these the curvatures are calculated and, with the inclusion of transformation to Cartesian curvatures, the element stiffness matrix is given by three congruent transformations.[14] Once the stiffness matrices have been summed static condensation may be used to eliminate the freedoms at the internal node and finally coordinate transformation to the global freedoms w, $\partial w/\partial x$, $\partial w/\partial y$, and $\partial w/\partial n$ is required.

The foregoing formulation involves protracted computation, but Chan and Kabaila[15] obtain a simplified formulation of the element by directly using Cartesian coordinates and, rather than using normal slopes, the midside slopes are taken as either $\partial w/\partial x$ or $\partial w/\partial y$, depending upon the orientation of each side of the element. Finally, the element can be reduced to twelve freedoms by constraining out the midside slopes by assuming a linear variation of the normal slope along each side of the element (see eqn (8.25)).

A twelve-freedom conforming quadrilateral was also obtained by Clough and Felippa,[16] again by combining four triangular subdomains. In this instance, though, the four normal slope freedoms are specified on the diagonals of the quadrilateral and are thus able to be condensed out, but once again triangular geometry has had to be used to develop quadrilateral elements. A more direct approach is achieved by the basis transformation technique described in Section 4.4,[17] and this can also be applied to triangular elements (see Sections 8.10 and 10.6).

8.4 Non-conforming triangle with nine freedoms

As noted in Section 5.2 development of a nine-freedom triangular bending element (see Fig. 8.4) presents some difficulties and suitable incomplete Cartesian displacement functions cannot be determined. Yet it is found that using the full ten-term cubic expansion (that is, eqn (5.2)), including a tenth freedom w_4 at the centroid, does not lead to good results.[18] However, by using area coordinates Bazeley *et al.*[19] obtained an economical non-conforming nine-freedom element which, although it fails the patch test for some irregular mesh patterns,[19,20] gives results of satisfactory accuracy.

The original explicit formulation of the element stiffness matrix in Cartesian coordinates is somewhat complicated[21] and thus the present treatment simplifies the formulation by using the technique of Argyris *et al.*[22] in which the global Cartesian slopes are transformed to obtain local natural slopes parallel to the element sides, as shown in Fig. 8.4. The 10×10 transformation matrix required to achieve this was given in eqn (5.67). The tenth row and column of this should be deleted because, in what follows, we delete the displacement freedom at the centroid, that is, w_4.

Observing eqns (5.73) the interpolation function for the central freedom involves only the term $L_1 L_2 L_3$ and this term vanishes on the element boundary and gives zero natural slopes along each side of the element. (The reader can verify this using eqns (5.68)–(5.70).) With the omission of the central freedom, adjustments to the remaining interpolation functions are made using the $L_1 L_2 L_3$ function and the final interpolation functions are obtained as[11]

$$f_1 = L_1 + L_1^2 L_2 + L_1^2 L_3 - L_2^2 L_1 - L_3^2 L_1 \quad \text{(for } w_1)$$

$$f_2 = L_2 + L_2^2 L_3 + L_2^2 L_1 - L_1^2 L_2 - L_3^2 L_2 \quad \text{(for } w_2)$$

$$f_3 = L_3 + L_3^2 L_1 + L_3^2 L_2 - L_1^2 L_3 - L_2^2 L_3 \quad \text{(for } w_3)$$

(8.13)

$$f_4 = L_1^2 L_2 + \tfrac{1}{2} L_1 L_2 L_3, \qquad f_5 = - L_1^2 L_3 - \tfrac{1}{2} L_1 L_2 L_3$$

$$\text{(for } \phi_{a1}, \phi_{c1})$$

$$f_6 = L_2^2 L_3 + \tfrac{1}{2} L_1 L_2 L_3, \qquad f_7 = - L_2^2 L_1 - \tfrac{1}{2} L_1 L_2 L_3$$

$$\text{(for } \phi_{b2}, \phi_{a2})$$

$$f_8 = L_3^2 L_1 + \tfrac{1}{2} L_1 L_2 L_3, \qquad f_8 = - L_3^2 L_2 - \tfrac{1}{2} L_1 L_2 L_3$$

$$\text{(for } \phi_{c3}, \phi_{b3}).$$

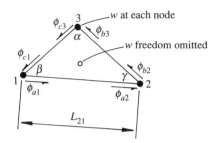

Fig. 8.4 Nine-freedom triangular bending element

Here the last function f_{10} for the centroidal freedom is, of course, omitted. The terms in $L_1 L_2 L_3$ are removed in the functions f_1, f_2, and f_3 and modified in the remaining six functions. This term is sometimes referred to as the 'bubble function' because it describes the shape of a triangular membrane resulting from the application of uniformly distributed moments along the edges.[23] Because the bubble function vanishes everywhere on the element boundary the coefficients of this term in the functions $f_1, \ldots, f_9$ may take any value without affecting the other terms in these functions. It can be shown, however, that the $L_1 L_2 L_3$ term must be multiplied by the factor $1/2$ to ensure the element has a constant twisting curvature mode.[24]

Applying eqns (5.94), (5.95), and (5.96) to these functions the curvature–displacement matrix for the natural curvatures χ_a, χ_b, and χ_c is given as

$$
B^t = \begin{bmatrix}
2(L_3 + b_1) & -4L_1 & 2(L_2 - b_3) \\
2(L_3 - b_1) & 2(L_1 + b_2) & -4L_2 \\
-4L_3 & 2(L_1 - b_2) & 2(L_2 + b_3) \\
b_1 - 1 & -L_1 & L_2 \\
-L_3 & L_1 & b_3 + 1 \\
L_3 & b_2 - 1 & -L_2 \\
b_1 + 1 & -L_1 & L_2 \\
-L_3 & L_1 & b_3 - 1 \\
L_3 & b_2 + 1 & -L_2
\end{bmatrix} \tag{8.14}
$$

where $b_1 = 3(L_2 - L_1)$, $b_2 = 3(L_3 - L_2)$, $b_3 = 3(L_1 - L_3)$ and the columns must be divided by L_{21}^2, L_{32}^2, and L_{13}^2, respectively.

These natural curvatures cannot be used directly to evaluate the strain energy because they 'overlap' (at least two sides will contain a component of χ_x). Thus they are first expressed in terms of *artificial curvature components*

κ_a, κ_b, and κ_c (which are independent and not physically measurable) by the transformation

$$\{\chi_n\} = \begin{Bmatrix} \chi_a \\ \chi_b \\ \chi_c \end{Bmatrix} = \begin{bmatrix} 1 & \cos^2\gamma & \cos^2\beta \\ \cos^2\gamma & 1 & \cos^2\alpha \\ \cos^2\beta & \cos^2\alpha & 1 \end{bmatrix} \begin{Bmatrix} \kappa_a \\ \kappa_b \\ \kappa_c \end{Bmatrix} = W\{\kappa_n\}. \quad (8.15)$$

Now the total natural curvatures are related to the Cartesian curvatures by the transformation

$$\{\chi_n\} = \begin{bmatrix} c_{ax}^2 & c_{ay}^2 & 2c_{ax}c_{ay} \\ c_{bx}^2 & c_{by}^2 & 2c_{bx}c_{by} \\ c_{cx}^2 & c_{cy}^2 & 2c_{cx}c_{cy} \end{bmatrix} \begin{Bmatrix} \chi_x \\ \chi_y \\ \chi_{xy} \end{Bmatrix} = C\{\chi_c\} \quad (8.16)$$

where c_{ax}, c_{ay} are the direction cosines of the first side, that is, $c_{ax} = (x_2 - x_1)/L_{21}$, $c_{ay} = (y_2 - y_1)/L_{21}$ and similarly for c_{bx}, c_{by}, c_{cx} and c_{cy}. Then we have

$$\{\chi_c\} = C^{-1}\{\chi_n\} = C^{-1}W\{\kappa_n\} = G^{-1}\{\kappa_n\}. \quad (8.17)$$

Corresponding to $\{\kappa_n\}$ we also define a set of *artificial moments* $\{m_n\}$ related to the Cartesian moments by

$$\{m_c\} = G^{-1}\{m_n\}. \quad (8.18)$$

Using the artificial moments and curvatures the strain energy per unit area (the strain energy density) can be calculated as

$$2U = \{m_n\}^t\{\kappa_n\} = \{m_n\}^t D_n^{-1}\{m_n\} \quad (8.19)$$

where by definition $\{m_n\} = D_n\{\kappa_n\}$.[25] Now writing the unit strain energy in terms of Cartesian values and using the Cartesian modulus matrix D gives

$$2U = \{m_c\}^t\{\chi_c\} = \{m_c\}^t D^{-1}\{m_c\}$$
$$= (G^{-1}\{m_n\})^t D^{-1}(G^{-1}\{m_n\}). \quad (8.20)$$

Comparing eqns (8.19) and (8.20), also noting that $W = W^t$, one obtains Argyris' natural modulus matrix D_n as[25]

$$D_n = GDG^t = W^{-1}CDC^tW^{-1}. \quad (8.21)$$

The effect of this result is that the matrix W is inverted,[25] rather than matrix C which may be singular for some element geometries. Now the element stiffness is given by three-point numerical integration (at the middle of each side) as

$$k = T^t(\Sigma(\tfrac{1}{3}\Delta)B^t D_n B)T. \quad (8.22)$$

A more direct approach is to form a transformation matrix $H^\dagger$ linking $\{\chi_c\}$ directly to $\{\chi_n\}$, that is,

$$H = G^t = C^t W^{-1} \qquad (8.23)$$

in the same fashion as the Jacobian matrix is used in isoparametric elements. The element stiffness and stress matrices are given as

$$k = T^t(\Sigma(\tfrac{1}{3}\Delta)S^t DS)\,T, \qquad N = DST$$

where $S = HB$ and N here yields the Cartesian moments directly.

An alternative transformation is simply $H = C^{-1}$ *which follows directly from eqn* (8.16) *and involves less computation.* That $H = C^t W^{-1}$ and $H = C^{-1}$ are equivalent can be shown by replacing the entries $2c_{ax}c_{ay}$ in eqn (8.16) by $\sqrt{2}c_{ax}c_{ay}$ on the understanding that D_{xy} in the modulus matrix will be doubled to retain the correct energy calculation (see Section 2.6). Then we require

$$C^{-1} = C^t W^{-1} \qquad \text{or} \qquad W = CC^t \qquad (8.24)$$

and the $\sqrt{2}$ substitution is necessary to obtain the correct result for CC^t. Observing the three relations of the form

$$\gamma = \cos^{-1}(c_{ax}) - \cos^{-1}(c_{bx})$$

eqn (8.24) is easily proved. With the use of natural strains or curvatures in eqn (8.23), in fact, the derivation of such triangular elements is more convenient than that of quadrilateral elements. Of particular note also is the use of the natural approach to triangular thick plate elements and this is discussed in Section 9.3.

8.5 Conforming cubic triangular elements

To obtain conforming cubic triangular elements three alternative approaches are:

(1) the combination of three 10 df triangular elements, each with one midside normal slope freedom, followed by static condensation to remove the internal freedoms;[18]

(2) the use of Lagrange multiplier constraints at the middle of each side;[26,27]

$\dagger$ This matrix could be called a *Hessian matrix*, that is, a matrix of second derivatives that are here evaluated as the squares of direction cosines (which are first derivatives); cf. the connecting matrix for first derivatives which is the Jacobian matrix.

(3) addition of normal slope freedoms at the middle of each side using the *singularity functions* described by Zienkiewicz and Taylor.[11] These slopes can then be suppressed in the manner described below (eqn (8.25)).

Clough and Tocher's twelve-freedom 'high compatibility triangle'[18] (HCT) element is derived using the first approach and has been widely used. The element is reduced to nine freedoms by restricting the normal slope on each side to a linear variation. Referring to Fig. 5.16 one writes for node 4

$$\left(\frac{\partial w}{\partial n}\right)_4 = \tfrac{1}{2}\left\{\left(\frac{\partial w}{\partial n}\right)_1 + \left(\frac{\partial w}{\partial n}\right)_2\right\}. \tag{8.25}$$

As this implies constant moments normal to the edge this restriction reduces the rate of convergence to h^2 (as shown in Section 3.9). The same rate is obtained with the non-conforming element of Section 8.4, which is more economical because three 'subelements' are not required.

Applied to shell problems the second approach has been found to yield excellent displacement solutions.[27] The Lagrange multiplier constraints increase the number of freedoms required in the solution and this is sometimes cited as a disadvantage. As shown in Chapter 23, however, such formulations can be condensed to obtain equivalent penalty factor formulations with fewer freedoms.

In the third procedure the singularity functions needed are quite complicated. For side 12, for example, one uses the function

$$f_{10} = \frac{L_1^2 L_2^2 L_3(1 + L_3)}{(L_2 + L_3)(L_1 + L_3)}$$

because the use of simple quartic modes such as $L_1^2 L_2^2$ added to the cubic modes leads to a singular interpolation matrix A^{-1}. Calculation of the appropriate entries in matrix B for such functions is a tedious exercise sometimes undertaken numerically.

Another disadvantage of the singularity function approach is that the singularity functions give singular second derivatives (and hence curvatures) at the vertices so that a high-order numerical integration is needed to give convergence.[20] The overcome this difficulty Irons and Razzaque[28] introduce 'substitute' shape functions (for example, $f_{10}^* = L_3(2L_3 - 1)(L_3 - 1)/6$ for side 12) and the least squares method is used to give a best fit of the linear polynomial given by the second derivatives of the substitute functions to the original singularity functions. Thus for the derivative $\partial^2 f_{10}/\partial L_1^2$, for example, one evaluates the expression for this at each integration point, giving

$$\frac{\partial^2 f_{10}}{\partial L_1^2} = F_i = a_1 + a_2 L_1^i + a_3 L_2^i \tag{8.26}$$

where L_3 is omitted as $L_3 = 1 - L_1 - L_2$. Now the unknown coefficients a_1, a_2, and a_3 are determined by the three conditions

$$\frac{\partial \left(\sum_{i=1}^{n} \{F_i - a_1 - a_2 L_1^i - a_3 L_2^i\}^2 \right)}{\partial a_j} = 0, \qquad j = 1, 2, 3 \qquad (8.27)$$

where n is the number of integration points. The fit is carried out numerically and Razzaque gives the coding necessary to obtain the element stiffness matrix.[20] The element, although now non-conforming, gives better results than the original conforming version.

Once again this latter element can be reduced to nine freedoms by using equations of the form of eqn (8.25) and Razzaque shows that the nine-freedom version gives significantly better nodal moment solutions than the element of Section 8.4. With the use of average nodal moments, though, the simpler element detailed in Section 8.4 gives comparable results in most cases[11] and although it does not pass the patch test the error in its displacement solutions is only of the order of 2 per cent, even in highly irregular meshes. Thus it is still widely used and provides an excellent introduction to the powerful natural curvature approach in the present text.

Clearly, cubic triangular displacement elements for the bending of thin plates have received much attention and accurate formulations involve considerable complications. Alternative nine- and twelve-freedom bending elements have been derived by the hybrid methods,[29,30,31] which are summarized in Chapter 3. As remarked by Zienkiewicz and Taylor[11] and Razzaque[20] the hybrid elements derived by Allwood and Cornes[30] (nine freedoms) and Allman[31] (twelve freedoms) are equivalent to the Irons–Razzaque element.

The advantages of the displacement formulation, though, are its greater simplicity, the ease with which all types of loadings are dealt with (not always so with hybrids), and its ready generalization to deal with other areas of application such as the analysis of viscous fluid flows.

Another approach to the nine-freedom bending triangle developed by Bergan and Hanssen[32] obtains the element stiffness matrix in three steps. First the rigid body displacement mode is written as

$$w_{rb} = c_1 + c_2 x + c_3 y \qquad (8.28)$$

and the solution for c_1, c_2, and c_3 is given by eqn (6.3). Now one writes the straining components of the slopes ϕ'_x and ϕ'_y as

$$\phi'_x = \phi_x - c_2, \qquad \phi'_y = \phi_y - c_3 \qquad (8.29)$$

where ϕ_x and ϕ_y are the global values. Applying eqns (8.29) at each vertex and expressing the results in matrix form one obtains[33]

$$
\begin{bmatrix} \phi'_{x1} \\ \phi'_{y1} \\ \phi'_{x2} \\ \phi'_{y2} \\ \phi'_{x3} \\ \phi'_{y3} \end{bmatrix} = \frac{1}{2\Delta}
\begin{bmatrix}
-y_{32} & 2\Delta & 0 & -y_{13} & 0 & 0 & -y_{21} & 0 & 0 \\
x_{32} & 0 & 2\Delta & x_{13} & 0 & 0 & x_{21} & 0 & 0 \\
-y_{32} & 0 & 0 & -y_{13} & 2\Delta & 0 & -y_{21} & 0 & 0 \\
x_{32} & 0 & 0 & x_{13} & 0 & 2\Delta & x_{21} & 0 & 0 \\
-y_{32} & 0 & 0 & -y_{13} & 0 & 0 & -y_{21} & 2\Delta & 0 \\
x_{32} & 0 & 0 & x_{13} & 0 & 0 & x_{21} & 0 & 2\Delta
\end{bmatrix}
\begin{bmatrix} w_1 \\ \phi_{x1} \\ \phi_{y1} \\ w_2 \\ \phi_{x2} \\ \phi_{y2} \\ w_3 \\ \phi_{x3} \\ \phi_{y3} \end{bmatrix}
$$

or

$$\{d'\} = T\{d\} \tag{8.30}$$

This transformation can be viewed as another example of basis transformation in which a modified interpolation is now able to be used. In this case this is the 'truncated' interpolation polynomial for the constant curvature modes which is

$$w_c = c_4 x^2 + c_5 xy + c_6 y^2. \tag{8.31}$$

The second derivatives of these modes are used to define boundary moments on the element sides which are transformed to the Cartesian axes and lumped at the nodes giving a vector of six nodal moments $\{M^*\}$. Then the constant straining mode contributions to a 6×6 kernel stiffness matrix k^* are determined by comparing $k^*\{d'\}$ with $\{M^*\}$.

Finally the cubic modes are introduced:

$$w = c_7 x^3 + c_8 x^2 y + c_9 xy^2 + c_{10} y^3 \tag{8.32}$$

and a patch test criterion is used to determine their contribution to the kernel stiffness matrix k^*. Applying the rigid body transformation of eqn (8.30) to k^* the final element stiffness matrix is given by

$$k = T^t k^* T. \tag{8.33}$$

The element is found to give superior results to the non-conforming model of Bazeley *et al.* (eqn. (8.14)) and the hybrid model of Allman.[31] As might be expected, the element gives slightly superior results to the smoothed derivative model of Irons and Razzaque as all the linear strain terms, not just the three singularity function terms are 'optimized' by the use of a patch test.

A similar element to the Bergan and Hanssen element, but now incorporating the natural strain technique, is derived by Argyris *et al.*[34] and both these elements, which are strictly hybrid elements, give excellent results, as shown in Figs 8.11–8.14. More recently a much more economical element has been obtained by basis transformation and this is detailed in Section 8.10.

8.6 Convergence of some basic thin plate elements

Figure 8.5 shows the boundary conditions required by basic thin plate elements (that is, those directly formulated in terms of vertex freedoms w, ϕ_x, and ϕ_y) for the analysis of a quadrant of a clamped square plate carrying a uniformly distributed load, and Table 8.1 compares the results obtained using three such elements. These are the ACM rectangle detailed in Section 8.1, the twelve-freedom bicubic rectangle given by eqns (5.45), (5.47), and (8.11), and the nine-freedom triangle detailed in Section 8.4.

The non-conforming ACM element converges rather slowly, but we may assume that it does so correctly, as it has been shown to pass the patch test.[2, 35] The bicubic element shows much better performance and its results are here of comparable accuracy to those obtained with the full sixteen-freedom bicubic element (that is, without incorporation of eqn (8.11)).[36]

The nine-freedom non-conforming triangular element is not as accurate as the bicubic element but the results are of acceptable accuracy for most practical purposes and it is still widely used, for example, coupled with the constant strain triangle of Section 6.1 to form a flat shell element. The second set of results given for the triangular element is obtained using centroidal moment solutions and averaging these between elements sharing a node. As might be expected the result for m_E is comparatively poor and in general average nodal values are recommended.

Then finally, for the economical analysis of plate bending, the non-conforming nine-freedom triangle of Bazeley *et al.* (eqn (8.14)) and the bicubic rectangle with twelve or sixteen freedoms are both straightforward displacement elements in which explicit interpolations for the whole element are available. These interpolations are not so complicated as to discourage applications to problems involving geometric non-linearity, for example, and both elements have been used extensively for this purpose.[11, 12]

Clearly, accurate triangular or quadrilateral thin plate elements become quite complex and the elements of Irons and Razzaque, Bergan and Hanssen, and Argyris *et al.*, for example, may be difficult to apply to large displacement

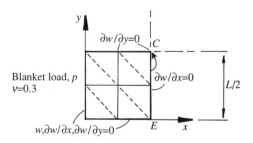

Fig. 8.5 Quadrant of a clamped plate with nine- and twelve-freedom plate elements

Table 8.1 Finite element solutions with nine-freedom triangular and twelve-freedom rectangular elements for the analysis of a quadrant of a clamped square slab with blanket load.

Mesh	$w_C D/pL^4$	m_C/pL^2	$-m_E/pL^2$
ACM rectangle[33], $v = 0.3$			
9 nodes	0.00148	0.0462	0.0355
25 nodes	0.00140	0.0278	0.0476
81 nodes	0.00130	0.0240	0.0503
12 freedom bicubic, $v = 0.3$			
9 nodes	0.00126	0.0235	0.0430
25 nodes	0.00127	0.0230	0.0487
81 nodes	0.00126	0.0226	0.0504
Triangle (eqn (8.14), $v = 0.3$			
9 nodes	0.00156	0.0186	0.0548
25 nodes	0.00135	0.0231	0.0553
81 nodes	0.00129	0.0239	0.0545
Triangle (eqn (8.14), $v = 0.2$[†]			
9 nodes	0.00156	0.0172	0.0211
25 nodes	0.00135	0.0213	0.0328
81 nodes	0.00129	0.0221	0.0408
Exact, $v = 0.3$	0.00127	0.0229	0.0514
Exact, $v = 0.2$	0.00127	0.0211	0.0514

† In this case the moment solutions are averaged centroidal moments

and shell problems. Alternatives are the higher-order elements discussed in the following sections, the nested interpolation technique of Section 8.10, or the thick plate elements of Chapter 9 with penalty factor constraints.

8.7 Constant moment element with six freedoms

Figure 8.6 shows a constant moment element with six freedoms and hence a quadratic displacement function. The element uses a complete quadratic interpolation and, although it is less accurate than some cubic elements, it is of special interest because it has been derived by the three main types of variational method outlined in Section 3.11, namely:

(1) as a hybrid element;[37]

(2) as an equilibrium element;[38]

(3) as a non-conforming displacement element.[39]

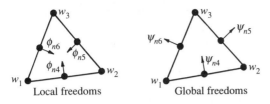

Fig. 8.6 Constant moment bending element

We shall here continue to follow the displacement method and although a simple explicit Cartesian solution is given by Morley,[39] we shall again use the natural strain approach introduced in Section 8.4 to provide an introduction to the higher-order elements detailed in Sections 8.8 and 8.9.

The main difficulty in formulating the element is the introduction of the normal slope freedoms at the middle of each side which are needed to allow the imposition of symmetry or clamped edge boundary conditions. The normal slopes are transformed to local dimensionless derivatives (that is, having the same dimension as w) using the signed multipliers m_a, m_b, and m_c of eqns (5.91).

The interpolation functions are established by applying eqns (5.77) and (5.79) to the quadratic area coordinate modes (eqn (5.87)) and are given in eqns (5.89). Applying eqns (5.94), (5.95), and (5.96) to the interpolation functions (for example eqn (5.94) is $\chi_a = \partial^2 w/\partial L_1^2 + \partial^2 w/\partial L_2^2 - 2\partial^2 w/\partial L_1 \partial L_2$ and the expressions for χ_b and χ_c follow by cyclic progression) the interpolation matrix for the natural curvatures is given by

$$B = \begin{bmatrix} 1 + e_c & 1 - e_b & -2 + e_b - e_c & 0 & -m_b & -m_c \\ -2 + e_c - e_a & 1 + e_a & 1 - e_c & -m_a & 0 & -m_c \\ 1 - e_a & -2 + e_a - e_b & 1 + e_b & -m_a & -m_b & 0 \end{bmatrix}$$

$$(8.34)$$

where the eccentricities e_a, e_b, and e_c are given by eqns (5.77) and (5.80). Noting that the actual curvature on side 12 is given by $\partial^2 w/\partial s_a^2 = \chi_a/L_{21}^2$ (and similarly for the other sides) the Cartesian curvatures are given by

$$S = H \begin{bmatrix} 1/L_{21}^2 & 0 & 0 \\ 0 & 1/L_{32}^2 & 0 \\ 0 & 0 & 1/L_{13}^2 \end{bmatrix} B \qquad (8.35)$$

where the matrix H is given by eqn (8.23), that is, $H = C^t W^{-1}$ or alternatively $H = C^{-1}$. Finally the element stiffness and stress matrices are given by

$$k = S^t DS(\Delta) \qquad (8.36)$$

$$N = DS \qquad (8.37)$$

that is, as the element has a constant moment character integration is not required in forming k. Matrix N gives the Cartesian unit moments acting throughout the element and these may be assigned to the element centroid or used to calculate average nodal values.

The quadratic modes could have been taken as either

$$\{L_1^2, L_2^2, L_3^3, L_1 L_2, L_2 L_3, L_3 L_1\} \tag{8.38}$$

or

$$\{L_1, L_2, L_3, L_1 L_2, L_2 L_3, L_3 L_1\} \tag{8.39}$$

both leading to the same result for the matrix B. The terms e_a, e_b, e_c (see eqns (5.77) and (5.80)) and the matrix H involve the vertex angles α, β, γ and these are calculated by standard trigonometrical formulae such as eqns (5.82) and (5.83).

The consistent loads for a blanket load of intensity p are given by

$$\{q_t\} = p \int \{f\} \, dA = p A^t \int \{M\} \, dA. \tag{8.40}$$

Using the area coordinate integration formula of eqn (6.32) to evaluate eqn (8.40) one obtains

$$\{q_t\} = p \Delta A^t \{1/6, 1/6, 1/6, 1/12, 1/12, 1/12\} \tag{8.41}$$

giving

$$\{q_t\} = \frac{p\Delta}{12} \{4 + e_a - e_c, 4 + e_b - e_a, 4 + e_c - e_b, m_a, m_b, m_c\} \tag{8.42}$$

again including the terms m_a, m_b, and m_c for the normal slopes.

Because of its simplicity the element is attractive but unfortunately it converges too slowly in some cases (see Table 8.2) and, in fact, a constant moment element developed by Mohr[40,41] with nine freedoms gives better results. This element is developed by using eqn (8.31) to include the straining portion of the w freedoms and a rigid body transformation, which performs a similar function to that in eqn (8.30), is used to exclude the rigid body actions. The slope freedoms are then used to define a second constant strain field and the two fields superposed. A disadvantage is that the weighting factors for this superposition are determined empirically but formulation of the element is otherwise simpler than for Morley's six-freedom element.

The nine-freedom constant moment element is marginally less accurate than the Bazeley *et al.* element in most applications (and hence details are not given here) but it does help demonstrate that the slow convergence of Morley's element is due to the inability of the element to model boundary conditions accurately and not just to the constant moment character of the element.[41]

8.8 Quartic element with fifteen freedoms

Figure 8.7 shows a non-conforming quartic thin plate bending element with fifteen freedoms, including normal slopes at the midside nodes. The transformation from global to local freedoms is given by the matrix T of eqn (5.90). The quartic modes are, with the omission of the rigid body modes,

$$\{M\} = \{L_1^3 L_2, L_1^3 L_3, L_2^3 L_3, L_2^3 L_1, L_3^3 L_1, L_3^3 L_2, L_1^2 L_2^2, L_2^2 L_3^3, \ldots ,$$

$$L_3^2 L_1^2, L_1^2 L_2 L_3, L_1 L_2^2 L_3, L_1 L_2 L_3^2\} \tag{8.43}$$

Applying eqns (5.68), (5.69), (5.70), (5.77), and (5.79) to these the truncated interpolation matrix A_s is given by eqn (5.92) (here truncation refers to the omission of the first three rigid body rows of the A matrix). Now applying eqns (5.94), (5.95), and (5.96) directly to these modes one obtains

$$
B^t =
\begin{bmatrix}
6L_1 L_2 - 6L_1^2 & 0 & 6L_1 L_2 \\
6L_1 L_3 & 0 & 6L_1 L_3 - 6L_1^2 \\
6L_2 L_3 & 6L_2 L_3 - 6L_2^2 & 0 \\
6L_2 L_1 - 6L_2^2 & 6L_2 L_1 & 0 \\
0 & 6L_3 L_1 & 6L_3 L_1 - 6L_3^2 \\
0 & 6L_3 L_2 - 6L_3^2 & 6L_3 L_2 \\
2L_1^2 + 2L_2^2 - 8L_1 L_2 & 2L_1^2 & 2L_2^2 \\
2L_3^2 & 2L_3^2 + 2L_2^2 - 8L_2 L_3 & 2L_2^2 \\
2L_3^2 & 2L_1^2 & 2L_3^2 + 2L_1^2 - 8L_3 L_1 \\
2L_2 L_3 - 4L_1 L_3 & -2L_1^2 & 2L_2 L_3 - 4L_1 L_2 \\
2L_1 L_3 - 4L_2 L_3 & 2L_1 L_3 - 4L_1 L_2 & -2L_2^2 \\
-2L_3^2 & 2L_1 L_2 - 4L_1 L_3 & 2L_1 L_2 - 4L_2 L_3
\end{bmatrix}.
\tag{8.44}
$$

Here the first three columns of matrix B, corresponding to the rigid body modes, are omitted. Now transforming to Cartesian curvatures using eqn (8.35) one obtains

$$
S = H
\begin{bmatrix}
1/L_{21}^2 & 0 & 0 \\
0 & 1/L_{32}^2 & 0 \\
0 & 0 & 1/L_{13}^2
\end{bmatrix}
B.
\tag{8.45}
$$

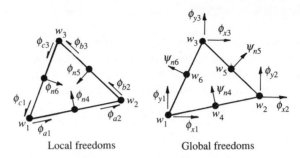

Local freedoms   Global freedoms

Fig. 8.7 Quartic bending element

Finally the element stiffness and force matrices are given as

$$k = T^t A_s^t (\Sigma S^t D(\omega_i \Delta) S) A_s T \qquad (8.46)$$

$$F = D S A_s T \qquad (8.47)$$

using the seven-point integration given in Table 5.3.

The consistent loads for a blanket load of intensity p are given by

$$\{q_t\} = p(A_s T)^t \iint \{M\} \, dx \, dy \qquad (8.48)$$

$$= T^t A_s^t \left(\frac{p\Delta}{180}\right) \{3, 3, 3, 3, 3, 3, 2, 2, 2, \ 1, 1, 1\} \qquad (8.49)$$

again using eqn (6.32) to evaluate this simple integration explicitly. In addition, loads $p\Delta/3$ must be added for the freedoms w_1, w_2, and w_3.

The element is non-conforming and this is not easily remedied. One remedy, for example, is to constrain w to a cubic variation on each side, eliminating the freedoms w_4, w_5, and w_6 using constraints of the form

$$w_4 = \frac{w_1 + w_2}{2} + \frac{\phi_{n1} - \phi_{n2}}{8}. \qquad (8.50)$$

The resulting element has the same freedoms as the conforming triangles derived by Irons and Razzaque[20,28] and Clough and Tocher.[18] Unfortunately, however, the element is overstiff when thus constrained and, to compound matters, has an even less convenient set of freedoms because the inconsistency in the number of degrees of freedom at the nodes increases.

An alternative means of enforcing conformity is to add three additional freedoms at the midside nodes. This is by no means easy: inclusion of the slope parallel to each side, for example, leads to a singular interpolation matrix. Another possibility is to include the cross derivative $\partial^2 w/\partial s \partial n$ at the midside nodes, adding three quintic modes

$$L_1^3 L_2 L_3, \quad L_1 L_2^3 L_3, \quad L_1 L_2 L_3^3$$

to eqn (8.43), but this still gives a singular A^{-1} matrix. This can be overcome by using the twist functions developed by Irons,[42] although these are relatively complicated and, as noted by Razzaque,[20] require a high-order numerical integration. The same procedure has, however, also been applied in a derivation in oblique coordinates.[43]

Another means of enforcing C_1 continuity is to follow the procedure used in the TRIB elements of Argyris *et al.*[22] and use two normal slopes at the third points of each side, treating these variables in the solution stage as though they acted at the midside nodes. Such an element would be similar in some respects to the SemiLoof elements (see Section 9.7) which allow normal slope freedoms at the Gauss points on the sides.

To obtain continuity a simpler course seems to be to use a quintic element with eighteen or twenty-one freedoms. Such an element is derived by Bell[44] using Cartesian polynomials and numerical inversion to obtain the interpolation matrix (for such purposes inversion routines with pivoting are required). The element has also been derived by Argyris *et al.*,[22] Butlin and Ford,[45] and Cowper *et al.*[46] The quintic element uses three curvature freedoms at the vertices, constraining merging elements to the same value. Often referred to as 'overcontinuity', it presents, in fact, no disadvantages[22] and the curvature boundary conditions usually required cause no more inconvenience than those for the sixteen-freedom bicubic rectangle. The interpolation matrix we use for this element was derived by Argyris, Fried, and Scharpf[22] and is given in Section 5.10. The reader may observe that it is no more complicated than that for the quartic element. Alternative explicit interpolation functions have been obtained for the element by Cowper *et al.*[47] using oblique coordinates and by Mitchell and Wait[2] using isoparametric mapping.

8.9 Quintic element with eighteen freedoms

Figure 8.8 shows a conforming quintic element with twenty-one freedoms. The transformation from global to local freedoms is given by eqn (5.107), in which the normal slopes are eliminated by constraining the normal derivative of w to a cubic variation. This does not reduce the accuracy of the element to any significant extent[44] but makes it a great deal more convenient to use, as the normal slope freedoms would cause much more inconvenience than the curvature freedoms in most applications.

Omitting the rigid body modes the modal vector is

$$\{M\} = \{L_1^4 L_2, L_1^4 L_3, L_2^4 L_3, L_2^4 L_1, L_3^4 L_1, L_3^4 L_2, L_1^3 L_2^2, L_1^3 L_3^2, L_2^3 L_3^2,$$

$$L_2^3 L_1^2, L_3^3 L_1^2, L_3^3 L_2^2, L_1^3 L_2 L_3, L_1 L_2^3 L_3, L_1 L_2 L_3^3, L_1^2 L_2^2 L_3, L_1 L_2^2 L_3^3,$$

$$L_1^2 L_2 L_3^2\} \tag{8.51}$$

and applying eqns (5.68), (5.69), (5.70), (5.77), (5.79), (5.94), (5.95) and (5.96) the truncated matrix A_s is given by eqn (5.108). Now applying eqns (5.94), (5.95), and (5.96) directly to the modal vector of eqn (8.51) the natural curvature components are expressed as dimensionless derivatives in the matrix

$$B = \begin{bmatrix} Z_1 + Z_2 - 2Z_4 \\ Z_2 + Z_3 - 2Z_5 \\ Z_3 + Z_1 - 2Z_6 \end{bmatrix} \quad (3 \times 18) \qquad (8.52)$$

where Z_i denotes row i of a matrix Z. Defining

$$z_1 = L_1^3, \qquad z_2 \quad = L_2^3, \qquad z_3 \quad = L_3^3$$

$$z_4 = L_1^2 L_2, \qquad z_5 = L_2^2 L_1, \qquad z_6 = L_2^2 L_3$$

$$z_7 = L_3^2 L_2, \qquad z_8 = L_1 L_3^2, \qquad z_9 = L_1^2 L_3$$

$$z_{10} = L_1 L_2 L_3$$

the matrix Z is given as

$$Z^t = \begin{bmatrix}
12z_4 & 0 & 0 & 4z_1 & 0 & 0 \\
12z_9 & 0 & 0 & 0 & 0 & 4z_1 \\
0 & 12z_6 & 0 & 0 & 4z_2 & 0 \\
0 & 12z_5 & 0 & 4z_2 & 0 & 0 \\
0 & 0 & 12z_8 & 0 & 0 & 4z_3 \\
0 & 0 & 12z_7 & 0 & 4z_3 & 0 \\
6z_5 & 2z_1 & 0 & 6z_4 & 0 & 0 \\
6z_8 & 0 & 2z_1 & 0 & 0 & 6z_9 \\
0 & 6z_7 & 2z_2 & 0 & 6z_6 & 0 \\
2z_2 & 6z_4 & 0 & 6z_5 & 0 & 0 \\
2z_3 & 0 & 6z_9 & 0 & 0 & 6z_8 \\
0 & 2z_3 & 6z_6 & 0 & 6z_7 & 0 \\
6z_{10} & 0 & 0 & 3z_9 & z_1 & 3z_4 \\
0 & 6z_{10} & 0 & 3z_6 & 3z_5 & z_2 \\
0 & 0 & 6z_{10} & z_3 & 3z_8 & 3z_7 \\
2z_6 & 2z_9 & 0 & 4z_{10} & 2z_4 & 2z_5 \\
0 & 2z_8 & 2z_5 & 2z_7 & 4z_{10} & 2z_6 \\
2z_7 & 0 & 2z_4 & 2z_8 & 2z_9 & 4z_{10}
\end{bmatrix}$$

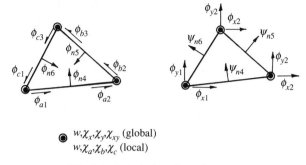

$w, \chi_x, \chi_y, \chi_{xy}$ (global)
$w, \chi_a, \chi_b, \chi_c$ (local)

Fig. 8.8 Quintic bending element

and finally the element stiffness matrix is given by eqns (8.45) and (8.46), now using a sixteen-point Gaussian quadrature. Using eqns (6.32) and (8.48) the consistent loads for blanket loading of intensity p are given by

$$\{q\} = T^t A_s^! \left(\frac{p\Delta}{630} \right) \{6, 6, 6, 6, 6, 6, 3, 3, 3, 3, 3, 3, 1.5, 1.5, 1.5, 1, 1, 1\} \qquad (8.53)$$

to which must be added values of $p\Delta/3$ for the freedoms w_1, w_2, and w_3. Without the use of consistent loads, in fact, the element does not perform as well as the quartic element.

Figure 8.9 shows the boundary conditions required for the analysis of a quadrant of a simply supported square plate. Here, along the side AB, for example, we have $\phi_x = 0$ at every node, that is

$$\frac{\partial \phi_x}{\partial x} = \chi_x = 0,$$

whereas along the side BC

$$\frac{\partial \phi_x}{\partial y} = \tfrac{1}{2} \chi_{xy} = 0$$

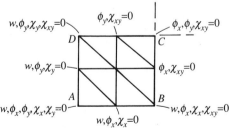

Fig. 8.9 Boundary conditions for quadrant of a simply supported plate with 18 df quintic elements

so that at point B one should include all boundary conditions appropriate to the intersecting edges AB and BC, that is, w, ϕ_x, $\chi_x = 0$ from side AB and $\chi_{xy} = 0$ from side BC.

With the quintic element one obtains C_1 continuity and uniquely defined curvatures for elements merging at a node. If the elements differ in thickness this causes moment discontinuities but this difficulty can be overcome by substituting moment freedoms m_x, m_y, m_{xy} for the curvature freedoms.[22]

8.10 Nine-freedom element using nested interpolations

Figure 8.10 shows a nine-freedom thin plate element with freedoms w, ϕ_x, ϕ_y at each vertice. We shall apply the method of nested interpolations to transform the global freedoms to local freedoms ϕ_x, ϕ_y at six points and, for these, complete quadratic interpolation can be employed.[48] To achieve this transformation we use a cubic interpolation to define the parallel slopes ϕ_{a4}, ϕ_{b5}, ϕ_{c6} on each side and linear interpolation to define the normal slopes ϕ_{na4}, ϕ_{nb5}, ϕ_{nc6}. Thus, using the first derivatives of eqns (5.24) and substituting $d = 0.5$, one obtains at node 4,

$$\phi_{a4} = -\frac{1.5w_1}{L_{21}} - \frac{\phi_{a1}}{4} + \frac{1.5w_2}{L_{21}} - \frac{\phi_{a2}}{4} \tag{8.54}$$

and

$$\phi_{na4} = \frac{\phi_{na1} + \phi_{na2}}{2} \tag{8.55}$$

using linear interpolation for the normal slope. Converting to Cartesian freedoms

$$\phi_{x4} = c_{ax}\phi_{a4} - c_{ay}\phi_{na4}$$

$$= A_a(w_2 - w_1) + B_a\phi_{x1} + C_a\phi_{y1} + B_a\phi_{x2} + C_a\phi_{y2} \tag{8.56}$$

$$\phi_{y4} = D_a(w_2 - w_1) + C_a\phi_{x1} + E_a\phi_{y1} + C_a\phi_{x2} + E_a\phi_{y2} \tag{8.57}$$

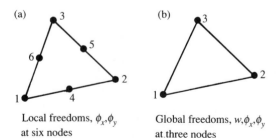

(a)

Local freedoms, ϕ_x, ϕ_y
at six nodes

(b)

Global freedoms, w, ϕ_x, ϕ_y
at three nodes

Fig. 8.10 Application of the method of nested interpolations to a triangular plate bending element

where

$$A_a = \frac{1.5c_{ax}}{L_{21}}, \qquad D_a = \frac{1.5c_{ay}}{L_{21}}, \qquad B_a = \frac{2c_{ay}^2 - c_{ax}^2}{4}$$

$$E_a = \frac{2c_{ax}^2 - c_{ay}^2}{4}, \qquad C_a = -\frac{3c_{ax}c_{ay}}{4}. \tag{8.58}$$

Repeating on the other two sides one obtains the complete transformation

$$\begin{bmatrix} \phi_{x4} \\ \phi_{y4} \\ \phi_{x5} \\ \phi_{y5} \\ \phi_{x6} \\ \phi_{y6} \end{bmatrix} = \begin{bmatrix} -A_a & B_a & C_a & A_a & B_a & C_a & 0 & 0 & 0 \\ -D_a & C_a & E_a & D_a & C_a & E_a & 0 & 0 & 0 \\ 0 & 0 & 0 & -A_b & B_b & C_b & A_b & B_b & C_b \\ 0 & 0 & 0 & -D_b & C_b & E_b & D_b & C_b & E_b \\ A_c & B_c & C_c & 0 & 0 & 0 & -A_c & B_c & C_c \\ D_c & C_c & E_c & 0 & 0 & 0 & -D_c & C_c & E_c \end{bmatrix} \{d\} = T_m\{d\}$$

$$\tag{8.59}$$

where A_b, B_b and so on are given by progression of subscripts in eqns (8.58). Thus the twelve local freedoms are obtained as

$$\begin{Bmatrix} \phi_{xi} \\ \phi_{yi} \end{Bmatrix} = \begin{bmatrix} T_c(6 \times 9) \\ T_m(6 \times 9) \end{bmatrix} \{d\} = T\{d\} \qquad i = 1 \to 6 \tag{8.60}$$

where T_c is a Boolean matrix with unit entries in the columns corresponding to the vertex slope freedoms. Defining the curvatures as

$$\chi_x = \frac{\partial \phi_x}{\partial x}, \qquad \chi_y = \frac{\partial \phi_y}{\partial y}, \qquad \chi_{xy} = \frac{\partial \phi_y}{\partial x} + \frac{\partial \phi_x}{\partial y} \tag{8.61}$$

the element stiffness matrix is given by

$$k = T^t k^* T \tag{8.62}$$

where the 12×12 stiffness matrix k^* is exactly that for the linear strain triangle (Section 7.3) except that $t^3/12$ is substituted for t in the plane stress modulus matrix.

An apparent deficiency is that eqns (8.61) do not satisfy the Kirchhoff constraints $\phi_x - \partial w/\partial x = 0$, $\phi_y - \partial w/\partial y = 0$ but as the assumptions $\phi_x = \partial w/\partial x$ and $\phi_y = \partial w/\partial y$ are implicit on each side in eqn (8.59) the element yields satisfactory results.

The same procedure may be used to derive a quadrilateral element with twelve freedoms. The k^* matrix for this is the stiffness matrix of the biquadratic membrane element (Section 7.2). The element performs reasonably well compared to other quadrilateral elements but is less accurate than the triangular element detailed above. (In fact it performs better than the triangle for distributed loads but overall is less accurate.[17])

8.11 Convergence of refined thin plate elements

Tables 8.2, 8.3, and 8.4 show the results obtained for the constant moment, quartic, and quintic elements for simply and clamped supported plates with blanket loading. The quadratic element has difficulty in modelling the clamped edge and here converges slowly. Morley[39] gives the displacement

Table 8.2 Solutions for blanket loaded square slabs using a six-freedom element. One quadrant analysed, $v = 0.3$.

Freedoms	$w_c D/pL^4$	m_c/pL^2	m_e/pL^2
Simply supported			
25	0.005367	0.04468	–
49	0.004650	0.04665	–
81	0.004394	0.04727	–
Exact	0.004062	0.04789	–
Clamped			
25	0.002921	0.02220	– 0.03507
49	0.002036	0.02306	– 0.04281
81	0.001708	0.02314	– 0.04644
Exact	0.001265	0.02290	– 0.05135

Table 8.3 Solutions for blanket loaded square slabs using a quartic element. One quadrant analysed, $v = 0.3$. Solutions obtained without the use of consistent loads are shown in parentheses.

Freedoms	$w_c D/pL^4$	m_c/pL^2	m_e/pL^2
Simply supported			
27	0.004182 (3395)	0.05824 (5059)	–
75	0.004077 (3872)	0.05047 (4808)	–
147	0.004068 (3976)	0.04900 (4782)	–
243	0.004065 (4013)	0.04850 (4779)	–
Exact	0.004062	0.04789	
Clamped			
27	0.001472 (1317)	0.03176 (3031)	– 0.04436 (3850)
75	0.001279 (1242)	0.02516 (2429)	– 0.05005 (4816)
147	0.001273 (1256)	0.02387 (2336)	– 0.04998 (4915)
243	0.001270 (1261)	0.02343 (2310)	– 0.05009 (4961)
Exact	0.001265	0.02290	– 0.05135

Table 8.4 Solutions for blanket loaded square slabs using a quintic element. One quadrant analysed, $v = 0.3$.

Freedoms	$w_c D/pL^4$	m_c/pL^2	m_e/pL^2
Simply supported			
24	0.004068	0.04815	–
54	0.004063	0.04791	–
96	0.004062	0.04789	–
150	0.004062	0.04789	–
Exact	0.004062	0.04789	–
Clamped			
24	0.001149	0.02257	– 0.03926 (4968)
54	0.001264	0.02295	– 0.04964 (5094)
96	0.001265	0.02291	– 0.05102 (5141)
150	0.001265	0.02291	– 0.05123 (5139)
Exact	0.001265	0.02290	– 0.05135

solution obtained with 289 nodes as 0.00138 and, using h^2 extrapolation, the element does appear to be converging correctly.

The quartic element converges rapidly and is more accurate than the cubic elements used in Table 8.1. The figures in parentheses are obtained using lumped loads of $p\Delta/9$ at the vertices and $2p\Delta/9$ at the midside nodes and overall are slightly better.

The quintic element proves the most accurate but, observing the first result, the clamped edge moment is not in keeping with the other figures. The remedy is to add the element *initial moment* to the moment solution at the clamped edge and in the first case one obtains

$$m_e = -0.03926 pL^2 - \frac{2p(L/2)^3}{24L} = -0.04968 pL^2 \qquad (8.63)$$

using eqns (8.9) applied to a pair of triangles and doubling the result to allow for symmetry. Repeating the process for the refined meshes the results are shown in parentheses in Table 8.4 and the convergence is much improved.

Indeed the adjustment of eqn (8.63) is, in principle, quite general, and was first encountered for a beam element in Section 1.2, but only the most rapidly convergent thin plate elements with coarse meshes are sufficiently accurate to take advantage of it. In general, therefore, body forces and surface tractions must be included in the same way as initial strains when the stresses are calculated.

Figures 8.11–8.14 show the convergence obtained with a number of alternative nine-freedom thin plate elements in square plates with simply or

Thin plate bending elements

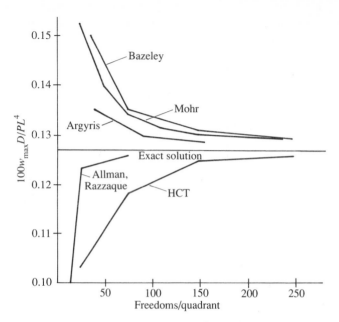

Fig. 8.11 Finite element solutions for a clamped plate with uniformly distributed load

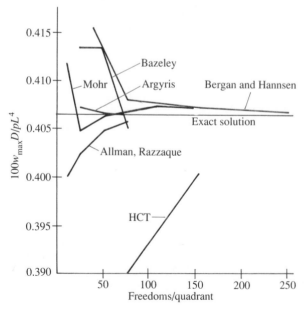

Fig. 8.12 Finite element solutions for a simply supported plate with uniformly distributed load

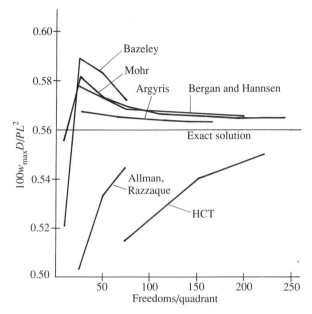

Fig. 8.13 Finite element solutions for a clamped plate with a central point load

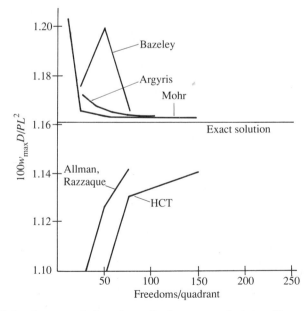

Fig. 8.14 Finite element solutions for a simply supported plate with a central point load

clamped supported edges and carrying blanket or centrally concentrated loads. Our interest in nine-freedom plate elements is that, provided convergence is reasonably fast, they are very economical, are easy to use, and are the most versatile (an example being their use to form flat elements for the approximate analysis of shells).

The element detailed in Section 8.4, that of Bazeley *et al.*,[19] performs as well as the conforming HCT element.[18] This is expected from the arguments introduced in Section 3.9, that is, they both have what we term D_2 *continuity* or continuity of the quadratic part of the displacement interpolation. The non-conforming Bazeley *et al.* element, however, is prone to somewhat erratic convergence.

The Allman–Razzaque element[20, 31] is disappointing with point loads but three more recent elements, those of Bergan and Hanssen,[32] Argyris *et al.*,[34] and Mohr[48] give excellent results. Of these newer elements the nested interpolation element detailed in Section 8.10 is much easier to derive than other elements of comparable accuracy.[49−51]

8.12 Closing remarks

In closing it should be remarked that the convergence plots of Figs 8.11–8.14 are for displacement solutions and that rates of convergence of stress solutions are generally less in displacement elements, as was shown in Section 3.9.

Many elements have been discussed and some closing recommendations are perhaps appropriate. When rectangular elements are able to model the problem at hand the bicubic element gives excellent results. When triangular elements are required the nine-freedom elements of Sections 8.4 and 8.10 are both very useful. The former also provides an excellent exercise in the natural strain approach and the latter is worth special remark as the same coding given in Section 15.8 for the six-node plane stress triangle can be used to calculate its kernel stiffness matrix. Following this by straightforward coding of the matrix T in eqn (8.60) the final element stiffness is given by $k_\cdot = T^t k^* T$.

References

1. Melosh, R. J. (1963). Basis for derivation of matrices for the direct stiffness method. *J. Am. Inst. Aeron. Astron.* **1** (7), 1631.
2. Mitchell, A. R. and Wait, R. (1977). *The Finite Element Method in Partial Differential Equations.* Wiley, London.
3. Oden, J. T. (1972). *Finite Elements of Nonlinear Continua.* McGraw-Hill, New York.
4. Deak, A. L. and Pian, T. H. (1967). Application of smooth surface interpolation to finite element analysis. *J. Am. Inst. Aeron. Astron.* **5**, 187.

5. Bogner, F. K., Fox, R. L., and Schmidt, L. A. (1965). The generation of inter-element compatible stiffness and mass matrices by the use of interpolation formulae. *Proceedings of the First Conference on Matrix Methods in Structural Mechanics*, Wright-Patterson AFB, Ohio.

6. Gallagher, R. H. (1975). *Finite Element Analysis Fundamentals*. Prentice-Hall, Englewood Cliffs, NJ.

7. Przemieniecki, J. S. (1968). *Theory of Matrix Structural Analysis*. McGraw-Hill, New York.

8. Wilson, R. R. and Brebbia, C. A. (1971). Dynamic behaviour of steel foundations for turbo-alternators. *J. Sound Vibn*, **18**, 405.

9. Brebbia, C. A. and Connor, J. J. (1973). *Fundamentals of Finite Element Techniques for Structural Engineers*. Butterworths, London.

10. Henshell, R. D., Walters, D., and Warburton, G. B. (1972). A new family of curvilinear plate bending elements for vibration and stability. *J. Sound Vibn*, **20**, 327.

11. Zienkiewicz, O. C. and Taylor, R. L. (1984). *The Finite Element Method*, (4th edn). McGraw-Hill, London.

12. Chan, Y. K. and Kabaila, A. P. (1973). A compatible element for buckling analysis of skew plates with parallel edges. *Conference on Finite Element Analysis*, Tokyo.

13. Sander, G. (1964). Bornes supérieures et inférieures dans l'analyse matricielle des plaques en flexion torsion. *Bull. Soc. Royale Sciences Liège*, **33**, 456.

14. Fraeijs de Veubeke, B. (1968). A conforming finite element for plate bending. *Int. J. Solids Struct*, **4**, 95.

15. Chan, Y. K. and Kabaila, A. P. (1973). A conforming quadrilateral element for buckling analysis of stiffened plates. *Civil Engineering Report R-121*, University of NSW, Sydney.

16. Clough, R. W. and Felippa, C. A. (1968). A refined quadrilateral element for the analysis of plate bending. *Proceedings of the Second Conference on Matrix Methods in Structural Mechanics*, Wright-Patterson AFB, Ohio.

17. Mohr, G. A. (1983). Finite element formulation by nested interpolations: application to a quadrilateral thin plate bending element. *Trans I.E. Aust.* **CE25** (3), 211.

18. Clough, R. W. and Tocher, J. L. (1965). Finite element stiffness matrices for analysis of plates in bending. *Proceedings of the First Conference on Matrix Methods in Structural Mechanics*, Wright-Patterson AFB, Ohio.

19. Bazeley, G. P., Cheung, Y. K., Irons, B. M., and Zienkiewicz, O. C. (1965). Triangular plates in bending – conforming and nonconforming solutions. *Proceedings of the First Conference on Matrix Methods in Structural Mechanics*, Wright-Patterson AFB, Ohio.

20. Razzaque, A. (1973). Program for triangular bending elements with derivative smoothing. *Int. J. Num. Meth. Engng*, **6**, 333.

21. Cheung, Y. K., King, I. P., and Zienkiewicz, O. C. (1968). Slab bridges with arbitrary shape and support conditions – a general method of analysis based on finite elements. *Proc. Inst. Civ. Eng.*, **40**, 9.

22. Argyris, J. H., Fried, I., and Scharpf, D. W. (1968). The TUBA family of plate elements for the matrix displacement method. *J. Aero. Sci.*, **72**, 701.

23. Timoshenko, S. P. and Woinowsky-Krieger, W. K. (1956). *Theory of Plates and Shells*, (2nd edn). McGraw-Hill, New York.

24. Meek, J. L. (1971). *Matrix Structural Analysis.* McGraw-Hill, New York.
25. Argyris, J. H. (1968). Three-dimensional anisotropic and inhomogeneous media–matrix analysis for small and large displacements. *Ingenieur Archiv.*, **31**, 33.
26. Harvey, J. W. and Kelsey, S. (1971). Triangular plate bending elements with enforced compatibility. *J. Am. Inst. Aeron. Astron.* **9**, 1023.
27. Thomas, G. R. and Gallagher, R. H. (1976). A triangular element based on generalized potential energy concepts. In *Finite Elements for Thin Shells and Curved Members*, (ed. D. G. Ashwell and R. H. Gallagher). Wiley, London.
28. Irons, B. M. and Razzaque, A. (1972). Shape function formulations for elements other than displacement models. *International Conference on Variational Methods in Engineering*, Southhampton.
29. Severn, R. T. and Taylor, P. R. (1966). The finite element method for flexure of slabs where stress distributions are assumed. *Proc. Inst. Civ. Eng.*, **34**, 153.
30. Allwood, R. J. and Cornes, G. M. M. (1969). A polygonal finite element for plate bending problems using assumed stress approach. *Int. J. Num. Meth. Engng*, **1**, 135.
31. Allman, D. J. (1970). Triangular finite elements for plate bending with constant and linearly varying bending moments. IUTAM Symposium on High Speed Computing for Elastic Structures, Liège.
32. Bergan, P. G. and Hanssen, L. (1976). A new approach for deriving good element stiffness matrices. In *The Mathematics of Finite Elements and Applications, Vol. 2*, (ed. J. R. Whiteman). Academic Press, London.
33. Rockey, K. C., Evans, H. R., Griffiths, D. W., and Nethercote, D. A. (1975). *The Finite Element Method: a Basic Introduction.* Crosby Lockwood Staples, London.
34. Argyris, J. H., Haase, M., and Mlejnek, H. P. (1980). On an unconventional but natural formation of a stiffness matrix. *Comp. Meth. Appl. Mech. Engng*, **22**, 1.
35. Strang, G. and Fix, G. J. (1973). *An Analysis of the Finite Element Method.* Prentice-Hall, Englewood Cliffs, NJ.
36. Faulkes, K. A. (1973). Application of finite elements to plate bending problems. In *Finite Element Techniques*, (ed. A. P. Kabaila and A. S. Hall). Clarendon Press, Sydney.
37. Herrmann, L. R. (1967). Finite element bending analysis for plates. *J. Eng. Mech. Div. ASCE*, **93**, 13.
38. Morley, L. S. D. (1968). The triangular equilibrium element in the solution of plate bending problems. *Aero. Quart.* **19**, 149.
39. Morley, L. S. D. (1970). The constant moment plate bending element. *J. Strain Analysis* **6**, 20.
40. Mohr, G. A. (1976). *Analysis and Design of Plate and Shell Structures using Finite Elements.* Ph.D. thesis, University of Cambridge.
41. Mohr, G. A. (1979). On triangular displacement elements for the bending of thin plates. *Proceedings of the Third Australian Conference on Finite Element Methods*, University of NSW, Sydney.
42. Irons, B. M. (1969). A conforming quartic triangular element for plate bending. *Int. J. Num. Meth. Engng*, **1**, 29.
43. Caramanlian, C., Selby, K. A., and Will, G. T. (1978). A quintic conforming plate bending triangle. *Int. J. Num. Meth. Engng*, **12**, 1109.

44. Bell, K. (1969). A refined triangular plate bending element. *Int. J. Num. Meth. Engng*, **1**, 101.
45. Butlin, G. A. and Ford, R. (1968). A compatible plate bending element. Engineering Department Report 68-15, University of Leicester.
46. Cowper, G. R., Kosko, E., Lindberg, G. M., and Olson, M. D. (1968). Formulation of a new triangular plate bending element. *Trans Canad. Aero-Space Inst.*, **1**, 86.
47. Cowper, G. W., Lindberg, G. M., and Olson, M. D. (1970). A shallow shell finite element of triangular shape. *Int. J. Solids Struct.*, **6**, 1133.
48. Mohr, G. A. (1981). Finite element formulation by nested interpolations: application to cubic elements. *Comput. Struct.*, **14**, 211.
49. Bathe, K. J., Brezzi, F., and Cho, S. W. (1989). The MITC7 and MITC9 plate bending elements. *Comput. Struct.*, **32**, 797.
50. Zienkiewicz, O. C., Taylor, R. L., Papadopoulos, P., and Onate, E. (1990). Mindlin-Reissner elements for plates. *Comput. Struct.* **35**, 505.
51. Robinson, J. and Lawson, J. (1990). An evaluation of the plate bending elements within MSC/pal 2. *Comput. Struct.*, **36**, 575.

9
Thick plate elements

The analysis of thick plates must allow for the additional deformations of the plate caused by transverse shearing, and appropriate equations for this were discussed in Section 2.7. These equations have been implemented in a variety of ways for rectangular finite elements but may meet with difficulties in triangular elements. These difficulties are reduced by the use of the natural strain technique of Argyris and Scharpf.[1] Such elements may still be used to analyse thin plates with the use of penalty factors to enforce the Kirchhoff constraints.

9.1 Introduction

In thick plate elements the deformations arising from transverse shearing are taken into account. Then the curvatures are calculated from eqns (2.25) and (2.28) and the average shear strains are calculated from eqns (2.38):

$$\chi_x = \frac{\partial \phi_x}{\partial x}, \quad \chi_y = \frac{\partial \phi_y}{\partial y}, \quad \chi_{xy} = \frac{\partial \phi_x}{\partial y} + \frac{\partial \phi_y}{\partial x} \tag{9.1a}$$

$$\gamma_x = \frac{\partial w}{\partial x} - \phi_x, \qquad \gamma_y = \frac{\partial w}{\partial y} - \phi_y \tag{9.1b}$$

where w is the total deflection, ϕ_x and ϕ_y are the slopes caused by flexing of the plate, and γ_x and γ_y are the additional slopes arising from deformation of the plate caused by transverse shearing. As shown in Section 2.7 it follows from an energy argument that γ_x and γ_y are also the average transverse shear strains.

If the transverse shear strains are suppressed by applying penalty factors such elements may also be used to analyse thin plates. These shear strains are suppressed by calculating the element stiffness matrix as

$$k = \int B_b^t D_b B_b \, \mathrm{d}V + \beta \int B_s^t D_s B_s \, \mathrm{d}V$$

where B_b is the strain matrix corresponding to eqns (9.1a) and B_s corresponds to eqns (9.1b).

In theory the *Kirchhoff constraints* $\gamma_x = \gamma_y = 0$ are more accurately enforced as $\beta \to \infty$. In practice, however, computation uses a limited word length and too large a value of β forces the curvatures out of the calculations, artificially increasing the flexibility of the plate.

Too small a value of β, on the other hand, also increases the flexibility of the plate by allowing shearing action to become artificially large. Thus, as we shall in Section 9.5, an optimum value of β should be estimated by using a sequence of trial values in a reasonably fine mesh.

Equations (9.1) omit only one of the three-dimensional strains, that of transverse compression. This leaves eqns (9.1b) without a third 'coupling' strain such as $\varepsilon_{xy} = \partial u/\partial y + \partial v/\partial x$ (for plane stress) as has appeared in all two-dimensional problems encountered thus far. As noted in Section 2.7 this sometimes leads to numerical difficulties when triangular thick plate elements are formulated. In Section 9.3, therefore, we derive a more stable triangular element by defining natural shear strains parallel to each side of the element.

9.2 Rectangular thick plate elements

The strain–displacement equations for thick plates are eqns (9.1) and the constitutive equations are obtained by combining eqns (2.29) and (2.37) to give

$$
\begin{bmatrix} m_x \\ m_y \\ m_{xy} \\ Q_x \\ Q_y \end{bmatrix} = \frac{Et}{12(1 - v^2)} \begin{bmatrix} t^2 & vt^2 & 0 & 0 & 0 \\ vt^2 & t^2 & 0 & 0 & 0 \\ 0 & 0 & \frac{1}{2}(1 - v)t^2 & 0 & 0 \\ 0 & 0 & 0 & 5(1 - v) & 0 \\ 0 & 0 & 0 & 0 & 5(1 - v) \end{bmatrix} \begin{bmatrix} \chi_x \\ \chi_y \\ \chi_{xy} \\ v_x \\ v_y \end{bmatrix}
$$
$$
= D\{\varepsilon\}
$$
(9.2)

where

$$
\chi_x = \frac{\partial \phi_x}{\partial x}, \quad \chi_y = \frac{\partial \phi_y}{\partial y}
$$

$$
\chi_{xy} = \frac{\partial \phi_x}{\partial y} + \frac{\partial \phi_y}{\partial x}
$$

$$
\gamma_x = \frac{\partial w}{\partial x} - \phi_x, \quad \gamma_y = \frac{\partial w}{\partial y} - \phi_y.
$$

These relations are successfully applied to rectangular elements by Griemann and Lynn,[2] yielding a convenient element with the same nodal freedoms as the simple non-conforming thin rectangle of Section 8.1. The element assumes the Cartesian displacement functions

$$w = a_1 + a_2 x + a_3 y + a_4 xy + a_5 x^2 + a_6 y^2 + a_7 x^2 y + a_8 xy^2$$

$$\phi_x = b_1 + b_2 x + b_3 y + b_4 xy$$

$$\phi_y = c_1 + c_2 x + c_3 y + c_4 xy \qquad (9.3)$$

and, restricting the element to constant shear strains in the directions of the axes, one applies the conditions[†]

$$\frac{\partial \gamma_x}{\partial x} = 0 = \frac{\partial^2 w}{\partial x^2} - \frac{\partial \phi_x}{\partial x}$$

$$\frac{\partial \gamma_y}{\partial y} = 0 = \frac{\partial^2 w}{\partial y^2} - \frac{\partial \phi_y}{\partial y} \qquad (9.4)$$

to eqns (9.3), obtaining

$$a_5 = \tfrac{1}{2} b_2, \qquad a_6 = \tfrac{1}{2} c_3, \qquad a_7 = \tfrac{1}{2} b_4, \qquad a_8 = \tfrac{1}{2} c_4. \qquad (9.5)$$

Hence the number of unknown coefficients is reduced to twelve so that freedoms w, ϕ_x, and ϕ_y are permitted at each of the vertices.

An alternative approach is that of Pryor et al.[3] in which the chosen displacement functions are

$$w = a_1 + a_2 x + a_3 y + a_4 x^2 + a_5 xy + a_6 y^2 + a_7 x^3 + a_8 x^2 y$$

$$\qquad + a_9 xy^2 + a_{10} y^3 + a_{11} x^3 y + a_{12} xy^3$$

$$\gamma_x = b_1 + b_2 x + b_3 y + b_4 xy, \qquad \gamma_y = c_1 + c_2 x + c_3 y + c_4 xy. \qquad (9.6)$$

These are exactly the functions of the non-conforming ACM rectangular thin plate element introduced in Section 8.1 and the bilinear plane stress rectangle introduced in Section 6.2. Rearranging eqns (9.1b) the slopes ϕ_x and ϕ_y are given by

$$\phi_x = \frac{\partial w}{\partial x} - \gamma_x, \qquad \phi_y = \frac{\partial w}{\partial y} - \gamma_y. \qquad (9.7)$$

Using these results the curvatures in eqns (9.1) can be calculated as

$$\chi_x = \frac{\partial \phi_x}{\partial x} = \frac{\partial^2 w}{\partial x^2} - \frac{\partial \gamma_x}{\partial x}$$

$$\chi_y = \frac{\partial \phi_y}{\partial y} = \frac{\partial^2 w}{\partial y^2} - \frac{\partial \gamma_y}{\partial y} \qquad (9.8)$$

$$\chi_{xy} = \frac{\partial \phi_x}{\partial y} + \frac{\partial \phi_y}{\partial x} = \frac{2 \partial^2 w}{\partial x \partial y} - \frac{\partial \gamma_x}{\partial y} - \frac{\partial \gamma_y}{\partial x}$$

[†] A successful precedent for this is the restriction of the 8 df plane stress rectangle to constant shear strain (see Section 6.6).

and the derivation of the element stiffness matrix is carried out by super-imposing the shear stiffness matrix obtained from the shear terms of eqns (9.8) on the standard stiffness matrix for the non-conforming thin plate element described in Section 8.1.

Several other rectangular elements have been derived along similar lines to these.[4] An alternative and more direct approach is to assume the same interpolations for w, ϕ_x, and ϕ_y.[5]

$$w = \{f\}^t\{w\}, \qquad \phi_x = \{f\}^t\{\phi_x\}, \qquad \phi_y = \{f\}^t\{\phi_y\} \qquad (9.9)$$

and to calculate the curvatures and shear strains directly from eqns (9.1). This procedure often leads to numerical difficulties,[6] however, particularly in the case of triangular elements.[7]

The principal cause of this difficulty appears to be that there are only two expressions for the average shear strains in eqns (9.1b), that is, there is no third expression coupling the other two as is the case in plane stress and strain or in thin plates. To take a hypothetical case, if the plate is such that the transverse shearing effects are dominant then for these there are only two uncoupled or independent Cartesian actions.

In the case of triangular elements this anomaly can be overcome by calculating natural shear strains on each side of an element[1] and this approach is taken up in the following section.

9.3 Triangular thick plate elements

Only a few triangular thick plate elements have yet been developed. These include the nine-freedom element developed by Lynn and Dhillon[8] and the thirty-six-freedom element developed by Rao *et al.* and Thangham Babu *et al.*[9,10] Figure 9.1 shows a six-noded isoparametric triangular thick plate element with eighteen freedoms, w, ϕ_x, and ϕ_y at each node. Hence quadratic

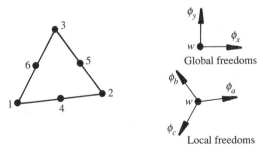

Fig. 9.1 Thick plate element with eighteen freedoms

interpolation may be used for the lateral displacement and the two independent flexural slopes:

$$w = \{f\}^t\{w\} = L_1(2L_1 - 1)w_1 + L_2(2L_2 - 1)w_2 + L_3(2L_3 - 1)w_3$$
$$+ 4L_1L_2w_4 + 4L_2L_3w_5 + 4L_3L_1w_6. \tag{9.10}$$

The same interpolation is used for ϕ_x and ϕ_y but first these Cartesian slopes must be transformed to natural slopes parallel to each side of the element, as shown in Fig. 9.1. In fact it is convenient to carry out all calculations on the element sides so that natural curvatures and shear strains are also calculated parallel to the element sides. The results are transformed to Cartesian values only in the final element stiffness matrix calculation.

Now defining natural curvatures and shear strains parallel to each side of the element, the strain–displacement relations are[1]

$$\chi_a = \frac{\partial \phi_a}{\partial s_a} = \frac{\partial \phi_a / \partial L_2 - \partial \phi_a / \partial L_1}{L_{21}}$$

$$\chi_b = \frac{\partial \phi_b}{\partial s_b} = \frac{\partial \phi_b / \partial L_3 - \partial \phi_b / \partial L_2}{L_{32}}$$

$$\chi_c = \frac{\partial \phi_c}{\partial s_c} = \frac{\partial \phi_c / \partial L_1 - \partial \phi_c / \partial L_3}{L_{13}}$$

$$\gamma_a = \frac{\partial w}{\partial s_a} - \phi_a = \frac{\partial w / \partial L_2 - \partial w / \partial L_1}{L_{21}} - \phi_a \tag{9.11}$$

$$\gamma_b = \frac{\partial w}{\partial s_b} - \phi_b = \frac{\partial w / \partial L_3 - \partial w / \partial L_2}{L_{32}} - \phi_b$$

$$\gamma_c = \frac{\partial w}{\partial s_c} - \phi_c = \frac{\partial w / \partial L_1 - \partial w / \partial L_3}{L_{13}} - \phi_c$$

where ϕ_a, ϕ_b, ϕ_c and s_a, s_b, s_c are respectively the slopes and coordinates referred to each side of the element. The natural slopes are given by

$$\phi_a = c_{ax}\phi_x + c_{ay}\phi_y, \qquad \phi_b = c_{bx}\phi_x + c_{by}\phi_y, \qquad \phi_c = c_{cx}\phi_x + c_{cy}\phi_y \tag{9.12}$$

where

$$c_{ax} = \frac{x_{21}}{L_{21}} = \frac{x_2 - x_1}{L_{21}}, \qquad c_{ay} = \frac{y_{21}}{L_{21}} = \frac{y_2 - y_1}{L_{21}}$$

$$c_{bx} = \frac{x_{32}}{L_{32}} = \frac{x_3 - x_2}{L_{32}}, \qquad c_{by} = \frac{y_{32}}{L_{32}} = \frac{y_3 - y_2}{L_{32}}$$

$$c_{cx} = \frac{x_{13}}{L_{13}} = \frac{x_1 - x_3}{L_{13}}, \qquad c_{cy} = \frac{y_{13}}{L_{13}} = \frac{y_1 - y_3}{L_{13}}.$$

Combining eqns (9.10)–(9.12) the 6×18 interpolation matrix B for the natural generalized strains is given by

$$\{\varepsilon\} = \{\chi_a, \chi_b, \chi_c, \gamma_a, \gamma_b, \gamma_c\} = B\{d\} \qquad (9.13)$$

where B may be schematically written in the form

$$B = \begin{bmatrix} B_b \\ B_s \end{bmatrix} = \begin{bmatrix} w \text{ columns} & \phi_x \text{ columns} & \phi_y \text{ columns} \\ \{0\}^t & c_{ax}\{\dot{f}_a\}^t & c_{ay}\{\dot{f}_a\}^t \\ \{0\}^t & c_{bx}\{\dot{f}_b\}^t & c_{by}\{\dot{f}_b\}^t \\ \{0\}^t & c_{cx}\{\dot{f}_c\}^t & c_{cy}\{\dot{f}_c\}^t \\ \{\dot{f}_a\}^t & -c_{ax}\{f\}^t & -c_{ay}\{f\}^t \\ \{\dot{f}_b\}^t & -c_{bx}\{f\}^t & -c_{by}\{f\}^t \\ \{\dot{f}_c\}^t & -c_{cx}\{f\}^t & -c_{cy}\{f\}^t \end{bmatrix}$$

and

$$\{\dot{f}_a\} = \frac{\{1 - 4L_1, 4L_2 - 1, 0, 4L_1 - 4L_2, 4L_3, -4L_3\}}{L_{21}} = \frac{\partial\{f\}}{\partial s_a} \qquad (9.14)$$

$$\{\dot{f}_b\} = \frac{\{0, 1 - 4L_2, 4L_3 - 1, -4L_1, 4L_2 - 4L_3, 4L_1\}}{L_{32}} = \frac{\partial\{f\}}{\partial s_b} \qquad (9.15)$$

$$\{\dot{f}_c\} = \frac{\{4L_1 - 1, 0, 1 - 4L_3, 4L_2, -4L_2, 4L_3 - 4L_1\}}{L_{13}} = \frac{\partial\{f\}}{\partial s_c} \qquad (9.16)$$

$$\{f\} = \{L_1(2L_1 - 1), L_2(2L_2 - 1), L_3(2L_3 - 1), 4L_1 L_2, 4L_2 L_3, 4L_3 L_1\}. \qquad (9.17)$$

Transformation of the curvatures given by B_b to Cartesian values is accomplished using the Hessian matrix of eqn (8.23) and Argyris and Scharpf[1] obtain the transformation required for the shears by writing

$$\gamma_a = c_{ax}\gamma_x + c_{ay}\gamma_y$$
$$\gamma_b = c_{bx}\gamma_x + c_{by}\gamma_y \qquad (9.18)$$

or

$$\gamma_b/c_{bx} = \gamma_x + c_{by}\gamma_y/c_{bx}$$
$$\gamma_a/c_{ax} = \gamma_x + c_{ay}\gamma_y/c_{ax} \qquad (9.19)$$

giving

$$\gamma_y(c_{ax}c_{by} - c_{bx}c_{ay}) = \gamma_b c_{ax} - \gamma_a c_{bx}. \qquad (9.20)$$

Now the area of the triangle is given by

$$2\Delta = x_{21}y_{32} - x_{32}y_{21} = L_a L_b(c_{ax}c_{by} - c_{bx}c_{ay}) \qquad (9.21)$$

where $L_a = L_{21}$, $L_b = L_{32}$, so that eqn (9.20) may be written

$$\gamma_y = \frac{L_a L_b(-c_{bx}\gamma_a + c_{ax}\gamma_b)}{2\Delta} \tag{9.22a}$$

Repeating, this time solving eqns (9.18) for γ_x, one obtains

$$\gamma_x = \frac{L_a L_b(c_{by}\gamma_a - c_{ay}\gamma_b)}{2\Delta}. \tag{9.22b}$$

Now by cyclic progression of the subscripts a, b, c one obtains

$$\gamma_x = \frac{L_b L_c(c_{cy}\gamma_b - c_{by}\gamma_c)}{2\Delta}$$

$$\gamma_y = \frac{L_b L_c(-c_{cx}\gamma_b + c_{bx}\gamma_c)}{2\Delta} \tag{9.23}$$

and

$$\gamma_x = \frac{L_c L_a(c_{ay}\gamma_c - c_{cy}\gamma_a)}{2\Delta}$$

$$\gamma_y = \frac{L_c L_a(-c_{ax}\gamma_c + c_{cx}\gamma_a)}{2\Delta}. \tag{9.24}$$

Averaging eqns (9.22), (9.23), and (9.24) one obtains

$$\{\gamma_x, \gamma_y\} = T_G\{\gamma_a, \gamma_b, \gamma_c\} \tag{9.25}$$

where

$$T_G^t = \frac{1}{6\Delta}\begin{bmatrix} c_{by}L_aL_b - c_{cy}L_cL_a & c_{cx}L_cL_a - c_{bx}L_aL_b \\ c_{cy}L_bL_c - c_{ay}L_aL_b & c_{ax}L_aL_b - c_{cx}L_bL_c \\ c_{ay}L_cL_a - c_{by}L_bL_c & c_{bx}L_bL_c - c_{ax}L_cL_a \end{bmatrix}.$$

Hence the transformed strain–displacement matrix for both plate flexing and shearing is given as

$$B = \begin{bmatrix} H B_b \\ T_G B_s \end{bmatrix} \tag{9.26}$$

and finally the element stiffness matrix and the element stresses are calculated using

$$k = \Sigma B^t D B(\omega_i \Delta), \qquad \{\sigma\} = DB\{d\} = N\{d\} \tag{9.27}$$

where D is given in eqn (9.2). Three-point integration is used and this reduced integration exactly computes all energy terms except those associated with the appearance of the flexural slopes ϕ_a, ϕ_b, and ϕ_c in the expressions for the average shear strains γ_a, γ_b, and γ_c.

Table 9.1 Deflection and stress resultant maxima for blanket loaded square slabs ($v = 0.2$). m_{max} and Q_{max} are sampled at vertices and midsides, respectively and eqn (5.135) is used for integration.

Net freedoms/quadrant	$w_{max}D/pL^4$	m_{max}/pL^2	Q_{max}/pL
Simply supported slab ($t/L = 0.2$)			
12	0.00359	0.0347	0.167
48	0.00470	0.0451	0.252
108	0.00478	0.0451	0.281
192	0.00479	0.0449	0.295
Exact solution[11]	0.00478	0.0442	0.338
Thin plate solution[12]	0.00406	0.0442	0.338
Clamped slab ($t/L = 0.1$)			
8	0.00104	0.0150	0.167
40	0.00136	0.0355	0.289
96	0.00144	0.0421	0.331
176	0.00146	0.0453	0.354
Argyris and Scharpf[1]	–	0.0443	0.365
Griemann and Lynne[2]	0.00148	0.0475	–
Thin plate solution[12]	0.00126	0.0513	0.338

Table 9.1 shows that satisfactory results are obtained using the element to analyse a quadrant of a square plate carrying a blanket load and having simply or clamped supported edges.

Argyris and Scharpf[1] derive four triangular thick plate elements using displacement functions one order higher for the lateral displacement freedoms than for the slopes. The first of these is the twelve-freedom element obtained by suppressing the midside slopes in the element of Fig. 9.1 (using equations of the form of eqn (4.29) and this element converges monotonically but more slowly than the eighteen-freedom element.[7] Argyris and Scharpf's three higher-order elements converge still more rapidly but involve complicated deployment of the nodal freedoms in contrast to the convenient eighteen-freedom element given by eqn (9.13).

Figure 9.2 shows the central deflection predicted by the eighteen-freedom element as the span/thickness ratio of a plate is increased. Here eighteen elements are used for a quadrant of the plate. When the plate is thick ($L/t \leq 20$) the finite element solution is within 3 percent of the exact (Reissner) solution. For thinner plates the error increases to about 5 per cent but begins to reduce again for very thin plates ($L/t \simeq 10\,000$) whereas results for an eight-node rectangular element diverge.[5] The same behaviour is observed

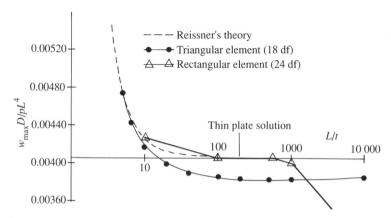

Fig. 9.2 Maximum deflection predicted in simply supported square plate with uniformly distributed load using triangular and rectangular finite elements which include the shearing stiffness

when the plate edges are clamped but the rectangular element diverges earlier.[5]

Pugh *et al.*[5] report results for five rectangular elements, all with separate interpolations for the displacements and slopes and their results are briefly summarized in Fig. 9.3. Element *A*, which is bilinear, converges satisfactorily though more slowly than the higher-order elements, provided a selective integration is used (four points for flexure and one point for shear). The authors report that this element is widely used for shell analysis[13] but conclude that the quadratic or cubic elements (cases *C* and *E*) would be more satisfactory.

The eighteen-freedom triangular element detailed in the foregoing is of comparable accuracy to the second quadratic rectangle (case *C*) in Fig. 9.3, although the two elements require three- and nine-point integrations.

9.4 Reduced integration in thick plate elements

The most obvious alternative integration for the 18 df triangle of Fig. 9.1 is *selective integration* with three points for flexure and one point for shear but this does not lead to an improvement in results. Another alternative is to use reduced interpolation for the slopes ϕ_x and ϕ_y in computing the shear strains but to retain the full interpolation in calculating the curvatures. Such procedures have proved successful in thin shell elements[14,15] where reduced interpolation is applied to the w/R strain components (see Section 11.5) but do not lead to an improvement in the triangular thick plate element considered here.

Element	Case	Integration points	Convergence error	
			Simple support	Clamped support
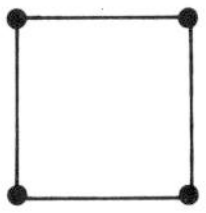	A	4/4	Divergent	Divergent
		4/1	0.5%	1.0%
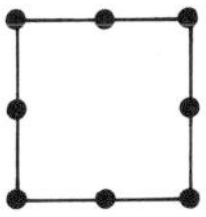	B	4/4	Divergent	Divergent
		9/9	Divergent	Divergent
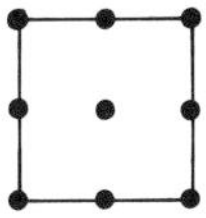	C	4/4	0.1%	0.1%
		9/9	0.5%	5.0%
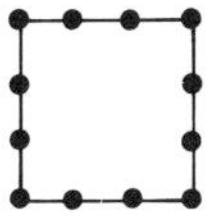	D	4/4	Divergent	Divergent
		9/9	Divergent	Divergent
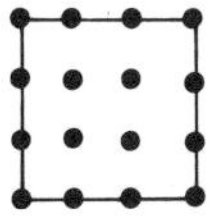	E	4/4	0.1%	0.1%
		9/9	0.1%	0.1%

Fig. 9.3 Convergence behaviour of various rectangular thick plate elements

Some insight into the effects of reduced integration in the 18 df triangle can be gained by examining the terms of the form $\{f\}\{f\}^t$ given by the ϕ–ϕ contributions to the energy product. Exact integration gives

$$\int \{f\}\{f\}^t \mathrm{d}A = \frac{\Delta}{14\,580} \begin{bmatrix} 486 & -81 & -81 & 0 & -324 & 0 \\ & 486 & -81 & 0 & 0 & -324 \\ & & 486 & -324 & 0 & 0 \\ & & & 2592 & 1296 & 1296 \\ & & & & 2592 & 1296 \\ \text{Symmetric} & & & & & 2592 \end{bmatrix}$$

(9.28)

whereas integration at the midside nodes (eqn (5.135)) gives a sparse matrix with the only non-zero entries being (4, 4), (5, 5), and (6, 6) = $\Delta/3$. This is an unsatisfactory result in shell applications (for which the element is designed) as no stiffness will be attributed to the w/R strain components (see eqns (2.40)) at the vertex nodes, leading to difficulty when these are loaded in the radial direction.

A better result is obtained using the integration at internal points of eqn (5.136) and this gives

$$\int \{f\} \{f\}^{t} dA = \frac{\Delta}{14\,580} \begin{bmatrix} 360 & -180 & -180 & 180 & -360 & 180 \\ & 360 & -180 & 180 & 180 & 360 \\ & & 360 & -360 & 180 & 180 \\ & & & 1980 & 1440 & 1440 \\ \text{Symmetric} & & & & 1980 & 1440 \\ & & & & & 1980 \end{bmatrix}.$$

$$(9.29)$$

Figure 9.4 shows the improvement in convergence obtained by using the alternative integration (eqn (5.136)) and this internal integration gives much improved results in shell applications.[16] The same phenomenon as observed here, namely the use of that reduced integration which gives the best approximation of the coupling strains, is also observed with an arch element in Section 11.1 and such behaviour is further discussed in Section 15.11.

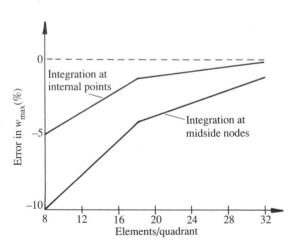

Fig. 9.4 Convergence of maximum deflection in clamped blanket loaded slab using an 18 df quadratic element with alternative integrations

9.5 Penalty factor techniques

Thick plate elements can be applied to the analysis of thin plates through the use of penalty factors to suppress the transverse shear.[5, 17] Without the use of these the eighteen-freedom triangle gives poor results when $L/t \geq 10$ if seven-digit computation is used. (Eleven-digit computation was used in Fig. 9.2.)

Denoting the penalty factor by β the element stiffness matrix is computed as

$$k = k_b + \beta k_s \qquad (9.30)$$

where

$$k_b = \Sigma B_b^t H^t(\omega_i \Delta) D_b H B_b \qquad (9.31)$$

$$k_s = \Sigma B_s T_G^t (\tfrac{5}{6}\omega_i \Delta G t) I_2 T_G B_s \qquad (9.32)$$

where D_b is the thin plate modulus matrix (eqn (2.29)), B_b and B_s are given in eqn (9.13), and I_2 denotes a unit matrix of order 2.

Table 9.2 shows results for blanket loaded square plates. Firstly various penalty factors are applied to a clamped plate of fixed thickness to ascertain

Table 9.2 Deflection maxima in square plates with $q = 1\ \text{kN/m}^2$, $E = 10^6\ \text{kN/m}^2$, $v = 0.2$, and $L = 10\ \text{m}$ with various thicknesses and penalty factors, using thirty-two elements for a quadrant mesh.

	t	β	w_{max}
Clamped	0.010	1000×10^{-4}	0.2386×10^3
	0.010	100×10^{-4}	0.1605×10^3
	0.010	50×10^{-4}	0.1461×10^3
	0.010	5×10^{-4}	0.1446×10^3
	0.010	1×10^{-4}	0.1676×10^3
	0.010	0.1×10^{-4}	0.3673×10^3
Exact solution			$0.1457 \times 10^{-3}/t^3$
	1.000	5	0.1447×10^{-3}
	0.100	5×10^{-2}	0.1446
	0.010	5×10^{-4}	0.1446×10^3
	0.001	5×10^{-6}	0.1446×10^6
Simply supported	1.000	5	0.4874×10^{-3}
	0.100	5×10^{-2}	0.4861
	0.010	5×10^{-4}	0.4861×10^3
	0.001	5×10^{-6}	0.4866×10^6
Exact solution			$0.4683 \times 10^{-3}/t^3$

which value gives the best approximation to the thin plate solution, that is, the underlined value is chosen as the closest lower bound. Now observing eqns (9.31) and (9.32) one sees that the numerical error arising from the disparate orders of magnitude of the shearing and flexural stiffnesses is prevented by using

$$\beta = \frac{(\text{const.})Gt^2}{1.2Eh^2}. \tag{9.33}$$

Note that h^2 appears in the denominator because the flexural strains are of the form $\partial\phi/\partial x$ whilst the shearing strains are of the form $\partial w/\partial x$ or ϕ. The derivatives here are evaluated over the element side length h. Using the value determined in Table 9.2 for β one writes

$$\beta = \frac{\alpha t^2}{\Delta} = 5 \times 10^{-4} = \frac{\alpha(0.01)^2}{(25/32)} \tag{9.34}$$

giving

$$\beta = \frac{125t^2}{32\Delta} \tag{9.35}$$

as the required penalty factor. Using this result for simply and clamped supported plates of various thicknesses in Table 9.2 consistent approximations to the thin plate solution are obtained regardless of the plate thickness. As β is mesh dependent it can be used to improve the rate of convergence as is seen by observing that[18]

(1) globally, rather than in the element sense, the magnitudes of the shearing and flexing strains in a plate depend on the ratio

$$\frac{Gt^2}{1.2EL^2}; \tag{9.36}$$

(2) the discretization error depends upon

$$\left(\frac{h}{L}\right)^N. \tag{9.37}$$

Assuming that the present element gives h^2 convergence (which Fig. 9.4 supports) and taking β proportional to the ratio of eqns (9.36) and (9.37) tends to give a cancellation of the shear and discretization errors:

$$\beta = (\text{const.})\frac{Gt^2}{1.2EL^2}\left(\frac{L}{h}\right)^2 = (\text{const.})\frac{Gt^2}{1.2Eh^2} \tag{9.38}$$

as in eqn (9.33). Table 9.3 gives the results for various plates using both a constant penalty factor (the value determined in Table 9.2) and a mesh dependent value (from eqn (9.35)). In each case reduction of the penalty factor

Table 9.3 Solution for maximum deflection in a plate with $L/t = 10$ with various boundary conditions, loads, meshes, and penalty factors. Key: CL = clamped edges, SS = simply supported edges, UDL = blanket load, PL = point load.

Nodes	β	CL/UDL	SS/UDL	CL/PL	SS/PL
		\multicolumn{4}{c}{$w_{max}D/QL^2$, $Q = qL^2$, or P}			
Exact solution		0.001265	0.004062	0.005605	0.011600
81	5.000	0.001277	0.004269	0.005727	0.012185
49	5.000	0.001200	0.004102	0.005388	0.011640
25	5.000	0.000996	0.003870	0.004501	0.010745
9	5.000	0.000493	0.003551	0.003551	0.009313
49	2.813	0.001262	0.004113	0.005721	0.011845
25	1.250	0.001280	0.004081	0.005887	0.011906
9	0.313	0.001534	0.004324	0.007172	0.013226

with coarser meshes accelerates convergence, particularly in the clamped plates.

The original use of penalty factors in the solution of non-linear optimization problems[19, 20] is worth noting because eqn (9.30) is indeed a constrained minimization problem, minimizing the total potential energy. Here the objective is to obtain the thin plate solution and, as in optimization problems, the penalty factor is increased as the objective is approached to force the solution to satisfy the constraints. Thus, in the finite element context the finer the mesh the larger the penalty factor that should be used and eqn. (9.35) ensures this. As to the largest value that the penalty factor may take, this depends upon the machine precision and this question is taken up in Section 11.1, but the procedure used in Table 9.2 used in conjunction with a fine mesh suffices to determine an acceptable value.

9.6 Applications to sandwich plates

A useful application of thick plate elements is to the analysis of sandwich plates. As shown in Fig. 9.5 the flexural stiffness of the weak core may be neglected, giving the stress–strain relations in the form

$$m_x = \frac{E_f d^2 t(\chi_x + v_f \chi_y)}{4(1 - v_f^2)} \tag{9.39}$$

$$Q_x = \tfrac{5}{6} G d \gamma_x. \tag{9.40}$$

In cases where $G \ll E_f/2(1 + v_f)$ thick plate elements give excellent results.[1] For modelling of cracked reinforced concrete slabs the model of Fig. 9.5 may

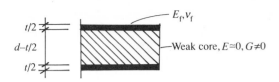

$t/2$

$d-t/2$

$t/2$

E_f, v_f

Weak core, $E \dot{=} 0, G \neq 0$

Fig. 9.5 Weak core sandwich slab

again be useful but it may be appropriate to assume $v_f = 0$, that is, that the reinforcement acts independently in orthogonal directions.[21]

9.7 Other elements for thick plates

An alternative approach to the analysis of thick plates is through the use of three-dimensional elements.[22] With the expedient of reduced integration[23, 24, 25] or constraint equations that coalesce the nodes at the neutral surface,[26] (see Section 7.5) these become more economical. Another element that relies upon the use of constraint equations is Irons' SemiLoof element[27] which employs normal slope freedoms at the Gauss points on each side (that is, at the Loof nodes).[28] Here the Kirchhoff constraints ($\partial w/\partial s - \phi_s = 0$) are enforced at the Gauss points and around the perimeter of the element.[27] A similar type of element is proposed by Lyons.[29] As Strang and Fix remark[30] such constraints possess only heuristic justification. Nevertheless the SemiLoof element, augmented with extensional freedoms (giving a total of 32 df), yields a useful shell element.[31]

The formulation of the SemiLoof element is quite elaborate,[32] however, Irons[27] cites the case of a point loaded clamped square plate where the rate of convergence is slower than that for the ACM rectangle. The much simpler eighteen-freedom element detailed in Section 9.3 is another alternative and extension to deal with shells is straightforward. The resulting thirty-freedom shell element also experiences difficulty with point loads (Section 11.6) but nevertheless proves an economical alternative to higher-order elements such as those based on plate elements using 5 df ($w, \partial w/\partial x, \partial w/\partial y, \phi_x$, and ϕ_y) at each node.[33]

References

1. Argyris, J. H. and Scharpf, D. W. (1971). Finite element theory of plates and shells including transverse shear strain effects. *IUTAM Symposium on High Speed Computation for Elastic Structures, Vol. 1*, University de Liège.

2. Griemann, L. F. and Lynn, P. P. (1970). Finite element analysis of plate bending with transverse shear deformation. *Nucl. Engng Des.* **14**, 223.

3. Pryor, C. W., Barker, R. M., and Frederick, D. (1970). Finite element bending analysis of Reissner plates. *J. Engng Mech. Div. ASCE*, **96**, 974.

4. Speare, P. R. S. and Kemp, K. O. (1976). Shear deformation in elastic homogeneous and sandwich slabs. *Proc. Inst. Civ. Eng., Part 2*, **61**, 697.

5. Pugh, E. D. L., Hinton, E., and Zienkiewicz, O. C. (1978). A study of quadrilateral plate bending elements with reduced integration. *Int. J. Num. Meth. Engng*, **12**, 1059.

6. Brebbia, C. A. and Connor, J. J. (1973). *Fundamentals of Finite Element Techniques for Structural Engineers*. Butterworths, London.

7. Mohr, G. A. (1978). A triangular finite element for thick slabs. *Comput. Struct.*, **9**, 595.

8. Lynne, P. P. and Dhillon, B. S. (1970). Triangular thick plate bending element. *Proceedings of the First International Conference on Structural Mechanics in Reactor Technology*. Vol. 6, Paper No. M61-365, Berlin.

9. Rao, G. V., Venkataramana, J., and Raju, I. S. (1974). A high precision triangular plate bending element for the analysis of thick plates. *Nucl. Engng Des.*, **30**, 408.

10. Thangham Babu, P. V., Reddy, D. V., and Sodhi, D. S. (1979). Frequency analysis of thick orthotropic plates on elastic foundations using a high precision triangular plate bending element. *Int. J. Num. Meth. Engng*, **14**, 531.

11. Reissner, E. (1947). On bending of elastic plates. *Quart. Appl. Math.*, **5**, 55.

12. Timoshenko, S. P. and Goodier, G. N. (1951). *Theory of Elasticity*, (3rd edn). McGraw-Hill, New York.

13. Zienkiewicz, O. C., Taylor, R. L., and Too, J. M. (1971). Reduced integration techniques in general analysis of plates and shells. *Int. J. Num. Meth. Engng*, **3**, 275.

14. Mohr, G. A. (1981). A doubly curved isoparametric triangular thin shell element. *Comput. Struct.*, **14**, 9.

15. Mohr, G. A. (1980). Numerically integrated triangular element for doubly curved thin shells. *Comput. Struct.*, **11**, 565.

16. Mohr, G. A. (1982). On displacement finite elements for the analysis of shells. *Fourth Australian International Conference on Finite Element Methods*. University of Melbourne.

17. Fried, I. (1973). Shear in C_0 and C_1 bending elements. *Int. J. Solids Struct.*, **9**, 449.

18. Fried, I. and Yang, S. K. (1973). Triangular nine degree of freedom C_0 plate bending element of quadratic accuracy. *Quart. Appl. Math.*, **31**, 303.

19. Bracken, J. and McCormick, G. P. (1968). *Selected Applications of Nonlinear Programming*. Wiley, New York.

20. Whittle, P. (1971). *Optimisation Under Constraints*. Wiley, London.

21. Mohr, G. A. (1979). Elastic and plastic predictions of slab reinforcement requirements. *Trans. I. E. Aust.*, **CE21**(1), 16.

22. Ergatoudis, J. B., Irons, B. M., and Zienkiewicz, O. C. (1968). Three-dimensional analysis of arch dams and their foundations. *Symposium on Arch Dams and their Foundations, Inst. Civ. Eng.*, London.

23. Gallagher, R. H. (1975). *Finite Element Analysis Fundamentals*. Prentice-Hall, Englewood Cliffs NJ.

24. Pawsey, S. F. and Clough, R. W. (1971). Improved numerical integration of thick shell elements. *Int. J. Num. Meth. Engng*, **3**, 575.

25. Cook, R. D. (1972). Some elements for analysis of plate bending. *J. Eng. Mech. Div. ASCE*, **98**, 1452.

26. Owen, D. R. J. and Li, Z. H. (1987). A refined analysis of laminated plates by finite element displacement methods – I. Fundamentals and static analysis. *Comput. Struct.*, **26**, 907.

27. Irons, B. M. (1976). The SemiLoof shell element. In *Finite Elements for Thin Shells and Curved Members*, (ed. D. G. Ashwell and R. H. Gallagher). Wiley, London.

28. Loof, H. W. (1966). The economical solution of stiffness of large structural elements. *International Symposium on Computers in Structural Engineering*, Univ. of Newcastle upon Tyne.

29. Lyons, L. P. R. (1977) A general finite element system with special reference to the analysis of cellular structures. Ph.D. thesis, Imperial College, London.

30. Strang, G. and Fix, G. J. (1973). *An Analysis of the Finite Element Method.* Prentice-Hall, Englewood Cliffs NJ.

31. Knowles, N. C., Razzaque, A., and Spooner, J. B. (1976). Experience of finite element analysis of shell structures. In *Finite Elements for Thin Shells and Curved Members*, (ed. D. G. Ashwell and R. H. Gallagher). Wiley, London.

32. Irons, B. M. and Ahmad, S. (1980). *Techniques of Finite Elements.* Ellis-Horwood, Chichester.

33. Hinton, E., Iossifidis, L., and Ren. J. G. (1987). Higher order plate bending analysis using the modified Razzaque–Irons triangle. *Comput. Struct.*, **26**, 681.

10
Flat elements
for shell analysis

Flat elements based on combining the plane stress and thin plate elements described in previous chapters provide the simplest means of approximate analysis of shell structures. In this chapter a variety of such elements are introduced. In addition line elements for the analysis of axisymmetric shells are discussed as these involve a special case in which straight elements are joined together and rotated about an axis of revolution to form an approximate model of an axisymmetric shell.

10.1 Introduction

The analysis of structural shells is perhaps the most difficult finite element problem. The governing differential equations are relatively complicated, if known at all for some cases, so that the energy approach of preceding chapters is now virtually obligatory, and combination of the extensional and flexural strains into a single element formulation presents few immediate difficulties. If the element is curved, however, the extensional and flexural actions are coupled by radial strain components of the form w/R_x, where R_x is the radius of curvature of the shell measured in a plane parallel to the ZY-plane.

This coupling gives rise to a phenomenon similar to the parasitic shear experienced in plane stress elements and reduces the rate of convergence considerably. Partly as a result of this, curved shell elements generally have at least thirty degrees of freedom and thus involve considerable computation.

An alternative approach to shell analysis is to use flat triangular elements (sometimes called 'facet elements') in which, in the same way as in the six-freedom beam/column element introduced in Chapter 1, the extensional and flexural effects are coupled only by coordinate transformation. Then the element stiffness matrix is given by

$$k = T^t \begin{bmatrix} k_a & 0 \\ 0 & k_b \end{bmatrix} T$$

where T is a coordinate transformation matrix and k_a and k_b are the extensional and flexural stiffness matrices. The matrices k_a and k_b are calculated in the same way as the stiffness matrices for plane stress and plate bending in previous chapters. A common choice, for example, is to couple the six-freedom constant strain triangle with a nine-freedom thin plate element. We then have freedoms u, v, w, ϕ_x, and ϕ_y at each node but to permit coordinate transformation of these in three dimensions we must add a third rotation ϕ_z. This could be assigned zero stiffness but this leads to zero stiffness for this *drilling freedom* at nodes where all adjoining elements are coplanar. The usual remedy is to allow a small artificial stiffness to ϕ_z. At the close of the chapter we show how to include ϕ_z rigourously in flat shell elements.

In the following section we derive a 6 df axisymmetric element using the same coordinate transformation matrix as used for the 6 df beam/column element introduced in Chap. 1, providing a useful introduction to shells.

10.2 Conical frustum element

Figure 10.1 shows an axisymmetric thin shell element, the surface of revolution being a conical frustum. The element has three degrees of freedom per node and the global freedoms (node 1 in Fig. 10.1) are transformed to local freedoms (node 2) using

$$\{d'\} = \begin{bmatrix} u'_1 \\ w'_1 \\ \phi'_1 \\ u'_2 \\ w'_2 \\ \phi'_2 \end{bmatrix} = \begin{bmatrix} \cos\alpha & \sin\alpha & 0 & 0 & 0 & 0 \\ -\sin\alpha & \cos\alpha & 0 & 0 & 0 & 0 \\ 0 & 0 & 1 & 0 & 0 & 0 \\ 0 & 0 & 0 & \cos\alpha & \sin\alpha & 0 \\ 0 & 0 & 0 & -\sin\alpha & \cos\alpha & 0 \\ 0 & 0 & 0 & 0 & 0 & 1 \end{bmatrix} \begin{bmatrix} u_1 \\ w_1 \\ \phi_1 \\ u_2 \\ w_2 \\ \phi_2 \end{bmatrix} = T\{d\}.$$

$$(10.1)$$

The interpolation functions for the local displacements are exactly those for the skeletal beam/column element introduced in Section 1.2, that is,

$$u' = (1 - d)u'_1 + du'_2 \tag{10.2}$$

$$w' = (1 - 3d^2 + 2d^3)w'_1 + (d - 2d^2 + d^3)L\phi_1 + (3d^2 - 2d^3)w_2'$$
$$+ (d^3 - d^2)L\phi_2. \tag{10.3}$$

Four strain–displacement equations are required and the first three are obtained from eqns (2.40) by substituting $r = R_y$, $R_x = \infty$, $v = 0$, $\partial(\ )/\partial y = 0$ and using the local dimensionless coordinate $d = x/L$, $d = 0 \to 1$. With these

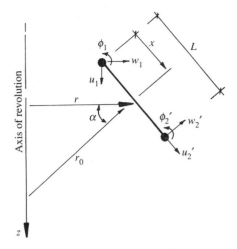

Fig. 10.1 Conical frustum shell element

substitutions we obtain the three relationships

$$\varepsilon_x = \frac{1}{L}\frac{\partial u'}{\partial d} \tag{10.4}$$

$$\varepsilon_c = \frac{w}{r} = \frac{w'\cos\alpha}{r} + \frac{u'\sin\alpha}{r} \tag{10.5}$$

$$\chi_x = \frac{1}{L^2}\frac{\partial^2 w'}{\partial d^2} \tag{10.6}$$

to which must be added[1]

$$\chi_c = \frac{-\sin\alpha}{rL}\frac{\partial w'}{\partial d} \tag{10.7}$$

for the change in circumferential curvature which arises through a mechanism similar to that shown in Fig. 11.7.

Applying eqns (10.4)–(10.7) to (10.2) and (10.3) the strain-interpolation matrix is given by[2]

$$B^{t} = \begin{vmatrix} -1/L & (1-d)a & 0 & 0 \\ 0 & (1-3d^2+2d^3)b & (12d-6)/L^2 & (6d^2-6)a/L \\ 0 & L(d-2d^2+d^3)b & (6d-4)/L & (1-4d+3d^2)a \\ 1/L & da & 0 & 0 \\ 0 & (3d^2-2d^3)b & (6-12d)/L^2 & (6d-6d^2)a/L \\ 0 & L(d^3-d^2)b & (6d-2)/L & (3d^2-2)a \end{vmatrix} \tag{10.8}$$

where $a = \sin \alpha/r$ and $b = \cos \alpha/r$. Integrating through the element thickness to obtain the stress resultants, the modulus matrix is given by

$$\{\sigma\} = \left\{ \begin{array}{c} N_x \\ N_c \\ m_x \\ m_c \end{array} \right\} = \frac{Et}{1-v^2} \begin{bmatrix} 1 & v & 0 & 0 \\ v & 1 & 0 & 0 \\ 0 & 0 & t^2/12 & vt^2/12 \\ 0 & 0 & vt^2/12 & t^2/12 \end{bmatrix} \left\{ \begin{array}{c} \varepsilon_x \\ \varepsilon_c \\ \chi_x \\ \chi_c \end{array} \right\} = D\{\varepsilon\}. \quad (10.9)$$

Now the circumferential strains depend upon $1/r$ and the radius of revolution at any point can be obtained from

$$r = r_1 + \frac{(r_2 - r_1)x}{L} \quad (10.10)$$

where r_1 and r_2 are the radii at the nodes and the integration for the stiffness matrix can be carried out explicitly.[3] The algebra required is tedious and a simpler result is obtained by setting r equal to the average radius in each element[4] (which avoids singularity when a node lies on the axis of revolution).

A more useful solution is obtained by using numerical integration to approximate the reciprocal terms.[5] Seeking to integrate all other contributions to the energy product exactly we note that it will involve terms in d^6 (neglecting the reciprocal terms) and therefore the stiffness matrix is integrated using four-point Gaussian quadrature

$$k = T^t(\Sigma B^t(\omega_i 2\pi rL)DB)T \quad (10.11)$$

where

$$d_1 = 0.0694318, \qquad \omega_1 = 0.1739274$$

$$d_2 = 0.3300095, \qquad \omega_2 = 0.3260726$$

$$d_3 = 0.6699905, \qquad \omega_3 = 0.3260726$$

$$d_4 = 0.9305682, \qquad \omega_4 = 0.1739274.$$

The results obtained for the analysis of a fixed base cylinder shown in Fig. 10.2 are given in Fig. 10.3, the radial expansion and circumferential stress converging to the expected values (pr^2/Et and pr/t respectively) at a distance along the cylinder approximately equal to 30 per cent of its diameter. Suppose the base of the cylinder is replaced by annular plate elements (for which $\alpha = 90$ degrees). To which internal pressure is also applied (that is, a complete pressure vessel is formed) the longitudinal wall stress also converges rapidly to the expected value ($pr/2t$) but at the junction of the cylinder substantial stress discontinuity develops[1] and the numerical results become nonsensical in this region. This difficulty remains even when the finite element

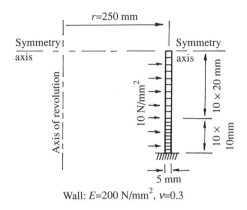

Fig. 10.2 Internally pressurized cylinder with axisymmetric elements

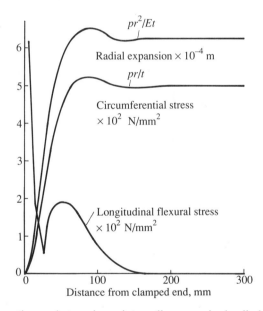

Fig. 10.3 Deformation and stress in an internally pressurized cylinder with clamped ends

model is carefully radiused at the junction, and it is necessary to use solid axisymmetric elements for the base.[6]

Figure 10.4 shows the element applied to the analysis of a circular plate with a simply supported edge and Table 10.1 shows the results obtained for the deflection and moments at the centre and edge of the plate when this is both simply supported and clamped. Also given, in parentheses, are the

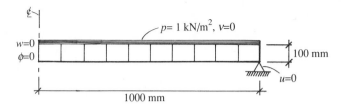

Fig. 10.4 Analysis of a circular slab using axisymmetric elements

Table 10.1 Finite element solutions for a uniformly loaded circular slab with simply supported or encastered edge

	Simply supported		Encastered	
	FEM	Exact	FEM	Exact
$w_{max} D/pR^4$	0.0782	0.0781	0.0157	0.0156
	(0.0777)		(0.0158)	
Unit moments/pR^2				
Radial at centre	0.188	0.188	0.0631	0.0625
	(0.191)		(0.0625)	
Radial at edge	0.009	0.0	− 0.1160	− 0.1250
	(0.017)		(− 0.1080)	
Circumferential	0.128	0.125	0.0031	0.0
at edge	(0.125)		(0.0060)	

results obtained with the element using an explicit 'average radius' integration. In general the numerically integrated solution gives results superior to the explicit solution.

Many other axisymmetric shell elements have been developed, including elements based upon Bessel and other non-polynomial functions,[7] curved isoparametric elements,[2] and elements including shearing deformations.[8, 9] A curved element of the last type is given in Section 11.2.

10.3 Rectangular folded plate elements

Flat rectangular shell elements can be used for certain shell problems, including barrel vault roofs,[2] folded plate roofs,[3] and box girders. Such elements can be formed by combining the 12 df ACM plate element with the bilinear plane stress element[3, 10] or the bicubic plate element with the bilinear[11] or biquadratic plane stress elements. Observing the folded plate structure of Fig. 10.5 one sees that in the vertical element a local in-plane

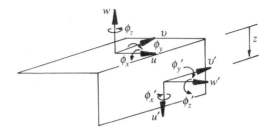

Fig. 10.5 Influence of the drilling freedom upon folded plate elements

rotation ϕ_z' (the drilling freedom) arises from ϕ_x in the horizontal element and it is desirable to ensure continuity of these two actions. This can be achieved by including ϕ_z in the plane stress part of the formulation and this question is taken up in Sections 10.5 and 10.6

10.4 Flat triangular elements

Figure 10.6 shows an eighteen-freedom flat triangular element for shell analysis arbitrarily oriented in space. Such 'facet' elements may be combined to model shell structures approximately.

To model the extensional behaviour we require only two in-plane translations and for the flexural behaviour a transverse deflection and two associated slopes. To permit coordinate transformation in three dimensions, however, we must have the six freedoms $u, v, w, \phi_x, \phi_y, \phi_z$, that is, without ϕ_z, components of the other rotations would be lost on transformation.

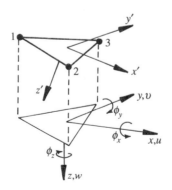

Fig. 10.6 Flat triangular shell element

Then, assuming rows and columns are suitably rearranged, the local element stiffness matrix may be written in the form

$$k' = \begin{bmatrix} k_a & 0 & 0 \\ 0 & k_b & 0 \\ 0 & 0 & \beta I_3 \end{bmatrix} \tag{10.12}$$

where k_a is the 6×6 stiffness matrix for the constant strain triangle (see Section 6.1) and k_b is the 9×9 stiffness matrix for the triangular plate element introduced in Section 8.10. Because of the flat nature of the element the extensional and flexural actions are uncoupled and the 3×3 unit matrix introduces the ϕ'_z freedoms with a small stiffness β.

To obtain the transformation from the global to the local axes we begin by writing a linear interpolation for the elevation of the element surface

$$z = a + bx + cy. \tag{10.13}$$

Substituting the nodal elevations and solving for b and c yields the solutions for the tangents of the slopes of the local x'- and y'-axes. These are the same solutions used for coefficients in the constant strain triangle, that is, using rows two and three in eqn (6.3):

$$t_x = \frac{\partial z}{\partial x} = b = \frac{-(y_{32}z_1 + y_{13}z_2 + \ + y_{21}z_3)}{2\Delta} \tag{10.14}$$

$$t_y = \frac{\partial z}{\partial y} = c = \frac{(x_{32}z_1 + x_{13}z_2 + x_{21}z_3)}{2\Delta} \tag{10.15}$$

where Δ is the area of the horizontal projection of the element. Hence a 3×3 matrix $[t]$ carries out the coordinate transformation as[12]

$$\begin{Bmatrix} x' \\ y' \\ z' \end{Bmatrix} \begin{bmatrix} c_x & -s_x t_y & s_x \\ -s_y t_x & c_y & s_y \\ -c_y s_x & -c_x s_y & c_x c_y \end{bmatrix} \begin{Bmatrix} x \\ y \\ z \end{Bmatrix} = [t]\{x\} \tag{10.16}$$

where

$$s_x = \sin(\tan^{-1} t_x), \qquad c_x = \cos(\tan^{-1} t_x)$$

$$s_y = \sin(\tan^{-1} t_y), \qquad c_y = \cos(\tan^{-1} t_y).$$

The first row of the matrix is obtained by applying the standard transformation

$$x' = c_x x + s_x z_x \tag{10.17}$$

for the rotation from the x- to x'-axis, where

$$z_x = z - t_y y \tag{10.18}$$

allows for the rotation from the y- to the y'-axis. The second row is obtained in like fashion whilst the third is obtained as the direction cosines of the vector cross product of the two unit vectors parallel to the x'- and y'-axes. If $\hat{\mathbf{i}}, \hat{\mathbf{j}}, \hat{\mathbf{k}}$ are unit vectors parallel to the global axes then these two unit vectors are

$$\hat{\mathbf{v}}_x = c_x \cdot \hat{\mathbf{i}} + 0 \cdot \hat{\mathbf{j}} + s_x \cdot \hat{\mathbf{k}} \qquad \hat{\mathbf{v}}_y = 0 \cdot \hat{\mathbf{i}} + c_y \cdot \hat{\mathbf{j}} + s_y \cdot \hat{\mathbf{k}} \qquad (10.19)$$

Then

$$\hat{\mathbf{v}}_z' = \hat{\mathbf{v}}_x' \times \hat{\mathbf{v}}_y' \quad \text{(vector cross product)}$$

$$= -c_y s_x \cdot \hat{\mathbf{i}} - c_x s_y \cdot \hat{\mathbf{j}} + c_x c_y \cdot \hat{\mathbf{k}} \qquad (10.20)$$

is the unit vector parallel to the z'-axis and its components are the required direction cosines. The element stiffness matrix is given as $k = T^t k' T$ where matrix T is formed by repeating $[t]$ six times along the diagonal. Finally the extreme fibre stresses are given by

$$\sigma_x = \frac{N_x}{t} \pm \frac{m_x}{Z}, \qquad \sigma_y = \frac{N_y}{t} \pm \frac{m_y}{Z}, \qquad \tau_{xy} = \pm \frac{m_{xy}}{Z}, \qquad Z = \frac{t^2}{6}. \qquad (10.21)$$

Note that eqn (10.16) fails when the elements lie in a vertical plane. This can be avoided by appropriate coding. An alternative coordinate transformation is obtained by applying eqn (1.60) to one side of the triangle and to the corresponding altitude of the triangle. These directions can then be used as local coordinates in the element formulation, as shown in Fig. 5.10. Parekh uses a similar element to that described above, but the element of Section 8.4 (rather than that of Section 8.10) is used for the bending, and again the drilling freedom is assigned a small arbitrary stiffness.[13] Clough and Johnson also use arbitrary stiffness for ϕ_z and the constant strain triangle, but the HCT plate element is used for the bending.[14]

10.5 The drilling freedom

Strictly speaking the local stiffness matrix of eqn (10.12) should not include the ϕ_z' freedoms, that is, there will be only five equations for each node. Coordinate transformation using eqn (10.16) then gives six equations for each node but the transformed global element stiffness matrix will still be of rank five for each node. Hence if all elements meeting at a node are coplanar (and thus have the same coordinate transformation) the node will introduce a singularity into the structure stiffness matrix. The arbitrary stiffness β in eqn (10.12) will generally prevent this, as will a similar scheme suggested by Zienkiewicz[2] in which a 3×3 matrix with small stiffnesses replaces the entry βI_3 in eqn (10.12) but computationally this change is of negligible significance.

Another approach is to omit the ϕ'_z freedom everywhere and to assemble the equations in local coordinates. This is the usual procedure in curved shell elements (see Chapter 11) but in flat shell elements it involves difficulties in that the local coordinates abruptly change at nodes, exactly where a unique definition is needed to allow omission of the drilling freedom.

Alternatively, rather than 'waste' the drilling freedom, it can be included in the plane stress part of the formulation by defining it as[15]

$$\phi_z = \tfrac{1}{2}\left(\frac{\partial u}{\partial y} - \frac{\partial v}{\partial x} \right) \tag{10.22}$$

and a similar freedom appears in fluid mechanics applications where it represents the vorticity of fluid flow.

The drilling freedom is logically included in a twelve-freedom plane stress element by Mohr, allowing complete quadratic interpolations and yielding marginally better results than the eight-freedom bilinear element in applications to plane stress (see Table 6.1) and folded plate problems.[16] The element cannot be generalized into quadrilateral shape, however, and is thus of limited application.

Tocher and Hartz[17] formulated an eighteen-freedom triangular membrane element by allowing the freedoms

$$u, v, \frac{\partial u}{\partial x}, \frac{\partial v}{\partial y}, \frac{\gamma_{xy}}{2}, \phi \tag{10.23}$$

where

$$\gamma_{xy} = \frac{\partial u}{\partial y} + \frac{\partial v}{\partial x}, \qquad \phi = \tfrac{1}{2}\left(\frac{\partial u}{\partial y} - \frac{\partial v}{\partial x} \right) \tag{10.24}$$

at each vertex. The element is derived by applying cubic expansions to u and v in three subtriangles of the element, so that condensation from a total of fifty-four freedoms to the final eighteen must be achieved by matrix transformations and this is a considerable disadvantage.[18,19]

A more economical element, the hybrid equivalent of which was first obtained by Dungar and Severn,[20] is used by Olson and Bearden[21] to form a flat shell element. It is based on an incomplete nine-term cubic displacement function proposed by Holand,[22] giving a total of eighteen in-plane freedoms (u, v, $\partial u/\partial x$, $\partial u/\partial y$, $\partial v/\partial x$, $\partial v/\partial y$ at each node). This is reduced to nine by the application of various constraints.

Olsen and Bearden, however, conclude that in shell applications this yields little or no improvement over results obtained using the constant strain triangle.[21] In addition it suffers from the disadvantage of requiring inversion of a 12×12 matrix, that is, the solution is not explicit, and involves the risk of failure with special element geometries. It also involves more computation. In the following section, however, we introduce a more economical element, the formulation of which is completely explicit.

10.6 Triangular element including the drilling freedom

To establish an expression for ϕ_z which might be more useful in a triangular element one defines, after the fashion of Argyris,[23] natural extensional freedoms α, β, γ parallel to each side, as shown in Fig. 10.7 for the constant strain triangle. Corresponding to these one defines coordinates a, b, c on the sides. With the use of these natural freedoms and coordinates the basis transformation exercise which follows is made considerably easier than it would otherwise be.

The natural extensional freedoms are given by

$$\alpha = c_{ax}u + c_{ay}v, \qquad \beta = c_{bx}u + c_{by}v, \qquad \gamma = c_{cx}u + c_{cy}v \qquad (10.25)$$

where the direction cosines c_{ax}, c_{ay}, and so on, are given by eqns (5.65). Now calculating the derivatives $\partial\alpha/\partial c$ and $\partial\gamma/\partial a$ using

$$\frac{\partial\alpha}{\partial c} = \frac{\partial x}{\partial c}\frac{\partial\alpha}{\partial x} + \frac{\partial y}{\partial c}\frac{\partial\alpha}{\partial y} = c_{cx}\frac{\partial\alpha}{\partial x} + c_{cy}\frac{\partial\alpha}{\partial y} \qquad (10.26)$$

$$\frac{\partial\gamma}{\partial a} = \frac{\partial x}{\partial a}\frac{\partial\gamma}{\partial x} + \frac{\partial y}{\partial a}\frac{\partial\gamma}{\partial y} = c_{ax}\frac{\partial\gamma}{\partial x} + c_{ay}\frac{\partial\gamma}{\partial y} \qquad (10.27)$$

one obtains, combining (10.22) and (10.25)–(10.27),

$$\frac{\partial\alpha}{\partial c} - \frac{\partial\gamma}{\partial a} = \frac{4\Delta\phi_1}{L_{21}L_{13}} \qquad (10.28)$$

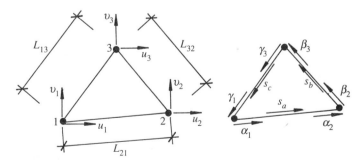

Fig. 10.7 Constant strain triangle with Cartesian and natural freedoms

where Δ is the area of the triangle.[24] Similarly

$$\frac{\partial \beta}{\partial a} - \frac{\partial \alpha}{\partial b} = \frac{4\Delta\phi_2}{L_{32}L_{21}}, \qquad \frac{\partial \gamma}{\partial b} - \frac{\partial \beta}{\partial c} = \frac{4\Delta\phi_3}{L_{13}L_{32}} \qquad (10.29)$$

where ϕ_1, ϕ_2, ϕ_3 are the Cartesian drilling freedoms to be included.

Following the nested interpolation method (Section 4.4) the global freedoms (Fig. 10.8(b)) will be redefined by interpolations on each side of the element to yield the local element freedoms shown in Fig. 10.8(a), which are exactly those of the LST element. Using orthogonal transformation at vertex 1 and denoting $\alpha_n, \beta_n, \gamma_n$ as the normal displacements,

$$\gamma = -\alpha_n \sin B - \alpha \cos B \qquad \text{(on side 12)} \qquad (10.30)$$

$$\alpha = \gamma_n \sin B - \gamma \cos B \qquad \text{(on side 13)}. \qquad (10.31)$$

Substituting these results into eqn (10.28) yields

$$\frac{4\Delta\phi_1}{L_{21}L_{13}} = \frac{\partial \gamma_n}{\partial c} \sin B - \frac{\partial \gamma}{\partial c} \cos B + \frac{\partial \alpha_n}{\partial a} \sin B + \frac{\partial \alpha}{\partial a} \cos B. \qquad (10.32)$$

Constraining the angle at the vertex to remain constant and assuming linear variation of the parallel displacements the rotation at node 1 is given as

$$\psi_1 = \left(\frac{\partial \gamma_n}{\partial c}\right)_1 = \left(\frac{\partial \alpha_n}{\partial a}\right)_1$$

$$= \frac{4\Delta\phi_1/L_{21}L_{13} + (\gamma_1 - \gamma_3)\cos B/L_{13} - (\alpha_2 - \alpha_1)\cos B/L_{21}}{2\sin B} \qquad (10.33)$$

and ψ_2, ψ_3 are obtained in like fashion. These nodal rotations are not to be confused with the drilling freedoms and are not referred to Cartesian axes, that is, the nodal rotation given by eqn (10.33) is the rotation of the sides meeting at this node, and eqn (10.33) assumes that these 'side rotations' are

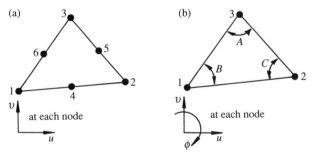

Fig. 10.8 Application of method of nested interpolations: (a) local freedoms and (b) global freedoms

equal. Now on side 12, for example, a quadratic variation of α_n, is allowed which gives

$$\alpha_{n4} = \frac{\alpha_{n1} + \alpha_{n2}}{2} + \frac{L_{21}(\psi_1 - \psi_2)}{8}. \tag{10.34}$$

Then substituting $\alpha_4 = (\alpha_1 + \alpha_2)/2$ and eqn (10.34) in $u_4 = c_{ax}\alpha_4 - c_{ay}\alpha_{n4}$, and also using eqns (10.25), one obtains

$$u_4 = \frac{u_1 + u_2}{2} + \frac{c_{ay}L_{21}(\psi_2 - \psi_1)}{8}. \tag{10.35}$$

Similarly

$$v_4 = \frac{v_1 + v_2}{2} + \frac{c_{ax}L_{21}(\psi_1 - \psi_2)}{8}. \tag{10.36}$$

Obtaining equations for u_5, v_5, u_6, v_6 in a similar fashion the transformation of variables required in Fig. 10.8 is accomplished by the equation

$$\{u_4, v_4, u_5, v_5, u_6, v_6\} = \tfrac{1}{2}I_6\{u_i, v_i\} + \tfrac{1}{8}N\{\psi_i\} \qquad i = 1 \rightarrow 3 \tag{10.37}$$

where

$$N^t = \begin{bmatrix} -y_{21} & x_{21} & 0 & 0 & y_{13} & -x_{13} \\ y_{21} & -x_{21} & -y_{32} & x_{32} & 0 & 0 \\ 0 & 0 & y_{32} & -x_{32} & -y_{13} & x_{13} \end{bmatrix}.$$

The vector $\{\psi_i\}$ is given by the three permutations of eqn (10.33), giving

$$\begin{Bmatrix} 2L_{21}L_{13}\tan B\,\psi_1 \\ 2L_{32}L_{21}\tan C\,\psi_2 \\ 2L_{13}L_{32}\tan A\,\psi_3 \end{Bmatrix} = G\{u_1, v_1, \phi_1, u_2, v_2, \phi_2, u_3, v_3, \phi_3\} = G\{d\} \tag{10.38}$$

where

$$G^t = \begin{bmatrix} L_{13}c_{ax} + L_{21}c_{cx} & -L_{32}c_{ax} & -L_{32}c_{cx} \\ L_{13}c_{ay} + L_{21}c_{cy} & -L_{32}c_{ay} & -L_{32}c_{cy} \\ 4\Delta/\cos B & 0 & 0 \\ -L_{13}c_{ax} & L_{21}c_{bx} + L_{32}c_{ax} & -L_{13}c_{bx} \\ -L_{13}c_{ay} & L_{21}c_{by} + L_{32}c_{ay} & -L_{13}c_{by} \\ 0 & 4\Delta/\cos C & 0 \\ -L_{21}c_{cx} & -L_{21}c_{bx} & L_{32}c_{cx} + L_{13}c_{bx} \\ -L_{21}c_{cy} & -L_{21}c_{by} & L_{32}c_{cy} + L_{13}c_{by} \\ 0 & 0 & 4\Delta/\cos A \end{bmatrix}$$

and appropriate coding avoids division by the cosines in eqn (10.38). The element stiffness matrix is given as $k = T^t k^* T$ where k^* is the LST stiffness matrix (Section 7.3), whilst T is a 12×9 matrix such that

$$\{u_i, v_i\} = T\{d\} \qquad i = 1 \rightarrow 6 \qquad (10.39)$$

and is obtained by combining eqns (10.37) and (10.38).

For the analysis of Fig. 6.3, Table 10.2 compares the results of the present nine-freedom element (DFT) with those of the CST and LST elements. Note that with the DFT element, boundary conditions for ϕ are not essential and have negligible effect upon the present results.[24] The DFT results appear to converge as rapidly as those of the LST and both the direct and shear stress results converge sensibly, regardless of whether the latter are centroidal or nodal values.

Applying eqn (3.62) the rate of convergence is approximately h^3, which is between the rates for the CST and LST elements, so that inclusion of ϕ has increased the rate of convergence and eliminated the shear stress difficulties. Thus the element is the first to justify inclusion of the drilling freedom. An improved version with less stringent constraints[25] at the vertices has been proposed[24] but results have not yet been reported.

Results obtained for the analysis of shallow shells by combining the DFT element with the plate element of Section 8.10 are of comparable accuracy to

Table 10.2 Finite element results for the analysis of Fig. 6.3

Element	Nodes	$-v_C$ (mm)	$(\sigma_x)_A$ (MPa)	$(\tau_{xy})_B$ (MPa)
CST	3×3	3.99	7.5	3.70
	4×4	4.77	12.7	4.06
	5×5	5.70	16.4	5.00
	6×6	6.29	18.7	5.16
	7×7	6.69	20.2	5.58
LST	3×3	6.78	25.2	3.08
	5×5	7.77	25.2	7.33
	7×7	7.86	25.2	5.74
DFT	3×3	6.98	19.8	8.00
	5×5	7.51	23.2	6.62
	7×7	7.84	24.5	6.52
	3×5	7.31	22.3	5.74
	2×2	3.68	–	–
	4×4	7.26	–	6.13[†]
	6×6	7.71	–	6.26[†]
Analytical		7.93	25.2	6.30

[†] centroidal values

those obtained with the doubly curved shell formulations described in Sections 11.5 and 11.6. As with a similar flat element formulation tested by Olson and Bearden,[21] large numbers of elements are needed to deal with deeper shells, whilst, as with most flat shell elements,[6] numerical difficulties are encountered in cylindrical shells.

10.7 Concluding remarks

Flat shell elements remain a useful means of deriving an approximate solution of shell problems. In the case of axisymmetric shells good results can be obtained with remarkably simple and economical elements.[3,4,9] For generally shaped shells, triangular elements are much more versatile than quadrilateral elements and inclusion of the drilling freedom can enhance their general performance.[24,26-28] Although less effective in deep shells, flat elements have been frequently used for the approximate analysis of shells[14,29] and perform well for relatively shallow shells such as hyperbolic paraboloids.[30,31]

References

1. Timoshenko, S. P. and Woinowsky-Krieger, W. K. (1956). *Theory of Plates and Shells, (2nd edn)*. McGraw-Hill, New York.
2. Zienkiewicz, O. C. (1977). *The Finite Element Method, (3rd edn)*. McGraw-Hill, London.
3. Rockey, K. C., Evans, H. R., Griffiths, D. W. and Nethercote, D. A. (1975). *The Finite Element Method: a Basic Introduction*. Crosby Lockwood Staples, London.
4. Grafton, P. E. and Strome, D. R. (1963). Analysis of axisymmetric shells by the direct stiffness method. *J. Amer Inst. Aeron. Astron.*, **1**, 2342.
5. Percy, J. H., Pian, T. H. H., Klein, S., and Navaratna, D. R. (1965). Application of matrix displacement method to linear elastic analysis of shells of revolution. *J. Amer Inst. Aeron. Astron.*, **3**, 2138.
6. Knowles, N. C., Razzaque, A., and Spooner, J. B. (1976). Experience of finite element analysis of shell structures. In *Finite Elements for Thin Shells*, (ed. D. G. Ashwell and R. H. Gallagher). Wiley, London.
7. Popov, E. P., Penzien, J., and Lu, Z. (1964). Finite element solution for axisymmetric shells. *J. Eng. Mech. Div. ASCE*, **90**, No. EM5.
8. Venkataramana, J. and Venkateswara Rao, G. (1975). Finite element analysis of moderately thick shells. *Nucl. Eng. Des.*, **33**, 398.
9. Zienkiewicz, O. C., Bauer, J., Morgan, K., and Onate, E. (1977). A simple element for axisymmetric shells with shear deformation. *Int. J. Num. Meth. Engng*, **11**, 1545.
10. Bose, G. K., McNiece, G. M., and Sherbourne, A. N. (1972). Column webs in steel beam to column connections. *Comput. Struct.*, **2**, 253.

11. Milner, H. R. and Grayson, R. (1977). *Computer User Manual for Non-linear Analysis of Stiffened Plates.* Simon Engineering Laboratories, University of Manchester.
12. Mohr, G. A. (1979). Design of shell shape using finite elements. *Comput. Struct.,* **10**, 745.
13. Parekh, C. J. (1969). *Finite Element Solution System.* Ph.D. thesis, University of Wales, Swansea.
14. Clough, R. W. and Johnson, C. P. (1968). A finite element approximation for the analysis of thin shells. *Int. J. Solids Struct.,* **4**, 43.
15. Aldstedt, E. (1969). Shell analysis using plane triangular elements. In *Finite Element Methods in Stress Analysis,* (ed. I. Holand and K. Bell). Tapir, Trondheim.
16. Mohr, G. A. (1981). A simple rectangular membrane element including the drilling freedom. *Comput. Struct.,* **13**, 483.
17. Tocher, J. L. and Hartz, B. J. (1967). Higher order finite element for plane stress. *J. Eng. Mech. Div. ASCE,* **93**, 5402.
18. Fraeijs de Veubeke, B. M. (1968). Discussion on reference 17 above. *J. Eng. Mech. Div. ASCE,* **94**, 359.
19. Holand, I. and Bergan, P. G. (1968). Discussion on reference 17 above. *J. Eng. Mech. Div. ASCE,* **94**, 698.
20. Dungar, R. and Severn, R. T. (1969). Triangular finite elements of variable thickness and their application to plate and shell problems. *J. Strain Analysis,* **4**, 10.
21. Olson, M. D. and Bearden, T. W. (1979). A simple triangular shell element revisited. *Int. J. Num. Meth. Engng,* **14**, 51.
22. Holand, I. (1969). The finite element method in plane stress analysis. In *Finite Element Methods in Stress Analysis,* (ed. I. Holand and K. Bell). Tapir, Trondheim.
23. Argyris, J. H. (1968). Three-dimensional anisotropic and inhomogeneous media, matrix analysis for small and large displacements. *Ingenieur Archiv.,* **31**, 33.
24. Mohr, G. A. (1982). Finite element formulation by nested interpolations: application to the drilling freedom problem. *Comput. Struct.,* **15**, 185.
25. Irons, B. M. and Ahmad, S. (1980). *Techniques of Finite Elements.* Ellis-Horwood, Chichester.
26. Cook, R. D. (1987). A plane hybrid element with rotational d.o.f. and adjustable stiffness. *Int. J. Num. Meth. Engng,* **24**, 1499.
27. Hughes, T. J. R. and Brezzi, F. (1989). On drilling degrees of freedom. *Comp. Meth. Appl. Mech. Engng,* **72**, 105.
28. Cook, R. D. (1991). Modified formulations for nine dof plane triangles that include vertex rotations. *Int. J. Num. Meth. Engng,* **31**, 825.
29. Megard, G. (1969). Planar and curved elements. In *Finite Element Methods in Stress Analysis,* (ed. I. Holand and K. Bell). Tapir, Trondheim.
30. Argyris, J. H., Dunne, P. C., Malehannakis, G. A., and Shelke, F. (1977). A simple triangular facet shell element with application to linear and non-linear equilibrium and elastic stability problems. *Comp. Meth. Appl. Mech. Engng,* **10**, 371.
31. Bathe, K. J. and Ho, L. W. (1981). A simple and effective element for analysis of general shell structures. *Comput. Struct.,* **13**, 673.

11
Curved shell elements

Several curved finite elements are described, beginning with an arch element and an axisymmetric thick shell element which is an extension of this. Rectangular finite elements for the analysis of cylindrical shells are then discussed, in particular the *strain element* of Ashwell and Sabir. Finally two triangular doubly curved shell elements are discussed, the second of which includes transverse shearing effects but is able to be used for thin shells with the application of penalty factors.

11.1 Quadratic arch element

A curved nine-freedom 'thick' arch element is shown in Fig. 11.1. The element is termed thick because a transverse shear strain is included but thin arches can also be dealt with by using a penalty factor to suppress this strain. As we shall see the element also provides a useful insight into the effects of reduced integration in coupled systems of equations.

The strain–displacement equations are a simplification of those given in Section 11.3 for an axisymmetric shell element (eqns (11.16)) and are

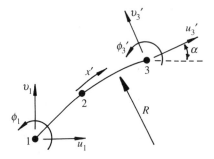

Fig. 11.1 Arch element, showing global freedoms at node 1 and local freedoms at node 3

$$\varepsilon = \frac{\partial u'}{\partial x'} + \frac{v'}{R} \tag{11.1}$$

$$\chi = \frac{1}{R}\frac{\partial u'}{\partial x'} + \frac{\partial \phi}{\partial x'} \tag{11.2}$$

$$\gamma = \frac{\partial v'}{\partial x'} - \phi \tag{11.3}$$

where x' is a curvilinear coordinate along the element arc. Using quadratic interpolation with a dimensionless coordinate $s = 2x'/L$, $s = -1 \rightarrow 1$ (where L is the element arc length) for the local freedoms u', v', and ϕ', the strain interpolation matrix B is obtained as

$$\begin{Bmatrix} \varepsilon \\ \chi \\ \gamma \end{Bmatrix} = \begin{bmatrix} \overset{u \text{ columns}}{a\{\dot{f}\}^{\mathrm{t}} - b\{f\}^{\mathrm{t}}/R} & \overset{v \text{ columns}}{b\{f\}^{\mathrm{t}} + a\{f\}^{\mathrm{t}}/R} & \overset{\phi \text{ columns}}{0} \\ a\{\dot{f}\}^{\mathrm{t}}/R & b\{f\}^{\mathrm{t}}/R & -\{\dot{f}\}^{\mathrm{t}} \\ -b\{\dot{f}\}^{\mathrm{t}} & a\{\dot{f}\}^{\mathrm{t}} & -\{f\}^{\mathrm{t}} \end{bmatrix}\{d\}$$

$$= \begin{Bmatrix} B_{\mathrm{m}} \\ B_{\mathrm{f}} \\ B_{\mathrm{s}} \end{Bmatrix}\{d\} \tag{11.4}$$

and

$$\{f\} = \{(s^2 - s)/2, 1 - s^2, (s^2 + s)/2\} \qquad s = -1 \rightarrow 1 \tag{11.5}$$

$$\{\dot{f}\} = \{(2s - 1), -4s, (2s + 1)\}/L \qquad s = -1 \rightarrow 1 \tag{11.6}$$

where $a = \cos \alpha$, $b = \sin \alpha$ transform to the local freedoms and L is the element arc length. The element stiffness matrix is best obtained by a two-point Gauss integration (see Table 11.2) as

$$k = \sum \omega_i L\{(EA)B_{\mathrm{m}}^{\mathrm{t}}B_{\mathrm{m}} + (EI)B_{\mathrm{f}}^{\mathrm{t}}B_{\mathrm{f}} + \beta(\tfrac{5}{6}GA)B_{\mathrm{s}}^{\mathrm{t}}B_{\mathrm{s}}\} \tag{11.7}$$

where β is a penalty factor and the element is here assumed to be of constant thickness. For applications to rib stiffened shells a linearly varying torsional strain can also be included without difficulty.

Table 11.1 shows results obtained for the analysis of a cantilever carrying an end load P with various penalty factor values. With $\beta = 1$ the result is close to the deep beam solution whilst with $\beta = 100$ it is closest to the shallow beam solution. This suggest that the penalty factor should be evaluated as $\beta = 100 t^2$ (not $\beta = \mathrm{const.}t^2/h^2$ as in eqn (9.33) as it now shifts the shearing stiffnesses towards the membrane stiffnesses, which are at least in part of the same dimensionality $(\partial u'/\partial x)$ as those for the shear strains $(\partial v'/\partial x - \phi)$ and this is confirmed in Table 11.1.

Table 11.1 Maximum transverse displacement in a cantilever with $P = 1$, $L = 1$, $E = 10^6$, $v = 0$, $b = 1$, $t = 1$, ten elements, and various penalty factors (seven-digit numerical computation)

t	β	Finite element	From eqn (11.8)
1.000	5000	5.143×10^{-6}	4.000×10^{-6}
	500	4.086×10^{-6}	4.005×10^{-6}
	100	4.044×10^{-6}	4.024×10^{-6}
	50	4.057×10^{-6}	4.048×10^{06}
	5	4.481×10^{-6}	4.480×10^{-6}
	1	6.401×10^{-6}	6.400×10^{-6}
0.100	1.0000	4.046×10^{-3}	–
0.010	0.0100	4.047	–
0.001	0.0001	4.000×10^3	–
Analytical, deep beam		6.400×10^{-6} $(t = 1)$	
Analytical, slender beam		$4.000 \times 10^{-6}/t^3$	

Table 11.1 also shows the numerical solutions expected in view of the penalty factors used, that is,

$$v_{\max} = \frac{PL^3}{3EI} + \frac{1.2\,PL}{\beta GA}. \tag{11.8}$$

Comparing the computer solutions with those expected one observes that the round-off error increases markedly when $\beta = 5000$, that is, when the expected solution is

$$v_{\max} = (4.00000 + 0.00048) \times 10^{-6}. \tag{11.9}$$

Noting that seven-digit computation is being used and that beam elements involve terms of the form $12EI/L^3$ and $4EI/L$, so that approximately four digits are needed to recover the flexural component of $v_{\max}$ (that is, 4×10^{-6}), one sees that the shearing component of $v_{\max}$ cannot be recovered and the coupling of the transverse displacements to the rotations in eqn (11.3) is lost, resulting in a poor solution when $\beta = 5000$.

If higher-precision computation is used with $\beta = 100\,t^2$ one still obtains the same results, and this has been verified by testing the axisymmetric element of Section 11.3 with both seven- and sixteen-digit computation. Higher precision will allow larger penalty factors to be used successfully but it proves instructive to use minimum precision here as this more quickly reveals the mechanisms of penalty factor behaviour.

Table 11.2 shows the results obtained when ten arch elements are used to analyse one half of a clamped circular arch carrying a transverse load at the

Table 11.2 Solutions for a circular clamped arch with unit central load and $E = 10^6, v = 0, b = 1, R = 160$ and $t = 4$. V_A and T_A are abutment shear and thrust

	v_{max}	V_A	T_A
Simpson's rule	0.416	1.044	0.577
Two-point Gauss	0.465	0.446	0.537
Three-point Gauss	0.393	0.889	0.027
Exact[2]	0.450	0.470[†]	0.543[†]

[†]at nearest element centroid

crown[1]. Two-point Gauss gives best results, though three points integrate eqn (11.7) exactly. This parallels the situation in Section 9.4 for a thick plate element and can again be explained by evaluating the $\{f\}\{f\}^t$ terms that were not exactly evaluated by the reduced integration (note that all the integrations considered here give the exact result for $\{\dot{f}\}\{f\}^t$ and $\{f\}\{\dot{f}\}^t$). Three-point Gauss yields the exact result which is

$$\int \{f\}\{f\}^t dL = \frac{L}{90} \begin{bmatrix} 12 & 6 & -3 \\ 6 & 48 & 6 \\ -3 & 6 & 12 \end{bmatrix}. \qquad (11.10)$$

Simpson's rule gives a similar result to that had by lumping the stiffnesses;

$$\int \{f\}\{f\}^t dL = \frac{L}{90} \begin{bmatrix} 15 & 0 & 0 \\ 0 & 60 & 0 \\ 0 & 0 & 15 \end{bmatrix} \qquad (11.11)$$

whilst two-point Gauss gives

$$\int \{f\}\{f\}^t dL = \frac{L}{90} \begin{bmatrix} 10 & 10 & -5 \\ 10 & 40 & 10 \\ -5 & 10 & 10 \end{bmatrix} \qquad (11.12)$$

which is clearly a better approximation to the exact result, maintaining coupling through the off-diagonal terms but reducing the diagonal stiffnesses, thereby accelerating convergence (see also Section 15.11).

11.2 Numerical calculation of arch element geometry

To calculate the arc length of the arch element of Section 11.1 (which is needed in eqns (11.6) and (11.7)) one writes an interpolation of the nodal elevations as

$$y = \tfrac{1}{2}(s^2 - s)y_1 + (1 - s^2)y_2 + \tfrac{1}{2}(s^2 + s)y_3 = \sum f_i y_i$$

where $s = -1 \rightarrow 1$ is the dimensionless coordinate along the plan projection of the element and the nodes are equispaced in plan. Then the arc length of the element is given by N-point one-dimensional numerical integration as

$$L = P \sum_{j=1}^{N} \omega_j \left\{ 1 + \left(\frac{2}{P}\right)^2 \left[\sum_{i=1}^{3} \left(\frac{\mathrm{d}f_i}{\mathrm{d}s}\right)_j y_i \right]^2 \right\}^{1/2} \tag{11.13}$$

where $P = X_3 - X_1$ is the plan length of the element and $(\mathrm{d}f_i/\mathrm{d}s)_j$ denotes the derivative of f_i at integration point s_j. As for the stiffness matrix and stress calculation the cubic Gauss points $\pm 1/\sqrt{3}$ give best results.[1]

Accurate calculation of element curvature (which is needed in eqn (11.4)) requires quintic interpolation (see eqn (5.33)) but this requires nodal slopes as well as elevations as data—a considerable disadvantage in the case of general shell applications. An alternative scheme is to fit a circle to the nodes, as shown in Fig. 11.2. The curvature is then given by[2]

$$\frac{1}{R} = \frac{4A}{abc} \tag{11.14}$$

and this result is given a sign depending upon whether the midside node lies inside or outside the chord c. The lengths a, b, c are easily estimated using Pythagoras' theorem and the triangular area which they enclose is given by the semi-perimeter formula

$$A = [p(p - a)(p - b)(p - c)]^{1/2} \tag{11.15}$$

where $p = (a + b + c)/2$. Equation (11.14) gives virtually exact results around

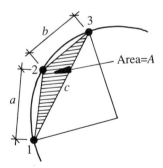

Fig. 11.2 Fitting a circle to the nodes

a circular arc, as might be expected, but in general is approximate. Equations (11.13) and (11.14) are also applied to the sides of triangular shell elements in Section 11.7 in conjunction with the natural strain approach.

11.3 Curved axisymmetric shell element

Figure 11.3 shows a nine-freedom curved axisymmetric shell element which includes transverse shearing strains. Cook[3] also derives a similar element but condenses out the internal freedoms. The strain–displacement equations are an extension of eqns (10.4)–(10.7),[4]

$$\varepsilon_x = \frac{\partial u'}{\partial x'}, \qquad \varepsilon_c = \frac{w}{r}$$

$$\chi_x = \frac{-\partial \phi}{\partial x'} + \frac{\partial(u'/R)}{\partial x'}, \qquad \chi_c = \frac{\sin \alpha\{(u'/R) - \phi\}}{r} \qquad (11.16)$$

$$\gamma = \frac{\partial w'}{\partial x'} - \phi,$$

obtained by adding the shear strain and an additional term for χ_x which incorporates the correction shown in Fig. 11.7. The modulus matrix is given by including the shear stiffness in eqn (10.9), yielding

$$
\begin{bmatrix} N_x \\ N_c \\ m_x \\ m_c \\ Q \end{bmatrix} = \frac{Et}{1 - v^2}
\begin{bmatrix}
1 & v & 0 & 0 & 0 \\
v & 1 & 0 & 0 & 0 \\
0 & 0 & t^2/12 & vt^2/12 & 0 \\
0 & 0 & vt^2/12 & t^2/12 & 0 \\
0 & 0 & 0 & 0 & 5(1 - v)/12
\end{bmatrix}
\begin{bmatrix} \varepsilon_x \\ \varepsilon_c \\ \chi_x \\ \chi_c \\ \gamma \end{bmatrix} = D\{\varepsilon\}.
$$

$$(11.17)$$

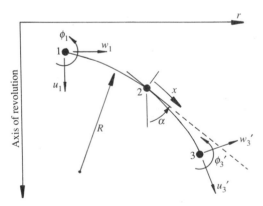

Fig. 11.3 Curved axisymmetric shell element

Using quadratic interpolation the strain interpolation matrix is given by

$$
\begin{array}{ccc}
u \text{ columns} & w \text{ columns} & \phi \text{ columns}
\end{array}
$$

$$
\begin{bmatrix} \varepsilon_x \\ \varepsilon_c \\ \chi_x \\ \chi_c \\ \gamma \end{bmatrix} = \begin{bmatrix} c\{\dot{f}\}^t & s\{\dot{f}\}^t & \{0\}^t \\ \{0\}^t & \{f\}^t/r & \{0\}^t \\ c\{\dot{f}\}^t/R & s\{\dot{f}\}^t/R & -\{\dot{f}\} \\ cs\{f\}^t/Rr & s^2\{f\}^t/Rr & -s\{f\}^t/r \\ -s\{\dot{f}\}^t & c\{\dot{f}\}^t & -\{f\}^t \end{bmatrix} = B\{d\} \quad (11.18)
$$

where $\{f\}$, $\{\dot{f}\}$ are given by (11.5) and (11.6) and $c = \cos\alpha$, $s = \sin\alpha$. Using a penalty factor to suppress the transverse shearing the stiffness matrix is given by

$$
k = \sum (2\pi Lr\omega_i)\left[B_f^t D_f B_f + \beta B_s^t \left\{ \frac{5Et}{12(1+v)} \right\} B_s \right] \quad (11.19)
$$

where B_f is given by the first four rows of (11.18), D_f is the corresponding 4×4 partition of (11.17), and B_s is given by the last row of (11.18). The consistent loads for distributed transverse pressure (p) on an element are obtained as

$$
\begin{array}{cc}
u \text{ columns} & w \text{ columns}
\end{array}
$$

$$
\{q\}^t = \sum (2\pi p Lr\omega_i)\left[-s\{f\}^t \quad c\{f\}^t \right]. \quad (11.20)
$$

As for the arch element of Fig. 11.1 two-point Gauss with $\beta = 100\,t^2$ is found to give best results.[5]

Figure 11.4 shows the element used for the analysis of an internally pressurized sphere with $\beta = 0.01$ and in Table 11.3 the results are compared to those with the flat six-freedom element given in Section 10.2. With a comparable mesh the nine-freedom element is much more accurate and performs reasonably well with only three elements (that is, seven nodes) whereas the flat element gives nonsensical results with so few nodes. Figure 11.5 shows the results with the nine-freedom element if the sphere is subjected to pinching loads at each pole. Here thirty-three nodes are used so that, as expected from Table 11.3, satisfactory agreement with the analytical solution is obtained.[5]

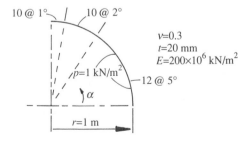

Fig. 11.4 Internally pressurized sphere with axisymmetric elements

Table 11.3 Solutions for the internally pressurized sphere shown in Fig. 11.4 using the six-freedom conical frustum element of Section 10.2 and a nine-freedom curved element including transverse shear and penalty factors

α°	34 nodes[†]	33 nodes[‡]	11 nodes[‡]	7 nodes[‡]
Radial displacement, m $\times 10^{-7}$				
0	0.8710	0.8737	0.8676	0.8563
50	0.8782	0.8737	0.8788	0.8835
90	0.9046	0.8768	0.8722	0.7610
Exact	0.8750			
Meridional force/unit thickness, kN/m				
0	0.4983	0.5019	0.5093	0.5223
50	0.4998	0.5019	0.4954	0.4890
90	0.5059	0.5004	0.4994	0.4861
Exact	0.5000			
Circumferential force/unit thickness, kN/m				
0	0.4995	0.5001	0.4999	0.4992
50	0.5010	0.5001	0.5001	0.4999
90	0.5059	0.5004	0.5005	0.5123
Exact	0.5000			

[†]six-freedom flat element   [‡]nine-freedom curved element

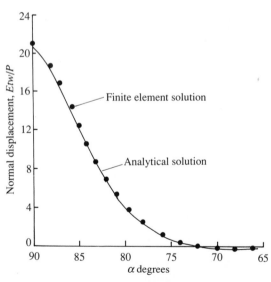

Fig. 11.5 Displacement at pole of pinched sphere

11.4 Cylindrical shell elements

The simplest choice for a rectangular cylindrical shell element is to couple the ACM plate element (eqns (8.3)) with the bilinear plane stress element (eqn 7.2c), giving a total of twenty freedoms.[6] Another alternative is to replace the ACM element with the bicubic, giving twenty-four freedoms.[7] Cantin and Clough[8] find that the convergence is improved, however, if the exact rigid body modes for the shell are included in the displacement functions. The strain–displacement equations are obtained from eqns (2.40) by substituting $1/R_{xy} = 1/R_x = 0$, $R_y = R$, and $y = \phi R$, giving

$$\varepsilon_x = \frac{\partial u}{\partial x} \tag{11.21a}$$

$$\varepsilon_\phi = \frac{1}{R}\left(\frac{\partial v}{\partial \phi} + w\right) \tag{11.21b}$$

$$\varepsilon_{x\phi} = \frac{1}{R}\frac{\partial u}{\partial \phi} + \frac{\partial v}{\partial x} \tag{11.21c}$$

$$\chi_x = \frac{-\partial^2 w}{\partial x^2} \tag{11.21d}$$

$$\chi_\phi = \tfrac{1}{2}\left(\frac{\partial v}{\partial \phi} - \frac{\partial^2 w}{\partial \phi^2}\right) \tag{11.21e}$$

$$\chi_{x\phi} = \frac{1}{R}\left(\frac{\partial v}{\partial x} - \frac{\partial^2 w}{\partial x \partial \phi}\right). \tag{11.21f}$$

To determine the exact rigid body modes for u, v, w the left sides of these equations are set equal to zero and the right sides integrated. Eliminating $\partial v/\partial \phi$ from eqns (b) and (e) one obtains $\partial^2 w/\partial \phi^2 + w = 0$, the general solution for which may be written as

$$w_{\text{rb}} = -(c_1 + c_2 x)\cos \phi - (c_3 + c_4 x)\sin \phi \tag{11.22}$$

as the coefficients of $\cos \phi$ and $\sin \phi$ may be linear in x and still satisfy (d).

Equation (b) requires $\partial v/\partial \phi = -w$ and, combining this result with eqn (11.22) and integrating, yields

$$v_{\text{rb}} = (c_1 + c_2 x)\sin \phi - (c_3 + c_4 x)\cos \phi + c_6 \tag{11.23}$$

and this also satisfies (f). Equation (c) requires $\partial u/\partial \phi = -R\partial v/\partial x$, and this yields

$$u_{\text{rb}} = Rc_2 \cos \phi + Rc_4 \sin \phi + c_5 \tag{11.24}$$

which also satisfies (a). With the inclusion of the exact rigid body modes the interpolation matrix must be obtained numerically. Hence complexity of the interpolation functions is no longer a difficulty and the approach of Ashwell and Sabir[9] in which assumed strain interpolations are integrated, adding the results to the rigid body modes, is feasible. Their element is shown in Fig. 11.6, having five nodal degrees of freedom and using local *curvilinear* co-ordinates x and $y = \phi R$ on the shell surface.

The freedom $(\partial w / \partial \phi - v)/R$ arises from integrating eqn (11.21e), and is used to represent the true slope[10] at each point in a deep shell, as shown in Fig. 11.7. This freedom is not inconvenient here as numerical inversion is used to obtain the interpolation matrix.

The key step is to choose the assumed strain functions

$$\varepsilon_x = c_7 + c_8 x, \qquad \varepsilon_\phi = c_9 + c_{10}\phi, \qquad \varepsilon_{x\phi} = c_{11},$$

$$\chi_x = c_{12} + c_{13}x + c_{14}\phi + c_{15}x\phi$$

$$\chi_\phi = c_{16} + c_{17}x + c_{18}\phi + c_{19}x\phi, \qquad \chi_{x\phi} = c_{20}. \qquad (11.25)$$

These correspond to the 8 df plane stress element with constant shear and the ACM plate element restricted to constant twist. Ashwell and Sabir equate the

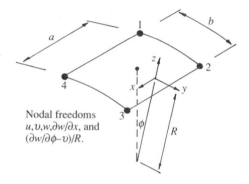

Nodal freedoms
$u, v, w, \partial w/\partial x$, and
$(\partial w/\partial\phi - v)/R$.

Fig. 11.6 Thin shell element with cylindrical coordinates $x, \phi = y/R$

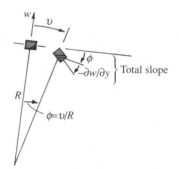

Fig. 11.7 Calculation of effective slope in cylindrical shell elements

right sides of eqns (11.21) and (11.25) and by integration determine the displacement functions corresponding to the assumed strains. Augmenting these with eqns (11.22)–(11.24) and with corrections obtained by integrating the compatibility conditions[11] the displacement functions are strongly coupled and numerical inversion is needed to obtain a 20×20 interpolation matrix C. For this purpose Gauss–Jordan reduction with pivotal condensation should be used. The kernel stiffness matrix $k*$ is able to be evaluated explicitly using eqn (6.18) and the final element stiffness matrix is given by $k = C^t k* C$.[9]

Although the element is somewhat specialized (so that only outline details are given here) some aspects of the formulation may find wider application, particularly the step leading to the choice of eqns (11.25). The element gives excellent displacement solutions[12] provided double precision computation is used (doubtless this is needed because of the use of transcendental functions in the interpolations). Although it is highly non-conforming, satisfactory stress solutions can be obtained by averaging at the nodes or restriction to centroidal values (which seems obligatory for N_ϕ)[13] but the element of Section 11.6 gives equally good stresses.

11.5 Doubly curved triangular shell elements

Flat triangular shell elements typically employ a cubic interpolation for the normal displacement and linear interpolations for the tangential displacements. The latter allow only constant extensional strains and improved flat shell elements generally seek to remedy this.[14] A number of curved shell elements have also employed linear interpolations for u, v[15-18] but these perform poorly when rigid body motions are significant.[19] The fifteen-freedom triangular element of Utku,[15] for example, includes transverse shear but as all strains are constant it converges very slowly.

Many curved elements, therefore, use cubic interpolations for u, v and, like the flat elements, employ a still higher-order interpolation for the normal displacement, w. An example is the triangular element of Cowper, Lindberg and Olson[20-22] which couples the 18 df quintic plate element with the cubic plane stress element of Fig. 6.4(b), condensing the centroidal freedom out of the latter so that the shell element has thirty-six freedoms. This element performs well, though greater difficulty is found with deep shells.[23]

Argyris and Scharpf's SHEBA element,[24] on the other hand, uses the quintic interpolations of the TUBA plate elements[25] for u, v and w. The element gives excellent results,[26] the rate of convergence being not much reduced from that of the quintic plate element.[27] The element requires sixty-three freedoms and, although this can be reduced to fifty-four,[28,29] it is costly to use.

Gallagher and Thomas[30] obtain a more economical element using cubic interpolations for u, v, and w together with Lagrange multiplier constraints in order to enforce C_1 continuity. Including these constraints the element requires thirty-six freedoms and gives excellent displacement solutions, though the same degree of accuracy is not obtained in stress calculations.[31] Hybrid elements have also been developed for doubly curved shells[32-34] but none of these prove more accurate than the best displacement elements.

An approximate approach to shells, which avoids the complexities of shell theory, is the use of three-dimensional elements[35] in which simple constraints can be used to coalesce the nodes at the neutral surface (see Section 7.5). Irons' SemiLoof shell element is an extension of this approach in which transverse shear constraints are also included, resulting in a quadrilateral element with thirty-two freedoms, some of these at the 'Loof nodes' which are located at the two cubic Gauss points on each side.[36,37]

In the remainder of this section a twenty-seven freedom quartic/quadratic thin shell element[38] is described, combining the quartic plate element of Section 8.8 with the LST element. Simplifying eqns (2.40),[29,38] the strain–displacement equations are

$$\chi_a = \frac{\partial^2 w}{\partial s_a^2}, \qquad \chi_b = \frac{\partial^2 w}{\partial s_b^2}, \qquad \chi_c = \frac{\partial^2 w}{\partial s_c^2}$$

$$\varepsilon_x = \frac{\partial u}{\partial x} + \frac{w}{R_x}, \qquad \varepsilon_y = \frac{\partial v}{\partial y} + \frac{w}{R_y} \tag{11.26}$$

$$\varepsilon_{xy} = \frac{\partial u}{\partial y} + \frac{\partial v}{\partial x} + \frac{2w}{R_{xy}}$$

using natural curvatures and Cartesian strains, and the strain matrix is

$$S = \begin{bmatrix} 0 & S_b & 0 \\ S_{ba} & 0 & S_a \end{bmatrix} \begin{Bmatrix} \{a\} \\ \{u\} \\ \{v\} \end{Bmatrix} \tag{11.27}$$

where S_a is obtained from eqn (6.30), S_b is obtained from eqn (8.45), and $\{a\}$ are the amplitudes of the quartic modes of eqn (8.43). Using

$$S_{ba} = \{L_1, L_2, L_3\}\{1/R_x, 1/R_y, 1/R_{xy}\}^t \tag{11.28}$$

the membrane and bending actions are coupled through the rigid body plate bending terms only.[38] Such limited coupling has been found by Wennerström to yield satisfactory results in applications to the analysis of large deflections of plates.[39] The 27×27 stiffness matrix is given by

$$k = T_s^t A_s^t \left\{ \sum S^t D_s S(\omega_i \Delta) \right\} A_s T_s \tag{11.29}$$

where

$$
T_s = \begin{bmatrix} T(15 \times 15) & 0 \\ 0 & I_{12} \end{bmatrix} \qquad A = \begin{bmatrix} A(15 \times 15) & 0 \\ 0 & I_{12} \end{bmatrix}
$$

$$
D_s = \begin{bmatrix} D & 0 \\ 0 & D_a \end{bmatrix} \qquad\qquad D_a = \left(\frac{12}{t^2}\right)D \tag{11.30}
$$

and T is given by eqn (5.90), A by eqn (5.92), and D by eqn (2.29). Thus the kernel stiffness matrix is given as

$$
S^t D_s S = \begin{bmatrix} S_{ba}^t D_a S_{ba} & 0 & S_{ba}^t D_a S_a \\ 0 & S_b^t D S_b & 0 \\ S_a^t D_a S_{ba} & 0 & S_a^t D_a S_a \end{bmatrix} \tag{11.31}
$$

and $S_b^t D S_b$, $S_a^t D_a S_a$ give the stiffness matrices for the separate bending and membrane elements. Without this limited coupling the element performs poorly, whether the formulation is explicitly or numerically integrated.[38] With full coupling terms in L_i^8 must be integrated, which requires twenty-five Gauss points. However, with limited coupling a selective three/seven-point integration (the latter being needed only for the $S_b^t D S_b$ term) exactly integrates eqn (11.31). Indeed, as well as reducing computation times, this reduced coupling approach gives much better results.

The element has one degree of freedom less at the midside nodes than at the vertices and some programs do not deal with this situation. One simple solution is to allow a fifth 'dummy freedom' at the midside nodes with no stiffness and skip this freedom during solution of the system equations to prevent pivotal division overflow or, as is done with the drilling freedom in many flat shell elements, to allow a small artificial stiffness.

The quartic interpolation used for flexure is, of course, non-conforming and this can be remedied by the techniques discussed in Chapter 4, an excellent example of such correction being the Lagrange multiplier cubic/cubic formulation developed by Gallagher and Thomas.[30] However, as the quartic thin plate element results of Table 8.3 suggest, the quartic part of the present formulation will, in fact, provide acceptable results and the quadratic part of the formulation has the same level of continuity.

The element provides a useful example of thin shell element formulation and the 'limited coupling' procedure, like reduced integration for example, represents yet another development in finite element techniques. However, the thick shell formulation of the following section can, with the use of penalty factors, deal with thin shells and is thus more widely applicable whilst yielding in general more accurate results than the element above. Moreover it is a good deal more economical.

11.6 Quadratic triangular shell element

To extend the thick plate element of Section 9.3 to deal with shells the LST element is incorporated, giving what would seem to be the minimum representations of moment and extensional stress, namely linear. The extensional strains are written as[40]

$$\varepsilon_x = \frac{\partial u}{\partial x} + \frac{w}{R_x}$$

$$\varepsilon_y = \frac{\partial v}{\partial y} + \frac{w}{R_y}$$

$$\varepsilon_{xy} = \frac{\partial u}{\partial y} + \frac{\partial v}{\partial x} + \frac{2w}{R_{xy}} \tag{11.32}$$

and these are evaluated using the Jacobian transformation

$$\begin{Bmatrix} \partial(\ )/\partial x \\ \partial(\ )/\partial y \end{Bmatrix} = \frac{1}{2\Delta} \begin{bmatrix} -y_{32} & -y_{13} \\ x_{32} & x_{13} \end{bmatrix} \begin{Bmatrix} \partial(\ )/\partial L_1 \\ \partial(\ )/\partial L_2 \end{Bmatrix} \tag{11.33}$$

to obtain the Cartesian derivatives, restricting attention to straight sided elements (in the local curvilinear coordinate system).

Combining eqns (9.13)–(9.17) and eqns (11.32) and (11.33) the complete strain–displacement matrix for the thirty freedom shell element is obtained as

$$\{\varepsilon\} = \{ \chi_a, \chi_b, \chi_c, \gamma_a, \gamma_b, \gamma_c, \varepsilon_x, \varepsilon_y, \varepsilon_{xy} \} = B\{d\} \tag{11.34}$$

where

$$
B = \begin{bmatrix} B_b \\ B_s \\ B_m \end{bmatrix} =
\begin{array}{ccccc}
u & v & w & \phi_x & \phi_y \\
\text{columns} & \text{columns} & \text{columns} & \text{columns} & \text{columns} \\
\end{array}
$$

$$
\begin{bmatrix}
\{0\}^t & \{0\}^t & \{0\}^t & c_{ax}\{\dot{f}_a\}^t & c_{ax}\{\dot{f}_a\}^t \\
\{0\}^t & \{0\}^t & \{0\}^t & c_{bx}\{\dot{f}_b\}^t & c_{bx}\{\dot{f}_b\}^t \\
\{0\}^t & \{0\}^t & \{0\}^t & c_{cx}\{\dot{f}_c\}^t & c_{cy}\{\dot{f}_c\}^t \\
\{0\}^t & \{0\}^t & \{\dot{f}_a\}^t & -c_{ax}\{f\}^t & -c_{ay}\{f\}^t \\
\{0\}^t & \{0\}^t & \{\dot{f}_b\}^t & -c_{bx}\{f\}^t & -c_{by}\{f\}^t \\
\{0\}^t & \{0\}^t & \{\dot{f}_c\}^t & -c_{cx}\{f\}^t & -c_{cy}\{f\}^t \\
\{\dot{f}_x\}^t & \{0\}^t & \{f\}^t/R_x & \{0\}^t & \{0\}^t \\
\{0\}^t & \{\dot{f}_y\}^t & \{f\}^t/R_y & \{0\}^t & \{0\}^t \\
\{\dot{f}_y\}^t & \{\dot{f}_x\}^t & 2\{f\}^t/R_{xy} & \{0\}^t & \{0\}^t \\
\end{bmatrix}
$$

where

$$\{\dot{f}_a\} = \frac{\{1 - 4L_1, 4L_2 - 1, 0, 4L_1 - 4L_2, 4L_3, -4L_3\}}{L_{21}}$$

$$\{\dot{f}_b\} = \frac{\{0, 1 - 4L_2, 4L_3 - 1, -4L_1, 4L_2 - 4L_3, 4L_1\}}{L_{32}}$$

$$\{\dot{f}_c\} = \frac{\{4L_1 - 1, 0, 1 - 4L_3, 4L_2, -4L_2, 4L_3 - 4L_1\}}{L_{13}}$$

$$\left\{\begin{matrix} \{\dot{f}_x\}^t \\ \{\dot{f}_y\}^t \end{matrix}\right\} = \frac{1}{2\Delta} \begin{bmatrix} -y_{32} & -y_{13} \\ x_{32} & x_{13} \end{bmatrix} \left\{\begin{matrix} \{\partial f/\partial L_1\}^t \\ \{\partial f/\partial L_2\}^t \end{matrix}\right\}$$

$$\{\partial f/\partial L_1\} = \{4L_1 - 1, 0, 4L_1 + 4L_2 - 3, 4L_2, -4L_2, 4 - 8L_1 - 4L_2\}$$

$$\{\partial f/\partial L_2\} = \{0, 4L_2 - 1, 4L_1 + 4L_2 - 3, 4L_1, 4 - 4L_1 - 8L_2, -4L_1\}$$

and $\{\dot{f}_a\}$ denotes $\{\partial f/\partial a\}$, $\{\dot{f}_x\}$ denotes $\{\partial f/\partial x\}$, and so on.

Here a schematic representation is used for matrix B. In fact the entries in each row must correspond with the sequence of freedoms in the element displacement vector which is

$$\{d\} = \{u_1, v_1, w_1, \phi_{x1}, \phi_{y1}, \ldots, u_6, v_6, w_6, \phi_{x6}, \phi_{y6}\}. \quad (11.35)$$

Now the natural curvatures and shearing strains must be transformed to Cartesian values and this is accomplished by the transformations

$$\{\chi_x, \chi_y, \chi_{xy}\} = H\{\chi_a, \chi_b, \chi_c\} \quad (11.36)$$

where $H = C^t W^{-1}$ or C^{-1} with W, C defined by eqns (8.15) and (8.16) and

$$\{\gamma_x, \gamma_y\} = T_G\{\gamma_a, \gamma_b, \gamma_c\} \quad (11.37)$$

where T_G is given by eqn (9.25).

Finally the element stiffness matrix is given by numerical integration, including a penalty factor to suppress the shears in thin shells,

$$k = \sum (\omega_i \Delta) [B_b^t D_b^* B_b + \beta B_s^t D_s^* B_s + B_m^t D_m B_m] \quad (11.38)$$

where the penalty factor β is given by eqn (9.35) (but for thick shells one uses $\beta = 1$) and the 'natural' modulus matrices D_b^* and D_s^* are given by the congruent transformations

$$D_b^* = H^t D_b H, \qquad D_s^* = T_G^t D_s T_G \quad (11.39)$$

with

$$D_{\rm m} = \frac{Et}{(1 - v^2)}\begin{bmatrix} 1 & 0 & 0 \\ 0 & 1 & 0 \\ 0 & 0 & (1 - v)/2 \end{bmatrix}$$

$$D_{\rm b} = \left(\frac{t^2}{12}\right)D_{\rm m}$$

$$D_{\rm s} = \frac{5Et}{12(1 + v)}\begin{bmatrix} 1 & 0 \\ 0 & 1 \end{bmatrix}.$$

As with the quartic/quadratic element of Section 11.5 the interaction between the flexural and membrane actions can be reduced, using only a linear interpolation for the terms of the form $\{f\}^t/R_x$ in eqn (11.34). Indeed this is found necessary when integration at the midside nodes is used[41,42] but, in fact, as shown for the parent thick plate element in Section 9.4, three-point integration at internal points yields better results and with this such reduced interpolation is not required.[42] Moreover this shift of the integration points, without change in number, demonstrates that integrations with a Gaussian character (that is, those involving symmetrically placed internal points) generally yield best results in finite elements.

As shown in Fig. 9.2 the parent thick plate element can be applied to thin plates with reasonable accuracy without the use of penalty factors and thus the shell element proves reliable in both thin and thick shell analysis. When applied to a thin cylinder with uniform radial pressure, provided the consistent loads ($q\Delta/3$ at the midside nodes) are used, the element yields the exact solution regardless of the number of elements used or their thickness, whereas flat shell elements are needed in large numbers for such problems and frequently fail to yield a solution, particularly when the shell is reasonably thin.[37] Analytically, of course, this is a trivial problem and results for a more demanding problem are given in the following section.

11.7 Quadratic triangular shell element with global coordinate system

To extend the shell element of Section 11.6 to deal with arbitrary shells the LST element is now incorporated with a 'natural' formulation. Unlike the formulation of Section 11.6 in which local nodal coordinates on the shell surface and curvatures $1/R_x$, $1/R_y$, and $1/R_{xy}$ are specified by the user, the following formulation calculates this data for shells of arbitrary shape from Cartesian nodal coordinate data. Now, including the effect shown in Fig. 11.7

in both the curvature and shear deformation calculations one writes

$$\chi_a = -\frac{1}{R_a}\left(\frac{\partial u_a}{\partial s_a}\right) + \frac{\partial \phi_a}{\partial s_a} \qquad a \to b, c$$

$$\gamma_a = \frac{\partial w}{\partial s_a} - \phi_a + \frac{u_a}{R_a} \qquad a \to b, c \qquad (11.40)$$

$$\varepsilon_a = \frac{\partial u_a}{\partial s_a} + \frac{w}{R_a} \qquad a \to b, c$$

where u_a, s_a, R_a, and ϕ_a are respectively curvilinear displacement, curvilinear coordinate, radius of curvature, and flexural rotation referred to side 'a' of the element (that is, that joining nodes 1 and 2). These equations are repeated for sides b and c as $a \to b, c$ denotes. The displacements u_a are given by

$$u_a = c_{ax}u + c_{ay}v \qquad a \to b, c \qquad (11.41)$$

where the direction cosines for the first side, for example, are given by

$$c_{ax} = \frac{x_2 - x_1}{L_a}, \qquad c_{ay} = \frac{y_2 - y_1}{L_a}. \qquad (11.42)$$

Here x, y are curvilinear coordinates whose plan projections are the global Cartesian axes X, Y, and L_a is the arc length of the element side.

The use of natural strains throughout eqns (11.40) (cf. eqns (11.26) and (11.32)) is convenient as the arc length and curvature calculation schemes described in Section 11.2 can now be applied on each side of the element. Indeed this seems a necessary approach because, although the curvatures $1/R_x$ and $1/R_y$ present little difficulty, no explicit formula is known for the twist curvature and the classical theory of surfaces[43] would be needed to determine this curvature. Direct use of natural strains avoids these difficulties.

Using quadratic areal coordinate interpolation on the shell surface the 9×30 strain interpolation matrix for the element follows from eqns (11.40) as

$$B = \left\{ \begin{array}{c} B_b \\ B_s \\ B_m \end{array} \right\} =$$

u columns	v columns	w columns	ϕ_x columns	ϕ_y columns

$$\left[\begin{array}{ccccc} c_{ax}\{f_a\}^t/R_a & c_{ay}\{f_a\}^t/R_a & \{0\}^t & -c_{ax}\{f_a\}^t & -c_{ay}\{f_a\}^t \\ c_{ax}\{f\}^t/R_a & c_{ay}\{f\}^t/R_a & \{f_a\}^t & -c_{ax}\{f\}^t & -c_{ay}\{f\}^t \\ c_{ax}\{f_a\}^t & c_{ay}\{f_a\}^t & \{f\}^t/R_a & \{0\}^t & \{0\}^t \end{array} \right]$$

$$(11.43)$$

where

$$\{f\} = \{L_1(2L_1 - 1),\ L_2(2L_2 - 1),\ L_3(2L_3 - 1),\ 4L_1L_2,\ 4L_2L_3,\ 4L_3L_1\}$$
(11.44)

$$\{f_a\} = \frac{(\partial\{f\}/\partial L_j - \partial\{f\}/\partial L_i)}{L_a} \qquad i = 1, 2, 3 \quad j = 2, 3, 1 \quad a \to b, c.$$
(11.45)

Then the element stiffness matrix is given by eqn (11.38) but now one also uses a natural modulus matrix $D_m^* = H^t D_m H$.

The arc lengths required in eqn (11.45) are given by applying eqn (11.13) to each side of the element and for this purpose it should be noted that, when the nodal elevations are specified, nodes 4, 5, and 6 are assumed to be half way between the vertices in the plan projection of the element, as shown in Fig. 11.8. The curvilinear element area Δ, required in eqn (11.38) and for calculating β, is then calculated by using $p = (L_a + L_b + L_c)/2$ in eqn (11.15) and the curvatures $1/R_a$, $1/R_b$ and $1/R_c$ required in eqn (11.43) are calculated by applying eqn (11.14) on each side of the element.

The matrices B, H, and T_G all require knowledge of the direction cosines c_{ax}, c_{ay}, and so on, on the shell surface and these are now calculated from the projection of the element upon the plane containing the corner nodes, as shown in Fig. 11.8. The tangents of the slopes of this plane relative to the global Cartesian axes are given by eqns (10.14) and (10.15):

$$t_x = \frac{\partial Z}{\partial X} = -\frac{(Y_{32}Z_1 + Y_{13}Z_2 + Y_{21}Z_3)}{2\Delta_p}$$
(11.46)

$$t_y = \frac{\partial Z}{\partial Y} = \frac{X_{32}Z_1 + X_{13}Z_2 + X_{21}Z_3}{2\Delta_p}$$
(11.47)

where $X_{32} = X_3 - X_2$, and so on, and Δ_p is the plan area of the element which is $\frac{1}{2}(X_{21}Y_{32} - X_{32}Y_{21})$. Using t_x and t_y the direction cosines are

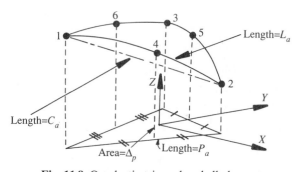

Fig. 11.8 Quadratic triangular shell element

given by

$$c_{ax} = \frac{X_{21}[1 + t_x^2]^{1/2}}{C_a}, \qquad c_{ay} = \frac{Y_{21}[1 + t_y^2]^{1/2}}{C_a} \qquad a \to b, c \quad (11.48)$$

where the numerators are projections onto the plane formed by the vertices and C_a is the chord length between nodes 1 and 2.

Table 11.4 shows results for the displacement under the load in the pinch loaded cylinder of Fig. 11.9, an octant of which is discretized into finite elements. The first column of results is obtained using data-specified curvatures and local coordinates (that is, one imagines a flat domain of size $\pi R/2$ in the x-direction in Fig. 11.9 and size $L/2$ in the y-direction and divides this into equisized elements). The specified curvatures are 0.01 for the circumferential element sides, 0.0 for the longitudinal sides, and 0.007116 for the diagonal sides (the last value follows from consideration of Mohr's circle) and a constant penalty factor of 0.00796 (the value given by eqn (9.35) for the finest mesh) is used.

Table 11.4 $wEt/100P$ under the loads in Fig. 11.9

Elements	Nodes	Local coordinate basis (a)[†]	(b)	Global coordinate basis (c)	(d)[††]
2	9	0.2473	0.3462	0.3380	0.3381
8	25	0.7906	1.2269	1.5773	1.5688
16	49	1.3279	1.6337	1.7038	1.6844
32	81	1.7173	1.7157	1.7284	1.7009
	Exact	1.7340	1.7340	1.7340	Unknown

[†]constant β [††]u_a/R terms in eqns (11.40) included

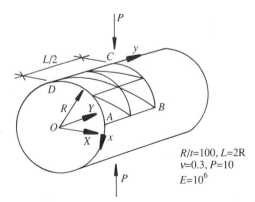

Fig. 11.9 Pinch loaded cylinder with closed ends

The second column of results uses eqn (9.35) to calculate the penalty factor; the convergence is clearly accelerated, as expected.

The third column uses the geometry calculations of eqns (11.13), (11.14), and (11.48). For this purpose one specifies a mesh of elements equisized in the plane XOY shown in Fig. 11.9, giving the nodal elevations for each node. In this situation eqn (11.14) yields relatively poor curvature estimates (about 10 per cent too low) on the diagonal sides of the steeply inclined elements near the 'waist' of the shell, adjacent to line AB in Fig. 11.9. Hence this scheme of curvature calculation strictly holds only for relatively shallow shells. Thus the load is applied at point C, as shown, where the curvatures are more accurately estimated. Because of the mesh refinement that then occurs in the xy-system near the load the results are slightly improved.

Until now the correction terms involving u_a in χ_a and γ_a in eqns (11.40) have not been included. (Note in passing that the first of these, that for χ_a, assumes R_a is constant, cf. eqns (2.40)). Including these the fourth column of results in Table 11.4 is obtained. As expected the shell is slightly stiffer but for most practical purposes these additional strain terms can evidently be neglected.

The element gives satisfactory stress solutions[41] and integration point values (using the integration of eqn (5.136)) are recommended for these. As these integration points are close to the vertex nodes they can be taken as nodal values and averaged between elements. This procedure also yields satisfactory stress solutions at more convenient sampling points; that is, their coordinates are known.

Section 15.7 gives a complete program listing of the local coordinate version of the element described in Section 11.6. This program should give the same results for the problem of Fig. 11.9 as in columns (a) and (b) of Table 11.4 and the data for this problem is given in Section 15.7. Stress results may be compared with the curves given by Mohr.[41]

11.8 Differential geometry of shells

In the preceding section the arc lengths and curvatures in a shell element were calculated using numerical integration and a circle fitting approximation respectively. As was noted, the latter strictly applies only to relatively shallow shells. This is because the fitted circle will always lie in a vertical plane, implying that the surface normal is always vertical which, of course, is not generally the case.

To deal with deep shells the differential theory of surfaces must be employed. An elegant example is provided by Argyris and Scharpf[24] where nine elevations are specified near each vertex of a triangular element and cubic interpolation used to evaluate first and second derivatives of the elevation. Then quintic interpolation can be applied to the elevations and their first and second derivatives at the vertices to determine the 'shape curvatures' at any point in the element.

However, this scheme requires twenty-seven elevations to be specified in each element, confronting the user with a formidable data preparation exercise when large numbers of elements are used. We therefore describe a simple scheme for calculating shell curvatures and apply this to the six nodal elevations of the quadratic triangular shell element shown in Fig. 11.8.

Because the 'circle fitting' approximation used in Section 11.7 experiences greatest difficulty when the elements are steeply inclined, as when they are close to AB in Fig. 11.9, we consider such an element. Considering a cylindrical shell of unit radius with four rows of elements equally spaced along OX in Fig. 11.9, the coordinates of an element with one edge lying on AB are

$$(x_1, y_1, z_1) = (0.75, 0, 0.6614), \quad (x_2, y_2, z_2) = (1, 0, 0)$$

$$(x_3, y_3, z_3) = (1, 0.25, 0), \quad (x_4, y_4, z_4) = (0.875, 0, 0.4841)$$

$$(x_5, y_5, z_5) = (1, 0.125, 0), \quad (x_6, y_6, z_6) = (0.875, 0.125, 0.4841).$$

The curvature on side 12, for example, is given by[24]

$$\frac{1}{R_a} = -\left(\frac{1}{m_a^2}\right) \hat{\mathbf{n}} \cdot \left(\frac{\partial^2 \mathbf{r}}{\partial \bar{s}_a^2}\right) \tag{11.49}$$

where $\partial^2 \mathbf{r}/\partial \bar{s}_a^2$ is the dimensionless derivative, that is $\bar{s}_a$ is taken as a dimensionless coordinate along the horizontal projection of side 12, of the radius vector $\mathbf{r} = x\hat{\mathbf{i}} + y\hat{\mathbf{j}} + z\hat{\mathbf{k}}$, $\hat{\mathbf{n}}$ is the unit normal vector to the surface and m_a is the 'modulus' of the surface, and is a measure of the arc length on side 12 compared to the dimensionless coordinate $\bar{s}_a$.

The second derivative in eqn (11.49) is given by eqn (5.94):

$$\frac{\partial^2 \mathbf{r}}{\partial \bar{s}_a^2} = \frac{\partial^2 \mathbf{r}}{\partial L_1^2} + \frac{\partial^2 \mathbf{r}}{\partial L_2^2} - \frac{2\partial^2 \mathbf{r}}{\partial L_1 \partial L_2} \tag{11.50}$$

and the modulus m_a is given by

$$m_a = \left| \frac{\partial \mathbf{r}}{\partial L_2} - \frac{\partial \mathbf{r}}{\partial L_1} \right|$$

where L_1, L_2, and L_3 are taken to apply to the horizontal projection of the element where nodes are exactly positioned at the middle of each side, as required.

The unit normal at any point in an infinitesimal element is given by

$$\hat{\mathbf{n}} = \frac{\mathbf{e}_a \times \mathbf{e}_b}{|\mathbf{e}_a \times \mathbf{e}_b|} = \frac{\mathbf{e}_b \times \mathbf{e}_c}{|\mathbf{e}_b \times \mathbf{e}_c|} = \frac{\mathbf{e}_c \times \mathbf{e}_a}{|\mathbf{e}_c \times \mathbf{e}_a|} \tag{11.51}$$

where e_a, e_b and e_c are natural tangent vectors given by cyclic permutation of

$$e_a = \frac{\partial \mathbf{r}}{\partial \bar{s}_a} = \left(\frac{\partial L_1}{\partial \bar{s}_a}\right)\left(\frac{\partial \mathbf{r}}{\partial L_1}\right) + \left(\frac{\partial L_2}{\partial \bar{s}_a}\right)\left(\frac{\partial \mathbf{r}}{\partial L_2}\right) = \left(\frac{\partial \mathbf{r}/\partial L_2 - \partial \mathbf{r}/\partial L_1}{C_a}\right)$$

(11.52)

because, on side 12, $\partial \bar{s}_a/\partial L_1 = -\partial \bar{s}_a/\partial L_2 = C_a$, a constant which need not be determined.

Using the usual quadratic interpolation functions of eqn (11.44) the reader can verify that at node 4 one obtains

$$\hat{\mathbf{n}} = \frac{e_a \times e_b}{|e_a \times e_b|} = 0.9355\hat{\mathbf{i}} + 0.3537\hat{\mathbf{j}}$$

(11.53)

but because of the 'large' nature of the element the three values for $\hat{\mathbf{n}}$ given by eqn (11.51) will not be equal. But in fact we know here that the normal at node 4 is given by

$$\hat{\mathbf{n}} = 0.875\hat{\mathbf{i}} + 0.4841\hat{\mathbf{k}}.$$

(11.54)

Applying eqn (11.50) the quadratic interpolation functions, $\mathbf{r} = \sum f_i \mathbf{r}_i$ where $f_1 = L_1(2L_1 - 1), \ldots, f_6 = 4L_3 L_1$, yield, at node 4,

$$\frac{\partial^2 \mathbf{r}}{\partial \bar{s}_a^2} = 4(\mathbf{r}_1 + \mathbf{r}_2 - 2\mathbf{r}_4).$$

(11.55)

Using the exact normal of eqn (11.54), evaluating eqn (11.55), and substituting these results into eqn (11.49) the curvature on side 12 is given by

$$\frac{1}{R_a} = \frac{-4(0.4841)(0.6614 - 0.9682)}{(0.7227)^2} = 1.1375$$

(11.56)

using the known arc length of side 12 (0.7227) in place of m_a. This value is also obtained with reasonable accuracy by numerical integration.

This is a poor result, as we know $1/R_a = 1/R = 1$ and eqn (11.14) gives an equally poor result. Clearly at least a cubic interpolation is needed to implement this scheme with reasonable accuracy. Indeed, it should be noted that we have in such elements used quadratic interpolations to calculate curvatures in the form $\chi_x = \partial \phi_x/\partial x$. As we saw in previous chapters, cubic interpolation is needed to provide reasonable accuracy in second derivative calculations in finite elements, in which case the foregoing scheme gives accurate results.[24]

11.9 Concluding remarks

The nine-freedom arch element introduced in Section 11.1 is a useful performer and is as accurate as the most accurate of the arch elements.[1]

Moreover most arch elements are extensions of the 6 df beam/column element and permit only constant extensional strain. We shall see in Section 12.8, however, that in problems where the structure deforms into a shell-like shape, incorporation of linear extensional strain into the element gives much improved results. Indeed, except in very simple problems, this 'at least linear strain' philosophy is advisable and certainly constant strain curved shell elements converge extremely slowly.[15]

The nine-freedom axisymmetric shell element of Section 11.3 is a straight-forward extension of the arch element of Section 11.1 and gives much better results than the more widely used 6 df straight axisymmetric element given in Section 10.2.

The shell element introduced in Section 11.6 is a good general-purpose element; it can be used for plane stress, thin plate, thick plate, thin shell, and thick shell problems, and has been included in the package program STRUDL where it has been found to compare well with other available shell elements. This type of element, based on *discrete Kirchhoff constraints*, is of increasing interest[44,45] and therefore a BASIC program for the element of Section 11.6 is included in Section 15.7.

Finally, it should be noted that some practical shell problems can be adequately approximated by ignoring the flexural effects, greatly simplifying the analysis.[46]

References

1. Mohr, G. A. and Garner, R. (1983). Reduced integration and penalty factors in an arch element. *Int. J. Struct.*, 3(1), 9.
2. Mohr, G. A. and Paterson, N. B. (1984). A natural numerical differential geometry scheme for a doubly curved shell element. *Comput. Struct.*, **18**, 433.
3. Cook, R. D. (1974). *Concepts and Applications of Finite Element Analysis*. Wiley, New York.
4. Timoshenko, S. P. and Woinowsky-Krieger, S. (1959). *Theory of Plates and Shells*, (2nd edn). McGraw-Hill, New York.
5. Mohr, G. A. (1982). Application of penalty factors to a curved isoparametric axisymmetric thick shell element. *Comput. Struct.*, **15**, 685.
6. Connor, J. J. and Brebbia, C. A. (1967). A stiffness matrix for a shallow rectangular shell element. *J. Eng. Mech. Div. ASCE*, **93**, 43.
7. Gallagher, R. H. (1966). The Development and Evaluation of Matrix Methods for Thin Shell Analysis. Ph.D. thesis, University of New York, Buffalo.
8. Cantin, G. and Clough, R. W. (1968). A curved cylindrical shell finite element. *J. Amer. Inst. Aeron. Astron.*, **6**, 1057.
9. Ashwell, D. G. and Sabir, A. B. (1972). A new cylindrical shell finite element based on simple independent strain functions. *Int. J. Mech. Sci.*, **14**, 171.
10. Gallagher, R. H. (1976). Problems and progress in thin shell analysis. In *Finite Elements for Thin Shells*, (ed. D. G. Ashwell and R. H. Gallagher). Wiley, London.

11. Timoshenko, S. P. and Goodier, J. N. (1951). _Theory of Elasticity_, (3rd edn). McGraw-Hill, New York.
12. Ashwell, D. G. (1976). Strain elements with applications to rings and cylindrical shells. _Finite Elements for Thin Shells_, (ed. D. G. Ashwell and R. H. Gallagher). Wiley, London.
13. Mohr, G. A. (1976). _Analysis and Design of Plates and Shells using Finite Elements_. Ph.D. thesis, University of Cambridge.
14. Olson, M. D. and Bearden, T. W. (1979). A simple triangular shell element revisited. _Int. J. Num. Meth. Eng._, **14**, 51.
15. Utku, S. (1967). Stiffness matrices for triangular elements of nonzero Gaussian curvature. _J. Amer. Inst. Aeron. Astron._, **10**, 1659.
16. Megard, G. (1969). Planar and curved shell elements. _Finite Elements in Stress Analysis_, (ed. I. Holand and K. Bell). Tapir, Trondheim.
17. Ford, R. (1969). TRIF–a triangular element for shell structures. CEGB Report No. RD/C/N359. London.
18. Strikland, G. and Loden, W. (1968). A doubly curved triangular shell element. _Proceedings of the Second Conference on Matrix Methods in Structural Mechanics_, Ohio.
19. Brebbia C. A. and Connor, J. J. (1973). _Fundamentals of Finite Element Techniques for Structural Engineers_. Butterworths, London.
20. Lindberg, G. M., Olson, M. D., and Cowper, G. R. (1969). New developments in the finite element analysis of shells. _DME/NAE Quart. Bull._, **4**, Nat. Res. Council Canada.
21. Cowper, G. R., Kosko, E., Lindberg, G. M., and Olson, M. D. (1968). Formulation of a new triangular plate bending element. _Trans. Canad. Aero-Space Inst._, **1**, 86.
22. Cowper, G. R., Lindberg, G. M., and Olson, M. D., (1970). A shallow shell finite element of triangular shape. _Int. J. Solids Struct._, **6**, 1133.
23. Olson, M. D. (1974). Analysis of arbitrary shells using shallow shell finite elements. _In Thin Shells_, (ed Y. C. Fung and E. E. Sechler). Prentice-Hall, Englewood Cliffs, NJ.
24. Argyris, J. H. and Scharpf, D. W. (1968). The SHEBA family of shell elements for the matrix displacement method. _Aero. J._ **72**, 873.
25. Argyris, J. H., Fried, I., and Scharpf, D. W. (1968). The TUBA family of plate elements for the matrix displacement method. _Aero. J._, **72**, 701.
26. Argyris, J. H. (1970). The impact of the digital computer on engineering sciences. _Aero. J._, **74**, 13–41, 111–127.
27. Ciarlet, P. G. (1975). Conforming finite element methods for shell problems. In _The Mathematics of Finite Elements and Applications_, Vol. 2, (ed. J. R. Whiteman). Academic Press, London.
28. Dupuis, G. and Goel, J. J. (1970). A curved finite element for thin elastic shells. _Int. Solids Struct._, **6**, 1413.
29. Dawe, D. J. (1975). High order triangular finite element for shell analysis. _Int. Solids Struct._, **11**, 1097.
30. Gallagher, R. H. and Thomas, G. R. (1976). A triangular element based on generalized potential energy concepts. In _Finite Elements for Thin Shells_, (ed. D. G. Ashwell and R. H. Gallagher). Wiley, London.

31. Harvey, J. W. and Kelsey, S. (1971). Triangular plate bending elements with enforced compatibility. *J. Amer. Inst. Aeron. Astron.*, **9**, 1023.
32. Henshell, R. D., Neale, B. K., and Warburton, G. B. (1971). A new hybrid cylindrical shell finite element. *J. Sound. Vib.*, **16**, 519.
33. Edwards, G. and Webster, J. J. (1976). Hybrid cylindrical shell finite elements. In *Finite Elements for Thin Shells*, (ed. D. G. Ashwell and R. H. Gallagher). Wiley, London.
34. Alayioglu, H. and Ali, R. (1977). A hybrid stress doubly curved shell finite element. *Comput. Struct.*, **7**, 477.
35. Ergatoudis, J. G., Irons, B. M., and Zienkiewicz, O. C. (1968). Three-dimensional analysis of arch dams and their foundations. *Symposium on Arch Dams*, Institute of Civil Engineers, London.
36. Irons, B. M. (1976). SemiLoof shell element. In *Finite Elements for Thin Shells.*, (ed. D. G. Ashwelll and R. H. Gallagher). Wiley, London.
37. Knowles, N. C., Razzaque, A., and Spooner, J. B. (1976). Experience of finite element analysis of shell structures. In *Finite Elements for Thin Shells.*, (ed. D. G. Ashwell and R. H. Gallagher). Wiley, London.
38. Mohr, G. A. (1980). Numerically integrated triangular element for doubly curved thin shells. *Comput. Struct.*, **11**, 565.
39. Wennerström, H. (1978). Nonlinear shell analysis performed with flat elements. In *Finite Elements in Nonlinear Mechanics*, Vol. 1, (ed. P. G. Bergan, I. Holand and K. Bell). Tapir, Trondheim.
40. Mohr, G. A. (1981). A doubly curved isoparametric triangular shell element. *Comput. Struct.*, **14**, 9.
41. Mohr, G. A. (1981). Application of penalty factors to a doubly curved quadratic shell element. *Comput. Struct.*, **14**, 15.
42. Mohr, G. A. (1982). On displacement finite elements for the analysis of shells. *Proceedings of the Fourth Australian International Conference on Finite Element Methods*, University of Melbourne.
43. Dym, C. L. (1974). *Introduction to the Theory of Shells.*, Pergamon, Oxford.
44. Murthy, S. S. and Gallagher, R. H. (1986). A triangular thin shell element based on discrete Kirchhoff theory. *Comput. Meth. Appl. Mech. Engng*, **54**, 197.
45. Saetta, A. V. and Vitaliani, R. V. (1990). A finite element formulation for shells of arbitrary geometry. *Comput. Struct.*, **37**, 781.
46. Hamid, M. S., Sabbah, H. N., and Stein, P. D. (1985). Comparison of finite element stress analysis of aortic valve leaflet using either membrane elements or solid elements. *Comput. Struct.*, **20**, 955.

Part III
Non-linear and time-dependent problems, and FEM programming

12
Non-linear materials
and large displacements

The present chapter provides a brief introduction to techniques for the solution of non-linear finite element problems. The two main types of non-linear problem are those where the magnitudes of the stresses, displacements velocities, and so on, are sufficient to alter the nature of the constitutive relationships or to make non-linear terms in the governing strain–displacement relationships or differential equations significant. In the mechanics of solids we call these non-linear material and large displacement problems respectively, and both types of problem are discussed in this chapter.

12.1 Introduction

The two main types of non-linear problem encountered in the mechanics of solids are those where the range of the solution variables involved (that is, the displacements or stresses) are sufficiently large to:

(1) alter the nature of the constitutive relationships, that is *non-linear material problems*;
(2) render non-linear terms in the strain–displacement relationships significant, that is, *large displacement problems*.

With non-linear materials, Hooke's law is no longer valid but it is generally possible to obtain a relationship between small increments of stress and strain in the form

$$d\{\sigma\} = D_T d\{\varepsilon\} \tag{12.1}$$

where D_T is called the *tangent modulus matrix*. Approximate procedures for dealing with materials with a relatively smooth stress–strain relationship can easily be devised but formal derivation of D_T for elasto-plastic problems is difficult. However, once the appropriate relationship has been found finite element solutions of such problems proceed without difficulty.

In problems where large geometry changes occur strains become non-linear relationships of displacements or, more correctly, displacement gradients. The usual measure of finite strains is Green's strain tensor in which

$$\varepsilon_x = \frac{\partial u}{\partial x} + \tfrac{1}{2}\left[\left(\frac{\partial u}{\partial x}\right)^2 + \left(\frac{\partial v}{\partial x}\right)^2 + \left(\frac{\partial w}{\partial x}\right)^2\right] \tag{12.2a}$$

$$\varepsilon_{xy} = \frac{\partial u}{\partial y} + \frac{\partial v}{\partial x} + \tfrac{1}{2}\left[\frac{\partial u}{\partial x}\frac{\partial u}{\partial y} + \frac{\partial v}{\partial x}\frac{\partial v}{\partial y} + \frac{\partial w}{\partial x}\frac{\partial w}{\partial y}\right] \tag{12.2b}$$

with cyclic permutation to obtain the other strain components. The existence of non-linear terms such as $(\partial u/\partial x)^2$ and $(\partial u/\partial x)(\partial u/\partial y)$ causes a non-linear relationship between loads and displacements in the displacement finite element method even when the material obey's Hooke's law.

The formulation of problems involving these non-linear effects may still be approached by the theorem of virtual work, that is, it is still correct to write for a non-linear system in equilibrium

$$\int \{\delta\varepsilon\}^{\mathrm t}\{\sigma\}\,\mathrm{d}V - \int \{\delta\bar u\}^{\mathrm t}\{T\}\,\mathrm{d}S = 0. \tag{12.3}$$

The reason that eqn (12.3) remains valid is because its derivation requires only the use of Green's theorem (that is, integration by parts) and the equations of equilibrium, neither of which are influenced by the non-linearity effects.

If we now introduce the finite element interpolations into eqn (12.3) this leads, after cancellation of a common factor $\{\delta d\}^{\mathrm t}$ and element summation, to

$$\sum\int B^{\mathrm t}\{\sigma\}\,\mathrm{d}V = \int F\{T\}\,\mathrm{d}S = \sum\{q_{\mathrm t}\}. \tag{12.4}$$

On the other hand, if the system is not in equilibrium then

$$\sum\int B^{\mathrm t}\{\sigma\}\,\mathrm{d}V - \sum\{q_{\mathrm t}\} = \{\dot R\} \tag{12.5}$$

where $\{R\}$ is an out of balance force vector (the *residual loads*), and the matrix $B = B(\{d\})$ and the stresses $\{\sigma\} = \{\sigma(\{\varepsilon\})\}$ are non-linear functions of the displacements and strains.

Then non-linear finite element problems can be exactly solved by calculating $\int B^{\mathrm t}\{\sigma\}\,\mathrm{d}V$, the *element reactions*, at each step of an iterative procedure such as Newton's method until the residual loads $\{R\}$ vanish and eqn (12.4) is satisfied. This process is called *equilibrium iteration* but approximate procedures such as *load stepping* may also be used and these are much easier to implement. Both types of procedure are described in Sections 12.2 and 12.3.

12.2 Load stepping techniques

The simplest approach to the analysis of non-linear problems in the finite element method is load stepping with the total load $\{Q_n\} = \Sigma\{\delta Q_i\}$, being built up in a series of steps $i = 1 \to n$. Then the displacement solution after application of the ith load increment is obtained as

$$\{D\}_{i+1} = \{D\}_i + K_T^{-1}\{\delta Q_i\} \tag{12.6}$$

and this is often referred to as *Euler's method*. We refer to it as load stepping, however, because, like the time stepping techniques introduced in Chapter 14, it is equivalent to a *linear extrapolation* along a tangent. To demonstrate this consider the single degree of freedom problem $q = ax/(b + x)$ which is a rectangular hyperbola passing through the origin and having an asymptote at $q = a$, as shown in Fig. 12.1. Using only equal load steps δq we can write a linear interpolation in a typical step as

$$x = (1 - \alpha)x_i + \alpha x_{i+1} \tag{12.7}$$

where $\alpha = dq/\delta q$ is a dimensionless measure in this interval.

Denoting $\dot{x} = dx/dq$ we can determine this by differentiating eqn (12.7):

$$\delta q \dot{x} = \delta q \left(\frac{dx}{dq}\right) = \delta q \frac{dx}{d\alpha}\frac{d\alpha}{dq} = \frac{dx}{d\alpha}$$

$$= -x_i + x_{i+1} \tag{12.8}$$

that is, we have

$$x_{i+1} = x_i + \delta q(dx/dq) = x_i + \delta q/(dq/dx) \tag{12.9}$$

which is indeed Euler's formula. Then using eight equal steps $q = 0.1a$ and calculating dq/dx from $q = ax/(b + x)$ eqn (12.9) yields

$$x_{i+1} = x_i + \frac{\delta q}{(dq/dx)_i} = x_i + \frac{\delta q(b + x_i)^2}{ab}, \qquad i = 0 \to 8 \tag{12.10}$$

giving the results shown in Table 12.1. Clearly the process is divergent but results can be improved by progressively diminishing the step size[1] and good approximations can usually be obtained with 10–20 load steps. In 'softening' structures, in which $\{Q\}$ decreases as $\{D\}$ increases, one can solve the problem using specified displacement increments rather than load steps and by such means post buckling phenomena, for example, can be analysed.[2]

An improvement over the latter method can be obtained by using *quadratic extrapolation*, rather than linear. For the first step one uses eqn (12.9) but thereafter the required formula is obtained by writing the quadratic interpolation

$$x = \frac{(\alpha^2 - \alpha)x_{i-1}}{2} + (1 - \alpha^2)x_i + \frac{(\alpha^2 + \alpha)x_{i+1}}{2} \tag{12.11}$$

where α interpolates dimensionlessly across two load increments of δq. On differentiation with respect to α,

$$\delta q \dot{x} = \frac{(2\alpha - 1)x_{i-1}}{2} - 2\alpha x_i + \frac{(2\alpha + 1)x_{i+1}}{2} \tag{12.12}$$

and

$$\delta q \dot{x}_i = -\frac{x_{i-1}}{2} + \frac{x_{i+1}}{2} \quad (\alpha = 0) \quad (12.13)$$

where $\dot{x}_i = (\mathrm{d}x/\mathrm{d}q)_i$, $\dot{x}_{i+1} = (\mathrm{d}x/\mathrm{d}q)_{i+1}$. Eliminating x_{i-1} one obtains the *second-order Runge–Kutta* predictor–corrector formula[3,4]

$$x_{i+1} = x_i + \delta q\left(1 - \frac{1}{2\alpha}\right)\dot{x}_i + \delta q\left(\frac{1}{2\alpha}\right)\dot{x}_{i+1}. \quad (12.14)$$

With $\alpha = \infty$ one obtains the Euler formula which must be used as a *predictor* for the first load increment. Equation (12.13) provides the *predictor* for subsequent increments and $\alpha = 1.0$ in eqn (12.14) provides the most convenient *corrector* formula, as only two tangents must be calculated at each step. Then we have

$$x_1 = x_0 + \delta q\left(\frac{\mathrm{d}x}{\mathrm{d}q}\right)_0 \qquad\qquad \text{First predictor} \qquad (12.15)$$

$$x_{i+1} = x_{i-1} + 2\delta q\left(\frac{\mathrm{d}x}{\mathrm{d}q}\right)_i \qquad\qquad \text{Subsequent predictor} \quad (12.16)$$

$$x_{i+1} = x_i + \tfrac{1}{2}\delta q\left(\frac{\mathrm{d}x}{\mathrm{d}q}\right)_i + \tfrac{1}{2}\delta q\left(\frac{\mathrm{d}x}{\mathrm{d}q}\right)_{i+1} \qquad \text{Corrector} \qquad (12.17)$$

Using four predictor–corrector steps (so that again eight tangents must be evaluated) the results are compared against those of the Euler method in Table 12.1 showing that the second-order Runge–Kutta method (that is, quadratic extrapolation) gives much better results than the Euler method.

By the procedure followed in eqns (12.11)–(12.17) higher-order extrapolation formulas can be obtained and the fourth-order Runge–Kutta formulae

Table 12.1 Comparison of Euler and second-order Runge–Kutta methods

q	x/b Euler	x/b Runge–Kutta	x/b Exact
$0.1a$	0.1000	—	0.1111
$0.2a$	0.2210	0.2440	0.2500
$0.3a$	0.3701	—	0.4286
$0.4a$	0.5578	0.6607	0.6667
$0.5a$	0.8005	—	1.0000
$0.6a$	1.1247	1.4874	1.5000
$0.7a$	1.5761	—	2.3333
$0.8a$	2.2397	3.8164	4.0000

have been much used in non-linear analysis.[3, 5] Although the corrector steps can be iterated this is likely to give divergence[3] and thus it is advisable to try various numbers of load steps and examine the convergence of the solution in order to ensure a reliable result. When a solution for a single load is required, and not the complete load history, it may be more economical to use equilibrium iteration, as described in Section 12.3.

12.3 Equilibrium iteration

The most widely used technique for solving systems of non-linear equations is Newton's method. Usually the total load is applied immediately and an initial solution $\{D\}_1 = K_0^{-1}\{Q\}$ obtained. Then eqn (12.5) is used to calculate residual loads $\{R\}$ from which corrections in the solution are calculated iteratively as $\{D\}_{i+1} = \{D\}_i + K_T^{-1}\{R\}$ where K_T and $\{R\}$ are calculated using the displacements $\{D\}_i$, until convergence is obtained.

Returning to the single-freedom example of Section 12.2 (that is, $q = ax/(b+x)$) and applying the load $q = 0.8a$ in a single-step Newton's method yields the iteration formula

$$x_{i+1} = x_i + \frac{R}{(dq/dx)_i} = x_i + \frac{R(b + x_i)^2}{ab} \qquad (12.18)$$

where

$$R = q - q' = 0.8a - \frac{ax_i}{b + x_i}. \qquad (12.19)$$

For the successive equilibrium iterations, this yields the results

$x/b = 0.8000, \quad 1.9521, \quad 3.1613, \quad 3.8593, \quad 3.9960, \quad 3.9998, \quad 4.0000$

and the solution is correct to four decimal places after six iterations. (Note that four decimal place calculation was used to obtain these results.)

Where estimation of the tangent stiffness is difficult or uneconomical the *modified Newton* or *initial stiffness method*[6, 7] may be used. Here the initial stinffness is used for each equilibrium iteration:

$$x_{i+1} = x_i + \frac{R}{(dq/dx)_0} = x_i + \frac{bR}{a} \qquad (12.20)$$

in our single-freedom example. Again applying the load $q = 0.8a$ in one step, and thus using eqn (12.19) to calculate R, we obtain the results

$x/b = 0.8000\,(1), \quad 1.8126\,(5), \quad 2.4367\,(10), \quad 3.0477\,(20), \quad 3.7588\,(50)$

the numbers in parentheses indicating the number of iterations used.

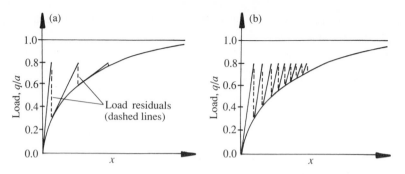

Fig. 12.1 (a) Newton's tangent stiffness method, and (b) initial stiffness method

Clearly the convergence is a good deal slower, as illustrated in Fig. 12.1. This is because the initial stiffness method has only first-order convergence ($\delta x = x_{i+1} - x_i = $ (const.) $(x_i - x_{\text{exact}})$) whereas Newton's method has second-order convergence ($\delta x = $ (const.) $(x_i - x_{\text{exact}})^2$).[8] Such rates of convergence are defined by the order of the first non-vanishing derivative in MacClaurin's formula (eqn (1.15)). If, for example, we seek the root of $f(x) = 0$ then Newton's formula is $g(x) = x - f(x)/f'(x)$. It is then easy to show that $g'(x) = -f(x)f''(x)/f'(x)^2 = 0$ at the root and $g''(x) \neq 0$ so that Newton's method yields second-order convergence.

Convergence of the initial stiffness method can be accelerated by using a convergence factor[9] or *Aitken acceleration* which is a h^1 extrapolation obtained by putting $h_i = d_i - d_{i-1}$, $h_j = d_{i+1} - d_i$ in eqn (3.60) with $N = 1$, giving

$$d_{i+1}^{\text{extrap}} = d_{i+1} - \frac{(d_{i+1} - d_i)^2}{(d_{i+1} - 2d_i - d_{i-1})}. \tag{12.21}$$

A similar formula can be derived for Newton's method using $n = 2$. Typically iteration ceases when the *Euclidean norm* $(\{\delta D\}^t \{\delta D\})^{1/2}$ of the displacement corrections is less than some specified tolerance, although energy or force criteria may also be used.[10]

Consideration of the results of Table 12.1 suggests a useful modification of Newton's method when applied with several load steps, namely to use eqn (12.16) as a predictor after the first step and thereafter Runge–Kutta (eqn (12.17)) for the first correction, all subsequent corrections being carried out with residuals using eqn (12.18). Table 12.2 compares the results thus obtained for our hyperbola problem with those with Newton's method. Here only the prediction and first correction are given and it is assumed that the correct solution is obtained by further corrections before allowing the next load step. Sufficient improvement in results is obtained to save one Newton iteration at the expense of keeping $\{D_{i-1}\}$.

Table 12.2 Comparison of second-order Runge–Kutta, Newton, and a modified Newton method.

q		x/b Runge–Kutta	x/b Newton	x/b Newton
0.2a	p	0.2000	0.2000	0.2000
	c	0.2440	0.2480	0.2480
	e		0.2500	
0.4a	p	0.6250	0.5625	0.6250[†]
	c	0.6703	0.6602	0.6657
	e		0.6667	
0.6a	p	1.3612	1.2223	1.3611[†]
	c	1.5020	1.4692	1.4923
	e		1.5000	
0.8a	p	3.1667	2.7500	3.1667[†]
	c	3.8611	3.6875	3.8611
	e		4.0000	

Key: p = prediction, c = first correction e = extract solution
[†]using eqn (12.16) for first correction

In matrix form Newton's method is sometimes called the Newton–Raphson method and we write

$$\{R\} = \{Q\} - \{Q'\} = K_T\{\delta D\}, \quad K_T = \sum k_T, \quad \{Q'\} = \sum\{q'\} \quad (12.22)$$

where it follows from eqn (12.5) that the element reactions are given by

$$\{q'\} = \int B^t\{\sigma\}\,dV \quad (12.23)$$

where $\{\sigma\} = D_T B\{d\}$. The element tangent stiffness matrices are given by

$$k_T = \int B^t D_T B\,dV \quad (12.24)$$

with $D_T = f(\{\sigma\})$ and $B = B(\{d\})$, and in turn $\{\sigma\}$ is evaluated using the last available displacement solutions $\{d\}$.

Equations (12.23) and (12.24) can be vertified by writing the changes at any point in an element resulting from equilibrium iteration as

$$\{\sigma_{i+1}\} = \{\sigma_i\} + D_i\{\varepsilon_{i+1} - \varepsilon_i\}, \quad \{\varepsilon_{i+1}\} = \{\varepsilon_i\} + B_i\{d_{i+1} - d_i\}(12.25)$$

Eliminating $\{\varepsilon_{i+1}\}$, premultiplying both sides by B_i^t and integrating over the element volume (that is, using eqn (3.13) with $C = B_i^t$—see eqn (3.19))

$$\int B_i^t\{\sigma_{i+1}\}\,dV - \int B_i^t\{\sigma_i\}\,dV = \int B_i^t D_i B_i\{\delta d\}\,dV. \quad (12.26)$$

Here the first term is a *forward projection* which will equal the applied loads if the next step is to yield the correct solution. Thus, eqn (12.26) justifies eqn (12.22) at element level.

Many alternative methods for non-linear problems derive from optimization theory[11,12] as will be evident to the reader of Chapter 22.

12.4 Non-linear materials

The behaviour of non-linear materials under three-dimensional stress can be described by calculating some stress, strain, or energy quantity and equating this to a limiting yield or fracture value of this, observed at a stress level F_y (yield), F_t (tensile cracking), or F_c (compressive crushing) in uniaxial tension. Some basic examples are[13]

(1) *Rankine's criterion*, that is, limiting direct stress,

$$\sigma_1 = F_y \quad \text{or} \quad |\sigma_3| = F_y. \tag{12.27}$$

(2) *Saint-Venant's criterion*, that is, limiting direct strain,

$$\sigma_1 - v(\sigma_2 + \sigma_3) = F_y \quad \text{or} \quad |\sigma_3 - v(\sigma_1 + \sigma_2)| = F_y. \tag{12.28}$$

(3) *Tresca's criterion*, limiting shear stress,

$$\sigma_1 - \sigma_3 = F_y \tag{12.29}$$

(4) *Beltrami's criterion*, limiting strain energy per unit volume,

$$\tfrac{1}{2}\{\varepsilon_i\}^t\{\sigma_i\} = \frac{F_y^2}{2E} \tag{12.30}$$

and substituting $\varepsilon_i = (\sigma_i - v\sigma_j - v\sigma_k)/E$ one obtains

$$\sigma_1^2 + \sigma_2^2 + \sigma_3^2 - 2v(\sigma_1\sigma_2 + \sigma_2\sigma_3 + \sigma_3\sigma_1) = F_y^2. \tag{12.31}$$

(5) *von Mises' criterion*, limiting distortion energy, which can be calculated by subtracting the strain energy of the isotropic component of stress, $p = \Sigma\sigma_i/3$, which is, writing $\sigma_i = p$, given by

$$3\left(\frac{p}{2E}\right)(p - vp - vp) = \frac{(1 - 2v)(\sigma_1^2 + \sigma_2^2 + \sigma_3^2)}{6E}. \tag{12.32}$$

Subtracting this from $\{\sigma_i\}^t\{\sigma_i - v\sigma_j - v\sigma_k\}/2E$ one obtains

$$\frac{(1 + v)(\sigma_1^2 + \sigma_2^2 + \sigma_3^2 - \sigma_1\sigma_2 - \sigma_2\sigma_3 - \sigma_3\sigma_1)}{3E}. \tag{12.33}$$

Equating this to the strain energy at yield under uniaxial tension, which, from eqn (12.33), is $(1 + v)F_y^2/3E$, leads finally to

$$\sigma_1^2 + \sigma_2^2 + \sigma_3^2 - \sigma_1\sigma_2 - \sigma_2\sigma_3 - \sigma_3\sigma_1 = F_y^2. \tag{12.34}$$

(6) *The Mohr–Coulomb criterion*, that is, weak tension. In two dimensions this may be expressed as

$$\frac{\sigma_1}{F_t} - \frac{\sigma_2}{F_c} = 1 \qquad (12.35)$$

where $F_t \leq F_c$ in general. A special case is *no-tension materials*[14] in which $F_t = 0$; in some cases a limit on F_c is of no concern.

The two-dimensional failure envelopes for all but Beltrami's criterion, which, like the von Mises' criterion, gives an ellipse as comparison of eqns (12.31) and (12.34) indicates, are shown in Fig. 12.2.

Two of the most widely used rules in finite element analysis of non-linear materials are the von Mises' rule, which is much used for ductile metals, and the Mohr–Coulomb rule which is much used for brittle vitreous-like materials.[14]

In applying rules (1)–(5) as 'failure' criteria, yield is taken to be the failure condition and the material is assumed to be elastic prior to yield. In elasto-plastic analysis, however, we wish to study the post yield behaviour and these rules are used as yield criteria. After yield the material is no longer elastic and new constitutive laws are required, and an additional failure criterion, usually of different form from the yield criterion, is required.

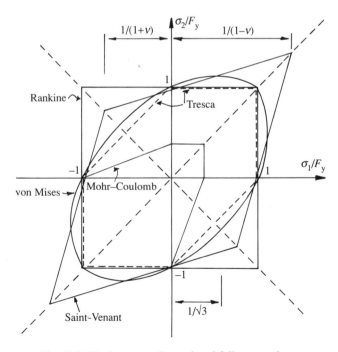

Fig. 12.2. Various two-dimensional failure envelopes

To clarify the discussion which follows it is worth summarizing some of the basic assumptions drawn upon in elasto-plastic analysis.

1. In uniaxial stress states the material obeys a law of form similar to that shown in Fig. 12.3. In a perfectly plastic material the slope H is zero.

2. Prior to yield the material is elastic, frequently being assumed to obey Hooke's law both on loading and unloading as shown in Fig. 12.3.

3. Under multiaxial stress states it is useful to take the view that there exists an equivalent uniaxial stress $\sigma^* = f(\{\sigma\})$ such that yielding occurs when $\sigma^* = F_y$, the yield stress in simple tension.

4. In a strain hardening material the stress level is regarded as a function of the plastic work $W_p = \{\sigma\}^t\{d\varepsilon_p\}$ and strain hardening takes place whenever positive plastic work is done on the material.

5. In the post yield range the plastic strains are usually regarded as not recoverable so that stress–strain laws in the plastic range are of necessity incremental in nature.

6. Plastic strains are usually assumed to be unaffected by the hydrostatic stress components, that is, they are related only to the deviatoric components of the direct stresses $\sigma_i' = \sigma_i - (\sigma_x + \sigma_y + \sigma_z)/3$ and to the shear stresses.

7. When the stress state moves from point P to P_2 in Fig. 12.4(a) it can do so via P_1 by elastic unloading and reloading followed by plastic strain under constant stress conditions at P_2. Thus in general strain increments are given by $\{d\varepsilon\} = \{d\varepsilon_e\} + \{d\varepsilon_p\}$ where $\{d\varepsilon_e\}$ and $\{d\varepsilon_p\}$ are the elastic and plastic increments, respectively.

8. As plastic strain increments involve less strain energy than partially elastic increments it is postulated that plastic increments occur normal to the yield surface as shown in Fig. 12.4(b). This follows from the *method of steepest descent* (see Section 22.6) where in unconstrained minimization

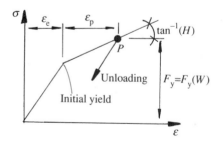

Fig. 12.3 Elasto-plastic behaviour

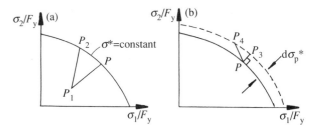

Fig. 12.4 (a) Perfectly plastic behaviour, and (b) strain hardening behaviour

problems $f(x)$ is minimized by moving in the direction $-\{\partial f/\partial x_i\}$, that is, normal to the contours $f(x) = \text{const}$. Thus, in Fig. 12.4(b), purely plastic straining occurs in passing from P to P_3 and elasto-plastic straining in passing from P to P_4 and the latter would involve greater energy.

9. It follows from 8 that the progression past the yield surface $d\sigma_p^*$ shown in Fig. 12.4(b) is given by

$$d\sigma_p^* = \{\partial f/\partial \sigma\}^t\{d\sigma\} = F^t\{d\sigma\} \qquad (12.36)$$

where $f(\{\sigma\})$ defines the equivalent uniaxial stress σ^* and, as it is obtained from yield criteria such as those of eqns (12.27)–(12.34), also defines the shape of the yield surface. In eqn (12.36) $\{\partial f/\partial \sigma\}$ denotes the vector formed by differentiating this function with respect to each stress component in turn.

10. The plastic strain increments are often assumed to be given by

$$\{d\varepsilon_p\} = F d\varepsilon_p^* \qquad (12.37)$$

where $d\varepsilon_p^*$ corresponds to $d\sigma_p^*$ and this is called an *associated flow rule*.

11. If the material strain hardens it is usual to assume that the size of the yield surface increases but that it does not change in shape and this is referred to as *isotropic hardening*.

Application of the von Mises' rule to elasto-plastic analysis is a convenient example because its two-dimensional envelope is a continuous elliptical curve, $\sigma_1^2 + \sigma_2^2 - \sigma_1\sigma_2 = F_y^2$, allowing more compact coding than some discontinuous envelopes.

For the purpose of calculating the tangent modulus matrix D_T we write the von Mises' rule for two dimensions in terms of the Cartesian stresses:

$$\sigma_x^2 + \sigma_y^2 - \sigma_x\sigma_y + 3\sigma_{xy}^2 = F_y^2. \qquad (12.38)$$

Taking the square root of the left-hand side gives the *yield function*

$$f(\sigma_x, \sigma_y, \sigma_{xy}) = (\sigma_x^2 + \sigma_y^2 - \sigma_x\sigma_y + 3\sigma_{xy}^2)^{1/2} \qquad (12.39)$$

and when the numerical value of this, the *equivalent uniaxial stress* σ^*, equals or exceeds F_y, yielding occurs. The rate of progression past the yield surface $d\sigma_p^*$ (Fig. 12.4(b)) can now be calculated using a *plastic flow law* obtained by chain differentiation of the *yield function* f of eqn (12.39) giving

$$d\sigma_p^* = \left(\frac{\partial f}{\partial \sigma_x}\right)d\sigma_x + \left(\frac{\partial f}{\partial \sigma_y}\right)d\sigma_y + \left(\frac{\partial f}{\partial \sigma_{xy}}\right)d\sigma_{xy}$$

$$= \left\{\frac{\partial f}{\partial \sigma}\right\}^{t}\{d\sigma\} = F^{t}\{d\sigma\} \tag{12.40}$$

where, for the Von Mises criterion, $F = \{2\sigma_x - \sigma_y, 2\sigma_y - \sigma_x, 6\sigma_{xy}\}/2\sigma^*$. Now using the elastic modulus matrix D one can write

$$\{d\sigma\} = D\{d\varepsilon_e\} = D\{d\varepsilon\} - D\{d\varepsilon_p\} \tag{12.41}$$

where $\{d\varepsilon_p\}$ are the plastic components of the strain increments. Using eqn (12.40) in eqn (12.41) we obtain

$$d\sigma_p^* = F^{t}D\{d\varepsilon\} - F^{t}D\{d\varepsilon_p\}. \tag{12.42}$$

Now a *hardening rule* based on uniaxial behaviour is introduced:

$$d\sigma_p^* = Hd\varepsilon_p^* \tag{12.43}$$

where H is the slope of the uniaxial stress–strain curve in the plastic range. This will be constant in bilinear models, which are applicable to ductile metals[7,15] and plastic clays,[16] but will vary continuously in hyperbolic models which are applicable to concrete[17,18] and rock.[19] Assuming also an *associated flow rule* for the plastic strain increments:

$$\{d\varepsilon_p\} = Fd\varepsilon_p^*. \tag{12.44}$$

This is sometimes referred to as the *Prandtl–Reuss flow rule*, and use of F for both the stress and strain flow rules gives rise to isotropic hardening, that is, the yield surface grows in size with the same shape as in the initial yield conditions. Now substituting eqns (12.43) and (12.44) into eqn (12.42) we obtain

$$Hd\varepsilon_p^* = F^{t}D\{d\varepsilon\} - F^{t}DF\,d\varepsilon_p^* \tag{12.45}$$

which is solved for $d\varepsilon_p^*$, giving

$$d\varepsilon_p^* = \frac{F^{t}D\{d\varepsilon\}}{H + F^{t}DF}. \tag{12.46}$$

Substituting the associated flow rule (eqn (12.44)) into eqn (12.41) one obtains

$$\{d\sigma\} = D\{d\varepsilon\} - DF\,d\varepsilon_p^* \tag{12.47}$$

and using eqn (12.46) in this last result gives

$$\{d\sigma\} = D\{d\varepsilon\} - \frac{DFF^tD\{d\varepsilon\}}{H + F^tDF} = D_T\{d\varepsilon\} \tag{12.48}$$

so that a symmetric tangent modulus matrix is obtained as

$$D_T = D - \frac{DFF^tD}{H + F^tDF}. \tag{12.49}$$

This replaces the elastic modulus matrix when $\sigma_i^* \geq F_y$ in the load increment *i*. If this causes a relatively large increment in σ^* then the proportion of the increment lying outside the yield surface can be estimated as

$$\alpha = \frac{\sigma_i^* - F_y}{\sigma_i^* - \sigma_{i-1}^*} \tag{12.50}$$

and revised stress increments calculated as

$$\{d\sigma\} = (1 - \alpha)D\{d\varepsilon\} + \alpha D_T\{d\varepsilon\}. \tag{12.51}$$

D_T is calculated using the values of F and H applying after $\{d\sigma\} = \alpha D_T\{d\varepsilon\}$ is added to the stresses prior to penetration of the yield surface. To this and succeeding load increments, of course, equilibrium iteration should be applied (and if non-linearity is permitted in the elastic range equilibrium iteration will also be required at the outset of the analysis).

The form of D_T in eqn (12.49) is such that 'perfect plasticity' (that is, $H = 0$) can be dealt with and *non-associative flow rules* for the strain increments can be used. In these, F in eqn (12.44) is replaced by a different function Q (sometimes referred to as the plastic potential) and Cormeau,[20] for example, investigates the use of a Mohr–Coulomb rule in which different parameters are used to evaluate F and Q.

This plastic potential corresponds to the plastic work W_p and to include this we write the yield function in the form $f = \sigma^* - R_y = 0$, where R_y is the 'radius' of the expanding yield surface. Then we write, for an increment in stress,

$$df = \left\{\frac{\partial f}{\partial \sigma}\right\}^t \{d\sigma\} + \frac{\partial f}{\partial W_p} dW_p = F^t\{d\sigma\} + \frac{\partial f}{\partial W_p} \frac{dW_p}{d\varepsilon_p^*} d\varepsilon_p^* = 0 \tag{12.52}$$

Substituting eqn (12.44) into eqn (12.41) we have

$$\{d\sigma\} = D\{d\varepsilon\} - F d\varepsilon_p^*. \tag{12.53}$$

Premultiplying by F^tD and rearranging leads to

$$F^t\{d\sigma\} = F^tD\{d\varepsilon\} - F^tDF d\varepsilon_p^*. \tag{12.54}$$

Substituting eqn (12.54) into eqn (12.52) provides an expression for $d\varepsilon_p^*$:

$$d\varepsilon_p^* = F^t D \{d\varepsilon\} \left[F^t DF - \frac{\partial f}{\partial W_p} \frac{dW_p}{d\varepsilon_p^*} \right]^{-1}. \qquad (12.55)$$

Substituting eqn (12.55) into eqn (12.53) gives

$$\{d\sigma\} = \left[D - DFF^t D \left(F^t DF - \frac{\partial f}{\partial W_p} \frac{dW_p}{d\varepsilon_p^*} \right)^{-1} \right] \{d\varepsilon\} \qquad (12.56)$$

which is the required constitutive relationship.

For a perfectly plastic material $\partial f / \partial W_p = 0$ but in general

$$dW_p = \{\sigma\}^t \{d\varepsilon_p\} = d\varepsilon_p^* \{\sigma\}^t F, \qquad \text{i.e.} \quad \frac{dW_p}{d\varepsilon_p^*} = \{\sigma\}^t F \qquad (12.57)$$

so that the final result for D_T is

$$D_T = D - DFF^t D \left(F^t DF - \frac{\partial f}{\partial W_p} \{\sigma\}^t F \right)^{-1}. \qquad (12.58)$$

If we determine $\partial f / \partial W_p$ from the value in simple tension then at yield $dW_p = R_y d\varepsilon_p$ so that

$$\frac{\partial f}{\partial W_p} = \frac{1}{R_y} \frac{\partial f}{\partial \varepsilon_p} = -\frac{1}{R_y} \frac{\partial R_y}{\partial \varepsilon_p} = -\frac{H}{R_y}. \qquad (12.59)$$

From eqn (12.40) it follows that $\sigma_p^* = \{\sigma\}^t F = R_y$ so that

$$\frac{\partial f}{\partial W_p} \{\sigma\}^t F = -H \qquad (12.60)$$

in which case eqn (12.58) reduces to eqn (12.49). When alternative criteria are used to calculate $\partial f / \partial W_p$, however, eqn (12.58) provides a generalization of eqn (12.49).

Many combined and modified yield criteria have been developed and the Mohr–Coulomb/von Mises' criterion for concrete proposed by Buyukozturk[21] is a useful example as it seeks a best fit to test data.

Using the first and second *stress invariants* J_1 and J_2 the yield criterion is written as

$$3J_2 + F_y J_1 + \frac{J_1^2}{5} = \frac{F_y^2}{9} \qquad (12.61)$$

where the yield stress is taken to be $F_c/3$ (F_c being the crushing strength of the concrete) and J_1 and J_2, the mean and deviatoric stress invariants, are given by

$$J_1 = \sum \sigma_i, \quad 3J_2 = \sum (\sigma_i - \tfrac{1}{3} J_1)^2 + \sum \sigma_{ij}^2 \qquad i = x, y, z \qquad (12.62)$$

where σ_{ij} is permuted to provide six terms. (Note in passing that $3J_2 = F_y^2$ is a concise expression for the von Mises criterion.)

In the case of plane stress, substituting for J_1 and J_2 in eqn (12.61) one obtains the yield function as

$$f = 3 \left\{ \frac{3(2\sigma_x^2 + 2\sigma_y^2 - \sigma_x \sigma_y + 5\sigma_{xy}^2)}{15} + F_y(\sigma_x + \sigma_y) \right\}^{1/2}. \quad (12.63)$$

Equating this to F_y (with $\sigma_{xy} = 0$) one obtains an ellipse which, using $F_y = F_c/3$, is scaled to provide the yield surface shown in Fig. 12.5. A modified hardening rule is obtained by using chain differentiation of eqn (12.63) to give

$$d\sigma_p^* = \frac{\partial f}{\partial \sigma_x} d\sigma_x + \frac{\partial f}{\partial \sigma_y} d\sigma_y + \frac{\partial f}{\partial \sigma_{xy}} d\sigma_{xy} + \frac{\partial f}{\partial \sigma^*} d\sigma^*. \quad (12.64)$$

Rearranging eqn (12.64) yields

$$d\sigma_p^* \left[1 - \frac{3(\sigma_x + \sigma_y)}{2\sigma^*} \right] = \{\partial f / \partial \sigma\}^t \{d\sigma\} = F^t\{d\sigma\}. \quad (12.65)$$

Substituting the hardening rule (eqn 12.43) into this last result gives

$$H' d\varepsilon_p^* = F^t\{d\sigma\} \quad (12.66)$$

where the gradient vector F is given by

$$F = \frac{3}{2\sigma^*} \left\{ \begin{array}{c} 2.4\sigma_x - 0.6\sigma_y + \sigma^* \\ 2.4\sigma_y - 0.6\sigma_x + \sigma^* \\ 6\sigma_{xy} \end{array} \right\}. \quad (12.67)$$

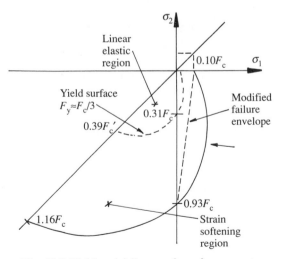

Fig. 12.5 Yield and failure surfaces for concrete

$H' = H(1 - 3J_1/2\sigma^*)$ is the modified hardening coefficient and, with the use of this in place of H, eqn (12.49) still holds.

As shown in Fig. 12.5, eqn (12.63) is also used as a compression failure criterion in the third quadrant (now using $F_y = F_c$) and a Mohr–Coulomb criterion with $F_t = F_c/10$ is used for the remaining quadrants. This failure criterion models the compression–compression behaviour more realistically (namely the intercept $1.16F_c$ for $\sigma_1 = \sigma_2$). Note that the intercept $0.93F_c$ for $\sigma_1 = 0$ arises from a 'best-fit' procedure[21] used to establish eqn (12.61). Linear elastic behaviour is used in the first quadrant and elsewhere up to 'yield' (in fact caused by microcracking) and thereafter H is based upon uniaxial compression test curves.

Until now discussion has related to plane stress or strain but extension of such elasto-plastic analyses with the von Mises' criterion, for example, to deal with plates is straightforward. Numerical integration, of course, is essential as D_T will vary throughout the elements and, as the strains vary non-linearly, increments in these must be calculated at each iteration and added to a stored total strain value. Then to deal with plates or shells a further numerical integration at points throughout the thickness is required to calculate the increments in the stress resultants. Assuming plane sections remain plane one can write

$$\left\{ \begin{matrix} dN \\ dm \end{matrix} \right\} = \Sigma \begin{bmatrix} D_T & yD_T \\ yD_T & y^2D_T \end{bmatrix} \left\{ \begin{matrix} d\varepsilon_0 \\ d\kappa \end{matrix} \right\} \omega_i t \tag{12.68}$$

where $d\varepsilon_0$ is the strain increment at the neutral surface, $d\kappa$ is the curvature increment, y is the distance from the neutral surface, and D_T must be revalued at each integration station through the thickness (that is, using strain increments calculated as $d\varepsilon = d\varepsilon_0 + yd\kappa$).[22]

To deal with composite materials alternative yield criteria can be used. In reinforced concrete, for example, one can use a Mohr–Coulomb model for the concrete and a von Mises' model for the reinforcement, 'smearing' the latter into a layer, and using an anisotropic modulus matrix. Then a weighted average tangent modulus matrix $\bar{D}_T$ is calculated at each integration point through the thickness as[21]

$$\bar{D}_T = \alpha(D_T)_{\text{concrete}} + \beta(D_T)_{\text{steel}} \tag{12.69}$$

where α and β are respectively the proportions of steel and concrete associated with each point.

The elasto-plastic procedure described above can be readily modified to include creep effects to form an *elasto-visco-plastic* model by generalizing eqn (12.44) to[23]

$$\dot{\varepsilon}_{vp} = \frac{d\varepsilon}{dt} = |f|^H \beta \left(t, \frac{\sigma^*}{f_y} \right) F \tag{12.70}$$

where $|f|^H$ is a Heavyside step function which is zero when f is less than a specified value and equal to unity otherwise. β is a function giving the rate of creep as a function of time and stress or strain level (see for example, eqn (14.24). Equation (12.70) is an associative flow rule and such models are stable if the function F is convex.[24] This is generally advisable as stability proofs are difficult and sometimes impossible to obtain for a non-associative rule.[24, 25]

The analysis must now include *time stepping*, such techniques being discussed in Chapter 14, and calculation of the effective stress levels must, of course take into account both initial strains and creep strains.

Some creep problems may result in the onset of large strains or displacements and these are discussed in the remainder of the present chapter. With the inclusion of these effects very general procedures for the analysis of non-linear and time-dependent behaviour in materials is possible with the aid of the finite element method.

Finally as creep, like vibration, may be classified as a *propagation problem* (see Section 1.1) the special case of elastic creep is discussed in Chapter 14 as this may be dealt with entirely as a time stepping analysis in which inertia effects are negligible (or not present if the loads are static).

12.5 Geometric non-linearity

There are two basic approaches to problems involving geometric non-linearity

(1) fixed frame of reference (Eulerian) methods; and
(2) moving frame of reference (Lagrangian) methods.

In the Eulerian methods the displacements u, v, w are functions of a fixed frame of reference (x, y, z). In the Lagrangian methods, if (x, y, z) is the position of a particle which moves to (a, b, c), thus the displacements u, v, w are functions of (a, b, c).

Both Eulerian and Lagrangian formulations may be incremental, the total load being built up gradually, and equilibrium iteration can be used to ensure that the residual loads at any stage of loading are small.

In the following sections we describe the following.

1. Stepwise moving frame of reference techniques. These are Lagrangian in the sense that the coordinates are updated after each load step (which corresponds to the use of the initial stiffness method for each load step) or are updated after each equilibrium iteration used for each load step (which corresponds to the tangent stiffness method). Each re-analysis with updated coordinates, however, has a Eulerian basis. (Section 12.6).

2. Eulerian methods in which corrections for the effects of large displacements upon extensional and flexural stresses are taken into account. (Sections 12.7 and 12.8).

3. Lagrangian methods in which the effects of large displacements upon extensional and flexural stresses are taken into account. (Section 12.9).

First, however, we examine the non-linear terms which significantly affect the strains when large displacements occur. Consider for example an infinitesimal element $dx = AB$ of a beam displaced by deformation to the position $A'B'$ shown in Fig. 12.6. As shown this element is displaced both laterally and longitudinally.

Putting $u_2 = u_1 + (\partial u/\partial x)\,dx$ and $v_2 = v_1 + (\partial v/\partial x)\,dx$ the length of the element after displacement is given by Pythagoras' theorem as

$$A'B' = [(dx + u_2 - u_1)^2 + (v_2 - v_1)^2]^{1/2}$$

$$= \left[\left(1 + \frac{\partial u}{\partial x} \right)^2 + \left(\frac{\partial v}{\partial x} \right)^2 \right]^{1/2} dx$$

$$= \left[1 + 2\frac{\partial u}{\partial x} + \left(\frac{\partial u}{\partial x} \right)^2 + \left(\frac{\partial v}{\partial x} \right)^2 \right]^{1/2} dx. \qquad (12.71)$$

Using the binomial approximation $(1 + \delta)^{1/2} \simeq 1 + \frac{1}{2}\delta$ and ignoring the term $(\partial u/\partial x)^2$ as 'second-order small' because lateral displacements caused by flexure are generally of much greater magnitude than the extensional displacements, eqn (12.71) reduces to

$$A'B' = \left[1 + \frac{\partial u}{\partial x} + \frac{1}{2}\left(\frac{\partial v}{\partial x} \right)^2 \right] dx. \qquad (12.72)$$

Then the extensional strain in the element is given by

$$\varepsilon_x = \frac{A'B' - AB}{dx} = \frac{A'B'}{dx} - 1 = \frac{\partial u}{\partial x} + \frac{1}{2}\left(\frac{\partial v}{\partial x} \right)^2. \qquad (12.73)$$

Hence the strain energy is a prismatic homogeneous finite element of length L

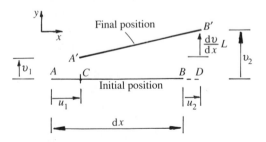

Fig. 12.6 Infinitesimal element of a beam

is given by

$$U_e = \frac{EA}{2} \int_0^L \left[\left(\frac{\partial u}{\partial x} \right)^2 + \frac{\partial u}{\partial x} \left(\frac{\partial v}{\partial x} \right)^2 + \frac{1}{4} \left(\frac{\partial v}{\partial x} \right)^4 \right] dx. \tag{12.74}$$

If we put $E(\partial u/\partial x) = \sigma$ in the second term then, applying the theorem of minimim potential energy, we obtain the usual elastic stiffness matrix, the *geometric* stiffness matrix,[26] and a 'large displacement' stiffness matrix:

$$k_0 = \int B_0^t D B_0 \, dV \quad \text{where } B_0 = \{\partial f_u/\partial x\}^t \text{ is applied to } \{u\} \tag{12.75}$$

$$k_G = \int B_G^t \sigma B_G \, dV \quad \text{where } B_G = \{\partial f_v/\partial x\}^t \text{ is applied to } \{v\} \tag{12.76}$$

$$k_L = \int B_L^t D B_L \, dV \quad \text{where } B_L = \dot{v}\{\partial f_v/\partial x\}^t \text{ is applied to } \{v\} \tag{12.77}$$

$\{f_u\}$ and $\{f_v\}$ are respectively the interpolation functions for the freedoms u and v, and $\dot{v} = \{\partial f_v/\partial x\}^t\{v\}$ is an interpolated value used as a scalar multiplier at each integration point used to compute k_L, numerical integration being essential in this case. In the case of the beam matrix D reduces to the scalar EA, $\int dV$ is replaced by $\int dx$, and the usual elastic stiffness matrix for flexure, $k = \int \{\partial^2 f_v/\partial x^2\}(EI) \{\partial^2 f_v/\partial x^2\} dx$ must be added. Application of these matrices is described in the following sections.

12.6 Stepwise moving frame of reference techniques

Moving frame of reference techniques provide a useful approach to large displacement analysis[1] (which may involve non-linear materials) and, coupled with equilibrium iteration, very large displacements may be dealt with. The procedure is illustrated in Fig. 12.7 where

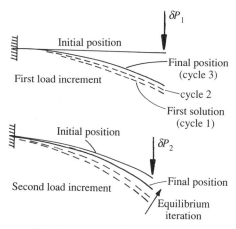

Fig. 12.7 Scheme for load stepping

(1) For each new load increment the preceding loads are ignored (if they were not one would obtain the solution for the total loading applied to an initially bent cantilever);

(2) equilibrium iteration is used after each increment in load;

(3) the current frame of reference is always that prior to application of the current load increment. The element coordinate transformation matrix for this frame is T_{i0} and if the initial stiffness method is used this is used in all calculations for this load increment;

(4) if the tangent stiffness method is used one uses $k_T = T_{ij}^t\, k_0'\, T_{ij}$ where T_{ij} is for the latest geometry after j equilibrium iterations and i load steps.

Applying the 'tangent' method (that is, updating the coordinate transformation after each equilibrium iteration) one obtains for the ith load increment

$$\{D_i\}_0 = \left(\sum T_{i0}^t k_0' T_{i0}\right)^{-1}\{Q_i\} \tag{12.78}$$

for the first linear trial solution. For the jth equilibrium iteration with this load increment the correction to $\{D_i\}_{j-1}$, the previous value, is

$$\{\delta D_i\}_j = \left(\sum T_{ij}^t k_0' T_{ij}\right)^{-1}\left(\{Q_i\} - \sum\{q_j'\}\right) \tag{12.79}$$

where

$$\{q_j'\} = T_{i0}^t \sum B_j^{*t} D\bar{B}_j^t(\omega_i L) T_{i0}\{d_i\}_{j-1} \tag{12.80}$$

gives the element reactions. k_0' is the local stiffness matrix and the matrices B_j^*, $\bar{B}_j$ are defined in Sections 12.7 and 12.8. In eqn (12.79) T_{ij} is based on the latest geometry which is given by

$$\{x, y\}_{ij} = \{x, y\}_{i0} + \{D_i\}_{j-1} \tag{12.81}$$

where

$$\{D_i\}_{j-1} = \{D_i\}_0 + \sum_{n=1}^{j-2}\{\delta D_i\}_n. \tag{12.82}$$

In eqn (12.80) T_{i0} is defined by $\{x, y\}_{i0} = \{x, y\}_{(i-1)M}$ where M refers to the last equilibrium iteration in the preceding load increment.

If calculation of the residuals and thus equilibrium iteration is difficult to achieve, simple load stepping can be used. With the use of a sufficient number of steps good approximations can be achieved (see Table 12.3) and such approximations can be improved by quadratic (see Section 12.2) or higher-order extrapolations.

12.7 Eulerian solution for large displacements

Considering the six-freedom beam element as an example, large displacement analysis commences with a first *linear trial solution* $\{d\}$ for the element displacements, after which the element curvatures are calculated as $\chi = \partial^2 v/\partial x^2$ and the extensional strains are calculated from eqn (12.73). Then the generalized element stresses are given by

$$\{\sigma\} = D\begin{Bmatrix} \varepsilon \\ \chi \end{Bmatrix} = \begin{bmatrix} EA & 0 \\ 0 & EI \end{bmatrix}(B_0 + \tfrac{1}{2}B_L)\{d\} = D\bar{B}\{d\} \qquad (12.83)$$

where $(B_0 + B_L)$ is obtained using linear and cubic interpolation with $s = 0 \to 1$. Underlining the B_L terms for emphasis, we have

$$B^* = \frac{1}{L}\begin{bmatrix} -1 & \dot{v}(6s^2 - 6s) & \dot{v}(3s^2 - 4s + 1)L & 1 & \dot{v}(6s - 6s^2) & \dot{v}(3s^2 - 2s)L \\ 0 & (12s - 6)/L & (6s - 4) & 0 & (6 - 12s)/L & (6s - 2) \end{bmatrix}$$

$$(12.84)$$

with

$$\dot{v} = \{0, 6s^2 - 6s, L(3s^2 - 4s + 1), 0, 6s - 6s^2, L(3s^2 - 2s)\}^{\mathrm{t}}\frac{\{d\}}{L} (12.85)$$

evaluated as a scalar multiplier at each integration point. Consider a *virtual* variation in the total extensional strain as

$$\delta\varepsilon^* = \delta\left(\frac{\partial u}{\partial x}\right) + \delta\left(\frac{1}{2}\left(\frac{\partial v}{\partial x}\right)^2\right) = \delta\left(\frac{\partial u}{\partial x}\right) + \frac{\partial v}{\partial x}\delta\left(\frac{\partial v}{\partial x}\right). \qquad (12.86)$$

Then the element reactions are calculated by numerical integration using

$$\{q'\} = \sum B^{*\mathrm{t}}D\bar{B}\{d\}(\omega_i L) \qquad (12.87)$$

where $B^* = B_0 + B_L$. Now considering a real increment $\{\delta\varepsilon\} = D\{\delta d\}$ (which will be the correction obtained by equilibrium iteration) the element tangent stiffness matrix is given by

$$\tfrac{1}{2}\int\{\delta\varepsilon^*\}D\{\delta\varepsilon\}\,\mathrm{d}V = \tfrac{1}{2}\{\delta d^*\}k_{\mathrm{T}}\{\delta d\} \qquad (12.88)$$

so that minimization will yield

$$k_{\mathrm{T}} = \int B^{*\mathrm{t}}DB^*\,\mathrm{d}V = \int\begin{bmatrix} B_0^{\mathrm{t}}DB_0 & B_0^{\mathrm{t}}DB_L \\ B_L^{\mathrm{t}}DB_0 & B_L^{\mathrm{t}}DB_L \end{bmatrix}\mathrm{d}V = k_0 + k_{\mathrm{L}}. \qquad (12.89)$$

Note that k_{L} evaluates the strain energy associated with the last term in eqn (12.74) and that in many applications the stress stiffening matrix should be included, that is,

$$k_{\mathrm{T}} = k_0 + k_{\mathrm{L}} + k_{\mathrm{G}} \qquad (12.90)$$

where

$$k_G = \int_0^1 \left\{ \frac{\partial f_v}{\partial s} \right\} \left(\frac{A\sigma^*}{L} \right) \left\{ \frac{\partial f_v}{\partial s} \right\}^t ds \tag{12.91}$$

with

$$\sigma^* = \frac{1}{L} E \left\{ \frac{\partial f_u}{\partial s} \right\}^t \{u\}. \tag{12.92}$$

σ^* is calculated on a small displacement basis using the latest displacement estimates which have been obtained with iterative improvement.

Omitting χ in the foregoing procedure, the method is readily extended to three dimensions,[6] but as it stands it is not correct for one or two dimensions. In both these cases large lateral displacements may induce large slopes which invalidate the usual small deflection calculation $\chi = \partial^2 v / \partial x^2$. Thus in matrix B^* of eqn (12.84) we have only made allowance for large displacements in the first row and clearly corrections in the second row should be considered.

To take account of large slopes the curvature in the beam should be calculated as

$$\chi = \frac{\partial^2 v}{\partial x^2} (1 + \dot{v}^2)^{-3/2} \tag{12.93}$$

and this is achieved merely by dividing row two of matrix $\bar{B}$ by the scalar quantity $(1 + \dot{v}^2)^{3/2}$ evaluated using eqn (12.85) at each integration point. Then, in the same manner as in eqn (12.86), we consider a curvature increment

$$\begin{aligned}
\delta\chi &= \frac{\delta(\partial^2 v / \partial x^2)}{(1 + \dot{v}^2)^{3/2}} + \frac{\partial^2 v}{\partial x^2} \delta \left\{ \frac{1}{(1 + \dot{v}^2)^{3/2}} \right\} \\
&= \frac{\delta(\partial^2 v / \partial x^2)}{(1 + \dot{v}^2)^{3/2}} - \frac{\partial^2 v}{\partial x^2} \left\{ \frac{3\dot{v}(\delta\dot{v})}{(1 + \dot{v}^2)^{5/2}} \right\}
\end{aligned} \tag{12.94}$$

Appealing to reciprocity, $\ddot{v}\delta\dot{v} = \delta\ddot{v}\,\dot{v}$, so that, to form the matrix B^*, each term of row two of matrix B^* is scalar multiplied by

$$\left(1 - \frac{3\dot{v}}{(1 + \dot{v}^2)} \right) (1 + \dot{v}^2)^{-3/2} \tag{12.95}$$

at each integration point and the curvature corrections require little additional coding above that required for the extensional corrections.

Figure 12.9 shows the results obtained using load stepping alone or load stepping in conjunction with equilibrium iteration for the cantilever of Fig. 12.8. Equilibrium iteration is carried out using either numerically integrated residual loads with and without the curvature correction included or the explicit *fictitious load* procedure of Kohnke.[9] This last procedure is

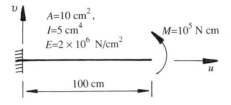

Fig. 12.8 Simple example of large displacement analysis with flexural elements

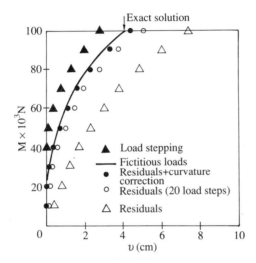

Fig. 12.9 Reduction in v(cm) from small displacement theory value

comparable to the use of s and c functions in stability analysis but, unlike the numerical integration solution described above, it is difficult to generalize to plate problems. In each case ten load steps are used and the final results for the displacement at the end of the cantilever are given in Table 12.3.[27]

The fictitious load result is statisfactory and even the load stepping result is reasonable. The residual load method with only the extensional correction gives a poor result; for the vertical displacement v the predicted departure from the linear solution is twice as great as it should be. Hence Fig. 12.9 shows that simple load stepping without equilibrium iteration yields less than half the error given by using the large displacement 'correction' for the extensional strains of eqn (12.84).

Hence the residual loads are actually damaging the solution, that is, when ten load steps are used one does better by dispensing with equilibrium iteration. But incorporating the curvature corrections of eqns (12.93) and 12.95 into the calculations the much improved results shown in Table 12.3 and Fig. 12.9 are obtained. Clearly these curvature corrections are important,

Table 12.3 Solutions for displacements at the free end of the cantilever of Fig. 12

	Load steps	Total cycles	Number of elements	v(cm)	u(cm)	ϕ (radians
Load stepping	10	10	10	47.2021	-14.6765	1.0076
Fictitious loads	10	30	10	45.9617	-15.7876	1.0000
Eulerian method[†]	10	40	10	42.6103	-13.1984	0.8967
Eulerian method[‡]	10	10	50	45.4878	-15.8529	1.0421
Lagrangian method[‡]	1	6	1	46.051	-15.730	1.000
Exact solution	–	–	–	45.9698	-15.8529	1.0
Linear solution	1	1	1	50.0	0.0	1.0

[†]without curvature correction, [‡]with curvature correction

therefore, in problems involving large slopes and the following section describes a Lagrangian technique which also includes these curvature corrections and yields excellent results.

In the case of finite strains in rubber-like materials the $(\partial u/\partial x)^2$ term in eqn (12.71) is more significant than the $(\partial v/\partial x)^2$ term but otherwise inclusion of this term follows the same general lines of approach used in eqns (12.84) and (12.86).

12.8 Lagrangian solution for large displacements

A Lagrangian solution for large displacements, in which the displacements u and v are functions of a coordinate s measured along the elastic curve, was developed by Milner.[28] This procedure is computationally very efficient but requires the use of an extra freedom $\partial u/\partial x$ at each node. This, however, allows the same interpolation to be used for the u and v freedoms. In arch and shell problems this is desirable and clearly it is also desirable in problems where the structure deforms into an arch or shell-like shape.

It follows from eqn (12.71) that the extensional strain can be expressed in terms of local displacements u' and v' as

$$\varepsilon = \frac{du'}{ds_0} + \tfrac{1}{2}\left[\left(\frac{du'}{ds_0}\right)^2 + \left(\frac{dv'}{ds_0}\right)^2\right] \tag{12.96}$$

using the same binomial approximation leading to eqn (12.73) but retaining all terms.

In Fig. 12.10(b) a point P lies on the centroidal x-axis and Q is at height z above P. Under load P and Q move to P' and Q', respectively, according to the usual assumption that plane sections remain plane. If ϕ is the slope of the

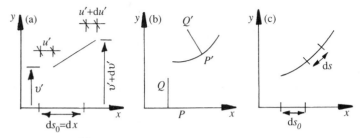

Fig. 12.10 Deformations of filament of a beam

beam the resulting displacements can be expressed as

$$u' = u - z \sin \phi, \quad v' = v - z(1 - \cos \phi) \tag{12.97}$$

where, if all variables are functions of s,

$$\frac{du'}{ds} = \frac{du}{ds} - z \cos \phi \frac{d\phi}{ds}$$

$$\frac{dv'}{ds} = \frac{dv}{ds} - z \sin \phi \frac{d\phi}{ds}. \tag{12.98}$$

Using Pythagoras' theorem it follows that the fibre ds shown in Fig. 12.10(c) originally had a length ds_0 given by

$$ds_0 = (ds^2 - dv^2)^{1/2} - du \tag{12.98}$$

from which we obtain

$$\frac{ds}{ds_0} = \left[\left(1 - \left(\frac{dv}{ds} \right)^2 \right)^{1/2} - \frac{du}{ds} \right]^{-1}. \tag{12.100}$$

Considering an infinitesimal element it is easy to show that

$$\sin \phi = \frac{dv}{ds}, \quad \cos \phi = \left[1 - \left(\frac{dv}{ds} \right)^2 \right]^{1/2} \tag{12.101}$$

from which it follows that

$$\frac{d\phi}{ds} = \frac{d^2 v}{ds^2} \left[1 - \left(\frac{dv}{ds} \right)^2 \right]^{-1/2}. \tag{12.102}$$

Substituting eqns (12.98) into eqn (12.96) leads to

$$\varepsilon_z = \frac{du}{ds} \frac{ds}{ds_0} - z \cos \phi \frac{d\phi}{ds} \frac{ds}{ds_0} + \frac{1}{2} \left[\left(\frac{du}{ds} \right)^2 + \left(\frac{dv}{ds} \right)^2 \right.$$

$$\left. - 2z \frac{d\phi}{ds} \left(\cos \phi \frac{du}{ds} + \sin \phi \frac{dv}{ds} \right) + z^2 \left(\frac{d\phi}{ds} \right)^2 \right] \left(\frac{ds}{ds_0} \right)^2 \tag{12.103}$$

which is without approximation but unwieldy to use in practice.

Neglecting the last three terms as 'second-order small' and assuming that $\cos\phi(\mathrm{d}s/\mathrm{d}s_0) \simeq 1$, which is equivalent to assuming $\mathrm{d}u/\mathrm{d}s \ll \cos\phi$ in eqn (12.100), the strain at distance z from the neutral axis is given by

$$\varepsilon_z = \frac{\mathrm{d}u}{\mathrm{d}s} - \frac{z\mathrm{d}\phi}{\mathrm{d}s} + \tfrac{1}{2}\left(\frac{\mathrm{d}u}{\mathrm{d}s}\right)^2 + \tfrac{1}{2}\left(\frac{\mathrm{d}v}{\mathrm{d}s}\right)^2 \qquad (12.104)$$

from which the stress resultants are obtained as

$$T = \int \sigma \,\mathrm{d}A = EA\varepsilon = EA\left[\frac{\mathrm{d}u}{\mathrm{d}s} + \tfrac{1}{2}\left(\frac{\mathrm{d}u}{\mathrm{d}s}\right)^2 + \tfrac{1}{2}\left(\frac{\mathrm{d}v}{\mathrm{d}s}\right)^2\right] \qquad (12.105)$$

$$M = -\int \sigma z \,\mathrm{d}A = EI\chi = EI\frac{\mathrm{d}\phi}{\mathrm{d}s} \qquad (12.106)$$

where the curvature χ is given by eqn (12.102).

Now using cubic interpolation for both u and v the strain interpolation matrix is given by

$$\left\{\begin{matrix} \varepsilon \\ \chi \end{matrix}\right\} = \left[\begin{matrix} B_1 + \tfrac{1}{2}\dot{u}B_1 + \tfrac{1}{2}\dot{v}B_2 \\ B_3/[1 - \dot{v}^2]^{1/2} \end{matrix}\right] = \bar{B}\{d\} \qquad (12.107)$$

where

$$B_1 = \frac{[6s^2 - 6s,\, 3s^2 - 4s + 1,\, 0,\, 0,\, 6s - 6s^2,\, 3s^2 - 2s,\, 0,\, 0]}{L} \qquad (12.108)$$

$$B_2 = \frac{[0,\, 0,\, 6s^2 - 6s,\, 3s^2 - 4s + 1,\, 0,\, 0,\, 6s - 6s^2,\, 3s^2 - 2s]}{L} \qquad (12.109)$$

$$B_3 = \frac{[0,\, 0,\, 12s - 6,\, 6s - 4,\, 0,\, 0,\, 6 - 12s,\, 6s - 2]}{L^2} \qquad (12.110)$$

$$\{d\} = \{u_1, \theta_1, v_1, \phi_1, u_2, \theta_2, v_2, \phi_2\}, \qquad \theta = \frac{\mathrm{d}u}{\mathrm{d}s}, \qquad \phi = \frac{\mathrm{d}v}{\mathrm{d}s} \quad (12.111)$$

and $\dot{u}$ and $\dot{v}$ are scalar multipliers at each integration point in $s = 0 \to 1$ calculated as $\dot{u} = B_1\{d\}$ and $\dot{v} = B_2\{d\}$.

Allowing a perturbation of the generalized strains we obtain

$$\delta\varepsilon^* = \delta\dot{u} + \dot{u}\delta\dot{u} + \dot{v}\delta\dot{v} \qquad (12.112)$$

$$\delta\chi^* = \frac{\delta\dot{v}}{[1 - \dot{v}^2]^{1/2}} + \frac{\ddot{v}\dot{v}\delta\dot{v}}{[1 - \dot{v}^2]^{3/2}} \qquad (12.113)$$

which can be expressed in matrix form as

$$\left\{\begin{matrix} \delta\varepsilon^* \\ \delta\chi^* \end{matrix}\right\} = \left[\begin{matrix} B_1 + \dot{u}B_1 + \dot{v}B_2 \\ B_3/[1 - \dot{v}^2]^{1/2} + \ddot{v}\dot{v}B_2/[1 - \dot{v}^2]^{3/2} \end{matrix}\right]\{\delta d^*\} = B^*\{\delta d^*\}$$
$$(12.114)$$

where $\{\delta d^*\}$ is a virtual or arbitrary increment in the displacements $\{d\}$ defined in eqn (12.111).

For such increments the residual loads are given by virtual work as

$$\int \{\delta\varepsilon^*\}^t\{\sigma\} \, dx - \{\delta d^*\}^t\{q\} = \{\delta d^*\}^t\{R\}. \qquad (12.115)$$

After substituting eqn (12.114) and eliminating $\{\delta d^*\}^t$ this gives

$$\int B^{*t}\{\sigma\} \, dx - \{q\} = \{R\} \qquad (12.116)$$

so that the element reactions are given by

$$\{q'\} = \int B^{*t}\{\sigma\} \, dx = \int B^{*t}D\bar{B} \, dx\{d\}. \qquad (12.117)$$

Note that the term $\{\delta d^*\}^t\{q\}$ in eqn (12.115) is given by

$$\{\delta d^*\}^t\{q\} = \delta u_1^* F_{u1} + \delta v_1^* F_{v1} + \delta\phi_1^* M_1 + \delta u_2^* F_{u2} + \delta v_2^* F_{v2} + \delta\phi_2^* M_2$$

$$= \{\delta d^*\}^t\{F_{x1}, 0, F_{v1}, M_1/\cos\phi, F_{u2}, 0, F_{v2}, M_2/\cos\phi\} \qquad (12.118)$$

where the loadings corresponding to the freedom $\theta = du/dx$ are taken to be zero.

The tangent stiffness matrix is obtained by considering a real increment $\{\delta d\}$ occurring during equilibrium iteration and corresponding increments $\{\delta\sigma\}$. Then the strain energy associated with $\{\delta d\}$ and $\{\delta d^*\}$ is

$$\delta U_e = \int \{\delta\varepsilon^*\}\{\delta\sigma\} = \{\delta d^*\}^t \left(\int B^{*t}DB^* \, dx\right)\{\delta d\} = \{\delta d^*\}^t k_T\{\delta d\} \qquad (12.119)$$

and the final equations for equilibrium iteration are obtained after element summation as

$$\left(\sum\int B^{*t}DB^* \, dx\right)\{\delta D\} = \left(\sum k_T\right)\{\delta D\} = \sum\{q\} - \sum\int B^{*t}D\bar{B}\{d\}$$

$$= \Sigma\{q\} - \Sigma\{q'\} = \{R\}. \qquad (12.120)$$

Note that in eqn (12.120) we have reversed the sign of eqn (12.116) to calculate the residual loads in the 'active' rather than 'reactive' sense, that is, if the applied loads $\{Q\} = \sum\{q\}$ exceed the summed element reactions $\{Q'\} = \sum\{q'\}$ then increases in the displacements must occur.

In cases where stress stiffening is significant the geometric or stress stiffness matrix k_G must also be included, as shown in eqn (12.90). In the problem of Fig. 12.8, however, this is not necessary as extensional stresses are not significant. Moreover, it is worth noting that in general k_T need only be an approximation (corresponding to the use of the initial stiffness method) but that the element reactions must be calculated exactly.

Applying eqn (12.120) with $\bar{B}$ given by eqn (12.107) and B^* given by eqn (12.114) to the problem of Fig. 12.8 yields the results shown in Table 12.3. Note that for this Lagrangian method only a single element and a single load step was used and yet the results are satisfactory. Much of this greater economy is due to the use of four freedoms and cubic interpolation to interpolate the extensional displacements u.

Thus in cases where large slopes as well as large displacement are involved the curvature corrections of eqns (12.93) and (12.95) (Eulerian method) or eqns (12.107) and (12.114) (Lagrangian method) are clearly of importance and these are readily included in plate bending problems. In the latter, however, formation of corrections for the twist curvature is very difficult and, as in Section 11.8, the use of the natural strain concept is the most expeditious means of solution.

Indeed it should be remarked that the large curvature correction introduced here is closely related to the *Lorentz transformation*[29] and clearly many further applications of the finite element method remain in the realm of relativistic physics, another example being the use of flat segments in Regge calculus[30] to approximate curved space–time, an approximation comparable to the use of flat shell elements in Chapter 10.

References

1. Oden, J. T. (1972). *Finite Elements of Nonlinear Continua.* Prentice-Hall, New Jersey.
2. Argyris, J. H. (1965). Continua and discontinua. *Proceedings of the First Conference on Matrix Methods in Structural Mechanics,* Wright-Patterson AFB, Ohio.
3. Beckett, R. and Hurt, J. (1967). *Numerical Calculations and Algorithms.* McGraw-Hill, New York.
4. Martin, H. C. and Carey, G. F. (1973). *Introduction to Finite Element Analysis.* McGraw-Hill, New York.
5. Connor, J. J. and Brebbia, C. A. (1976). *Finite Element Techniques for Fluid Flow.* Butterworths, London.
6. Zienkiewicz, O. C. (1977). *The Finite Element Method,* (3rd edn). McGraw-Hill, New York.
7. Zienkiewicz, O. C., Valliappan, S., and King, I. P. (1969). Elasto-plastic solutions of engineering problems, initial stress finite element approach. *Int. J. Num. Meth. Engg,* **1**, 75.
8. Livesley, R. K. (1975). *Matrix Structural Analysis,* (2nd edn). Pergamon, Oxford.
9. Kohnke, P. C. (1978). Large deflection analysis of frame structures with fictitious forces. *Int. J. Num. Meth. Engng* **12**, 1279.
10. Bathe, K. J. and Cimento, A. P. (1980). Some practical procedures for the solution of nonlinear finite element equations. *Comp. Meth. Appl. Mech. Engng,* **22**, 59.
11. Matthies, H. and Strang, G. (1979). The solution of nonlinear finite element equations. *Int. J. Num. Meth. Engng,* **14**, 1613.
12. Whittle, P. (1971). *Optimization under Constraints.* Wiley, London.
13. Timoshenko, S. P. (1956). *Strength of Materials part II,* (3rd edn). Van Nostrand Reinhold, New York.

14. Zienkiewicz, O. C., Valliappan, S., and King, I. P. (1968). Stress analysis of rock as a no-tension material. *Geotechnique*, **18**, 56.

15. Harrison, H. B. (1973). *Computer Methods in Structural Analysis*. Prentice-Hall, Englewood Cliffs, NJ.

16. Simpson, B. and Wroth, C. P. (1972). Finite element computation for a model retaining wall in sand. *Proceedings of the Fifth European Conference on Soil Mechanics and Foundation Engineering*, **1**, 85.

17. Hsu, T. R. and Bertels, W. M. (1974). Improved approximation of constitutive elastic-plastic stress-strain relationships for finite element analysis. *J. Amer. Inst. Aeron. Astron.*, **12**, 1450.

18. Grayson, R. and Stevens, L. K. (1979). Nonlinear analysis of structural systems of steel and concrete. *Proceedings of the Third Australian International Conference on Finite Element Methods*, University of NSW, Sydney.

19. Christian, J. T. and Desai, C. S. (1977). Constitutive laws for geological data. In *Numerical Methods in Geotechnical Engineering*, (ed. C. S. Desai and J. T. Christian). McGraw-Hill, New York.

20. Cormeau, I. C. (1975). Numerical stability in quasi-static elasto-viscoplasticity. *Int. J. Num. Meth. Engng*, **9**, 109.

21. Buyukozturk, O. (1977). Nonlinear analysis of reinforced concrete structures. *Comput. Struct.*, **7**, 149.

22. Grayson, R. and Milner, H. R. (1977). NASP: program for nonlinear analysis of stiffened plates. Report of Simon Engineering Laboratories, University of Manchester.

23. Malvern, L. E. (1969) *Introduction to the Mechanics of a Continuous Medium*. Prentice-Hall, Englewood Cliffs, NJ.

24. Drucker, D. C. (1956). On uniqueness in the theory of plasticity. *Quart. Appl. Math.*, **14**, 35.

25. Wright, J. P. and Baron, M. L. (1978). Dynamic deformation of materials and structures under explosive loadings. *IUTAM Symposium on High Velocity Deformation of Solids, Tokyo*. Springer-Verlag, Berlin.

26. Yang, T. Y. (1986). *Finite Element Structural Analysis*. Prentice-Hall, Englewood Cliffs, NJ.

27. Mohr, G. A. and Milner, H. R. (1981). Finite element analysis of large displacements in flexural systems. *Comput. Struct.*, **13**, 553.

28. Milner, H. R. (1981). Accurate analysis of large displacements in skeletal structures. *Comput. Struct.*, **14**, 205.

29. Misner, C. W., Thorne, K. S., and Wheeler, J. A. (1973). *Gravitation*. Freeman, San Francisco.

30. Regge, T. (1961). General relativity without coordinates. *Nuovo Cimento*, **19**, 558.

13
Eigenvalue analysis of vibration and stability

Many physical systems possess *eigenvalues*, that is, values of some important parameter at which the system falls into characteristic states of behaviour. Each of these states is described by an *eigenvector* which is associated with one of the eigenvalues. The eigenvalue problems most frequently encountered in the mechanics of solids are those of vibration and buckling and both of these are discussed in the present chapter. Techniques for solving linear eigenvalue problems are then discussed and these can also be applied to the eigenvalue problems of electromagnetic vibration and seiche motion which we shall encounter in Chapters 18 and 20.

13.1 Introduction

A wide variety of physical problems are governed by the *wave equation* (derived in Section 16.3.3) which for a single particle of mass m is

$$c \frac{\partial \phi}{\partial t} + m \frac{\partial^2 \phi}{\partial t^2} = \frac{\partial^2 \phi}{\phi x^2} + \frac{\partial^2 \phi}{\phi y^2} + \frac{\partial^2 \phi}{\partial z^2} = \nabla^2 \phi \tag{13.1}$$

where c is a damping parameter and ∇^2 is the Laplacian operator. Using separation of variables we assume a solution in the form $\phi = \psi(x, y, x) T(t)$ and substitute this into eqn (13.1), giving

$$T \nabla^2 \psi = \psi \dot{T} + m \psi \ddot{T}. \tag{13.2}$$

Dividing through by the function ϕ one obtains

$$\frac{\nabla^2 \psi}{\psi} = \frac{c \dot{T}}{T} + \frac{m \ddot{T}}{T} \tag{13.3}$$

and as the left and right sides are functions only of position (x, y, z) and time (t), respectively, each side may be equated to a constant $-k$, giving

$$m \ddot{T} + c \dot{T} + k T = 0 \tag{13.4}$$

and

$$\nabla^2 \psi + k\psi = 0. \tag{13.5}$$

The first of these is a linear ordinary differential equation which can be solved by time stepping techniques (see Chapter 14). The second is the *Helmholtz* equation which governs a wide range of eignevalue problems, including those of incipient structural stability, acoustic vibration, seiche motion, and electromagnetic waves. In the present chapter we shall restrict attention to structural examples but applications in other areas follow the same general lines.

The importance of eigenvalues in structural stability needs no elaboration whilst in vibration problems they can be used to predict any structural response by *modal superposition*. In relation to *harmonic excitation* (by loading of a cyclic nature) and *resonance* (which occurs when the frequency of excitation approaches a natural frequency or eigenvalue of the system) eignevalues are also of interest.

Equation (13.4) should in general include a forcing function on the right-hand side. This is most readily seen by applying Newton's second law to the single-mass systems of Fig. 13.1, giving

$$m\frac{\partial^2 x}{\partial t^2} = q(t) - kx - c\frac{\partial x}{\partial t}$$

or

$$kx + c\dot{x} + m\ddot{x} = q(t) \tag{13.6}$$

where $p(t)$ is the time-dependent forcing function which must be expressed in Newtons if the mass and displacement are in kilograms and metres respectively.

Generalizing to a system with several degrees of freedom one writes

$$M\{\ddot{D}\} + C\{\dot{D}\} + K\{D\} = \{Q(t)\} \tag{13.7}$$

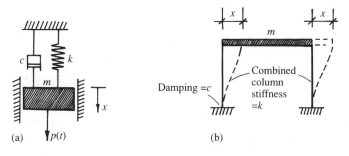

Fig. 13.1 Simple vibrational systems with one freedom

where M, C, and K are respectively the system mass, damping, and stiffness matrices, all of which may be assembled from element matrices in the same manner in which stiffness matrices were assembled in previous chapters.

In order to obtain formulae for the damping and mass matrices of eqn (13.7) we begin by writing the sum of the changes in statical and dynamical energy of a system undergoing a perturbation $\{\delta \bar{u}\} = \{\delta u, \delta v, \delta w\}$ as

$$\delta \Pi + \delta \Lambda = \tfrac{1}{2} \int \{\delta \varepsilon\}^{t} \{\sigma\} \, dV - \int \{\delta \bar{u}\}^{t} \{T\} \, dS - \int \{\delta \bar{u}\}^{t} \{X\} \, dV$$
$$+ \int \{\delta \bar{u}\}^{t} \left\{ \rho \frac{\partial^2 \bar{u}}{\partial t^2} \right\} dV + \int \{\delta \bar{u}\}^{t} \left\{ \mu \frac{\partial \bar{u}}{\partial t} \right\} dV \qquad (13.8)$$

where ρ is the density of the material and μ is a viscosity or 'internal friction' parameter which has a damping effect upon the motion of the material.

Introducing the same spatial interpolations used in static analysis and also applying these to the velocities and accelerations we have

$$\{\bar{u}\} = F^{t}\{d\}, \qquad \left\{ \frac{\partial \bar{u}}{\partial t} \right\} = F^{t} \left\{ \frac{\partial d}{\partial t} \right\}, \qquad \left\{ \frac{\partial^2 \bar{u}}{\partial t^2} \right\} = F^{t} \left\{ \frac{\partial^2 d}{\partial t^2} \right\}. \quad (13.9)$$

Using the first of these to form the strain interpolation $\{\varepsilon\} = B\{d\}$ and substituting this and eqns (13.9) into eqn (13.8) we obtain, after deleting a common factor $\{\delta d\}^{t}$, the equations for a finite element as

$$\int B^{t} DB \, dV \{d\} - \int F \{T\} \, dS - \int F \{X\} \, dV + \int F \rho F^{t} \, dV \{\ddot{d}\}$$
$$+ \int F \mu F^{t} \, dV \{\dot{d}\} = \{0\} \qquad (13.10)$$

neglecting the interelement reactions in anticipation that these will cancel after summation of the element equations. Then summing the element equations the resulting system equations can be expressed in the form of eqn (13.7):

$$M\{\ddot{D}\} + C\{\dot{D}\} + K\{D\} = \{Q(t)\} \qquad (13.11)$$

where

$$M = \sum m = \sum \int F \rho F^{t} \, dV \qquad (13.12)$$

$$C = \sum c = \sum \int F \mu F^{t} \, dV \qquad (13.13)$$

$$K = \sum k = \sum \int B^{t} DB \, dV \qquad (13.14)$$

$$\{Q(t)\} = \sum \int F \{T\} \, dS + \sum \int F \{X\} \, dV + \{Q_n(t)\} \qquad (13.15)$$

and $\{Q_n(t)\}$ are loadings applied directly at the nodes.

The eigenvalue problem corresponding to eqn (13.11) is obtained by deleting the damping and loading terms, giving

$$M\{\ddot{D}\} + K\{D\} = \{0\}. \qquad (13.16)$$

Separation of variables is then used to write the displacements as

$$\{D\} = \{e_i(x,y,x) \sin(\omega t + \alpha_i)\} \tag{13.17}$$

where $e_i(x,y,z)$ is a function only of position, ω is the frequency of periodic vibration of the system, and α_i is a phase angle which differs for each degree of freedom.

Then the accelerations can be expressed as $\{\ddot{D}\} = -\omega^2\{e_i \sin(\omega t + \alpha_i)\}$ and substituting this result and eqn (13.17) into eqn (13.16) we obtain

$$(K - \omega^2 M) \{e_i \sin(\omega t + \alpha_i)\} = \{0\} \tag{13.18}$$

To determine the amplitudes e_i of vibration of each degree of freedom we set $\sin(\omega t + \alpha_i) = 1$ in eqn (13.18) yielding the eigenvalue problem

$$(K - \lambda M)\{e_i\} = \{0\} \tag{13.19}$$

where $\lambda = \omega^2$. The various values of λ, λ_i, which satisfy eqn (13.19), are the eigenvalues of the system. In the case of vibration problems the corresponding values of ω are the natural frequencies of free vibration, that is, periodic vibration without loading.

If we solve eqn (13.19) as it stands we obtain the trivial solution $\{e_i\} = \{0\}$. But if we note that multiplication by the vector $\{e_i\}$ in eqn (13.19) corresponds to a linear combination of the columns of the matrix $(K - \lambda M)$ resulting in a null column, then it is clear that this matrix is singular, from which follows the *characteristic equation*

$$|K - \lambda M| = 0. \tag{13.20}$$

Then the Laplace expansion of the determinant in eqn (13.20) (that is, the sum of the entries in any row or column multiplied by their cofactors) yields a polynomial equation the roots of which are the eigenvalues λ_i.

In practice, formation of the characteristic polynomial is generally impractical and numerical methods, most of which have a search character are used to extract the eigenvalues. Once these have been extracted, however, the associated eigenvectors are determined from eqn (13.19) by substituting each eigenvalue and solving for the corresponding eigenvector.

First, however, at least one arbitrary value in $\{e_i\}$ must be specified. The need for this can be demonstrated by introducing an arbitrary translation of the axes of magnitude β, so that $\{e_i'\} = \{e_i\} + \beta\{I\}$ where $\{I\}$ is a unit vector. Then substituting for $\{e_i\}$ in eqn (13.19) we obtain

$$(K - \lambda M)\{e_i'\} - \beta(K - \lambda M)\{I\} = \{0\}. \tag{13.21}$$

From eqn (13.19) it follows that the first term in eqn (13.21) = $\{0\}$ so that $(K - \lambda M)\{\beta I\} = \{0\}$, corresponding to the *rigid body mode* $\{\beta I\}$. If eqn(13.19) permits such a rigid body mode, however, then at least one boundary condition (that is, one specified value in $\{e_i\}$) is required before it can be solved. This corresponds to setting an arbitrary datum for the magnitudes of the *eigenmodes* $\{e_i\}$.

A common practice is to set one entry in $\{e_i\}$ to unity. Solution for the remaining entries in $\{e_i\}$ then follows the procedure used for *imposed displacement problems*: one multiplies the corresponding column in the matrix $(K - \lambda_i M)$ by this specified value, deletes the corresponding row in this matrix, and uses Gauss reduction (typically) to solve the remaining equations. Equations (1.62) and (1.63) showed an example of this procedure and an equivalent technique which is more amenable to computer implementation is discussed in Section 15.4.

An alternative approach to solving eqn (13.20) for the eigenvalues is to write eqn (13.19) in the form

$$K\{e_i\} = \lambda M\{e_i\} \tag{13.22}$$

and eqn (13.22) is referred to as the *generalized eigenproblem*. Then if we begin with an initial guess for $\{e_i\}$ such as $\{e_i\}_1 = \{I\}$, for example, then we can iteratively solve the equation $\{e_i\}_{n+1} = K^{-1}M\{e_i\}_n$ until $\{e_i\}_{n+1} = \{e_i\}_n$ to obtain the eigenvector which satisfies eqn (13.22). The eigenvalue was, of course, omitted in forming this iteration as it is as yet unknown, but it therefore follows that λ_i is given by the ratio of any non-zero entry in the final eigenvector solution to its original guessed value. Alternatively if we premultiply both sides of eqn (13.22) by $\{e_i\}^t$ then we obtain each eigenvalue in terms of its *Rayleigh quotient*:

$$\lambda_i = \frac{\{e_i\}^t K\{e_i\}}{\{e_i\}^t M\{e_i\}}. \tag{13.23}$$

As the numerator of eqn (13.23) is twice the strain energy associated with displacements $\{e_i\}$, it is clear that the eigenvalues of rigid body modes, which involve zero strain energy, will be zero.

Many techniques for extracting eigenvalues, however, are based upon the canonical or *standard eigenvalue problem* which has the form

$$H\{e_i\} = \lambda I\{e_i\} = \lambda\{e_i\} \tag{13.24}$$

and to convert eqn (13.22) to this form matrix M is inverted to give $H = M^{-1}K$ or matrix K is inverted to give $H = K^{-1}M$.

Various techniques for eigenvalue extraction are discussed later in the chapter but it is worth noting, however, that eqn (13.24) leads to the definition of the eigenvalues of a matrix H as the roots of the equation $|H - \lambda I| = 0$. As we shall see in the following section this has useful application in investigating the properties of finite element matrices.

13.2 Eigenvalues of finite element stiffness matrices

Eigenvalues have immediate application in investigating finite element matrices. The eigenvalues of the stiffness matrix for a two-freedom spar element, for example, are given by

$$|k - \lambda I| = 0$$

or

$$\left| \frac{EA}{L} \begin{bmatrix} 1 & -1 \\ -1 & 1 \end{bmatrix} - \lambda \begin{bmatrix} 1 & 0 \\ 0 & 1 \end{bmatrix} \right| = 0. \tag{13.25}$$

Assuming $EA/L = 1$ for convenience and expanding this determinant, we find

$$(1 - \lambda)^2 - 1 = 0 \tag{13.26}$$

the roots of which are the eigenvalues $\lambda_1 = 0$, $\lambda_2 = 2$. Substituting these into the equation

$$[k - \lambda I]\{u_1, u_2\} = \{0, 0\}, \tag{13.27}$$

in each case assuming $u_1 = 1$ (an arbitrary reference value), and solving for u_2 one obtains the eigenvectors

$$\{e_1\} = \{1, \ 1\} \quad \text{for } \lambda_1 = 0 \tag{13.28}$$

$$\{e_2\} = \{1, \ -1\} \quad \text{for } \lambda_2 = 2 \tag{13.29}$$

which correspond to the rigid body and constant strain modes of the element, respectively.

As we saw in Chapter 5 a necessary condition for the convergence of finite element solutions with mesh refinement is that the elements exactly model the rigid body and constant strain modes, at least in the limit of mesh refinement. Thus determination of the eigenvalues of the stiffness matrix for an element is a useful way of testing to determine whether the element does provide the required rigid body and constant strain modes.

As another simple example, if we examine the stiffness matrix for the four-freedom beam element with freedoms v and ϕ at each end node (eqn (1.10)) we see that the third row is opposite to the first. This indicates that there is at least one zero eigenvalue corresponding in fact to the rigid body translation mode. With a little manipulation it is easy to show that it possesses a second zero eigenvalue corresponding to the rigid body rotation mode.

Eigenvalue examination is also useful in determining whether more complex finite elements possess spurious modes or too many rigid body modes.[1] In the six-node triangular element for Stokes' flow discussed in Section 19.5, for example, the element possesses six zero eigenvalues rather than the required three for two-dimensional motion unless the pressure freedoms at the midside nodes are suppressed or alternative constraints applied.

Although it is not important in the present application, it should be noted that the usual practice in the applications of the following sections is to *normalize* eigenvectors. The simplest means of doing this is to make the largest entry in $\{e_i\}$ unity and rescale the remaining entries accordingly. Alternatively one divides each entry by the Euclidean norm of the eigenvector, $(\{e_i\}^t \{e_i\})^{\frac{1}{2}}$. For example, normalizing eqns (13.28) and (13.29) in this way gives

$$\{\bar{e}_1\} = \{1/\sqrt{2},\ 1/\sqrt{2}\} \qquad \text{for } \lambda_1 = 0 \tag{13.30}$$

$$\{\bar{e}_2\} = \{1/\sqrt{2},\ -1/\sqrt{2}\} \qquad \text{for } \lambda_2 = 2. \tag{13.31}$$

Yet another alternative is to set $\{e_i\}^t M \{e_i\} = 1$ and this is referred to as M normalizing the eigenvector.

Finally it is worth noting that the eigenvalues of eqns (13.30) and (13.31) provide a simple example of the property

$$\sum \lambda_i = \text{trace}(k) = \sum k_{ii} \tag{13.32}$$

where k_{ii} are the diagonal entries in k and this provides a useful check of such eigenvalue calculations.

13.3   Mass and damping matrices for finite elements

For a spar element with extensional freedoms u_1 and u_2 the strain energy is

$$U_e = \tfrac{1}{2} \int EA \left(\frac{\partial u}{\partial x} \right)^2 dx \tag{13.33}$$

which leads to the stiffness matrix of eqn (13.25) and the kinetic energy is given as

$$\Lambda_e = \tfrac{1}{2} \int_0^L \rho A \left(\frac{\partial u}{\partial t} \right)^2 dx = \frac{L}{2} \int_0^1 \rho A \left(\frac{\partial u}{\partial t} \right)^2 ds. \tag{13.34}$$

(This is equivalent to using the fourth term on the right-hand side of eqn (13.8))

Applying linear interpolation to the velocity $\partial u / \partial t$, that is,

$$\frac{\partial u}{\partial t} = \{f\}^t \left\{ \frac{\partial u}{\partial t} \right\} = \{1 - s, s\}^t \left\{ \frac{\partial u}{\partial t} \right\}, \qquad s = 0 \to 1 \tag{13.35}$$

we obtain

$$\Lambda_e = \frac{L}{2} \{\dot{u}\}^t \left(\int_0^1 \{f\}\, (\rho A)\, \{f\}^t\, ds \right) \{\dot{u}\} = \tfrac{1}{2} \{\dot{u}\}^t\, m\{\dot{u}\} \tag{13.36}$$

from which

$$m = \rho A L \int_0^1 \{f\} \{f\}^t \, ds \qquad (13.37)$$

or

$$m = \rho A L \begin{bmatrix} 1/3 & 1/6 \\ 1/6 & 1/3 \end{bmatrix}. \qquad (13.38)$$

This is the *consistent mass matrix*[2,3,4] (a matrix is consistent if it satisfies the governing differential equations to an accuracy consistent with that of the assumed polynomials[3]). It clearly conserves mass as its entries sum to unity. The *lumped mass matrix* is given by assigning half the mass to each node:

$$m = \rho A L \begin{bmatrix} 1/2 & 0 \\ 0 & 1/2 \end{bmatrix}. \qquad (13.39)$$

and its use is comparable to the approximation of using lumped loads rather than consistent loads. The relative merits of using consistent or lumped masses are discussed in Section 13.5.

For a beam element with freedoms v_1, ϕ_1, v_2, ϕ_2 one uses in eqn (13.37):

$$\{f\} = \{1 - 3s^2 + 2s^3, L(s - 2s^2 + s^3), 3s^2 - 2s^3, L(s^3 - s^2)\} \qquad s = 0 \to 1$$
$$(13.40)$$

giving

$$m = \frac{\rho A L}{420} \begin{bmatrix} 156 & 22L & 54 & -13L \\ 22L & 4L^2 & 13L & -3L^2 \\ 54 & 13L & 156 & -22L \\ -13L & -3L^2 & -22L & 4L^2 \end{bmatrix} \qquad (13.41)$$

and in the lumped mass matrix for the beam the entries 156 become 210 and the others vanish. An advantage of the finite element method here is that the consistent mass matrix directly includes the *rotatory inertia*, and the effect of transverse shear upon the rotatory inertia (often greater than the statical shear effect) can also be included.[5-8]

As shown in Section 13.1 a damping matrix which takes the same form as the consistent mass matrix can be defined;

$$c = \int F \mu F^t \, dV \qquad (13.42)$$

where μ is a friction coefficient. Integration will in some cases be over the surface, or some part thereof, rather than the volume of each element.

For skeletal elements inclined to the horizontal the coordinate transformation of eqn (1.21) must be applied to the element stiffness, mass, and damping matrices before they are assembled to form the structure matrices.

The consistent mass matrix for the linear strain triangle is obtained by premultiplying eqn (9.28) by $\rho \Delta t$ and using this result for both the u, v freedoms. In general, though, consistent mass matrices are best obtained by numerical integration so that an interpolation for variable thickness, for example, can easily be included in eqn (13.37).

13.4 Natural frequencies of free vibration

Free vibrations are set up by releasing an unloaded system from a displaced position at rest. As shown in Section 13.1 the eigenvalues of free vibration are given by the characteristic equation

$$|K - \lambda M| = 0 \tag{13.43}$$

the solutions $\lambda_i = \omega_i^2$ yielding the natural frequencies of vibration. Then the eigenvectors are obtained by solving

$$(K - \lambda_i M)\{e_i\} = \{0\} \tag{13.44}$$

As an example consider extensional vibration of the cantilever shown in Fig. 13.2. Using two elements, consistent mass matrices, and deleting the freedom $u_3 = 0$ as a boundary condition, we obtain from eqn (13.44)

$$\left\{ \begin{bmatrix} 1 & -1 \\ -1 & 2 \end{bmatrix} \frac{EA}{L} - \begin{bmatrix} 1/3 & 1/6 \\ 1/6 & 2/3 \end{bmatrix} \rho A L \omega^2 \right\} \begin{Bmatrix} u_1 \\ u_2 \end{Bmatrix} = 0. \tag{13.45}$$

Then the characteristic equation is

$$\begin{vmatrix} 1 - 2\lambda & -1 - \lambda \\ -1 - \lambda & 2 - 4\lambda \end{vmatrix} = 0 \tag{13.46}$$

where $\lambda = \omega^2 \rho L^2/6E$. Expanding the determinant explicitly we obtain

$$(1 - 2\lambda)(2 - 4\lambda) - (1 + \lambda)^2 = 0 \tag{13.47}$$

giving the roots $\lambda_1 = 0.1082$, $\lambda_2 = 1.3204$, from which, if $S = 2L$,

$$\omega_1 S \left(\frac{\rho}{E} \right)^{1/2} = 2\sqrt{(6\lambda_1)} = 1.6114 = 1.0259 \left(\frac{\pi}{2} \right) \tag{13.48}$$

$$\omega_2 S \left(\frac{\rho}{E} \right)^{1/2} = 2\sqrt{(6\lambda_2)} = 5.6293 = 1.1946 \left(\frac{3\pi}{2} \right) \tag{13.49}$$

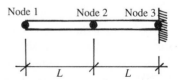

Fig. 13.2 Cantilever with three extensional freedoms

compared to the exact solutions which are $\pi/2$, $3\pi/2$.[9] Substituting ω_1^2, ω_2^2 and $u_1 = 1$ into eqn (13.45) yields the eigenvectors $\{1, 0.7071\}$ and $\{1, -0.7071\}$. If u_3 is unrestrained, however, we obtain $\lambda_1 = 0$, $\{e_1\} = \{1, 1, 1\}$ (the rigid body mode), $\lambda_2 = 0.5$, $\{e_2\} = \{1, 0, -1\}$ (the antisymmetric mode) and $\lambda_3 = 2$, $\{e_3\} = \{1, -1, 1\}$ (the symmetric mode).

With the finite element method, inclusion of flexural stiffness is accomplished without difficulty, allowing the solution of complex problems involving interaction of extensional and flexural vibrations.[9, 10]

13.5 Consistent masses versus lumped masses

Considering the example of Fig. 13.2 and using one, two, or three elements with either lumped or consistent masses the errors in the first three natural frequencies are shown in Table 13.1. For the fundamental frequency consistent and lumped masses give much the same accuracy but for the higher modes consistent loads are more accurate. The result for ω_2 with three elements and lumped masses is poor and shows that lumped masses do no guarantee monotonic convergence.

The diagonal lumped mass matrix, though, greatly simplifies eigenvalue extraction and it can be shown that, so long as the total mass is conserved, it will lead to convergent results.[11]

For some two-dimensional elements, in fact, good results can be obtained by scaling the diagonal terms of the consistent mass matrix, so that they alone account for the total mass, and omitting the off-diagonal terms.[12] In the linear strain triangle, for example, the lumped masses are $\rho\Delta t/9$ at the vertex nodes and $2\rho\Delta t/9$ at the midside nodes, whereas such scaling gives masses for the midside nodes five times those associated with the vertex nodes (see eqn (9.28) where a matrix of the same form as the consistent mass matrix is calculated for this element).

Table 13.1 Errors in finite element estimation of eigenvalues using consistent and lumped masses

Frequency	Elements	Error (%)	
		Consistent loads	Lumped loads
	1	10.3	− 10.0
ω_1	2	2.6	− 2.5
	3	1.1	− 1.1
	2	19.5	− 21.6
ω_2	3	10.3	− 55.0
ω_3	3	20.0	− 26.2

Generally the best recommendation, however, is to use lumped masses with low-order elements and consistent masses with high-order elemens.[13] The quintic triangle is a case in point where a fully consistent formulation should be used for both the loads and the masses. It is also worth noting that in flexural problems the lumped mass matrix omits the rotational inertia so that the rotational freedoms can be removed from the stiffness matrix by static condensation (that is, using eqn (6.42)). In typical thin plate elements with freedoms ω, ϕ_x, and ϕ_y at the nodes this reduces the number of equations to be solved by two thirds, a considerable reduction in computational effort.

The same condensation scheme can be used in the analysis of plates on elastic foundations, for example if the foundation is modelled by lumped or 'spring' stiffnesses at each node rather than using the contact stiffness matrix discussed in Section 3.7. It can also be used in incipient instability problems,[14] when in the case of beam elements, for example, the geometric stiffness matrix is approximated by the *string stiffness matrix*.[10] The string stiffness matrix is defined in Section 13.7 and, because of its sparse form, its use is directly comparable to the use of lumped mass matrices instead of consistent mass matrices.

13.6   Eigenvalues of plate vibration

As an example consider a quadrant of a simply supported square plate analysed using a single 12df rectangular bicubic element (eqns 5.45). First, as an introduction, consider static equilibrium of the plate under a uniformly distributed load. The boundary conditions are those shown for w, ϕ_x, ϕ_y in Fig. 8.9 and the loads are given by eqns (8.9), so that if the element occupies the 'bottom left' quadrant one obtains

$$\frac{Et^3}{12(1-v^2)} \begin{bmatrix} k_\phi & -k_{w\phi} & k_{\phi\phi} \\ -k_{w\phi} & k_w & -k_{w\phi} \\ k_{\phi\phi} & -k_{w\phi} & k_\phi \end{bmatrix} \begin{Bmatrix} \phi_{y2} \\ w_3 \\ \phi_{x4} \end{Bmatrix} = \begin{Bmatrix} pL_aL_b^2/24 \\ pL_aL^b/4 \\ pL_a^2L^b/24 \end{Bmatrix} \quad (13.50)$$

that is, nodes are taken anticlockwise with node 3 at the plate centre. Taking k_ϕ as an example, this is obtained using eqns (8.5) and (8.6) to calculate k_ϕ as row 6 of B^t times column 6 of DB which, denoting $f_{i,xx} = \partial^2 f_i/\partial x^2$, and so on, yields

$$k_\phi = \iint \{ (f_{6,xx})^2 + 2v(f_{6,xx}f_{6,yy}) + (f_{6,yy})^2 + 2(1-v)(f_{6,xy}) \} \, dx \, dy \quad (13.51)$$

which corresponds to eqn (2.31) and f_6 is given by eqns (5.45). After some algebra one obtains $k_\phi = (4\lambda^2/35 + 52\lambda^{-2}/35 + 8/25)\lambda = 1.92$ with $\lambda = L_b/L_a = 1$. Other entries in eqn (13.50) are obtained similarly and the complete stiffness matrix is given by Przemieniecki.[10] Solving eqn (13.50) with $L_a, L_b = L/2$ one obtains $w_3 Et^3/12pL^4 (1 - v^2) = 0.003605$ which is 10 per cent less than the exact solution, a satisfactory result with so few equations.

The characteristic equation for transverse vibration of the plate quadrant is obtained, if one quarter of the mass is lumped at each vertex, as

$$
\left| \frac{Et^3}{12(1 - v^2)} \begin{bmatrix} k_\phi & -k_{w\phi} & k_{\phi\phi} \\ -k_{w\phi} & k_w & -k_{w\phi} \\ k_{\phi\phi} & -k_{w\phi} & k_\phi \end{bmatrix} - \frac{\lambda \rho t L_a L_b}{4} \begin{bmatrix} 0 & 0 & 0 \\ 0 & 1 & 0 \\ 0 & 0 & 0 \end{bmatrix} \right| = 0.
$$

(13.52)

This gives $(\omega/2\pi^2)\sqrt{(\rho t L^4(1 - v^2)/Et^3)} = 1.0126$, where $\omega = \sqrt{\lambda}$, as the solution for the fundamental frequency, whereas the exact result is unity.[9] This is an excellent result with a single element but note that with a polynomial basis accuracy declines for higher frequencies, as Table 13.1 shows. The advantage of the finite element method, however, is the ease with which rotatory inertia, transverse shear, non-homogeneity, and similar effects can be included.

13.7 The initial stress matrix

The initial stress or geometric stiffness matrix, k_G, is used to take into account the effects of:

(1) stress stiffening in tension, for example by the development of membrane stresses in plates with large deflections;

(2) stress destiffening in compression in incipient instability (that is, buckling) problems.

The term initial stress matrix is perhaps more appropriate as this matrix is related to the strain energy associated with transverse displacement of an element with initial extensional stresses.

We have already discovered k_G in Section 12.5, where the extensional strain energy of a finite beam element of length L subjected to longitudinal and transverse displacements was shown to be

$$
U_e = \frac{EA}{2} \int_0^L \left[\left(\frac{\partial u}{\partial x}\right)^2 + \frac{\partial u}{\partial x}\left(\frac{\partial v}{\partial x}\right)^2 + \tfrac{1}{4}\left(\frac{\partial v}{\partial x}\right)^4 \right] dx. \qquad (13.53)
$$

In incipient stability problems 'large' (that is, finite) lateral displacements do not occur. Indeed we consider the state of instability or buckling to be one of *metastable equilibrium* in which the transverse displacements are arbitrary in magnitude but have a characteristic mode. Thus, at the onset of buckling the lateral displacements are infinitesimal so that now we neglect the last term of eqn (13.53).

Then considering an arbitrary or virtual displacement $\{\delta d\}$ into the buckled state and introducing the interpolations $u = \{f_u\}^t \{d\}$ and $v = \{f_v\}^t \{d\}$, where $\{d\}$ includes freedoms u, v, and ϕ, we require for equilibrium of a beam element

$$\delta U_e = \{\delta d\}^t \left[\int (B_a^t (EA)B_a + B_f^t(EI)B_f) \, \mathrm{d}x + k_G^* \right] \{d\} - \{\delta d\}^t \{q\} = 0 \quad (13.54)$$

where

$$B_a = \{\partial f_u/\partial x\}^t, \qquad B_f = \{\partial^2 f_v/\partial x^2\}^t \quad (13.55)$$

$$k_G^* = \int \{\partial f_v/\partial x\} \, (A\sigma^*) \, \{\partial f_v/\partial x\}^t \, \mathrm{d}x. \quad (13.56)$$

Equations (13.55) are the usual extensional and flexural strain matrices and k_G corresponds to the second term of eqn (13.53) which is linearized by writing $\sigma^* = EA(\partial u/\partial x)$. Here σ^* is the initial stress in the element and this is the value resulting from a loading $\{Q^*\}$ of arbitrary magnitude, that is, relative to one another the loads are correctly proportioned and our analysis seeks to determine the scale factors or eigenvalues by which $\{Q^*\}$ must be multiplied to cause buckling in various modes.

Deleting the common factor $\{\delta d\}^t$ in eqn (13.54), summing the matrices given by B_a and B_f to form the complete element stiffness matrix k, and then summing the element stiffness and initial stress matrices in the usual manner we obtain

$$(K + \lambda K_G^*) \{D^*\} = \lambda \{Q^*\} \quad (13.57)$$

where λ is the scale factor for the arbitrary loading $\{Q^*\}$ and $K_G^* = \Sigma k_G^*$ is computed using the element stresses σ^* corresponding to this loading.

In some special cases, such as those of statically determinate structures, the initial stresses in each element are obvious, but in general a preliminary elastic analysis is needed to determine the initial stresses.

If, in eqn (13.57), λ takes a value just sufficient to cause buckling and we write the displacements after buckling as $\{D\} = \{D^*\} + \{e_i\}$ then we obtain

$$(K + \lambda K_G^*) \{D^*\} + (K + \lambda K_G^*) \{e_i\} = \lambda \{Q^*\} + \{0\} \quad (13.58)$$

the null vector on the right-hand side appearing because buckling occurs without increase in loading.

Then substituting eqn (13.57) in eqn (13.58) the latter reduces to

$$(K + \lambda K_G^*) \{e_i\} = \{0\}. \quad (13.59)$$

In eqn (13.59) multiplication by the vector $\{e_i\}$ yields a linear combination of the columns of the augmented stiffness matrix $(K + \lambda K_G^*)$ which results in a zero column $\{0\}$, that is, the augmented stiffness matrix is singular. It follows that the characteristic equation is

$$|K + \lambda K_G^*| = 0. \tag{13.60}$$

The various values of λ, λ_i, which satisfy eqn (13.60) are the eigenvalues of incipient buckling and the corresponding loadings which cause buckling follow as $\{Q\} = \lambda\{Q^*\}$.

The associated buckling modes are then the eigenvectors $\{e_i\}$ which are found by solving eqn (13.59) with $\lambda = \lambda_i$ and, typically, with one entry in $\{e_i\}$ set to unity.

For a four-freedom beam element with freedoms v_1, ϕ_1, v_2, and ϕ_2 the initial stress matrix is given by differentiating eqns (13.40) to give

$$\left\{\frac{\partial f_v}{\partial x}\right\} = \left\{\frac{\partial f_v}{\partial s}\right\}\left(\frac{\mathrm{d}s}{\mathrm{d}x}\right) = \frac{1}{L}\left\{\frac{\mathrm{d}f_v}{\mathrm{d}s}\right\} = \{f_v'\}$$

$$= \frac{\{6s^2 - 6s,\, L(3s^2 - 4s + 1),\, 6s - 6s^2,\, L(3s^2 - 2s)\}}{L} \tag{13.61}$$

and substituting this result into eqn (13.56), yielding

$$k_G^* = A\sigma^* \int \{f_v'\}\{f_v'\}^{\mathrm{t}}\,\mathrm{d}x = A\sigma^* \int_0^1 \{f_v'\}\{f_v'\}^{\mathrm{t}}\,\mathrm{d}s \left(\frac{\mathrm{d}x}{\mathrm{d}s} = L\right)$$

$$= \frac{A\sigma^*}{30L}\begin{bmatrix} 36 & 3L & -36 & 3L \\ & 4L^2 & -3L & -L^2 \\ \text{Symmetric} & & 36 & -3L \\ & & & 4L^2 \end{bmatrix}. \tag{13.62}$$

Equation (13.62) is an approximation although, as noted in Section 1.2, the stiffness matrix for this element is exact. This is because cubic interpolation is just sufficient to yield the complementary integral of $EI\mathrm{d}^4y/\mathrm{d}x^4 = p$ and integration of this to obtain the consistent loads recovers the particular integral. An exact geometric stiffness for the four-freedom beam element can be obtained, however, using the transcendental s and c functions.[15, 16]

An approximate geometric stiffness matrix for the four-freedom beam element can be obtained by writing $\partial v/\partial x \simeq (v_2 - v_1)/L$. This is equivalent to neglecting the slope freedoms and applying linear interpolation to the freedoms v_1 and v_2. Then using $\{\partial f_v/\partial x\} = \{0, -1/L, 0, 1/L\}$ in eqn (13.56) we obtain

$$k_G^* = \frac{A\sigma^*}{L} \begin{bmatrix} 0 & 0 & 0 & 0 \\ 0 & 1 & 0 & -1 \\ 0 & 0 & 0 & 0 \\ 0 & -1 & 0 & 1 \end{bmatrix} \tag{13.63}$$

This is called the *string stiffness matrix*[10] and it is also the appropriate initial stress matrix for the four-freedom spar element with freedoms u_1, v_1, u_2, and v_2. Used as an approximation for beam elements, eqn (13.63) permits static condensation to be applied to the element stiffness matrices in order to eliminate the slope freedoms from the analysis, as noted at the close of Section 13.5.

Generalization of eqn (13.56) to deal with thin plate finite elements is easily achieved by writing[17]

$$k_G^* = \int B^t \begin{bmatrix} \sigma_x^* & \sigma_{xy}^* \\ \sigma_{xy}^* & \sigma_y^* \end{bmatrix} B \, dv \tag{13.64}$$

$$B = \begin{bmatrix} \partial\{f\}^t/\partial x \\ \partial\{f\}^t/\partial y \end{bmatrix} \tag{13.65}$$

where $w = \{f\}^t \{w_1, \phi_{x1}, \phi_{y1}, w_2, \ldots, w_n, \phi_{xn}, \phi_{yn}\}$ is the interpolation for the transverse displacement. Generally, of course, the integration of eqn (13.64) is carried out numerically.

Alternatively, comparable approximations to that of eqn (13.63) can be made for thin plate finite elements. The simplest procedure, for example, is to apply eqn (13.63) to each side of triangular elements, replacing σ^* by the natural stresses parallel to the sides so frequently used in previous chapters.

13.8   Calculation of the eigenpairs for incipient buckling

As a simple example of the calculation of the eigenpairs, (the eigenvalues and their associated eigenvectors) for incipient buckling we consider a pin-ended column subject to a compressive axial load, $-P$. If we use only a single element with the boundary conditions $v_1 = v_2 = 0$ then, deleting the first and third rows and columns of k (given by eqn (1.10)) and k_G^* (given by eqn (13.62)) and substituting the results in eqn (13.60) we obtain

$$|k + \lambda k_G^*| = \begin{vmatrix} (4EI/L - 4\lambda L/30) & (2EI/L + \lambda L/30) \\ (2EI/L + \lambda L/30) & (4EI/L - 4\lambda L/30) \end{vmatrix} = 0 \tag{13.66}$$

where P is taken as unity so that $\lambda A\sigma^* = -\lambda P = -\lambda$.

Expanding eqn (13.66) to obtain the characteristic polynomial, the roots are $\lambda_1 = 12EI/L^2$ and $\lambda_2 = 60EI/L^2$ and the corresponding eigenvectors are $(1/\sqrt{2}, -1/\sqrt{2})$ and $(1/\sqrt{2}, 1/\sqrt{2})$ which represent the symmetric and antisymmetric modes. The exact solutions are $\lambda_1 = \pi^2 EI/L^2$ and $\lambda_2 = 4\lambda_1$ so that, as shown in Table 13.1, we once again see the typical increase in error in the higher eigenvalue estimates expected when polynomial approximating techniques are used.

In general, however, coordinate transformation must be applied to both element stiffness and initial stress matrices of beam/column elements before they are assembled to form the system equations. Taking the system of two spar elements in Fig. 13.3 as an example the element stiffness matrices are gien by eqn (1.33):

$$ k = \begin{bmatrix} k_{uv} & -k_{uv} \\ -k_{uv} & k_{uv} \end{bmatrix} \quad \text{where } k_{uv} = \frac{EA}{L} \begin{bmatrix} s^2 & cs \\ sc & c^2 \end{bmatrix}. \quad (13.67) $$

Applying the same coordinate transformation to k_G^* in eqn (13.63), that is $T^t k_G^* T$ with matrix T of eqn (1.21) reduced to omit slope freedoms, we obtain

$$ k_G^* = \begin{bmatrix} g & -g \\ -g & g \end{bmatrix} \quad \text{where } g = \frac{\sigma^* A}{L} \begin{bmatrix} s^2 & -cs \\ -sc & c^2 \end{bmatrix} \quad (13.68) $$

where $c = \cos\alpha = (x_j - x_i)/L_{ij}$, $\sin\alpha = (y_j - y_i)/L_{ij}$ for a member between nodes i and j.

Then, in Fig. 13.3, $\cos\alpha = 0$, $\sin\alpha = -1$ and $\cos\alpha = -1/\sqrt{2}$, $\sin\alpha = -1/\sqrt{2}$ for the vertical and inclined members respectively. Using these values in eqns (13.67) and (13.68) and assembling the resulting element matrices the characteristic equation obtained is, if we take $\sigma^* A = -P$,

$$ \left| \frac{\sqrt{2}EA}{4L} \begin{bmatrix} 1 & 1 \\ 1 & 1 + 2\sqrt{2} \end{bmatrix} - \frac{P}{L} \begin{bmatrix} 1 & 0 \\ 0 & 0 \end{bmatrix} \right| = 0 \quad (13.69) $$

after deletion of rows and columns corresponding to the boundary conditions $u_2 = v_2 = u_3 = v_3 = 0$. Solving eqn (13.69) we obtain $P = EA/(1 + 2\sqrt{2})$ as

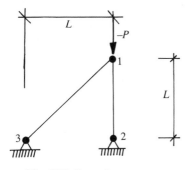

Fig. 13.3 Two-element truss

the magnitude of the buckling load and the associated eigenvector is $\{1, \ -1/(1 + 2\sqrt{2})\}$.

Such analyses are applicable to space truss structures, for example, but will only predict overall buckling of the structure and will not predict local buckling of a single member. To deal with the latter phenomenon eqns (13.62) and (13.63) should be combined to obtain a 6×6 initial stress matrix corresponding to freedoms u, v, and ϕ at each node. This corresponds to the stiffness matrix of eqn (1.12) and, with the use of these, Euler-type buckling in individual members can also be modelled. If the truss is 'pin jointed', however, the element stiffness matrices should be modified to permit only a small rotational stiffness at each joint.[15]

13.9 Numerical determination of eigenpairs.

The writing of numerical routines to extract eigenvalues and eigenvectors is very much a specialized task and many of the nuances involved require detailed study. Herein we present only a few of the available techniques, not so much to provide a definitive statement of the various problems but more to provide a general background enabling readers to gain an appreciation of the processes involved. To this end we consider the following techniques:

(1) transformation techniques such as the Jacobi[18] and Givens methods;

(2) vector iteration methods;[19-23]

(3) polynomial iteration methods;[24]

(4) determinant search methods[16] and the Sturm sequence;

(5) Composite methods.

13.9.1 Transformation techniques

Jacobi diagonalization is one of a number of transformation techniques and is useful for small problems. It simultaneously extracts all eigenvalues and eigenvectors of the general eigenproblem $(K - \lambda M)\{e_i\} = \{0\}$. This is done by reducing the K and M matrices to diagonal form using successive transformations to eliminate their off-diagonal entries. To eliminate the entries in positions ij and ji the required transformations are

$$k_{n+1} = T_n^t k_n T_n, \qquad M_{n+1} = T_n^t M_n T_n \qquad (13.70)$$

where T_n is a unit matrix with entries α and β added in positions ij and ji, respectively, and α and β are given by

$$\alpha = -P/S, \qquad \beta = Q/S \qquad (13.71)$$

where

$$P = k_{ii}M_{ij} - m_{ii}k_{ij}, \qquad Q = k_{ji}m_{ij} - m_{jj}k_{ij}, \qquad R = k_{ii}m_{jj} - k_{jj}m_{ii}$$

$$S = \tfrac{1}{2}R + \mathrm{sgn}(R)\sqrt{(\tfrac{1}{4}R^2 + PQ)}. \tag{13.72}$$

Row i and column j may be chosen in order to eliminate the off-diagonal entry of largest magnitude. Search to find these, however, is time-consuming and it is common practice to sweep through the rows or columns progressively.

When $k_{ij}/(k_{ii}k_{jj})^{\frac{1}{2}}$ is less than a small tolerance value transformation is not applied to k_{ij} and iteration of eqns (13.70) ceases when all off-diagonal entries satisfy this test. When matrices K and M have been diagonalized (to sufficient accuracy) the eigenvalues follow as $\lambda_i = k_{ii}/m_{ii}$ and the eigenvectors are given by the matrix

$$E = T_1 T_2 T_3 \ldots T_n \tag{13.73}$$

in which column i contains $\{e_i\}$.

The *Givens method* is a special form of the Jacobi method in which the matrices are reduced to *tridiagonal* form, with the only non-zero entries in row i being k_{ii-1}, k_{ii}, and k_{ii+1}. Thus i and j are chosen in the sequence $(1, 3)$, $(1, 4), \ldots, (1, m), (2, 4), (2, 5), \ldots, (2, m)$, and so on. With this sequence, once an entry k_{ij} is eliminated it remains zero so that only one pass through the matrix is required.

Once the matrices have been reduced to tridiagonal form, extraction of the eigenvalues and eigenvectors is straightforward and the final eigenvectors are obtained as

$$\{e_i\} = T_1 T_2 T_3 \ldots T_n \{\bar{e}_i\} \tag{13.74}$$

where $\{\bar{e}_i\}$ are the eigenvectors of the tridiagonalized problem.

Such transformation techniques, however, involve a great deal of computation when the matrices are relatively large, that is, the amount of computation required to obtain a single eigenvalue is comparable to that required to solve the linear equilibrium problem $K\{D\} = \{Q\}$. Moreover we generally require only a few of the eigenvalues, usually the smallest. In such cases the methods described in the remainder of the present section are preferable, a these provide only the number of eigenpairs required.

13.9.2 Vector iteration techniques

Vector iteration techniques begin with an initial guessed eigenvector and use the eigenproblem $(K - \lambda M)\{e_i\} = \{0\}$ to form a recurrence relation that provides progressively improved estimates by iteration. If we seek the eigenvector associated with the highest eigenvalue then we write the recurrence relation as

$$\{e_i\}_{n+1} = M^{-1}K\{e_i\}_n \tag{13.75}$$

and if lumped masses are used, M is diagonal and M^{-1} is simply $\langle 1/m_{ii} \rangle$ where $\langle \ \rangle$ denotes a diagonal matrix with the non-zero entries indicated.

In the problem of eqn (13.45), for example, putting $EA/L = \rho AL\omega^2 = 1$ and using lumped masses we obtain

$$\{e_i\}_{n+1} = \begin{bmatrix} \frac{1}{2} & 0 \\ 1 & 1 \end{bmatrix}^{-1} \begin{bmatrix} 1 & -1 \\ -1 & 2 \end{bmatrix} \{e_i\}_n = \begin{bmatrix} 2 & -2 \\ -1 & 2 \end{bmatrix} \{e_i\}_n.$$

$$(13.76)$$

commencing with $\{e_i\} = \{1, \ 1\}$ the successive eigenvector estimates obtained are

$$\{e_i\} = \begin{Bmatrix} 0 \\ 1 \end{Bmatrix} = \begin{Bmatrix} -2 \\ 2 \end{Bmatrix} = \begin{Bmatrix} -8 \\ 6 \end{Bmatrix} = \begin{Bmatrix} -30 \\ 20 \end{Bmatrix} = \begin{Bmatrix} -100 \\ 70 \end{Bmatrix} = \begin{Bmatrix} -340 \\ 240 \end{Bmatrix}$$

$$(13.77)$$

Dividing the last result by -340 we obtain $\{e_i\} = \{1, \ -0.7059\}$ which is close to the solution obtained for the highest frequency mode from the roots of the characteristic polynomial in Section 13.4, that is, $\{e_2 = 1, \ -0.7071\}$.

The growth in magnitude of eigenvector estimates in eqn (13.77) should be prevented by normalization. As this growth occurs because λ is omitted in eqn (13.75) it follows that the ratio of the kth entries in successive normalized eigenvectors provides an estimate of the eigenvalue.† Normalizing the last two vectors in eqn (13.77) we obtain $\lambda_2 \simeq 1$ which is approaching the value 1.3204 obtained in Section 13.4. Alternatively the eigenvalue can be calculated as the Rayleigh quotient of the last eigenvector estimate and this procedure, together with *shifting* to obtain further eigenpairs, is used below.

If we seek the lowest eigenvalue and its associated eigenvector then instead of eqn (13.75) we use the iteration $\{e_i\}_{n+1} = k^{-1}M\{e_i\}_n$. Then, beginning with the initial trial vector $\{\bar{e}\}_0$ and separating operations involving M and K, we iterate the following steps:

$$\text{(a)} \quad \{e\}_{n+1} = K^{-1}\{\bar{e}\}_n \qquad (13.78)$$

$$\text{(b)} \quad \{g\}_{n+1} = M\{e\}_{n+1} \qquad (13.79)$$

$$\text{(c)} \quad \rho_{n+1} = \frac{\{e\}_{n+1}^t \{e\}_n}{\{e\}_{n+1}^t \{g\}_{n+1}} \qquad (13.80)$$

$$\text{(d)} \quad \{\bar{e}\}_{n+1} = \frac{\{e\}_{n+1}}{(\{e\}_{n+1}^t \{g\}_{n+1})^{\frac{1}{2}}}. \qquad (13.81)$$

† If we normalize simply by dividing by the first element then obviously $k \neq 1$.

substituting for $\{e\}_n$ and $\{g\}_{n+1}$ from eqns (13.78) and (13.79), eqn (13.80) becomes

$$\rho_{n+1} = \frac{\{e\}_{n+1}^t K \{e\}_{n+1}}{\{e\}_{n+1}^t M \{e\}_{n+1}}. \tag{13.82}$$

Comparing this with eqn (13.23) we see that eqn (13.82) is the Rayleigh quotient and hence ρ converges on λ_1 as $n \to \infty$.

Equation (13.81) normalizes with respect to the mass and the Euclidean norm is given by

$$\{\bar{e}\}_{n+1} = \frac{\{e\}_{n+1}}{(\{e\}_{n+1}^t \{e\}_{n+1})^{\frac{1}{2}}}. \tag{13.83}$$

Step (a), of course, is actually carried out by direct solution techniques such as Gauss reduction and the amount of computation can be considerably reduced by storing the upper triangular matrix formed by the first Gauss reduction and using this for subsequent computations. This 're-solution' procedure is described in Section 15.5.

As an example consider the eigenproblem $(K - \lambda M) \{e_i\} = \{0\}$ with

$$K = \begin{bmatrix} 6 & 2 & 1 \\ 2 & 4 & -2 \\ 1 & -2 & 3 \end{bmatrix}, \quad M = \begin{bmatrix} 5 & 0 & 0 \\ 0 & 3 & 0 \\ 0 & 0 & 2 \end{bmatrix}. \tag{13.84}$$

The exact solutions (to five decimal places) for the eigenvalues and normalized eigenvectors are

$$\{\lambda\} = \begin{Bmatrix} 0.22883 \\ 1.55229 \\ 2.25222 \end{Bmatrix}, \quad E = \begin{bmatrix} -0.39850 & 0.71969 & -0.09154 \\ 0.63815 & 0.32784 & -0.62744 \\ 0.65876 & 0.61202 & 0.77327 \end{bmatrix} \tag{13.85}$$

with the eigenvectors stored by column in matrix E.

Using inverse iteration and commencing with the initial vector $\{e\}_0 = \{1, 1, 1\}$ we obtain

$$\{e\}_1 = \{-0.33333, 0.95833, 1.08333\}, \quad \rho_1 = 0.30193 \tag{13.86a}$$

$$\{e\}_2 = \{-0.94007, 1.62105, 1.69768\}, \quad \rho_2 = 0.23051 \tag{13.86b}$$

so that ρ is approaching the lowest eigenvalue, $\lambda_1 = 0.22883$. Seeking now to find the second eigenvalue we introduce a *shift* $\gamma = 1.525$ in the 'origin' for λ so that $\lambda^* = \lambda - \gamma$. Then the characteristic equation becomes $|K - \gamma M - \lambda^* M| = 0$, that is we subtract γM from the stiffness matrix in each iteration of eqn (13.78). Then inverse iteration will yield an eigenvalue solution λ^* related to the shifted origin and again, beginning with $\{\bar{e}\}_0 = \{1, 1, 1\}$, we obtain

$$\{e\}_1 = \{12.04616, 5.27232, 10.03037\}, \quad \rho_1 = 0.02707 \quad (13.87a)$$

$$\{e\}_2 = \{3.75944, 1.64181, 3.20447\}, \quad \rho_2 = 0.027279. \quad (13.87b)$$

The Euclidean norm of the eigenvector and the final eigenvalue are

$$\{\bar{e}\}_2 = \{0.72219, 0.31541, 0.61559\}, \quad \lambda_2 = \rho_2 + 1.525 = 1.55228. \quad (13.88)$$

The eigenvalue solution is already correct to four decimal places because the shift was close to the actual value.

An alternative technique for shifting is based on Gram–Schmidt orthonormalization and involves eliminating components of eigenvectors already determined from the initial iterating vector. If k eigenvectors have been determined then the modified initial vector is

$$\{\bar{e}'\}_0 = \{\bar{e}\}_0 - \sum_{i=1}^{k} \{e_i\}^{\mathrm{t}} M \{\bar{e}\}_0 \{e_i\}. \quad (13.89)$$

A disadvantage of this procedure, however, is that errors in the approximate eigenvectors accumulate in the summation of eqn (13.89). Hence it is common practice to use this technique in conjunction with the 'eigenvalue shift' procedure described above, for example to separate nearly equal eigenvalues after shifting by using only a single eigenvector $\{e_i\}$ on the right-hand side of eqn (13.89).

13.9.3   Polynomial iteration

In cases where the determinant $|K - \lambda M|$ is very small it can be expanded using the Laplace expansion to obtain the characteristic polynomial explicitly. Then the roots of this can be extracted by an iterative technique such as Newton's method. In most cases, however, explicit evaluation of the characteristic polynomial is impractical and *implicit polynomial iteration* must be used. In this method trial values of λ are used to compute the value of $|K - \lambda M|$ numerically, for example by Chio's method[21] where the value of the determinant is obtained as the chain product of the pivots encountered during Gauss reduction of the combined matrix $(K - \lambda M)$. Then *secant iteration*, for example, can be used to search for the nearest root.

Figure 13.4 illustrates secant iteration from two trial values γ_{n-1} and γ_n for λ, which have corresponding values of $|K - \gamma M| = p_{n-1}$ and p_n. Then an improved estimate of the eigenvalue is obtained as

$$\gamma_{n+1} = \gamma_n - \frac{p_n(\gamma_{n-1} - \gamma_n)}{p_{n-1} - p_n}. \quad (13.90)$$

Considering the example of eqns (13.84), if we choose $\gamma_{n-1} = 1.21$ (our first

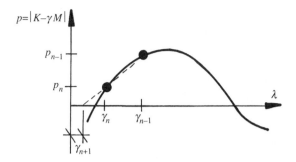

Fig. 13.4 Secant method of locating eigenvalues

choice was 1.2 but this resulted in a zero pivot, the simplest remedy being another choice) then we have

$$(K - 1.21M) = \begin{bmatrix} -0.05 & 2 & 1 \\ 2 & 0.37 & -2 \\ 1 & -2 & 0.58 \end{bmatrix}. \qquad (13.91)$$

Gauss reduction of eqn (13.91) encounters the pivots -0.05, 80.37, and 2.61 so that we have

$$p_{n-1} = |K - 1.21M| = -0.05 \times 80.37 \times 2.61 = -10.50. \qquad (13.92)$$

Similarly we obtain, for $\gamma_n = 1.3$, the value of the determinant as $p_n = -7.72$ so that eqn (13.90) yields

$$\gamma_{n+1} = 1.30 - \frac{(-7.72)(1.21 - 1.30)}{-10.50 + 7.72} = 1.55 \qquad (13.93)$$

which, as comparison with eqn (13.85) shows, is close to λ_2.

The process is then continued with the last two points (p, γ) used in eqn (13.90) and if the sign of the determinant changes (that is, a root has been passed) eqn (13.90) will linearly interpolate the last two points to approximate the root. When a change in sign in the determinant has been encountered, however, quadratic interpolation can be applied to the last three points to locate the root more closely and in general this will give better results than eqn (13.90), and eqns (13.96) and (13.97) can be used for this purpose.

13.9.4 Determinant search methods and the Sturm sequence

A Sturm sequence for the problem $(K - \lambda M)\{e_i\} = \{0\}$ is formed by taking a trial value γ of the eigenvalue and evaluating the cofactors of the diagonal entries of the resulting combined matrix. The resulting values $P_1, P_2, P_3, \ldots, P_m$, where m is the order of the matrices K and M, is called the Sturm sequence and the number of sign changes in his sequence is equal to the number of eigenvalues less than γ,

If the matrix $H = K - \gamma M$ is positive definite or positive semidefinite (so that $\{\Delta\}^t H \{\Delta\} \geq 0$ for any $\{\Delta\}$) then the number of eigenvalues less than γ also equals the number of negative pivots encountered during Gauss reduction of $K - \gamma M$. In the matrix of eqn (13.91), for example, Gauss reduction encountered one negative pivot, indicating that one eigenvalue was less than $\gamma = 1.21$.

The Sturm sequence property can be used in conjuction with *bisection* to locate a desired eigenvalue. To this end, regular steps in γ are taken until the number of sign changes corresponding to the required eigenvalue has occurred, that is, $k + 1$ if we require the kth eigenvalue. Then the last increment in γ is bisected by

$$\gamma_{n+1} = \tfrac{1}{2}(\gamma_n + \gamma_{n-1}) \tag{13.94}$$

where $p_n = |K - \gamma_n M|$ and $p_{n-1} = |k - \gamma_{n-1} M|$ are of opposite sign. Then if p_{n+1} is of opposite sign to p_n, for example, these two points are bisected and the process continued in this fashion until the root is located with sufficient accuracy.

Once regular stepping in γ has bracketed the required eigenvalue, however, it is generally more accurate to use linear interpolation with eqn (13.90) or the quadratic interpolation of eqn (13.95) to home in upon the solution.

13.9.5 Composite methods

As an example, one of the most widely used composite methods combines determinant search with inverse iteration when the required eigenvalue has been bracketed. The typical steps involved in such a 'search and iterate' scheme are as follows

1. Progressive stepping in the trial eigenvalue γ is used, together with the Sturm sequence property, to establish two lower bounds on the required eigenvalue γ_1 and γ_2 and one upper bound γ_3.

2. Quadratic extrapolation is then applied to these three values to obtain an improved estimate of the eigenvalue. If γ_1, γ_2 and γ_3 are equally spaced, for example, we write

$$p = \frac{p_1(s^2 - s)}{2} + p_2(1 - s^2) + \frac{p_3(s^2 + s)}{2} \qquad s = -1 \to 1 \tag{13.95}$$

where s interpolates dimensionlessly across the interval in γ. Then setting $p = 0$ in eqn (13.95) and solving for s we obtain

$$s = \frac{\sqrt{(b^2 - 4ac)} - b}{2a} \tag{13.96}$$

where

$$a = \frac{p_1 - 2p_2 + p_3}{2}, \qquad b = \frac{p_3 - p_1}{2}, \qquad c = p_2 \qquad (13.97)$$

and the improved eigenvalue estimate is given as

$$\gamma^* = \gamma_2 + s(\gamma_3 - \gamma_2). \qquad (13.98)$$

3. Commencing with the value γ^*, vector iteration (that is, eqns (13.78)–(13.81)) is used to find the eigenvector and associated eigenvalue. If reconvergence towards a previously determined eigenvalue is observed, as may occur when two or more eigenvalues are very close together, then eqn (13.89) is used to eliminate the associated eigenvector.

There are, of course, many other possible combinations of transformation, search, and/or iterative techniques to form composite methods. These include subspace iteration and Householder–QR–inverse iteration.[24] In the latter, for example, the Householder method is used to tridiagonalize the problem. It is then reduced to the standard form (eqn (13.24)) using factorization of M (see section 15.3) and vector iteration is then applied.

13.10 Extrapolation of eigenvalue solutions

The accuracy of eigenvalue solutions obtained by the finite element method depends in large measure upon the fineness of the finite element mesh used. To improve the accuracy of eigenvalue calculations based upon coarse meshes, Richardson extrapolation, that is, eqn (3.60) with $N = 2$, has been found to be effective.[25] Note however that in general the rate of convergence will not be $O(h^2)$ but $O(h^{2s-2m+2})$, *where s is the degree of displacement continuity.* When this is not known the rate of eigenvalue convergence with mesh refinement can be estimated empirically for given k and m matrices. When lumped masses are used, however, this corresponds to the use of constant strain elements to form K. As constant strain elements give $O(h^2)$ convergence the hypothesis implicit in Riehardson extrapolation that $N = 2$ is a reasonable first approximation when lumped masses are used.

An alternative to the use of eigenvalue extraction in incipient stability problems is to use a large displacement analysis and extrapolate the results using a Southwell plot.[26] Considering a single load q and a corresponding displacement x we describe the relationship between them as the rectangular hyperbola

$$q = \frac{q_{cr}x}{b + x} \qquad (13.99)$$

with asymptotes $x = -b$ and $q = q_{cr}$, the latter being the critical or buckling load. With a little rearrangement eqn (13.99) yields

$$\frac{x}{q} = \frac{b}{q_{cr}} + \frac{x}{q_{cr}} = b' + \frac{x}{q_{cr}}.$$ (13.100)

Thus, if we plot ordinates x/q and abscissae x we obtain a straight line (a Southwell plot), the inverse slope of which gives the critical load. Then if we have finite element solutions (q_i, x_i) and (q_j, x_j), $q_j > q_i$, for two loadings, substituting these results into eqn (13.100) and subtracting the two equations so obtained to eliminate b', we obtain

$$q_{cr} = \frac{x_j - x_i}{(x_i/q_i) - (x_j/q_j)}$$

$$= q_j + \frac{q_j - q_i}{(q_i x_j / q_j x_i) - 1}$$ (13.101)

and eqn (13.101) provides an estimate of the critical buckling load. Note also that if we replace q_i/x_i by h_i it reduces to eqn (3.60) (when $N = 1$) so that this is a h^1 extrapolation based on the line given by eqn (13.100).

13.11   BASIC program for eigenpairs

To demonstrate some of the techniques discussed in preceding sections it is useful to include a highly efficient interactive BASIC program (see Appendix 1 for details of the 'MEGABASIC' language used here) for eigenpair extraction.

The program uses the following strategy to extract eigenpairs.

1. Vector iteration of

$$\{e_i\}_{n+1} = M^{-1} K \{e_i\}_n = H \{e_i\}_n$$ (13.102)

is used to obtain the highest eigenvalue and associated eigenvector, the eigenvalue estimate being obtained as $e_{1,n+1}$ since $\{e_i\}_n$ is normalized by division by $e_{1,n}$. In the vector iteration the lumped mass matrix is used, so that formation of M^{-1} is trivial. Vector iteration is continued with $F = 1$ (line 165) until the first eigenpair has been obtained with sufficient accuracy (that is, when successive estimates closely agree).

2. Once the first eigenpair has been obtained with sufficient accuracy Gram–Schmidt orthogonalization is used to STEP (invoked by $F = 2$ in the program) to calculation of the next eigenvalue, that is, the H matrix is modified using

$$H^* = H - \frac{\lambda \{e_1\} \{e_1\}^t M}{\{e_1\}^t M \{e_1\}}$$ (13.103)

where, once again, computation is minimized by the use of the lumped mass matrix. By using eqn (13.103) after each vector iteration has produced an approximate result we can thus move down through the eigenvalues.

3. Inversion of a diagonal M, rather than of K, is, of course, computationally expedient but usually we require the lowest eigenvalue as a matter of priority. Thus when the lowest eigenvalue is that of most interest the program, after the highest eigenvalue has been obtained, SHIFTs to calculation of the lowest eigenvalue (invoked by $F = 3$ in the program) by modifying the stiffness matrix using[27]

$$K^* = K - \lambda_s M \qquad (13.104)$$

where λ_s is the last estimate of the highest eigenvalue (which need only be a fairly crude approximation).

```
10 ! "INPUT N";Input N     ;Rem SIZE OF EIGENVALUE PROBLEM
20 Dim SM(N,N),M(N),E(N),B(N),MS(N,N),H(N,N)
30 For I=1 to N
40 For J=1 to N
50 Read SM(I,J)            ;Rem READ STIFFNESS MATRIX -USE UNIT PIVOT AND ZERO
60 Next;Next               ;Rem ROW & COL FOR B.C.'s
70 For I=1 to N            ;Rem READ MASS MATRIX (LUMPED)
80 Read M(I);Next          ;Rem USE M(I)>100000 FOR B.C
82 For I=1 to N
83 Read E(I);Next          ;Rem READ INITIAL TRIAL EIGENVECTOR
84 If F=3 then SHIFT SM
85 For I=1 to N
86 For J=1 to N
87 H(I,J)=SM(I,J)/M(I)     ;Rem FORM ITERATION MATRIX H = INV(M)*K
88 Next;Next
110 For I=1 to N
115 EN(I)=0
120 For J=1 to N
130 EN(I)=EN(I)+H(I,J)*E(J)    ;Rem VECTOR ITERATION
140 Next J;Next I
145 X=EN(1)                ;Rem EIGENVALUE ESTIMATE
146 If X=0 then X=1        ;Rem CAN HAPPEN AT START IF SUM OF ROW OF K=0 AND
,147 ! "X = ",X            ;Rem INITIAL E( ) IS A UNIT VECTOR
150 For I=1 to N
155 EN(I)=EN(I)/X          ;Rem NORMALIZE NEW TRIAL EIGENVECTOR
160 Print EN(I)
164 Next
165 ! "TYPE 1 TO CONT, 2 TO STEP, 3 TO SHIFT";Input F
166 Rem 1 CONTINUES VECTOR ITERATION
167 Rem 2 STEPS TO NEXT EIGENPAIR USING SCHMIDT ORTHOGONALIZATION
168 Rem 3 SHIFTS ORIGIN BY AMOUNT OF CURRENT EIGENVALUE ESTIMATE
169 Rem NOTE THIS BEFORE CONTINUING - SEE ALSO NOTES AT LINE 400
170 If F=1 then Goto 175;If F=2 then Gosub 200
172 If F=3 then Goto 84
175 For I=1 to N
180 E(I)=EN(I);Next
190 Goto 110
199 End
200 For J=1 to N
215 B(J)=EN(J)*M(J)        ;Rem TEMPORARY MATRIX
220 Next
223 S=0
225 For I=1 to N
227 S=S+B(I)*EN(I)         ;Rem DENOMINATOR FOR ORTHOGONALIZATION
230 For J=1 to N
235 MS(I,J)=EN(I)*B(J)     ;Rem NUMERATOR MATRIX FOR ORTHOGONALIZATION
240 Next;Next
250 For I=1 to N
255 For J=1 to N
260 H(I,J)=H(I,J)-X*MS(I,J)/S   ;Rem ORTHOGONALIZATION CALCULATION
265 Next;Next
300 Return
```

```
310 Def proc SHIFT @SM        ;Rem SIMILAR TO A FORTRAN SUBROUTINE
320 For I=1 to N
330 SM(I,I)=SM(I,I)-X*M(I)     ;Rem ORIGIN SHIFT CALCULATION
335 If M(I)>100000 then SM(I,I)=1
337 Next
340 Return;Proc end
400 Rem Shift (F=3) can be used any time but suggested use is after a rough
410 Rem estimate of the first (highest) eigenvalue has been obtained - and
420 Rem NOTED! - this shifts calculation to near the origin for calculation
430 Rem of the fundamental (lowest) eigenvalue (& associated eigenvector)
440 Rem Then stepping (F=2) works upwards through the eigenvalues (rather
450 Rem than downwards as normally (i.e. if shift to origin not used).
460 Rem
500 Rem In conncection with the present technique note also the possibility
510 Rem of improved approximation with the lumped mass matrix by summing
520 Rem rows of the full M to form the lumped matrix as discussed in Sec. 13.5.
900 Data 1,-1,0              ;Rem STIFFNESS MATRIX FOR THE PROBLEM OF FIG. 13.2
910 Data -1,2,0              ;Rem NOTE ZERO ROW & COL. & UNIT PIVOT FOR B.C.
920 Data 0,0,1
930 Data 0.5,1,1000000       ;Rem DIAGONAL ENTRIES OF M - NOTE LARGE NO. FOR B.C.
940 Data 1,1,1               ;Rem INITIAL TRIAL EIGENVECTOR
950 Rem
1000 Data 6,2,1              ;Rem ANOTHER TEST PROBLEM FOR WHICH EIGENVALUES ARE
1010 Data 2,4,-2             ;Rem 2.25221,1.55229,0.22883 AND EIGENVECTORS ARE
1020 Data 1,-2,3             ;Rem 1,6.85418,-8.44728
1030 Data 5,3,2              ;Rem 1,0.45553,0.85039
1040 Data 1,1,1              ;Rem 1,-1.60138,-1.65310
```

When using the SHIFT subroutine the user should note the value of λ_s, which has been used as subsequent eigenvalues will be shifted in value by this amount, so that subsequent eigenvalues are given by

$$\lambda = \lambda^* + \lambda_s \qquad (13.105)$$

where λ^* is the value printed by the program.

When SHIFTing (to the fundamental) is used Gram–Schmidt STEPing allows us to move *up* through the eigenvalues.

The data for the problem of Fig. 13.2 is included in lines 900–950 and the program should yield the solutions given in Section 13.4. Note that in the K matrix (lines 900–920) the boundary condition $u_3 = 0$ has been set by zeroing row and column 3 and placing a 1 on he diagonal (this approach is more satisfactory than adding a penalty number to the diagonal when seeking eigenvalues).

Data for another hypothetical problem, along with the exact solutions, is given in lines 1000–1040.

Note that, in lines 940 and 1040, the initial trial eigenvector is a unit vector (not an uncommon practice) but that when the problem to be solved possesses rigid body modes (and associated zero eigenvalues) some other choice is needed, for example $\{1, 0.5, 0.75\}$ for the problem of Fig. 13.2.

13.12 Other eigenvalue problems

An eigenvalue problem of considerable interest is that given by the vibrations of frames affected by axial thrusts, when the characteristic equation becomes

$$|K_0 + \lambda K_G^* - \omega^2 M| = 0 \qquad \lambda < 0 \qquad (13.106)$$

The natural frequencies will decrease as λ approaches the (first) critical value, vanishing when $\lambda = \lambda_{cr}$. This phenomenon has useful application in the measurement of frequencies in structures under various loads to determine the buckling loads by extrapolation.[10]

It should also be noted that many eigenvalue problems are non-linear, with non-linear material or geometric behaviour, so that iterative analysis is needed to determine the correct *tangent stiffness matrix* and $\{\sigma*\}$ distribution at the prebuckling load level. In such cases a large displacement analysis with the inclusion of time stepping techniques when dynamical effects are present may prove more economical than an eigenvalue analysis.

References

1. Walters, R. A. and Carey, G. F. (1983). Analysis of spurious oscillation modes for the shallow mater and Navier-Stokes equations. *Comput. Fluids*, **11**, 51.
2. Leckie, F. A. and Linberg, G. M. (1969). The effect of lumped parameters on beam frequencies. *Aero. Quart.*, **14**, 234.
3. Archer, J. S. (1963). Consistent mass matrix for distributed systems. *Proc. Struct. Div. ASCE*, **89**, 161.
4. Dawe, D. J. (1965). A finite element approach to plate vibration problems. *J. Mech. Eng. Sci.* **7**, 28.
5. Timoshenko, S. P. (1921). On the correction for shear of the differential equation for transverse vibrations of prismatic bars. *Phil. Mag. Series 6*, **41**, 744.
6. Mindlin, R. D. (1951). Influence of rotatory inertia and shear on flexural motions of isotropic elastic plates. *J. Appl. Mech.*, **18**, 31.
7. Griemann, L. F. and Lynn, P. P. (1970). Finite element analysis of plate bending with transverse shear deformation. *Nucl. Engg Des.*, **14**, 233.
8. Hinton, E., Owen, D. R. J., and Shataram, D. (1976). Dynamic transient linear and nonlinear behaviour of thick and thin plates. *The Mathematics of Finite Elements and Applications, Vol. 2*, (ed. J. R. Whiteman). Academic Press, London.
9. Warburton, G. B. (1976). *The Dynamical Behaviour of Structures*, (*2nd edn*). Pergamon, Oxford.
10. Przemieniecki, J. S. (1968). *Theory of Matrix Structural Analysis*. McGraw-Hill New York.
11. Clough, R. W. (1971). Analysis of structure vibrations and response. In *Recent Advances in Matrix Methods of Structural Analysis and Design*, (R. H. Gallagher, Y. Yamada and J. T. Oden). Alabama Press.
12. Hinton, E., Rock, A., and Zienkiewicz O. C. (1976). A note on mass lumping in related processes in the finite element method. *Int. J. Earthquake Engg Struct. Dyn.*, **4**, 245.
13. Tong, P. and Rosettos, J. N. (1977). *Finite Element Method: Basic Technique and Implementation*. MIT Press, Mass.
14. Irons, B. M. (1965). Structural eigenvalue problems: elimination of unwanted variables. *J. Amer. Inst. Aeron. Astron.* **3**, 961.

15. Livesley, R. K. (1975). *Matrix Structural Analysis*, (*2nd edn*). Pergamon, Oxford.
16. Williams, F. W. and Howson, W. D. (1977). Compact calculation of natural frequencies and buckling loads for plane frames. *Int. J. Num. Meth. Engg*, **11**, 1067.
17. Bose, G. K., McNiece, G. M., and Sherbourne, A. N. (1972). Column webs in steel beam to column connections. *Comput. Struct.*, **2**, 254.
18. Connor, J. J. and Brebbia, C. A. (1977). *Finite Element Techniques for Fluid Flow.* Butterworths, London.
19. Cheung, Y. K. and King, I. P. (1971). Computer methods and computer programs. *The Finite Element Method in Engineering Science*, Chap. 20, (*2nd edn.*) (ed. O. C. Zienkiewicz). McGraw-Hill, London.
20. Wang, C. K. (1973). *Computer Methods in Advanced Structural Analysis*. Intext Press, New York.
21. Steinberg, D. I. (1974). *Computational Matrix Algebra*. McGraw-Hill, New York.
22. Taylor, R. L. (1977). Computer methods in finite element analysis. In *The Finite Element Method*, Chap. 24, (3rd edn) (ed. O. C. Zienkiewicz). McGraw-Hill, London.
23. Pipes, L. A. and Hovanessan, S. A. (1968) *Matrix Computer Methods in Engineering*. Wiley, New York.
24. Bathe, K. J. and Wilson, E. L. (1976). *Numerical Methods in Finite Element Analysis*. Prentice-Hall, Englewood Cliffs, NJ.
25. Müller, C. H. and Heise, U. (1980). Numerical calculation of eigenvalues of integral operators for plane elastostatic boundary value problems. *Comp. Meth. Appl. Mech. Eng.* **22**, 241.
26. Harvey, J. W. and Kelsey, S. (1971). Triangular plate bending elements with enforced compatibility. *J. Amer. Inst. Aeron. Astron.*, **9**, 1023.
27. Hughes, T. J. R. (1987). *The Finite Element Method: Linear Static and Dynamic Finite Element Analysis*. Prentice-Hall, Englewood Cliffs, NJ.

14
Time stepping analysis
of vibration and creep

Time-dependent *propagation problems* for which finite element methods are much used include those of structural vibrations, creep problems, and transient field problems such as those of seepage through porous media and heat conduction. The linear eigenvalue problem of dynamics has already been discussed in Chapter 13 and any dynamical analysis can be made by mode superposition once the eigenvectors have been determined. In this chapter we shall restrict attention to the application of time stepping techniques to analyse the creep and vibration of solids. Heat and fluid flow problems are deferred until Chapters 18, 19, and 20 following the introduction of the governing equations and weighted residual methods needed to deal with these.

14.1 Introduction

The most fundamental starting point for the study of the dynamics of a system whose configuration can be uniquely specified by a number of time-dependent coordinates x_i is *Hamilton's principle*. This states that if a system moves from one configuration at time t_0 to another at time t_1 then the path of the motion is such as to minimize the integral

$$F(x_i, t) = \int_{t_0}^{t_1} L \, dt = \int_{t_0}^{t_1} (\Pi - \Lambda) \, dt$$

where the *Lagrangian function L* is defined in terms of the potential energy (Π) and the kinetic energy (Λ).

Taking transverse vibrations of a string with constant mass per unit length ρ stretched by a tension T as an example we have

$$\Pi = \tfrac{1}{2} \int_0^L P \left(\frac{\partial y}{\partial x} \right)^2 dx, \qquad \Lambda = \tfrac{1}{2} \int_0^L \rho \left(\frac{\partial y}{\partial t} \right)^2 dx.$$

These results follow from eqns (12.73) and (13.34). Applying the first-order

Euler–Lagrange equation[1] to L gives

$$\frac{\mathrm{d}}{\mathrm{d}t}\left(\frac{\partial F}{\partial \dot{y}}\right) = \frac{\partial F}{\partial y},$$

that is, using the first two terms of eqn (17.13). Noting that as y does not appear explicitly, $\partial F/\partial y$ is replaced by $\partial(\partial F/\partial y_x)/\partial x$ where $y_x = \partial y/\partial x$, one obtains the wave equation

$$\frac{\partial^2 y}{\partial x^2} = c^2 \left(\frac{\partial^2 y}{\partial t^2}\right) \quad \text{with } c^2 = \rho/P$$

and this is the basis for many analytical solutions in dynamics.

As a governing energy principle is available, however, a more direct approach via the finite element method is to use the stiffness and mass matrices defined in previous chapters to evaluate Π and Λ. Then the time integration required by Hamilton's principle is replaced by the use of *time stepping techniques* and these are discussed in the following section.

Although we use finite difference approximations to form the recurrence relations for time-dependent problems, it is also shown that these approximations can be deduced from standard interpolation functions. Thus it is also possible to include a time coordinate at finite element nodes, and interpolation is also applied to these. The velocities and accelerations will then enter the formulation in the same way as do strains in equilibrium problems. This approach has many advantages but we restrict attention to the more conventional time stepping approach in the present text.

14.2 Time stepping techniques

In time stepping techniques we allow small increments in time and calculate the displacements in the system using a *recurrence relation*. Two types of problem which may be investigated in this way are first-order and second-order problems.

14.2.1 First-order problems

In these the governing differential equation involves first derivatives with respect to time (that is, velocities). These include creep, pore pressure dissipation, and transient heat conduction, for example. First-order problems are governed by matrix equations of the form

$$K_{\mathrm{D}}\{\dot{D}\} + K\{D\} = \{Q\} \tag{14.1}$$

where $\{\dot{D}\} = \partial\{D\}/\partial t$ and these derivatives can be replaced by finite difference approximations. In the *Crank–Nicolson scheme* one writes

$$\{D\} = \frac{\{D\}_{n+1} + \{D\}_n}{2} \tag{14.2a}$$

$$\{\dot{D}\} = \frac{\{D\}_{n+1} - \{D\}_n}{\delta} \tag{14.2b}$$

where δ is the time step length. Equation (14.2a) establishes the centre of this step as the point of reference and thus eqn (14.2b) is a *central difference*. The recurrence relation is obtained by substituting eqns (14.2a) and (14.2b) into eqn (14.1), giving

$$\left[K + \frac{2}{\delta}K_D\right]\{D\}_{n+1} = \{Q\}_{n+\frac{1}{2}} + \left[\frac{2}{\delta}K_D - K\right]\{D\}_n \tag{14.3}$$

and it is recommended that, for consistency, eqn (14.2a) should also be used for $\{Q\}$. If instead, we use the end of the interval as the point of reference, so that, $\{D\} = \{D\}_{n+1}$, then eqn (14.2b) is the *backward difference*, giving

$$\left[K + \frac{1}{\delta}K_D\right]\{D\}_{n+1} = \{Q\}_{n+1} + \frac{1}{\delta}K_D\{D\}_n. \tag{14.4}$$

This formula has also been frequently used (see references 1, 3, and 10 of Chapter 18) but seems to have little advantage over the unbiased Crank–Nicholson scheme. Another alternative is to use the beginning of the interval as the point of reference, that is, $\{D\} = \{D\}_n$. Then eqn (14.2) is the *forward difference*, giving

$$\frac{1}{\delta}K_D\{D\}_{n+1} = \{Q\}_n + \left[\frac{1}{\delta}K_D - K\right]\{D\}_n. \tag{14.5}$$

Now if the damping matrix K_D is diagonalized by lumping, which is a comparable approximation to using lumped masses in a vibration problem (for example eqn (13.39)), the recurrence relation is explicit, so that matrix inversion is not required to solve it. This is the most economical approach (but not the most accurate!) and examples of its use are given in Sections 18.3 and 20.4.

The three basic finite difference methods, in fact, represent special cases of a general weighted residual approach with weighting factors of (1,0,0), (0,1,0), and (0,0,1) at three consecutive points in time. Hence it is possible to construct, using other weightings, three-point schemes the performance of which is governable by variation of the weighting factors. Of the three basic methods, however, the Crank–Nicolson scheme has the smallest truncation error. This can be demonstrated by the use of Taylor's theorem or by considering the actual value of a nodal displacement u to be given by the infinite series

$$u = f(t) = c_1 + c_2 t + c_3 t^2 + c_4 t^3 + \cdots + c_n t^{n-1} \qquad n \to \infty. \quad (14.6)$$

Then the finite difference approximation for the velocity in a subsequent time interval h is

$$\frac{du}{dt} \simeq \frac{f(t+h) - f(t)}{h}$$

$$\simeq c_2 + c_3 h + 2c_3 t + c_4 h^2 + 3c_4 ht + 3c_4 t^2 + \cdots, \quad (14.7)$$

whereas at times t, $t + h/2$, and $t + h$ (that is, the backward, central, and forward points, respectively corresponding to the forward, central, and backward difference methods) the true velocities, obtained by substituting t, $t + h/2$, and $t + h$ for t in the derivative of eqn (14.6) with respect to t, are respectively obtained as

$$\frac{du}{dt} = c_2 + 2c_3 t + 3c_4 t + \cdots \qquad (14.8)$$

$$\frac{du}{dt} = c_2 + 2c_3(t + h/2) + 3c_4(t + h/2)^2 + \cdots \quad (14.9)$$

$$\frac{du}{dt} = c_2 + 2c_3(t + h) + 3c_4(t + h)^2 + \cdots \qquad (14.10)$$

For small h, the terms omitted by the finite difference solution (eqn (14.7)) when compared to these three results show that the errors for forward, central, and backward differences are respectively $c_3 h$, $c_4 h^2/4$, and $-c_3 h$ so that the central difference method has a truncation error of only $O(h^2)$ compared with $O(h)$ for the forward and backward difference methods. The reader can verify that the same conclusion holds for all higher-order differences and indeed this $O(h^2)$ result is the basis for Richardson extrapolation (eqn (3.60) with $N = 2$) for central finite differences. As is stressed in the present text such processes are also widely applicable in the finite element method.

14.2.2 Second-order problems

In these the governing differential equation involves second derivatives with respect to time (that is, accelerations) and vibration problems are predominant among this type. Second-order problems are governed by a matrix equation of the form

$$M\{\ddot{D}\} + K_{\mathrm{D}}\{\dot{D}\} + K\{D\} = \{Q\} \qquad (14.11)$$

and displacement vectors at three different times are required to approximate the accelerations $\{\ddot{D}\}$. Forward, backward, or central differences can again be used but, save in exceptional circumstances, central differences are the natural choice.

A variety of other methods exist, including the Fox–Goodwin,[2] Newmark β,[3] and Wilson θ methods.[4] Application of these other methods follows similar lines to the central difference method example in Section 14.3.

It is of more importance, perhaps, to demonstrate that time stepping formulae can be derived by rearranging interpolation formulae. This process can be interpreted as one of *extrapolation* and has very wide application. In the case of the central difference method, for example, the finite difference approximations for the velocity and acceleration are

$$\dot{u}_n = \left(\frac{\partial u}{\partial t}\right)_n = \frac{u_{n+1} - u_{n-1}}{2\delta} \qquad (14.12)$$

$$\ddot{u}_n = \left(\frac{\partial^2 u}{\partial t^2}\right)_n = \frac{u_{n+1} - 2u_n + u_{n-1}}{\delta^2} \qquad (14.13)$$

where δ is the time step length. These results can also be obtained by differentiating a quadratic interpolation in time. To this end differentiating eqns (5.10) once one obtains for the velocity

$$\dot{u}_n = \frac{(2s - 1)u_{n-1} - 4su_n + (2s + 1)u_{n+1}}{2\delta} \qquad (14.14)$$

where s is the time measured from the centre of the interval 2δ. Differentiating again yields eqn (14.13). Using $s = 0$, that is, the centre of the 'double interval' 2δ, in eqn (14.14) gives eqn (14.12) but one can also choose other values of s and hence obtain a range of alternative velocity approximations.

Using $s = 0.5$, for example, yields $\dot{u}_n = (u_{n+1} - u_n)/\delta = \dot{u}_{n+\frac{1}{2}}$ which sometimes leads to better results.

With the use of other interpolations a wide range of schemes for time stepping can be constructed. Using cubic Lagrangian interpolation, for example, a four-point formula is obtained and it is also possible to use cubic Hermitian interpolation, with velocities and accelerations at each point, to obtain a two-point formula.[5,6] Although these require more storage, fewer steps and thus less computation will be needed.

Returning to the quadratic (central difference) case, to obtain a recurrence relation one substitutes eqns (14.12) and (14.13) into eqn (14.11) and obtains[7]

$$(M + \tfrac{1}{2}\delta K_D)\{D\}_{n+1} = \delta^2\{Q\}_n + (2M - \delta^2 K)\{D\}_n + (\tfrac{1}{2}\delta K_D - M)\{D\}_{n-1} \qquad (14.15)$$

for the complete structure. The process, though often referred to as 'time integration' is clearly one of extrapolation (on a quadratic basis in this case) and thus corresponds to the second-order Runge–Kutta technique (see Section 12.2). Fourth-order Runge–Kutta techniques have also been much used for time stepping analysis.[8]

In the following section vibration problems, to which time stepping techniques have been frequently applied, are discussed, and the final section discusses a type of creep problem which has yet been given little attention in the finite element method.

14.3 Time stepping analysis of structural vibrations

The problem of Fig. 14.1 is one of transient boundary excitation, that is, the ground displacement u_3 is a transient function of time. To study the sway mode of vibration we consider only the lateral displacements at each floor level (whereas study of the 'rocking mode' also requires inclusion of vertical displacement freedoms). Neglecting compression in the floor and roof, only a single displacement freedom is needed at each level. The stiffness matrix for each column is thus a contraction of eqn (1.12) in which only the shearing entries k_4 are required and we sum the pair of column stiffness matrices at each level, giving

$$k_{ij} = 2\left(\frac{12EI}{h^3}\right)\begin{bmatrix} 1 & -1 \\ -1 & 1 \end{bmatrix}.$$

Assembling the 'element' matrices for each floor level, neglecting damping, and partitioning eqn (14.15) to distinguish the specified boundary displacements (subscript b) and the free displacements (subscript f) one obtains the recurrence relation

$$\begin{bmatrix} M_{ff} & M_{fb} \\ M_{fb}^t & M_{bb} \end{bmatrix}\begin{Bmatrix} D_f \\ D_b \end{Bmatrix}_{n+1} = \begin{bmatrix} 2M_{ff} - \delta^2 K_{ff} & 2M_{fb} - \delta^2 K_{fb} \\ 2M_{fb}^t - \delta^2 K_{fb}^t & 2M_{bb} - \delta^2 K_{bb} \end{bmatrix}\begin{Bmatrix} D_f \\ D_b \end{Bmatrix}_n$$
$$- \begin{bmatrix} M_{ff} & M_{fb} \\ M_{fb}^t & M_{bb} \end{bmatrix}\begin{Bmatrix} D_f \\ D_b \end{Bmatrix}_{n-1}. \tag{14.16}$$

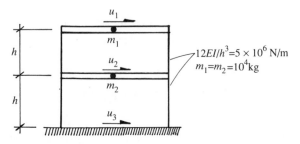

Fig. 14.1 Lumped mass analysis of a multistorey frame

As $\{D_b\}$ (in our example simply u_3) is specified we require only the first row of eqn (14.16) to determine $\{D_f\}$ at each time step, so that, we do not need to specify M_{bb}; even if we did it would be a very large number and thus act as a penalty factor to uncouple $\{D_b\}$ from the solution.

A further simplification arises if we neglect the masses of the columns or alternatively, lump them together with the 'naturally' lumped floor masses. In this case the system mass matrix is diagonalized so that M_{fb} is null. Then we have

$$M_{ff}\{D_f\}_{n+1} = -\delta^2 K_{fb}\{D_b\}_n + (2M_{ff} - \delta^2 K_{ff})\{D_f\}_n - M_{ff}\{D_f\}_{n-1} \quad (14.17)$$

where the first term on the right-hand side gives the loading effect of the specified boundary displacements. The matrices in eqn (14.17) are given by

$$K_{ff} = 10^7 \begin{bmatrix} 1 & -1 \\ -1 & 2 \end{bmatrix} \quad K_{fb} = 10^7 \begin{bmatrix} 0 \\ 1 \end{bmatrix} \quad M_{ff} = 10^4 \begin{bmatrix} 1 & 0 \\ 0 & 1 \end{bmatrix}. \quad (14.18)$$

Substituting these into eqn (14.17) and dividing through by a common factor of 10^4 one obtains

$$\begin{Bmatrix} u_1 \\ u_2 \end{Bmatrix}_{n+1} = \beta \begin{Bmatrix} 0 \\ 1 \end{Bmatrix} (u_3)_n + \begin{bmatrix} 2 - \beta & \beta \\ \beta & 2 - 2\beta \end{bmatrix} \begin{Bmatrix} u_1 \\ u_2 \end{Bmatrix}_n - \begin{Bmatrix} u_1 \\ u_2 \end{Bmatrix}_{n-1} \quad (14.19)$$

where $\beta = \delta^2 \times 10^3$.

If the transient boundary excitation at ground level is given by

$$u_3 = \begin{cases} 0.01 \sin\dfrac{\pi t}{t_0} & t \le t_0 \\ 0 & t > t_0 \end{cases}$$

with $t_0 = 0.2$ secs (that is, a single sinusoidal half wave 'shake' of peak amplitude 10 mm) and we use a time step length of $\delta = 0.02$ seconds (as this is one tenth of t_0 we should expect this to give reasonable results unless the mass/stiffness ratio were much lower) then we have

$$\beta = \delta^2 \times 10^3 = (0.02)^2 \times 10^3 = 0.4.$$

Substituting this value of β and the specified value of u_3 for $t \le t_0$ into eqn (14.19) (on the understanding that the term involving the latter drops out when $t > t_0$) we obtain the two recurrence relations

$$(u_1)_{n+1} = (1.6u_1 + 0.4u_2)_n - (u_1)_{n-1}$$

$$(u_2)_{n+1} = 0.004 \sin(5\pi t_n) + (0.4u_1 + 1.2u_2)_n - (u_2)_{n-1}. \quad (14.20)$$

Using these relations one obtains results close to the exact solution, which can be seen in Table 14.1 where the exact and central difference predictions of the peak values for u_1 and u_2 are compared.

Table 14.1 Comparison of finite element solution using the central difference method with exact solution of the problem in Fig. 14.1

Peak number	max $(u_1)_{abs}$		max $(u_2)_{abs}$	
	Exact[7]	Central difference	Exact[7]	Central difference
1	0.0196	0.0195	0.0131	0.0137
2	0.0200	0.0197	0.0123	0.0126
3	0.0206	0.0202	0.0115	0.0119
4	0.0202	0.0204	0.0120	0.0113

The results are of acceptable accuracy, though the time step used here is as large as advisable. If too large a time step is used the calculations become unstable and the predicted displacements oscillate with increasing amplitudes. For the central difference method this instability is avoided if[7]

$$\frac{\delta}{T_{min}} < \frac{1}{\pi} \qquad (14.21)$$

where T_{min} is the smallest natural period of vibration. But to obtain acceptable accuracy (that is, errors less than 10 per cent) Bathe and Wilson[4] show that one requires

$$\frac{\delta}{T_{eff}} \leq 0.18 \qquad (14.22)$$

where T_{eff} is the smallest period of the natural modes that play a significant part in the particular vibration situation considered. In the example considered here the periods of natural vibration are 0.3125 and 0.1229 seconds, so that if the second mode is significant we require $s \leq 0.022$. The value chosen here satisfies this limit so that the errors in Table 14.1 are of the expected proportions. If too small a time step is used, on the other hand, numerical difficulties may also arise. Hence optimum step sizes have been deduced for some problems.[9]

Warburton[7] compares the central difference and Newmark β methods for the problem of Fig. 14.1 and finds the two to be of comparable accuracy, although the Newmark method has the advantage that it is unconditionally stable for $\beta \geq 0.25$ whilst for $0 < \beta < 0.25$ the stability conditions are less stringent, for example $\delta/T_{min} \leq 0.551$ for $\beta = 1/6$.[7]

Another point of importance is the question of damping, the inclusion of which may considerably affect time stepping analyses [10,11] and reduce the risk of unstable solutions. The use of variable step lengths may also considerably reduce the incidence of instability.[12]

Time stepping analyses closely parallel those of load stepping because in both cases the stiffness matrices can be recalculated at each step to deal with non-linear problems and equilibrium iteration can be used at each load or time step.[12]

14.4 Time stepping analysis of creep

There are two types of creep problem worth remarking upon at the outset:

(1) those special cases of structural excitation in which inertia effects are negligible compared to damping effects, an example being supercritical damping (that is, a high M/K ratio) in a structure;[7]

(2) materials in which a time-dependent 'flow' occurs, examples being concrete and timber (in which long-term strains under dead loads are about three times the short-term values) and metals at high temperatures.

The first is relatively rare (an example being earthquake vibration of earth dams) and it is the second which we shall consider here.

As an example of elastic creep consider the three-bar truss shown in Fig. 14.2 for which the static solution for the stresses in the members is, if the members have equal modulus E and unit cross-sectional area,

$$X = \tfrac{4}{5}Q, \qquad Y = \tfrac{1}{5}Q \tag{14.23}$$

It is assumed that the creep behaviour is governed by a power law

$$\dot{\varepsilon}_0 = C\sigma^n \tag{14.24}$$

where the constant C has the dimensions of strain/(time × stress^n) and is here taken as unity. The exponent will be assumed to be 2 (in fact practical values in metals vary from 2 for rolled chromium steel at 600°C to 25 for forged chromium steel at 400°C).[13] Creep will cause a stress redistribution in the

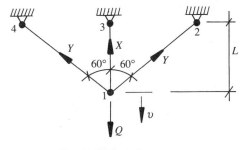

Fig. 14.2 Three-bar truss

truss until a steady state is reached (that is, no further redistribution of stress occurs but creep continues), when the stress in the vertical member is[14]

$$X = \frac{Q}{1 + \rho} = \frac{2Q}{3} \tag{14.25}$$

where $\rho = 2(\cos \alpha)^{1 + 2/n}$ and $\alpha = 60$ degrees. Owing to symmetry only the vertical displacement of the load need be considered and the steady state rate of increase in this is

$$\dot{v} = L \left(\frac{Q}{1 + \rho} \right)^n. \tag{14.26}$$

The stresses in the members are calculated by subtracting the cumulative creep strain ε_0 given by eqn (14.24) from the total strain ε_t, giving, after n equal time steps δt,

$$\sigma = E\varepsilon_t - E\varepsilon_0 = EBT\{d\} - E \sum_{i=1}^{n} (C\sigma_i^2 \delta t) \tag{14.27}$$

where

$$BT = \frac{1}{L_e} [-cu_i, \ -sv_i, \ cu_j, \ sv_j] \tag{14.28}$$

and c and s are the direction cosines of the member, and L_e is its length.

Considering a-single time increment δt, eqn (14.27) gives the member stresses as

$$X = E \left\{ \frac{v}{L} - \left(\frac{4Q}{5} \right)^2 \delta t \right\} \quad Y = E \left\{ \frac{v}{4L} - \left(\frac{Q}{5} \right)^2 \delta t \right\}. \tag{14.29}$$

Thus the vertical member sheds load to the inclined members, as expected from eqn (14.25). The displacement increments are calculated using the equilibrium iteration

$$\{\delta D\} = K^{-1}(\{Q\} - \{R\}) \tag{14.30}$$

where $\{Q\}$ are the applied loads and the summed element reactions are

$$\{R\} = \sum T^t B^t \sigma \tag{14.31}$$

with the stresses calculated using eqn (14.27). As eqns (14.27) and (14.29) show, these stresses have decreased from the initial values so that eqn (14.30) calculates the displacement increments resulting from creep. As the summation calculation for ε_0 already involves an approximation, only one or two equilibrium iterations are customary.[15]

With the use of such equilibrium iteration at a sufficient number of time steps the steady state solution, eqn (14.25) is reached so that in each time increment the change in stress in the vertical member is given by

$$\delta X = E \frac{\delta v}{L} - E X^n \delta t. \tag{14.32}$$

Using eqns (14.26) and (14.25) to evaluate δv and X respectively, one obtains

$$\frac{\delta X}{E} = \left[L \left(\frac{Q}{1 + \rho} \right)^n \middle/ L - \left(\frac{Q}{1 + \rho} \right)^n \right] \delta t = 0 \tag{14.33}$$

so that no further redistribution of stress occurs. However, the displacements steadily increase in eqn (14.26) as the creep components of the stresses from eqns (14.29) (which in turn are used to calculate $\{R\}$) continue to increase.

As creep problems can be seen to involve an initial strain basis (see eqn (14.27)) creep behaviour may also be included in the elasto-plastic procedure described in Section 12.4 by including the yield function in eqn (14.24) and the resulting elasto-visco-plastic procedure was discussed briefly at the close of Section 12.4.

14.5 Concluding remarks

Time stepping techniques, used in conjunction with the finite element method, provide a powerful tool with which vibrations arising from transient excitation, for example, present no real difficulties, whereas analytical solution of these can only be accomplished for certain very specialized problems.[7] Applications of time stepping techniques to creep problems are less widespread. This is straightforward but other alternative models are worthy of note, particularly formulation of creep as a diffusion problem,[16] which can be attacked in the manner described in Chapter 18, and the analysis of plastic flow in metal forming processes as a Stokes' flow problem, using the formulations described in Section 19.5.

With a combination of the techniques of Chapters 12 and 14, completely general analyses taking into account initial stresses and strains incurred during fabrication, thermal effects, non-linear elasticity, plasticity, creep, large displacements, and cracking are possible.

References

1. Riley, K. F. (1974). *Mathematical Methods for the Physical Sciences*. Cambridge University Press.
2. Fox, L. and Goodwin, E. T. (1949). Some new methods for the numerical integration of ordinary differential equations. *Proc. Camb. Phil. Soc.*, **49**, 373.

3. Newmark, N. M. (1959). A method for computation of structural dynamics. *Proc. ASCE*, **85** (EM5), 67.
4. Bathe, K. J. and Wilson, E. L. (1976). *Numerical Methods in Finite Element Analysis*. Prentice-Hall, Englewood Cliffs, NJ.
5. Argyris, J. H., Dunne, P. C. and Agepoulos, T. (1973). Nonlinear oscillations using the finite element method. *Comp. Meth. Appl. Mech. Engng*, **2**, 203.
6. Argyris, J. H., Dunne, P. C., and Agepoulos, T. (1973). Dynamic response by large step integration. *Earthquake Engng Struct. Dyn.*, **2**, 185.
7. Warburton, G. B. (1976). *The Dynamical Behaviour of Structures*, (2nd edn). Pergamon, Oxford.
8. Connor, J. J. and Brebbia, C. A. (1976). *Finite Element Techniques for Fluid Flow*. Butterworths, London.
9. Pugh, E. D. L., Hinton, E., and Zienkiewicz, O. C. (1978). A study of quadrilateral plate bending elements with reduced integration. *Int. J. Num. Meth. Engng*, **12**, 1059.
10. Smith, I. M. (1977). Some time dependent soil-structure interaction problems. In *Finite Elements in Geomechanics*, (ed. G. Güdehus). Wiley, London.
11. Park, K. C. and Underwood, P. G. (1980). A variable step central difference method for structural dynamics analysis – part I: theoretical aspects. *Comp. Meth. Appl. Mech. Engng*, **22**, 241.
12. Clough, R. W. and Petersson, H. (1976). Application of finite element method. In *The Mathematics of Finite Elements and Applications*, Vol. 2, (ed. J. R. Whiteman). Academic Press, London.
13. Odqvist, F. K. G. (1966). *Mathematical Theory of Creep and Creep Rupture*, Clarendon Press, Oxford.
14. Rabatnov, Y. N. (1969). *Creep problems in structural members*. (trans. Transcripta Service Ltd.) (ed. F. A. Leckie). North Holland, Amsterdam.
15. Zienkiewicz, O. C. (1977). *The Finite Element Method.*, (3rd edn). McGraw-Hill, London.
16. Kachanov, L. M. (1967). *The Theory of Creep.*, (trans. E. Bishop) (ed. A. J. Kennedy). Nat. Lending Lib. Sci. Tech., Boston Spa, Yorkshire.

15
Finite element programming

The present chapter introduces some practical aspects of modelling and programming with finite elements. Only a brief discussion is given to such aspects as equation solution techniques, of which there are many, and the reader is referred to specialist texts in these areas. The chapter includes programs for plane stress/strain, thin plate bending, and thick/thin doubly curved shells.

15.1 Mesh design

There are many factors worth considering in designing a finite element mesh, some of which are listed below.

1. Programmed mesh generation. Generation of regular meshes is a straightforward exercise and the program given in Section 15.7 gives an example of this. Generation of complex three-dimensional meshes, however, can be a difficult task requiring parametric mapping and similar techniques.

2. Mesh refinement. In regions of high stress gradient, for example, over-zealous refinement may result in numerical difficulties and in this context special crack tip elements (Section 7.7) or substructuring techniques are of interest.[1] To analyse a domain in which there is a hole, for example, detailed stress information in the region of the hole can be obtained by re-analysis of a substructure including the hole. This can be solved as an imposed displacement problem (see Section 15.4) using boundary displacements determined from a full analysis.

3. Bandwidth minimization. In rectangular domains, for example, numbering the nodes consecutively in the shorter direction gives the smallest bandwidth. Many programs renumber nodes to minimize the bandwidth and others use 'frontal' routines not dependent upon the bandwidth.

4. Diagonalization in triangular meshes. A common practice is to place all diagonals in the same direction (e.g. Fig. 6.3) rather than in a 'criss-cross' pattern, as this tends to distribute any element errors more evenly. (For example, in right-angled triangles assume stress errors δ at the right angle and 2δ at the other vertices and compare the pattern of summed errors in one-way diagonal and criss-cross meshes.) In some special cases optimal meshes can be determined[2] but generally experiment with trial meshes is required.

5. Mixtures of different element types. Different element types can be connected by simple penalty constraints. If, for example, we wish to connect a layer of plate elements with nodal freedoms w, ϕ_x, and ϕ_y to an underlying layer of brick elements with nodal freedoms u, v, and w this is accomplished by incorporating stiffness matrices

$$\begin{bmatrix} \beta & -\beta \\ -\beta & \beta \end{bmatrix} \quad \beta \gg |k| \tag{15.1}$$

to connect the common w freedoms. Where some of these are not in the same plan location the offending freedoms can be suppressed using constraints similar to those of eqn (4.29).[3] Note also that dummy elements with near zero stiffness can be used to 'place a hole' in an existing mesh.[4]

6. Moving meshes. Examples of moving meshes include mesh adjustment to accomodate a propagating crack,[5] to improve an initial guess of the location of the phreatic surface in unconfined seepage through porous media,[6] and to predict improved structural shapes.[7] In the case of spontaneous crack propagation (that is, without increase in loading) one writes the change in potential energy resulting from a mesh adjustment which propagates the crack as

$$\Pi_2 - \Pi_1 = \tfrac{1}{2}(\{D + \delta D\}^t(K + \delta K)\{D + \delta D\} + \{D + \delta D\}^t\{Q\}$$
$$- \tfrac{1}{2}\{D\}^t K\{D\} - \{D\}^t\{Q\}. \tag{15.2}$$

Neglecting second-order terms one obtains

$$\delta\Pi = \Pi_2 - \Pi_1 = \tfrac{1}{2}\{D\}^t(\delta K)\{D\} \tag{15.3}$$

so that, rather than needing a full analysis to find Π_2 and Π_1 (and hence the energy release rate $(\Pi_2 - \Pi_1)/\delta r$ where δr is the propagation distance of the crack), one requires only a single full analysis to find $\{D\}$ followed by calculation of δK for the mesh changes corresponding to propagation of the crack a distance δr. Then, when $\delta\Pi/\delta r$ exceeds a critical value, spontaneous crack propagation is predicted.[8]

15.2 Assembly and solution of the element equations

Figure 15.1 illustrates the assembly process for spar elements with four freedoms. Here the structure matrix is assembled by placing 2×2 blocks from the element matrices into it. For these blocks numerical subscripts refer to local element node numbers and alphabetical subscripts to structure or global node numbers. Because of its symmetry only that part of the structure matrix on and above the diagonal is required and this is stored as a rectangular array.

The program of Section 15.7 follows this approach, and further savings in storage are made by reducing the equations for each node in turn as soon as they have been assembled. With this approach the number of equations that can be solved is unlimited and the bandwidth determines storage requirements.

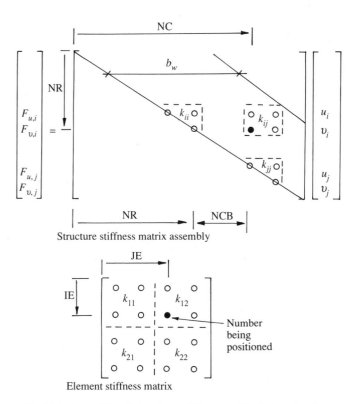

Structure stiffness matrix assembly

Element stiffness matrix

Fig. 15.1 Assembly of structure stiffness matrix in banded form

15.3 Solution of linear equations

There are many techniques for the solution of linear matrix equations including the following.

15.3.1 Gauss–Jordan reduction

The Gauss–Jordan method[9] applies successive column transformations T_n, $T_{n-1}, \ldots, T_2, T_1$ to reduce the *augmented matrix* matrix (K, I) to (I, K^{-1}). For the direct solution of the equations $\{Q\} = K\{D\}$ one writes $T_n T_{n-1} \ldots T_2 T_1(K, \{Q\}) = (I, \{D\})$. In practice Gauss–Jordan reduction is usually programmed using algorithms similar to those of eqns (15.6) and (15.7), rather than using transformation matrices T. As reduction is both above and below the diagonal approximately twice as much computation as in Gauss reduction is required but computation can be halved if the inverse matrix is not saved (as is done in the program of Section 1.11).

As Gauss–Jordan reduction destroys banding, however, it cannot be used for large finite element problems. It is useful for forming interpolation matrices, however, although pivoting is often needed to obtain these. Gauss–Jordan reduction is also much used in the *Simplex Method*[10] (which in the finite element method may be applied to plastic collapse problems)[11] and in *constrained linear estimation*[10] where in $\{Q\} = K\{D\}$ a range is specified for each entry in $\{Q\}$ and this can be cast as a linear programming problem. Indeed both linear and non-linear programming techniques play an important part in the finite element method and these are discussed in Chapter 22.

15.3.2 Gauss–Seidel iteration

This is a *relaxation* method where, beginning with an initial trial solution one rearranges each equation to obtain the recurrence relation[12]

$$D_i^m = \frac{\beta\left(Q_i - \sum_{j=1}^{n} (1 - \delta_{ij})K_{ij}D_j^{m-1}\right)}{K_{ii}} + (1 - \beta)D_i^{m-1} \qquad (15.4)$$

where the superscripts are iteration numbers, δ_{ij} is the Kronecker delta, and β is a relaxation factor, optimum values of which may decrease the number of iterations by an order of magnitude[13] (for example, $1.85 < \beta < 1.92$ has been suggested for plane stress[14] but in some non-linear problems $\beta_{\text{opt}} \to 1$).

For large numbers of equations (more than 2000) *successive overrelaxation* or *SOR* (that is, eqn (15.4) with $\beta > 1$) may be more economical than direct solution techniques.[13] It may also be competitive in non-linear problems[15] solved in steps, as the trial solutions from preceding steps are reasonably good and the non-linear corrections can be included as soon as $\{d\}$ has been recalculated for each element. Thus, one obtains successive rather than simultaneous corrections, as in the Jacobi iteration method.[13]

The principal difficulty of the Gauss–Seidel method is the assurance of convergence. (For example, try and solve the equations $x_1 + 2x_2 = 8$, $2x_1 + x_2 = 7$ in this and reversed order, it will be found that convergence is obtained only if the order is reversed. Dividing each row of the stiffness matrix by its pivot and writing the result as $K = I + L + U$ where U, L are the above and below diagonal parts, respectively, Gauss–Seidel iteration can be written as

$$(I + L)\{D\}^m = \{Q_i/K_{ii}\} - U\{D\}^{m-1} \tag{15.5}$$

and convergence is unconditionally guaranteed only if all the eigenvalues of the iteration matrix $(I + L)^{-1}U$ are less than one in magnitude. In fact the rate of convergence also depends upon the largest of these magnitudes (the *spectral radius*). A sufficient but more stringent condition is $|K_{ii}| \geq \sum |K_{ij}|$, $i \neq j$, and this is satisfied by finite difference methods (for which Gauss–Seidel iteration is much used) but is less easy to assure in variable geometry finite elements.

15.3.3 Gauss reduction

Here reduction of any column p (to zero entries below the pivot) is achieved by the *forward reduction* algorithms

$$K'_{pj} = \frac{K_{pj}}{K_{pp}} \quad \text{and} \quad Q'_p = \frac{Q_p}{K_{pp}}, \qquad j = p \to N \tag{15.6}$$

$$K'_{ij} = K_{ij} - K_{ip}(K'_{pj}), \qquad i = p + 1 \to N, \quad j = i \to N \tag{15.7}$$

$$Q'_i = Q_i - K_{ip}(Q'_p), \qquad i = p + 1 \to N. \tag{15.8}$$

The *back substitution* algorithms are

$$D_n = Q'_n \quad \text{(last row)} \tag{15.9}$$

and

$$D_m = Q'_m - K_{mj}D_j, \qquad m = n - 1 \to 1, \quad j = m + 1 \to n. \tag{15.10}$$

If the matrix is symmetric, $K_{ip} = K_{pi} = K_{pp}K'_{pi}$ and the below diagonal entries need not be stored. In the program given in Section 15.7 the equations for each node are reduced as soon as all element matrices contributing at this node have done so.[16, 17] Thus a rectangular array of dimensions equalling or exceeding the half bandwidth must be stored.

Such 'node by node' procedures are not be confused with *frontal routines*[18–21] in which the front width is the number of freedoms an element shares with other elements. Frontal schemes involve a good deal more 'housekeeping' but are more economical in 'ring structure' problems, for example.

15.3.4 Factorization techniques

In *Cholesky decomposition*, K is expressed as $K = L(I + U)$ where L, U are respectively lower and upper triangular matrices with null diagonals.[9] If K is symmetric the factorization can be expressed as $K = LL^t$, now populating the leading diagonal in L (compare eqn (15.5)). The solution is obtained as

$$\{D\} = (L^{-1})^t L^{-1} \{Q\}$$

and inversion of the triangular factor is quite straightforward, requiring an algorithm very similar to that for back substitution in Gauss reduction.[9]

In fact Gauss reduction achieves a similar result more directly.[20] This is observed by writing the stiffness matrix in the form

$$K = K_\Delta^t \langle K_{ii}' \rangle K_\Delta$$

where K_Δ is the upper triangular matrix (with unit leading diagonal) obtained after forward reduction (see eqn (15.21)), $\langle K_{ii}' \rangle$ denotes a diagonal matrix with entries K_{ii}' and K_{ii}' are the pivots encountered during reduction to obtain K_Δ.

Then Gauss reduction can be formally expressed as

$$\{D\} = (K_\Delta)^{-1} \left\langle \frac{1}{K_{ii}'} \right\rangle (K_\Delta^t)^{-1} \{Q\} \tag{15.11}$$

which closely resembles the Cholesky procedure.

Another useful factorization technique is LDL^t factorization, whilst an alternative Cholesky procedure is to use square factors, expressing K as $F^t F$ and then reducing F to upper triangular form.[3] Notable here is the *natural factor* technique of Argyris and Bronlund[22] in which the rigid body modes are excluded. This is advantageous in large displacement analysis, and K is directly assembled in factorized form. The numerical errors that arise during normal factorization are thus avoided. In general we have with numerical integration $k = \sum B^t DB(\omega_i V)$ and matrix D is factorized, giving

$$D = D_r^t D_r. \tag{15.12}$$

The element integration point contribution to F is then given as

$$F_e = D_r B \sqrt{(\omega_i V)}. \tag{15.13}$$

These are summed to give F, from which the rigid body rows may be excluded (as they are in A_s in Section 8.8, for example) and this improves the conditioning.[23] With the QR method and using the natural factors Johnsen and Roy[23] conclude that it is possible to solve almost all problems in linear structural analysis with single precision IBM 360 computation (that is, with 6.3 effective decimal digits).[24]

15.4 Boundary conditions

A boundary condition involving zero displacement can be dealt with by the technique described in Section 4.2, that is, by placing a very large number on the diagonal.[18] Non-zero specified or imposed displacements are also of considerable importance and they arise, for example, in earthquake dynamics, in the calculation of eigenvectors once the eigenvalues are known (where a single displacement is set to unity), and in the falling branch regimes of non-linear problems. To deal with such problems K can be partitioned thus:

$$\begin{bmatrix} K_{bb} & K_{bf} \\ K_{bf}^{t} & K_{ff} \end{bmatrix} \begin{Bmatrix} \{D\}_b \\ \{D\}_f \end{Bmatrix} = \begin{Bmatrix} \{Q\}_b \\ \{Q\}_f \end{Bmatrix} \tag{15.14}$$

where $\{D\}_b$, $\{D\}_f$ are the vectors of imposed boundary and unrestricted or free displacements, respectively. Rearranging this result gives

$$\begin{bmatrix} -K_{bb} & -K_{bf} \\ 0 & K_{ff} \end{bmatrix} \begin{Bmatrix} \{D\}_b \\ \{D\}_f \end{Bmatrix} = \begin{Bmatrix} -\{Q\}_b \\ \{Q\}_f - K_{bf}^{t}\{D\}_b \end{Bmatrix}. \tag{15.15}$$

In practice partitioning is not needed and appropriate modifications are made during Gauss reduction. We need to solve only the second row of eqn (15.15) so that during the forward pass the boundary rows are reversed in sign and reduction is undertaken only on the right-hand side using $\{D\}_b$ in place of $\{Q\}_b$. Thus the rearrangements in eqn (15.15) are achieved without partitioning.

In the non-boundary rows, Gauss reduction proceeds in the usual manner. Then during back substitution, boundary rows are recognized by their negative diagonal entry and the reactions associated with the specified displacements will be calculated at the same time as the solution for $\{D\}_f$ as

$$-\{Q\}_b + K_{bf}\{D\}_f + K_{bb}\{D\}_b. \tag{15.16}$$

Thus with minor modifications to standard Gauss reduction algorithms, $\{D\}_f$ and the reactions associated with $\{D\}_b$ are calculated simultaneously.

Skew boundary conditions involve specification of a displacement at an angle to the global xy-axes and can be enforced by applying a coordinate transformaton to the equations for the affected node. This is tedious and a much simpler solution is obtained by using a stiff spar element at the appropriate inclination, suppressing both displacements at its remote end.[25]

Until now we have discussed only boundary conditions enforced in a pointwise fashion at the nodes. *Blending functions* ensure continuous[26] satisfaction of inhomogeneous boundary values. As an example of their implementation consider a bilinear element with four freedoms $\{u\}$ and specified boundary functions $u = g_{12}(a)$, $g_{23}(b)$, $g_{43}(a)$, $g_{14}(b)$ on the four sides. Then

$$u = \frac{(1 - b)g_{12}(a)}{2} + \frac{(1 + a)g_{23}(b)}{2} + \frac{(1 + b)g_{43}(a)}{2} + \frac{(1 - a)g_{14}(b)}{2}$$

$$- f_1 g_{12}(-1) - f_2 g_{23}(-1) - f_3 g_{43}(1) - f_4 g_{14}(1) \qquad (15.17)$$

will produce a hyperbolic paraboloidal surface which blends with the boundary functions, the subtracted values preventing 'doubling up' of the corner values. Here the f_i are given by eqn (7.2c) and if g_{12} is parabolic, for example, one can use the corner values to write it in the form

$$g_{12}(a) = \frac{(u_1 + u_2 - 2c)}{2} + \frac{(u_2 - u_1)a}{2} + ca^2 \qquad (15.18)$$

where c is the amplitude of the a^2 shape on this boundary, which must be specified as data. Applying similar conditions on the other sides of the element eqn (15.17) can be written as

$$u = \frac{(1 - b)g_{12}(a, u_1, u_2)}{2} + \frac{(1 + a)g_{23}(b, u_2, u_3)}{2} + \frac{(1 + b)g_{43}(a, u_4, u_3)}{2}$$

$$+ \frac{(1 - a)g_{14}(b, u_1, u_4)}{2} - \sum f_i u_i \qquad (15.19)$$

and a stiffness matrix can now be formed in the usual way, some of the corner values being specified as boundary conditions at the equation solution stage. General implementation of this procedure with other elements is difficult and little improvement in results is obtained,[27] so that it may be more advisable to use higher-order elements which will allow the higher-order boundary interpolation which blending functions provide. A more consistent formulation will be achieved with improved approximation both within the domain and at the boundary (see for example the figures in parentheses in Table 8.4).

Such procedures are, however, useful in the mapping of shells[28] of specified shape with edge shapes that are not related to that of the shell and are more easily implemented in this context.

15.5 Spectrum loadings

Specification of direct nodal loadings requires no explanation (these having one-to-one correspondence with the nodal freedoms) whilst the question of distributed loading within an element is discussed in Section 3.5. In many practical situations loadings are *stochastic*[29, 30] so that analysis with a single mean or extremal loading is not realistic. In such cases one can repeat the analysis with a spectrum of load cases (which can be estimated by Monte Carlo simulation) and, in Gauss reduction, further reductions of K can be avoided by storing the pivots which occur before each application of eqn (15.6). Considering the equations

$$K\{D\} = \begin{bmatrix} 2 & -2 & 1 \\ -2 & 8 & -4 \\ 1 & -4 & 6 \end{bmatrix} \begin{Bmatrix} D_1 \\ D_2 \\ D_3 \end{Bmatrix} = \begin{Bmatrix} 2 \\ 3 \\ 1 \end{Bmatrix} = \{Q\}, \qquad (15.20)$$

Gauss reduction gives

$$K_\Delta\{D\} = \begin{bmatrix} 1 & -1 & 1/2 \\ 0 & 1 & -1/2 \\ 0 & 0 & 1 \end{bmatrix} \begin{Bmatrix} D_1 \\ D_2 \\ D_3 \end{Bmatrix} = \begin{Bmatrix} 1 \\ 5/6 \\ 5/8 \end{Bmatrix} = \{Q_\Delta\} \qquad (15.21)$$

and the vector of stored 'active' pivots (used in eqn (15.6) is

$$\{K'_{11}, K'_{22}, K'_{33}\} = \{2, 6, 4\}. \qquad (15.22)$$

Then the reduced load vector for any new loading $\{Q\}$ is given by

$$\langle K'_{11}, K'_{22}, K'_{33} \rangle \{Q_\Delta\} = K_\Delta^{t(-)}\{Q\} \qquad (15.23)$$

where $K_\Delta^{t(-)}$ refers to the transpose of the triangular stiffness matrix with reversed signs for the off-diagonal entries and the $\langle\ \rangle$ parentheses refer to a diagonal matrix with the entries indicated within them. Returning to the original load vector to verify eqn (15.23) one obtains

$$\begin{bmatrix} 2 & 0 & 0 \\ 0 & 6 & 0 \\ 0 & 0 & 4 \end{bmatrix} \{Q_\Delta\} = \begin{bmatrix} 1 & 0 & 0 \\ 1 & 1 & 0 \\ -1/2 & 1/2 & 1 \end{bmatrix} \begin{Bmatrix} 2 \\ 3 \\ 1 \end{Bmatrix} = \begin{Bmatrix} 2 \\ 5 \\ 5/2 \end{Bmatrix} \qquad (15.24)$$

as required. Note that, as in back substitution for displacements, the load vector must be progressively updated as each new term is evaluated, beginning from the top, and eqn. (15.23) applies only on this understanding. This forward reduction, once complete, is followed by the usual back substitution for the displacements (which is equivalent to inverting K_Δ).

Another incentive for storing the pivots is that their product gives the determinant of K (Chio's method; see Section 13.9).[31]

Such repeated solutions can be used to obtain solutions for axisymmetric structures with non-axisymmetric loadings, each load case being an amplitude of a Fourier series which approximates the true loading. The series of displacement and stress solutions so obtained is then progressively combined to yield the complete solution.[20]

15.6 Stress sampling points

Element stresses can be calculated using element stress matrices that are formed in the same subroutine as the element stiffness matrices. Stress calculations must take into account such effects as thermal strains and residual stresses[32] and, as shown in Chapter 3, will generally involve a larger error than the displacement solutions.

In some highly non-conforming shell elements stress calculations are restricted to the element centroids[33, 34] but usually stresses are sampled at the nodes or at the integration points. Because, in energy-based formulations, the stresses satisfy the equilibrium equations in an integral sense, it has been argued that the integration points are the best stress sampling points[35] and results with the arch element of Section 11.1, for example, bear this out.[36]

Nodal stresses generally require smoothing by averaging between elements[16] or by conjugate gradient techniques such as that proposed by Oden and Reddy.[37] Alternatively one can use extrapolation from integration point values to obtain nodal values[35] (which will also require averaging).[38]

Generally stress values calculated at the nodes are more convenient than integration point values and insistence that these be satisfactory, at least in the mean sense (that is, after averaging), in meshes of practical fineness is perhaps a more stringent test of an element than the patch test.

In shell problems, however, integration point stress sampling is widely recommended and we should also note that one must first be at pains to choose that numerical integration which gives the best displacement solutions. Further, when numerous integration points are available, those with the larger weights should generally give the best results and these points are normally those closest to the element centroid.

15.7 A BASIC microcomputer program for the quadratic shell element

The following section gives a BASIC microcomputer program for the thirty-freedom quadratic shell element developed in Section 11.6 (the element is denoted as SIP6R in STRUDL). The element calculations are those described in Section 11.6 and some relevant equation numbers are indicated alongside the corresponding coding.

Assembly of the element equations follows the *half-band* scheme shown in Fig. 15.1. The solution routine is based upon Parekh's FESS system[16] and reduces the equations on a node by node basis, allowing the use of the *block solution* technique described below. Boundary conditions are dealt with in the manner described in Section 15.4.

When the stiffness array SK(,) in subroutine SOLVE is full (that is, when the number of rows filled is equal to SIZ) the top segment of SK of size BUF × LB (where LB is greater than or equal to the half bandwidth for the problem) is written to disc and the remaining equations are moved up to make room for continuation of the node by node reduction process. This process is then reversed during back substitution. Note that in line 60 of SOLVE the size allocated for the stiffness 'block' ($\text{SIZ} \times \text{LB}$) is the minimum possible for the test problem, that is, $\text{SIZ} = $ half bandwidth $+ \text{NDF}$ (where $\text{NDF} = $ degrees of freedom per node) and $\text{LB} = $ half bandwidth, to fully test out the block solution system.

Introductory details of the *MEGABASIC* language used are given in Appendix 1, along with basic details of the organization of the Finite Element Microcomputer Solution System (*FEMSS*) used for the present and for five other programs in the text.

The element mesh is specified directly in the curvilinear coordinate system, as shown in Fig. 15.2. The half bandwidth b_w is specified by the user and is

Table 15.1 Data input for the quadratic shell element

Lines	Data
1	Number of nodes (NP); number of elements (NE); number of boundary condition nodes (NB); number of property sets (NT); half bandwidth (NBW)
1	PENF, POW, giving the penalty factor as $\beta = \text{PENF} \times 10^{\text{POW}}$
NT lines	Property sets $(N = 1 \to NT)$: $E/10^6$ (PROP(N, 1)); v (PROP(N, 2)); ρ (PROP (N, 3)); t (element thickness) (PROP(N, 4)); p (blanket load intensity in w-direction) (PROP(N, 5))
1	$1/R_x$, $1/R_y$, $1/R_{xy}$ (RX, RY, RXY): curvatures $\times 1000$
NP lines	x (CORD(N, 1)), y (CORD(N, 2)): curvilinear coordinates for each node read in ascending order
NE lines	Node numbers for each element (NOP(N, M), M $= 1 \to 6$), read in ascending order. Note that in line 130 the instruction IMAT(N) $= 1$ assigns property set 1 to every element. If the elements have different properties line 130 should be For N $= 1$ to NE; Read IMAT(N); For M $= 1$ to 6; Read NOP(N, M); Next; Next
NB lines	Boundary condition node number (NBC(I)); flags for boundary freedoms (NFIX(I)). If node 6, for example, had u, v, w suppressed (to zero) then the appropriate data line will be 6,11100
NL lines	NL $= $ number of nodes with concentrated loads; node number (NQ); concentrated loads corresponding to u, v, w, ϕ_x, ϕ_y freedoms (R(1), R(2), R(3), R(4), R(5)). This data is terminated by a data line with six zeros.

calculated as $(\text{Max} - \text{Min}) \times \text{NDF} + \text{NDF}$, where Max and Min are respectively the largest and smallest global node numbers in an element and b_w must be the largest value so encountered.

For the node by node solution routine elements must be in the order of nodal introduction, that is, taking the element numbers in ascending order these must introduce new node numbers in ascending order. In simple meshes such as those in rectangular domains this is automatically ensured by numbering nodes and elements first in the Y-direction and then in the X-direction, as is done in Fig. 15.2. When the elements are in arbitrary order, coding to re-order them for node by node solution takes only some 15 lines of code, including search for the half bandwidth.

The present program does not deal with non-zero specified displacements as boundary conditions. Coding to deal with these is included in the SOLVE routine of the program of Section 20.7.

The program listing includes the data (in lines 1000–1240) for the problem of Fig. 15.2, where for introductory purposes only two elements are used to analyse an octant of a pinch loaded cylinder with diaphragmatic ends. The data lines give the following information:

Line 1000	9 nodes, 2 elements, 8 b.c. nodes, 1 property set, half bandwidth $= 35$. For example, in element 1 the vertex node numbers are $1, 7, 3$ so that $b_w = 5(7 - 1) + 5 = 35$.
Line 1010	$\beta = 5 \times 10^{-4}$, the rounded-off value from eqn (9.35) (of course calculation of β can easily be included in the program).
Line 1020	Property set 1 has $E = 1 \times 10^6$, $v = 0.3$, $\rho = 0$, $t = 1$, $p = 0$.
Line 1025	$1/R_x = 1/R = 10/1000 = 0.01$, $1/R_y = 0$, $1/R_{xy} = 0$.

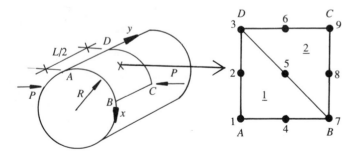

Fig. 15.2 Pinched cylinder problem of Fig. 11.9 with one octant analysed using two six-node elements

Line 1030–1110   Nodal curvilinear coordinates; the x-coordinate is circumferential and the y-coordinate longitudinal.

Line 1120–1130   Node numbers for each element, with the usual conventions of anticlockwise numbering taking the vertex nodes first.

Line 1150–1220   Boundary conditions are: AD and BC: $u = 0$, $\phi_x = 0$ (circumferential symmetry); AB: $u = 0$, $w = 0$, $\phi_x = 0$ (diaphragm at end of cylinder); CD: $v = 0$, $\phi_y = 0$ (longitudinal symmetry conditions); and these conditions 'overlap' at the corners A, B, C, and D.

Line 1230   Inward normal load of 25 at node 9 (one quarter of the actual load because we have biaxial symmetry at node 9).

The results output will be the displacements u, v, w at each node (from subroutine SOLVE) and the *average nodal stress resultants* m_x, m_y, N_x, N_y at each node (from subroutine STRESS). The latter are calculated using the *nodal valency* (the number of elements impingeing on any node) which is calculated in line 180 of the main (calling) subroutine. Generally the vertex node results are more reliable but note that these are actually obtained from the proximate integration points given by eqn (5.136).

With the above data, the displacement under the load should be that given in line 1, column 2 of Table 11.4. Naturally, to obtain more accurate results a finer mesh should be used (to obtain 'tolerable' stress results at least thirty-two elements, and thus eighty-one nodes, should be used). Results for displacement and stress distributions obtained using finer meshes should be in agreement with the approximate graphs given in references 38 and 41 of Chapter 11.

The same form of input data is equally applicable to the analysis of shallow spherical caps, for example, when $1/R_x = 1/R_y = 1/R$ and $1/R_{xy} = 0$, whilst for shallow shells of arbitrary shape the scheme outlined in Section 11.7 must be used.

The program uses *sequential access* (floppy) disc files to store the element stiffness matrices, the element stress matrices, and the blocks of the global stiffness produced by the solution routine. Note that retrieval of K_Δ in SOLVE for back substitution is achieved using a pair of Filepos( ) statements to read the blocks of K_Δ in reverse order.

Another useful test problem recommended to the reader is that of Fig. 15.2, with the pinching loads replaced by a uniform internal pressure loading (that is, $p = 1$ in line 1020, and line 1230 is omitted) when the exact analytical solutions for radial displacement and circumferential strain should be obtained with only two elements.

```
1 Include 'ELMAT','SOLVE','STRESS'          ;Rem LOAD SUBROUTINES
3 Access * from *                           ;Rem EVERYTHING ACCESS EVERYTHING
4 Rem Use CREATE "B:EMATS", CREATE "B:ELSTR", CREATE "B:STIFM" before running
5 Rem the program to create files for matrix storage on disc 2 if this has
6 Rem never been done previously (these files are defined immediately below)
7 Open £8,"B:EMATS"                          ;Rem FILE FOR ELE STIFFNESS MATRICES
8 Open £9,"B:ELSTR"                          ;Rem FILE FOR ELE STRESS MATRICES
9 Open £10,"B:STIFM"                         ;Rem FILE FOR GLOBAL STIFFNESS MATRIX
10 Dim FNE(100),NBC(50),NFIX(50),Q(500),IMAT(100),CORD(100,2),PROP(20,5)
20 Dim NOP(100,6),CI(7,2),WF(7),R(5)
25 Rem ***** DECLARE DATA TO BE USED BY OTHER SUBROUTINES
30 Def shared data NP,NE,NB,NT,NBW,NCN,NDF,PENF,FNE( )
32 Def shared data NBC( ),NFIX( ),Q( ),RX,RY,RXY,CI( ),WF( )
34 Def shared data IMAT( ),CORD( ),PROP( ),NOP( )
40 NDF=5;NCN=6
50 For I=1 to 3;CI(I,1)=1/6;CI(I,2)=1/6;WF(I)=1/6;CI(I+3,1)=0.5;CI(I+3,2)=0.5
60 WF(I+3)=0;Next                           ;Rem INTEG PT COORDS AND WEIGHTS
70 CI(1,1)=4/6;CI(2,2)=4/6;CI(5,1)=0;CI(6,2)=0
80 Read NP,NE,NB,NT,NBW;NBW=NBW-NDF          ;Rem PROBLEM SIZE
90 Read PENF,POW;PENF=PENF*10^POW            ;Rem PENALTY FACTOR, SEE EQN 9.35
100 For N=1 to NT;For I=1 to 5;Read PROP(N,I);Next;Next  ;Rem PROPERTY SETS
110 Read RX,RY,RXY                           ;Rem CURVATURES*1000
112 RX=RX/1000;RY=RY/1000;RXY=RXY/1000
120 For N=1 to NP; Read CORD(N,1),CORD(N,2);Next     ;Rem NODAL COORDINATES
125 Rem ***** NODE NUMBER SETS FOR EACH ELEMENT
130 For N=1 to NE;IMAT(N)=1;For M=1 to 6;Read NOP(N,M);Next;Next
140 For I=1 to NB;Read NBC(I),NFIX(I);Next   ;Rem BOUNDARY CONDITIONS
150 Read NQ,R(1),R(2),R(3),R(4),R(5)         ;Rem NODAL POINT LOADS
155 If NQ=0 then Goto 170                     ;Rem ADD LATTER TO LOAD VECTOR
160 For K=1 to NDF;IC=(NQ-1)*NDF+K;Q(IC)=Q(IC)+R(K);Next
165 Goto 150
170 Rem
180 For N=1 to NE;For II=1 to NCN;NN=NOP(N,II);FNE(NN)=FNE(NN)+1
185 Next;Next    ;Rem CALCULATE NO. OF ELEMENTS SHARING EACH NODE
190 For N=1 to NE
200 ELSUB N;Next                             ;Rem CALL ELEMENT STIFFNESS ROUTINE
205 Filepos(8)=0
206 Filepos(9)=0                             ;Rem BACKSPACE ELEMENT MATRIX FILES
210 SOLSUB                                   ;Rem CALL SOLUTION ROUTINE
220 STRUB                                    ;Rem CALL STRESS CALCULATION ROUTINE
300 End
1000 Data 9,2,8,1,35                         ;Rem DATA READ AS ABOVE
1010 Data 5,-4
1020 Data 1,0.3,0,1,0
1025 Data 10,0,0
1030 Data 0,0
1040 Data 0,50
1050 Data 0,100
1060 Data 78.5,0
1070 Data 78.5,50
1080 Data 78.5,100
1090 Data 157,0
1100 Data 157,50
1110 Data 157,100
1120 Data 1,7,3,4,5,2
1130 Data 3,7,9,5,8,6
1150 Data 1,10110
1160 Data 2,10010
1170 Data 3,11011
1180 Data 4,10110
1190 Data 6,01001
1200 Data 7,10110
1210 Data 8,10010
1220 Data 9,11011
1230 Data 9,0,0,-25,0,0
1240 Data 0,0,0,0,0,0                        ;Rem ZERO LINE TO TERMINATE PT LOADS
```

```
10 Rem ELMAT
20 Def shared proc ELSUB @N
30 Rem ***** Latter statement defines following as a routine (called in MAIN)
40 Rem ***** with argument being only the element number
50 Dim S(30,30),C(24,30),D(9,9),B(9,30),XY(6,2),F(6),TEMP(9,30)
60 Dim CT(3,2),DG(2,2),CG(2,3),DF(9,9),V(3,6),Z(2,3)
80 L=IMAT(N);E=PROP(L,1)*10^6;P=PROP(L,2)  ;Rem COLLECT ELEMENT PROPERTIES
90 DENS=PROP(L,3);TH=PROP(L,4);UDL=PROP(L,5)
95 For M=1 to 6;K=NOP(N,M);XY(M,1)=CORD(K,1);XY(M,2)=CORD(K,2);Next
100 X21=XY(2,1)-XY(1,1);X32=XY(3,1)-XY(2,1);X13=XY(1,1)-XY(3,1)
110 Y21=XY(2,2)-XY(1,2);Y32=XY(3,2)-XY(2,2);Y13=XY(1,2)-XY(3,2)
120 S21=sqrt(X21*X21+Y21*Y21)
130 S32=sqrt(X32*X32+Y32*Y32)                 ;Rem ELEMENT SIDE LENGTHS
140 S13=sqrt(X13*X13+Y13*Y13)
150 CT(1,1)=X21/S21;CT(1,2)=Y21/S21
160 CT(2,1)=X32/S32;CT(2,2)=Y32/S32           ;Rem DIRECTION COSINES OF SIDES
170 CT(3,1)=X13/S13;CT(3,2)=Y13/S13
180 A=X21*Y32-X32*Y21;AA=A*A                  ;Rem A = TWICE ELEMENT AREA
190 Z(1,1)=-Y32/A;Z(1,2)=-Y13/A;Z(1,3)=-Y21/A   ;Rem EXTENDED FORM OF EQN 11.33
200 Z(2,1)=X32/A;Z(2,2)=X13/A;Z(2,3)=X21/A       ;Rem WHEN L3 NOT ELIMINATED
210 CG(1,1)=CT(2,2)*S21*S32-CT(3,2)*S13*S21
220 CG(1,2)=CT(3,2)*S32*S13-CT(1,2)*S21*S32
230 CG(1,3)=CT(1,2)*S13*S21-CT(2,2)*S32*S13    ;Rem TG IN EQN 9.25
240 CG(2,1)=-CT(2,1)*S21*S32+CT(3,1)*S13*S21
250 CG(2,2)=-CT(3,1)*S32*S13+CT(1,1)*S21*S32
260 CG(2,3)=-CT(1,1)*S13*S21+CT(2,1)*S32*S13
270 For I=1 to 3;NI=I+3;NF=NOP(N,NI)*NDF-2     ;Rem CONSISTENT LOADS FOR
280 Q(NF)=Q(NF)+A*UDL/6;Next                   ;Rem RADIAL PRESSURE LOADING
285 Rem ***** Form modulus matrix as defined in eqn 11.39 for stress calcs
286 Rem ***** form DF which omits multipication by H and Tg in eqn 11.39
290 D(1,1)=E*TH^3/(12*(1-P*P));D(2,2)=D(1,1);D(1,2)=P*D(1,1);D(2,1)=D(1,2)
300 D(3,3)=0.5*(1-P)*D(1,1);D(1,3)=0;D(2,3)=0;D(3,1)=0;D(3,2)=0
310 For I=1 to 3;For J=1 to 3;DF(I+6,J+6)=D(I,J)*12/(TH*TH)
320 D(I+6,J+6)=D(I,J)*12/(TH*TH);Next;Next
330 DG(1,1)=5*E*TH/(12*(1+P));DG(2,2)=DG(1,1);DG(1,2)=0;DG(2,1)=0
335 Rem ***** Form matrix C in eqn 8.16
340 B(1,1)=CT(1,1)^2;B(1,2)=CT(1,2)^2;B(1,3)=2*CT(1,1)*CT(1,2)
350 B(2,1)=CT(2,1)^2;B(2,2)=CT(2,2)^2;B(2,3)=2*CT(2,1)*CT(2,2)
360 B(3,1)=CT(3,1)^2;B(3,2)=CT(3,2)^2;B(3,3)=2*CT(3,1)*CT(3,2)
370 AX=1;BB=1;CZ=1;AZ=(S21^2+S13^2-S32^2)/(2*S21*S13)
380 AZ=AZ*AZ;CX=AZ;DD=1-AZ;AY=1-DD*(S13/S32)^2  ;Rem MATRIX W IN EQN 8.15
390 BX=AY;BZ=1-DD*(S21/S32)^2;CY=BZ
400 DD=AX*(BB*CZ-BZ*CY)-AY*(BX*CZ-BZ*CX)+AZ*(BX*CY-BB*CX)
410 S(1,1)=BB*CZ-BZ*CY;S(1,2)=-(AY*CZ-AZ*CY);S(1,3)=AY*BZ-AZ*BB
420 S(2,1)=-(BX*CZ-BZ*CX);S(2,2)=AX*CZ-AZ*CX;S(2,3)=-(AX*BZ-AZ*BX);Rem INVERT W
430 S(3,1)=BX*CY-BB*CX;S(3,2)=-(AX*CY-AY*CX);S(3,3)=AX*BB-AY*BX
440 For I=1 to 3;For J=1 to 3;C(I,J)=0;For K=1 to 3  ;Rem NOTE EQN 8.24 FOR
450 C(I,J)=C(I,J)+D(I,K)*B(J,K);Next;Next ;Rem ALTERNATIVE TRANSFORMATION
460 For I=1 to 3;For J=1 to 3;D(I,J)=0;Next;Next ;For K=1 to 3
470 DF(I,J)=DF(I,J)+C(I,K)*S(K,J)/DD;D(I,J)=D(I,J)+B(I,K)*C(K,J)
480 Next;Next;Next                    ;Rem DF = Db IN EQN 11.39
490 For I=1 to 3;For J=1 to 3;C(I,J)=0;For K=1 to 3
500 C(I,J)=C(I,J)+D(I,K)*S(K,J)/DD;Next;Next;Next
510 For I=1 to 3;For J=1 to 3;D(I,J)=0;For K=1 to 3
520 D(I,J)=D(I,J)+S(I,K)*C(K,J)/DD;Next;Next;Next  ;Rem HtDbH IN EQN 11.39
530 For I=1 to 2;For J=1 to 3;S(I,J)=0;For K=1 to 2
540 S(I,J)=S(I,J)+DG(I,K)*S(K,J)/(9*AA);Next;Next;Next
550 For I=1 to 3;IA=I+3;For J=1 to 3;JA=J+3;For K=1 to 2
560 D(IA,JA)=D(IA,JA)+CG(K,I)*S(K,J);Next;Next;Next ;Rem ADD TGtDsTG TO D
570 For I=1 to 2;For J=1 to 3; DF(I+3,J+3)=S(I,J)*A*3;Next;Next ;Rem DsTG TO DF
580 For I=1 to 30;For J=1 to 30;S(I,J)=0;Next;Next
590 For II=1 to 6        ;Rem COMMENCE ELEMENT INTEGRATION LOOP £££££££££££££££££££
600 F1=4*CI(II,1);F2=4*CI(II,2)          ;Rem INTEGRATION POINT COORDS
610 For I=1 to 9;For J=1 to 30;B(I,J)=0;Next;Next ;Rem INITIALIZE STRAIN MATRIX
620 C1=CI(II,1);C2=CI(II,2);CC=C1+C2
630 F(1)=2*C1*C1-C1;F(2)=2*C2*C2-C2;F(3)=1-3*CC+2*CC*CC       ;Rem QUADRATIC
640 F(4)=4*C1*C2;F(5)=4*C2*(1-CC);F(6)=4*C1*(1-CC);F3=4-F1-F2;Rem INTERPOLATION
650 For I=1 to 3;For J=1 to 6;V(I,J)=0;Next;Next
```

```
660 V(1,1)=F1-1;V(1,4)=F2;V(1,6)=F3        ;Rem d(f)/dL1 - w/o L3 eliminated
670 V(2,2)=F2-1;V(2,4)=F1;V(2,5)=F3        ;Rem d(f)/dL2 -  "    "       "
680 V(3,3)=F3-1;V(3,5)=F2;V(3,6)=F1        ;Rem d(f)/dL3 -  "    "       "
690 For J=1 to 6;JA=5*J-2
700 B(4,JA)=(V(2,J)-V(1,J))/S21            ;Rem d(f)/da, d(f)/db, d(f)/dc for
710 B(5,JA)=(V(3,J)-V(2,J))/S32            ;Rem w columns in eqn 11.34
720 B(6,JA)=(V(1,J)-V(3,J))/S13
730 Next
740 For I=1 to 3;For J=1 to 6;IA=I+3;JA=5*J-1
750 B(IA,JA)=B(IA,JA)-CT(I,1)*F(J)                      ;Rem ENTRIES IN ROWS 4,5,6
760 B(IA,JA+1)=B(IA,JA+1)-CT(I,2)*F(J);Next;Next ;Rem FOR PHI COLUMNS
770 For J=1 to 6;JX=5*J-1;JY=JX+1
780 B(1,JX)=-CT(1,1)*(V(2,J)-V(1,J))/S21
790 B(1,JY)=-CT(1,2)*(V(2,J)-V(1,J))/S21
800 B(2,JX)=-CT(2,1)*(V(3,J)-V(2,J))/S32               ;Rem ENTRIES IN ROWS 1,2,3
810 B(2,JY)=-CT(2,2)*(V(3,J)-V(2,J))/S32               ;Rem FOR PHI COLUMNS
820 B(3,JX)=-CT(3,1)*(V(1,J)-V(3,J))/S13
830 B(3,JY)=-CT(3,2)*(V(1,J)-V(3,J))/S13
840 Next
850 For I=1 to 2;For J=1 to 6;TEMP(I,J)=0;For K=1 to 3
860 TEMP(I,J)=TEMP(I,J)+Z(I,K)*V(K,J);Next;Next;Next ;Rem FORM d(f)/dx, d(f)/dy
870 For J=1 to 6;JA=5*J-4;B(7,JA)=TEMP(1,J);B(9,JA)=TEMP(2,J)
880 B(8,JA+1)=TEMP(2,J);B(9,JA+1)=TEMP(1,J);Next      ;Rem PLACE THESE IN B
890 For J=1 to 6;JA=5*J-2;B(7,JA)=F(J)*RX;B(8,JA)=F(J)*RY
900 B(9,JA)=2*F(J)*RXY;Next                            ;Rem (f)t/R ENTRIES IN B
910 IA=4*(II-1)+1
920 For J=1 to 30;C(IA,J)=0;C(IA+1,J)=0;C(IA+2,J)=0;C(IA+3,J)=0
930 For K=1 to 9
940 C(IA,J)=C(IA,J)+DF(1,K)*B(K,J)
950 C(IA+1,J)=C(IA+1,J)+DF(2,K)*B(K,J)     ;Rem STRESS MATRIX TO CALCULATE
960 C(IA+2,J)=C(IA+2,J)+DF(7,K)*B(K,J)     ;Rem Mx,My,Nx,Ny AT EACH POINT
970 C(IA+3,J)=C(IA+3,J)+DF(8,K)*B(K,J)
980 Next K;Next J
990 If WF(II)=0 then Goto 1070
1000 For I=1 to 9;For J=1 to 30;TEMP(I,J)=0;For K=1 to 9
1010 TEMP(I,J)=TEMP(I,J)+D(I,K)*B(K,J);Next;Next;Next
1020 SMUL=WF(II)*A
1030 For I=1 to 30;For J=1 to 30;For K=1 to 3
1040 S(I,J)=S(I,J)+B(K,I)*TEMP(K,J)*SMUL                ;Rem ELEMENT STIFFNESS MATRIX
1050 S(I,J)=S(I,J)+B(K+3,I)*TEMP(K+3,J)*SMUL*PENF ;Rem ELEMENT STIFFNESS MATRIX
1060 S(I,J)=S(I,J)+B(K+6,I)*TEMP(K+6,J)*SMUL
1065 Next K;Next J;Next I
1070 Next II          ;Rem END OF ELEMENT INTEGRATION LOOP     ££££££££££££££££££
1080 For I=1 to 30;For J=1 to 30
1090 Write £8,S(I,J);Next;Next                         ;Rem FILE STIFFNESS MATRICES
1100 For I=1 to 24;For J=1 to 30
1110 Write £9,C(I,J);Next;Next                         ;Rem FILE STRESS MATRICES
1115 ! N                                               ;Rem REPORT COMPLETION FOR ELEMENT
1120 Return;Proc end
1130 End

10 Rem SOLVE
20 Def shared proc SOLSUB
45 Dim DIS(5,100)
50 Dim ESM(30,30),SK(40,35),SKP(35)
55 Def shared data DIS( )        ;Rem SHARE DISPLACEMENT SOLUTION WITH STRESS
60 SIZ=40;LB=35;BUF=SIZ-LB;NRW=0;NTW=NBW+NDF;NLOAD=NP*NDF;NBN=1
70 L=0;N=1                       ;Rem SIZ,LB ARE STIFFNESS BLOCK DEPTH & WIDTH
80 For I=1 to 30;For J=1 to 30
90 Read £8,ESM(I,J);Next;Next ;Rem READ FIRST k (ELEMENT STIFFNESS MATRIX)
100 L=L+1        ;Rem COMMENCE LOOP FOR NODE BY NODE FORWARD REDUCTION £££££££££££
110 For M=1 to 8
120 If N=(NE+1) then Goto 280
130 For I=1 to NCN
140 If NOP(N,I)=L then Goto 170                        ;Rem CHECK IF NEXT k NEEDED YET
150 Next
```

```
160 Goto 280
170 Rem
180 For I=1 to NCN;For J=1 to NCN
190 For IL=1 to NDF;IE=(I-1)*NDF+IL;NR=(NOP(N,I)-1)*NDF+IL
200 NRE=NR-NRW                          ;Rem NRW = NO. ROWS OF K FILED
210 For JL=1 to NDF;JE=(J-1)*NDF+JL;NC=(NOP(N,J)-1)*NDF+JL
220 NCB=NC-NR+1;If NR>NC then Goto 240
230 SK(NRE,NCB)=SK(NRE,NCB)+ESM(IE,JE)   ;Rem ASSEMBLY OF K (SEE FIG. 15.1)
240 Next;Next;Next;Next
245 If N=NE then Goto 265
250 For I=1 to 30;For J=1 to 30
260 Read £8,ESM(I,J);Next;Next           ;Rem READ NEXT k
265 N=N+1
270 Next M
280 Rem
290 NDIF=(NP-L+1)*NDF;If NDIF>NBW then LIM=NBW+NDF;JZ=0
300 If NBN=NB+1 then Goto 320;If L<>NBC(NBN) then Goto 320
310 JZ=NFIX(NBN);IZ=10^(NDF-1);NBN=NBN+1 ;Rem CHECK IF NODE HAS B.C.s
320 For ID=1 to NDF
330 LIM=LIM-1;IP=ID+NDF*(L-1);IPE=IP-NRW;R=Q(IP);NOB=0
340 If JZ=0 then Goto 370;If JZ<IZ then Goto 360      ;Rem CHECK FOR B.C.
350 RS=-R;R=0;NOB=1;JZ=JZ-IZ
360 IZ=IZ/10
370 Rem
380 If NOB=1 then Goto 410
385 ! L,ID                             ;Rem REPORT SOLUTION PROGRESS
390 XK=1/SK(IPE,1);Q(IP)=XK*R
400 Goto 430
410 Rem
420 Q(IP)=RS+SK(IPE,1)*R;XK=1;R=-R      ;Rem Q(IP) = BOUNDARY 'REACTION'
430 Rem
440 For J=1 to LIM;JA=J+1;SKP(J)=SK(IPE,JA);Next ;Rem STORE 'ROW MULTIPLIERS'
450 NC=LIM+1
460 For J=1 to NC;SK(IPE,J)=SK(IPE,J)*XK          ;Rem DIVIDE ROW BY PIVOT
470 If NOB=1 then SK(IPE,J)=-SK(IPE,J);Next       ;Rem NEGATE BOUNDARY ROW
480 If (L+ID-NP-NDF)=0 then Goto 660              ;Rem END TEST
490 For I=1 to LIM;NR=IP+I;NRE=IPE+I
500 If SKP(I)=0 then Goto 550;If NOB=1 then Goto 530;NC=LIM-I+1
510 For J=1 to NC;JP=J+I
520 SK(NRE,J)=SK(NRE,J)-SK(IPE,JP)*SKP(I);Next    ;Rem FORWARD REDUCTION
530 JP=I+1
540 Q(NR)=Q(NR)-SK(IPE,JP)*R                      ;Rem REDUCTION IN LOAD VECTOR
550 Next I
560 If (IPE+NTW)<SIZ then Goto 630                ;Rem TEST IF STIFFNESS BLOCK FULL
570 If (NLOAD-NRW)<=SIZ then Goto 630
580 For I=1 to BUF;For J=1 to LB
590 Write £10,SK(I,J);Next;Next                   ;Rem FILE PART OF STIFFNESS BLOCK
600 NRW=NRW+BUF                                   ;Rem NRW = NO. ROWS OF K FILED
610 For I=1 to LB;For J=1 to LB;IA=I+BUF
620 SK(I,J)=SK(IA,J);SK(IA,J)=0;Next;Next ;Rem SHIFT REMAINING ROWS UP
630 Rem
640 Next ID
650 Goto 100      ;Rem END NODE BY NODE FORWARD REDUCTION LOOP £££££££££££££££
660 Rem
670 NR=NDF*NP;NRE=NR-NRW;DIS(NDF,NP)=Q(NR);Rem LAST DISPLACEMENT NOW KNOWN
680 Q(NR)=0;I=NDF;L=NP
690 Goto 780
700 L=L-1                              ;Rem LOOP ON NODES FOR BACK SUBSTITUTION
710 I=I-1                              ;Rem LOOP ON D.F./NODE FOR BACK SUBSTITUTION
720 NR=NDF*(L-1)+I;NRE=NR-NRW
730 DIS(I,L)=Q(NR);Q(NR)=0
740 If LIM<(NBW+NDF-1) then LIM=LIM+1
750 For J=1 to LIM;JA=J+1
760 LJ=L+(J+I-1)/NDF;LJ=trunc(LJ);K=I+J-(LJ-L)*NDF
770 DIS(I,L)=DIS(I,L)-SK(NRE,JA)*DIS(K,LJ);Next   ;Rem BACK SUBSTITUTION
780 If SK(NRE,1)>0 then Goto 800
790 Q(NR)=DIS(I,L);DIS(I,L)=0                      ;Rem SET SUPPRESSED DISPL'T
800 Rem
```

```
810 If (NRE-NTW))>0 or NRW=0 then Goto 880
820 For II=1 to LB;For J=1 to LB
830 IA=SIZ-II+1;IB=LB-II+1
840 SK(IA,J)=SK(IB,J);Next;Next
850 NRW=NRW-BUF
855 Filepos(10)=filepos(10)-BUF*LB*5          ;Rem BACKSPACE K FILE
860 For II=1 to BUF;For J=1 to LB             ;Rem READ BACK FILED PARTS OF
870 Read £10,SK(II,J);Next;Next               ;Rem REDUCED K AS NEEDED
875 Filepos(10)=filepos(10)-BUF*LB*5          ;Rem BACKSPACE AGAIN - NOTE 5 IS
880 Rem                                       ;REM BYTES/NUMBER IN MEGABASIC
890 If (I+L-2)=0 then Goto 930                ;Rem END TEST
900 If I<>1 then Goto 710                     ;Rem END LOOP ON FREEDOMS/NODE
910 I=NDF+1
920 Goto 700                                  ;Rem END BACKSUB LOOP ON NODES
930 Rem
935 !;! "NODAL DISPLACEMENTS U,V,W";!
940 For N=1 to NP
950 ! %"I5",N,%"15E6",DIS(1,N),DIS(2,N),DIS(3,N)
960 Next
970 Return;Proc end
980 End


10 Rem STRESS
20 Def shared proc STRUB
30 Dim F(24),B(24,30),R(30),FORCE(100,4)
40 For N=1 to NE
50 For I=1 to 24;For J=1 to 30
60 Read £9,B(I,J);Next;Next                   ;Rem READ ELEMENT STRESS MATRIX
70 For I=1 to NCN;M=NOP(N,I)
80 If M=0 then Goto 120
90 K=(I-1)*NDF
100 For J=1 to NDF;IJ=J+K
110 R(IJ)=DIS(J,M);Next J                     ;Rem COLLECT ELEMENT DISPLACEMENTS
120 Next I
130 IA=K+NDF
140 For I=1 to 24;F(I)=0;For J=1 to IA
150 F(I)=F(I)+B(I,J)*R(J)                     ;Rem CALCULATE ELEMENT STRESSES
160 Next;Next
170 For II=1 to 6;NI=NOP(N,II)
180 IA=4*(II-1)+1;IB=IA+3
190 For I=1 to 4                              ;Rem CALCULATE AVERAGE NODAL STRESSES
200 FORCE(NI,I)=FORCE(NI,I)+F(IA+I-1)/FNE(NI)
210 Next;Next
220 Next N
225 !;! "Mx, My, Nx and Ny at each node - N.B. AVERAGED BETWEEN ELEMENTS"
226 !  " - note also that corner node values are taken from the proximate"
227 !  " - integration points as this gives better results";!
230 For N=1 to NP
240 ! %"I5",N,%"15E6",FORCE(N,1),FORCE(N,2),FORCE(N,3),FORCE(N,4)
250 Next
260 Return;Proc end
270 End
```

Note: In this simple program, many time-saving refinements are possible, for example:

(1) when a regular mesh re-uses the same two triangle ESMs only these two ESMs need be calculated to form the SSM (*element copying*);

(2) with large symmetric ESMs the below diagonal entries in *k* need not be calculated as they are not used in the assembly of *K* (see line 220 of SOLVE).

15.8 'FEMSS' programs for plane stress/strain and plate bending

The following section includes a program for plane stress and strain and for thin plate bending problems. Both use the same FEMSS (Finite Element Microcomputer Solution System) used in Section 15.7 and hence, except for trivial changes, the same solution routine.

The plane stress program uses the six-node isoparametric linear strain triangle, following the details given in Section 7.3 for element formulation, and reference to corresponding equation numbers in Chapters 6 and 7 is made within the program listing.

Both programs use the simple mesh generation scheme illustrated in Fig. 15.3 to discretize rectangular domains. As data this scheme requires simply the numbers NX, NY, XLIM, YLIM as data, where NX and NY are the numbers of nodes in the X- and Y-directions and XLIM and YLIM are the domain sizes in these directions. Then nodes and elements are automatically numbered first in the Y-direction and then in the X-direction, as illustrated in Fig. 15.3 for the six-node plane stress element. For the three-node plate element the same scheme is followed but, of course, the midside nodes are omitted.

Then in Fig. 15.3 the program's mesh generation will assign to the elements the node number sets:

Element 1   1, 7, 3, 4, 5, 2
Element 2   3, 7, 9, 5, 8, 6
Element 3   7, 13, 9, 10, 11, 8
Element 4   9, 13, 15, 11, 14, 12

for the six-node plane stress elements. Data input requirements for the plane stress program are:

Lines	Data
1	number of nodes (NP); number of elements (NE); number of boundary condition nodes (NB); number of property sets (NT); half bandwidth (NBW).
NT lines	E, v, ρ, t, p, as defined in Section 15.7, but note that in line 290 all elements are assigned the first property set. As noted in Section 15.7 this restriction is easily removed.
1	number of nodes in X-direction (NX); number of nodes in Y-direction (NY); domain size in X-direction (XLIM); domain size in Y-direction (YLIM).
NB lines	boundary condition node number (NBC(I)); boundary condition flags (NFIX(I)). For example, for both u and v suppressed (to zero) the boundary flag number is 11.
NL lines	NL = number of nodes at which concentrated loads are specified; node number (NQ); loads in X- and Y-directions (R(1), R(2)). This data is terminated by a line with three zeros.

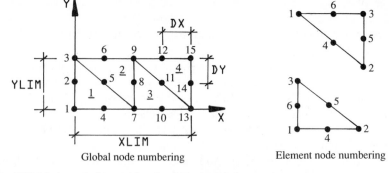

Global node numbering   Element node numbering

Fig. 15.3 Node and element numbering schemes for mesh generation in rectangular domains

The data given in lines 500–680 uses twenty-five six-node elements to analyse the plane stress problem of Fig. 6.3. Hence the data lines give the following information:

Line 500	25 nodes, 8 elements, 10 b.c. nodes, 1 property set, half bandwidth = 22 (for example, the node numbers given by the program for the first element will be 1, 11, 3, 6, 7, 2).
Line 510	$E = 20 \times 10^6$, $v = 0.2$, $\rho = 0$, $t = 0.5$, $p = 0$.
Line 520	NX = 5, NY = 5, XLIM = 2, YLIM = 1.
Line 530–620	$v = 0$ (simple support) at left-hand end, $u = 0$ (symmetry) at right-hand end in Fig. 6.3.
Line 630–670	line load split between nodes at right-hand end in Fig. 6.3.

Results from the program agree closely with those in Table 6.1. For example the vertical displacement at node 23 is − 7.849300, the shear stress at node 13 is 7.327, and the extreme fibre stresses at nodes 11 and 15 are 22.814 and 27.489.[†] (To compensate for the use of mesh diagonals all biased in one direction these last two results may be averaged). This need for averaging reminds us of the approximation involved in FEM whilst the slight difference in the displacement result at node 23 from that given in Table 6.1 arises only because of the different numerical precision used in the present program (eight-digit rather than seven).

[†] These are average nodal stresses, as in the program of Section 15.7, except that the vertex node values are taken at the nodes (not at the proximate integration points; the latter practice is only recommended for the quadratic shell element.

```
10 Include 'ELMAT','SOLVE','STRESS'        ;Rem SUBROUTINE 'MAIN' HERE INCLUDES
20 Access * from *                          ;Rem A SIMPLE MESH GENERATION CODE
30 Open £8,"B:EMATS"                        ;Rem OTHERWISE IT IS THE SAME AS FOR
40 Open £9,"B:ELSTR"                         ;Rem SEC. 15.7 EXCEPT FOR TRIVIAL
50 Open £10,"B:STIFM"                        ;Rem CHANGES, e.g. NDF IN LINE 110
60 Dim FNE(100),NBC(50),NFIX(50),Q(200),IMAT(100),CORD(100,2),PROP(20,5)
70 Dim NOP(100,6),CI(6,2),WF(6),R(5)
80 Def shared data NP,NE,NB,NT,NBW,NCN,NDF,FNE( )
90 Def shared data NBC( ),NFIX( ),Q( ),CI( ),WF( )
100 Def shared data IMAT( ),CORD( ),PROP( ),NOP( )
110 NDF=2;NCN=6
120 For I=1 to 3;CI(I,1)=0;CI(I,2)=0;WF(I)=0;CI(I+3,1)=0.5;CI(I+3,2)=0.5
130 WF(I+3)=1/6;Next               ;Rem INTEGRATION AT MIDSIDE NODES
140 CI(1,1)=1;CI(2,2)=1;CI(5,1)=0;CI(6,2)=0
150 Read NP,NE,NB,NT,NBW;NBW=NBW-NDF
160 For N=1 to NT;For I=1 to 5;Read PROP(N,I);Next;Next
170 Read NX,NY,XLIM,YLIM              ;Rem NX,NY ARE NO. NODES IN X,Y
180 NEX=NX-1;NEY=NY-1;RNX=NEX;RNY=NEY  ;Rem DIRECTIONS AND XLIM,YLIM ARE THE
190 DX=XLIM/RNX;DY=YLIM/RNY            ;Rem DOMAIN SIZES IN THOSE DIRECTIONS
200 For I=1 to NX;For J=1 to NY
210 RNDX=I-1;RNDY=J-1;NN=NY*(I-1)+J
220 CORD(NN,1)=RNDX*DX;CORD(NN,2)=RNDY*DY;Next;Next   ;Rem NODAL COORDINATES
230 NEX=(NX-1)/2;NEY=(NY-1)/2
240 For I=1 to NEX;For J=1 to NEY       ;Rem ELEMENT NODE NUMBERS ARE SET
250 NI=(I-1)*2*NY+(J-1)*2+1;NJ=NI+2*NY  ;Rem UP AS SHOWN IN FIG. 15.3
260 NS=NEY*(I-1)+J;NN=2*NS-1
270 NOP(NN,1)=NI;NOP(NN,2)=NJ;NOP(NN,3)=NI+2
280 NOP(NN,4)=NI+NY;NOP(NN,5)=NOP(NN,4)+1;NOP(NN,6)=NI+1
290 IMAT(NN)=1;NN=2*NS
300 NOP(NN,1)=NI+2;NOP(NN,2)=NJ;NOP(NN,3)=NJ+2
310 NOP(NN,4)=NOP(NN-1,5);NOP(NN,5)=NJ+1;NOP(NN,6)=NOP(NN-1,4)+2
320 IMAT(NN)=1;Next;Next               ;Rem ONLY ONE PROPERTY SET USED
330 For I=1 to NB;Read NBC(I),NFIX(I);Next
340 Read NQ,R(1),R(2)                    ;Rem READ POINT LOADS
350 If NQ=0 then Goto 380
360 For K=1 to NDF;IC=(NQ-1)*NDF+K;Q(IC)=Q(IC)+R(K);Next
370 Goto 340
380 Rem
385 For N=1 to NE;For II=1 to NCN;NN=NOP(N,II);FNE(NN)=FNE(NN)+1
387 Next;Next                          ;Rem 'NODAL VALENCY'
390 For N=1 to NE
400 ELSUB N;Next
410 Filepos(8)=0;Filepos(9)=0
415 SOLSUB
420 ! "INPUT 1 FOR STRESSES";Input FLAG  ;Rem TO INTERRUPT SCREEN OUTPUT ONLY
425 STRUB                                ;Rem WHEN PRINTER NOT USED FOR THIS
430 End
500 Data 25,8,10,1,22                    ;Rem DATA READ AS ABOVE
510 Data 20E6,0.2,0,0.5,0
520 Data 5,5,2,1
530 Data 1,01
540 Data 2,01
550 Data 3,01
560 Data 4,01
570 Data 5,01
580 Data 21,10
590 Data 22,10
600 Data 23,10
610 Data 24,10
620 Data 25,10
630 Data 21,0,-420
640 Data 22,0,-420
650 Data 23,0,-420
660 Data 24,0,-420
670 Data 25,0,-420
680 Data 0,0,0                           ;Rem ZERO LINE TO TERMINATE DATA
```

```
10 Rem ELMAT
20 Def shared proc ELSUB @N                    ;Rem ONLY ELEMENT NO. AS ARGUMENT
30 Dim S(12,12),C(18,12),D(3,3),B(3,12),XY(6,2)
40 Dim F(6),TEMP(3,12),TJ(2,2),DL(2,6),QF(12)
50 L=IMAT(N);E=PROP(L,1);P=PROP(L,2)           ;Rem COLLECT ELEMENT PROPERTIES
60 DENS=PROP(L,3);TH=PROP(L,4);UDL=PROP(L,5)
70 For M=1 to 6;K=NOP(N,M);XY(M,1)=CORD(K,1);XY(M,2)=CORD(K,2);Next
80 D(1,1)=E/(1-P*P);D(2,2)=D(1,1);D(1,2)=P*D(1,1);D(2,1)=D(1,2)
90 D(3,3)=0.5*(1-2*P)*D(1,1)/(1-P)             ;Rem MODULUS MATRIX
100 D(1,3)=0;D(2,3)=0;D(3,1)=0;D(3,2)=0
110 For II=1 to 6                              ;Rem START INTEGRATION LOOP £££££££££
120 F1=4*CI(II,1);F2=4*CI(II,2)
130 DL(1,1)=F1-1;DL(1,2)=0;DL(1,3)=F1+F2-3     ;Rem THE MATRIX OF EQN 7.15
140 DL(1,4)=F2;DL(1,5)=-F2;DL(1,6)=4-2*F1-F2
150 DL(2,1)=0;DL(2,2)=F2-1;DL(2,3)=F1+F2-3
160 DL(2,4)=F1;DL(2,5)=4-F1-2*F2;DL(2,6)=-F1
170 For I=1 to 2;For J=1 to 2;TJ(I,J)=0;For K=1 to 6
180 TJ(I,J)=TJ(I,J)+DL(I,K)*XY(K,J)            ;Rem JACOBIAN MATRIX, SEE EQN 7.16
190 Next;Next;Next
200 DJ=TJ(1,1)*TJ(2,2)-TJ(1,2)*TJ(2,1)
210 DD=TJ(1,1);TJ(1,1)=TJ(2,2)/DJ;TJ(2,2)=DD/DJ  ;Rem INVERT JACOBIAN MATRIX
220 TJ(1,2)=-TJ(1,2)/DJ;TJ(2,1)=-TJ(2,1)/DJ
230 For I=1 to 2;For J=1 to 6;TEMP(I,J)=0;For K=1 to 2
240 TEMP(I,J)=TEMP(I,J)+TJ(I,K)*DL(K,J);Next;Next;Next  ;Rem AS PER EQN 7.17
250 For J=1 to 6;JA=2*J-1
260 B(1,JA)=TEMP(1,J);B(3,JA)=TEMP(2,J)        ;Rem FILL STRAIN MATRIX AS
270 B(2,JA+1)=TEMP(2,J);B(3,JA+1)=TEMP(1,J);Next ;Rem PER EQN 6.30
280 For I=1 to 3;IA=3*(II-1)+I;For J=1 to 12;C(IA,J)=0
290 For K=1 to 3;C(IA,J)=C(IA,J)+D(I,K)*B(K,J)
300 Next;Next;Next   ;Rem STRESS MATRIX CONTRIBUTION FOR INTEGRATION POINT
310 If WF(II)=0 then Goto 390
320 For I=1 to 3;For J=1 to 12;TEMP(I,J)=0;For K=1 to 3
330 TEMP(I,J)=TEMP(I,J)+D(I,K)*B(K,J);Next;Next;Next
340 SMUL=WF(II)*TH*abs(DJ)
350 For I=1 to 12;For J=1 to 12;For K=1 to 3
360 S(I,J)=S(I,J)+B(K,I)*TEMP(K,J)*SMUL        ;Rem CONTRIBUTION TO k FOR THIS
370 Next;Next;Next                             ;Rem INTEGRATION POINT
380 QF(2*II)=QF(2*II)-DENS*SMUL                ;Rem ADD CONSISTENT LOADS TO Q
390 Next II                                    ;Rem END INTEGRATION LOOP £££££££££
400 For I=1 to 12;For J=1 to 12
410 Write £8,S(I,J);Next;Next                  ;Rem FILE ESMs
420 For I=1 to 18;For J=1 to 12
430 Write £9,C(I,J);Next;Next                  ;Rem FILE STRESS MATRICES
440 ! N                                        ;Rem PROGRESS REPORT ONLY
450 Return;Proc end
452 Rem Solution routine as in Sec. 15.7 with only necessary changes being
453 Rem dimensioning loops in lines 80 and 250 to 12. One should also
454 Rem change line 50 to, say, Dim ESM(12,12),SK(100,50),SKP(50) to handle
455 Rem larger problems more sensibly - the small dimensions of SK,SKP used
456 Rem in Sec. 15.7 are to test the 'buffering' of the solution routine.
460 End


10 Rem STRESS
20 Def shared proc STRUB
30 Dim F(18),B(18,12),R(12),FORCE(100,3)
40 For N=1 to NE
50 For I=1 to 18;For J=1 to 12
60 Read £9,B(I,J);Next;Next                    ;Rem READ ELEMENT STRESS MATRIX
70 For I=1 to NCN;M=NOP(N,I)
80 If M=0 then Goto 120
90 K=(I-1)*NDF
100 For J=1 to NDF;IJ=J+K
110 R(IJ)=DIS(J,M);Next J                      ;Rem COLLECT ELEMENT DISPLACEMENTS
120 Next I
130 IA=K+NDF
140 For I=1 to 18;F(I)=0;For J=1 to IA
150 F(I)=F(I)+B(I,J)*R(J)                      ;Rem CALCULATE ELEMENT STRESSES
160 Next;Next
```

```
170 For II=1 to 6;NI=NOP(N,II)
180 For I=1 to 3;IA=3*(II-1)+I
190 FORCE(NI,I)=FORCE(NI,I)+F(IA)/FNE(NI)   ;Rem AVERAGE NODAL STRESSES
200 Next;Next
210 Next N
215 !;! "NODAL AVERAGE DIRECT (X&Y) AND SHEAR STRESSES";!
220 For N=1 to NP
230 ! %"I5",N,%"15E6",FORCE(N,1),FORCE(N,2),FORCE(N,3)
240 Next
250 Return;Proc end
260 End
```

The program for plate bending uses the element presented in Section 8.10. This element uses basis transformation to transform the nine global freedoms (w, ϕ_x, and ϕ_y at each node) to twelve local freedoms (ϕ_x and ϕ_y at the vertices and midsides). As a result the kernel stiffness matrix k^* in eqn (8.62) is exactly that for the linear strain triangle used in the foregoing program. The plate bending element, however, cannot have curved sides and thus the isoparametric derivation of Section 7.3 is not used to form k^*. Instead, as noted in the program listing (line 270 of ELMAT), an extension of eqn (6.26) is used to transform from area coordinates to Cartesian derivatives.

The program uses the same mesh generation scheme illustrated in Fig. 15.3 except that the midside nodes are omitted. Data input requirements for the program are:

Lines	Data
1	number of nodes (NP); number of elements (NE); number of boundary conditions (NB); number of property sets (NT); half bandwidth (NBW).
NT lines	E, v, ρ, t, p, again all elements are assigned the same property set (line 320) to simplify data input.
1	number of nodes in X-direction (NX); number of nodes in Y-direction (NY); domain size in X-direction (XLIM); domain size in Y-direction (YLIM).
NB lines	boundary condition node number (NBC(I)); boundary condition flags (NFIX(I)). For example, for w, ϕ_x, ϕ_y all suppressed to zero the boundary flag number is 111.
NL lines	NL = number of point loaded nodes; node number (NQ); loads corresponding to w, ϕ_x, and ϕ_y. As in the two preceding programs this data is terminated by a data line of zeros.

```
10 Include 'ELMAT','SOLVE','STRESS'          ;Rem SUBROUTINE 'MAIN' HERE INCLUDES
20 Access * from *                           ;Rem A SIMPLE MESH GENERATION CODE
30 Open £8,"B:EMATS"                          ;Rem OTHERWISE IT IS THE SAME AS FOR
40 Open £9,"B:ELSTR"                          ;Rem SEC. 15.7 EXCEPT FOR TRIVIAL
50 Open £10,"B:STIFM"                         ;Rem CHANGES, e.g. NDF IN LINE 110
60 Dim FNE(100),NBC(50),NFIX(50),Q(300),IMAT(100),CORD(100,2),PROP(20,5)
70 Dim NOP(100,6),CI(6,2),WF(6),R(5)
80 Def shared data NP,NE,NB,NT,NBW,NCN,NDF,FNE( )
90 Def shared data NBC( ),NFIX( ),Q( ),CI( ),WF( )
100 Def shared data IMAT( ),CORD( ),PROP( ),NOP( )
110 NDF=3;NCN=3
120 For I=1 to 3;CI(I,1)=1/6;CI(I,2)=1/6;WF(I)=1/6;CI(I+3,1)=0;CI(I+3,2)=0
130 WF(I+3)=0;Next                            ;Rem INTEGRATION OF EQN 5.136
140 CI(1,1)=4/6;CI(2,2)=4/6;CI(4,1)=1;CI(5,2)=1
150 Read NP,NE,NB,NT,NBW;NBW=NBW-NDF
160 For N=1 to NT;For I=1 to 5;Read PROP(N,I);Next;Next
170 Read NX,NY,XLIM,YLIM                      ;Rem NX,NY ARE NO. NODES IN X,Y
180 NEX=NX-1;NEY=NY-1;RNX=NEX;RNY=NEY         ;Rem DIRECTIONS AND XLIM,YLIM ARE THE
190 DX=XLIM/RNX;DY=YLIM/RNY                   ;Rem DOMAIN SIZES IN THOSE DIRECTIONS
200 For I=1 to NX;For J=1 to NY
210 RNDX=I-1;RNDY=J-1;NN=NY*(I-1)+J
220 CORD(NN,1)=RNDX*DX;CORD(NN,2)=RNDY*DY;Next;Next    ;Rem NODAL COORDINATES
240 For I=1 to NEX;For J=1 to NEY             ;Rem ELEMENT NODE NUMBERS ARE SET UP
250 NI=(I-1)*NY+J;NJ=NY*I+J                   ;Rem IN SAME FASHION AS IN FIG. 15.3
260 NS=NEY*(I-1)+J;NN=2*NS-1                  ;Rem BUT OMITTING MIDSIDE NODES
270 NOP(NN,1)=NI;NOP(NN,2)=NJ;NOP(NN,3)=NI+1
290 IMAT(NN)=1;NN=2*NS
300 NOP(NN,1)=NI+1;NOP(NN,2)=NJ;NOP(NN,3)=NJ+1
320 IMAT(NN)=1;Next;Next                      ;Rem ONLY ONE PROPERTY SET USED
330 For I=1 to NB;Read NBC(I),NFIX(I);Next
340 Read NQ,R(1),R(2),R(3)                    ;Rem READ POINT LOADS
350 If NQ=0 then Goto 380
360 For K=1 to NDF;IC=(NQ-1)*NDF+K;Q(IC)=Q(IC)+R(K);Next
370 Goto 340
380 Rem
385 For N=1 to NE;For II=1 to NCN;NN=NOP(N,II);FNE(NN)=FNE(NN)+1
387 Next;Next                                 ;Rem 'NODAL VALENCY'
390 For N=1 to NE
400 ELSUB N;Next
410 Filepos(8)=0;Filepos(9)=0
415 SOLSUB
420 ! "INPUT 1 FOR STRESSES";Input FLAG      ;Rem TO INTERRUPT SCREEN OUTPUT ONLY
425 STRUB                                     ;Rem WHEN PRINTER NOT USED FOR THIS
430 End
1000 Data 9,8,8,1,12                          ;Rem DATA READ AS ABOVE.
1010 Data 12,0,0,1,0
1020 Data 3,3,0.5,0.5
1030 Data 1,111
1040 Data 2,101
1050 Data 3,101
1060 Data 4,110
1070 Data 6,001
1080 Data 7,110
1090 Data 8,010
1100 Data 9,011
1110 Data 9,0.25,0,0
1120 Data 0,0,0,0                             ;Rem ZERO LINE TO TERMINATE DATA
```

```
10 Rem ELMAT
20 Def shared proc ELSUB @N                    ;Rem ONLY ELEMENT NO. AS ARGUMENT
30 Dim S(12,12),C(9,12),D(3,3),B(3,12),XY(6,2),CT(3,2)
40 Dim TEMP(12,12),T(12,9),V(3,6),Z(2,3),SS(9,9),CC(9,9)
50 L=IMAT(N);E=PROP(L,1);P=PROP(L,2)          ;Rem COLLECT ELEMENT PROPERTIES
60 DENS=PROP(L,3);TH=PROP(L,4);UDL=PROP(L,5)
70 For M=1 to 3;K=NOP(N,M);XY(M,1)=CORD(K,1);XY(M,2)=CORD(K,2);Next
80 D(1,1)=E*TH^3/(12*(1-P*P));D(2,2)=D(1,1);D(1,2)=P*D(1,1);D(2,1)=D(1,2)
90 D(3,3)=0.5*(1-P)*D(1,1)                     ;Rem MODULUS MATRIX
100 D(1,3)=0;D(2,3)=0;D(3,1)=0;D(3,2)=0
110 X21=XY(2,1)-XY(1,1);X32=XY(3,1)-XY(2,1);X13=XY(1,1)-XY(3,1)
120 Y21=XY(2,2)-XY(1,2);Y32=XY(3,2)-XY(2,2);Y13=XY(1,2)-XY(3,2)
130 S21=sqrt(X21*X21+Y21*Y21)
140 S32=sqrt(X32*X32+Y32*Y32)                  ;Rem ELEMENT SIDE LENGTHS
150 S13=sqrt(X13*X13+Y13*Y13)
160 CT(1,1)=X21/S21;CT(1,2)=Y21/S21
170 CT(2,1)=X32/S32;CT(2,2)=Y32/S32            ;Rem DIRECTIONS COSINES OF SIDES
180 CT(3,1)=X13/S13;CT(3,2)=Y13/S13
190 A=X21*Y32-X32*Y21                          ;Rem A=TWICE ELEMENT AREA
200 For I=1 to 3;NF=NOP(N,I)*NDF-2
210 Q(NF)=Q(NF)+A*UDL/6;Next                   ;Rem ADD CONSISTENT LOADS TO Q
220 Z(1,1)=-Y32/A;Z(1,2)=-Y13/A;Z(1,3)=-Y21/A  ;Rem EXTENDED FORM OF EQN 6.28
230 Z(2,1)=X32/A;Z(2,2)=X13/A;Z(2,3)=X21/A     ;Rem WHEN L3 NOT ELIMINATED
240 For II=1 to 6      ;Rem COMMENCE INTEGRATION LOOP ££££££££££££££££££££££££££££
250 F1=4*CI(II,1);F2=4*CI(II,2);F3=4-F1-F2
260 V(1,1)=F1-1;V(1,4)=F2;V(1,6)=F3
270 V(2,2)=F2-1;V(2,4)=F1;V(2,5)=F3            ;Rem EXTENDED FORM OF EQN 6.26
280 V(3,3)=F3-1;V(3,5)=F2;V(3,6)=F1            ;Rem WHEN L3 NOT ELIMINATED
290 For I=1 to 2;For J=1 to 6;TEMP(I,J)=0;For K=1 to 3
300 TEMP(I,J)=TEMP(I,J)+Z(I,K)*V(K,J);Next;Next;Next
310 For J=1 to 6;JA=2*J-1
320 B(1,JA)=TEMP(1,J);B(2,JA)=TEMP(2,J)        ;Rem FILL STRAIN MATRIX AS
330 B(3,JA)=TEMP(2,J);B(3,JA+1)=TEMP(1,J);Next ;Rem PER EQN 6.30
340 If II<4 then Goto 370
350 For I=1 to 3;IA=3*(II-4)+I;For J=1 to 12;C(IA,J)=0;For K=1 to 3
360 C(IA,J)=C(IA,J)+D(I,K)*B(K,J);Next;Next;Next
370 Rem ***** ABOVE GIVES STRESS MATRIX FOR NODAL STRESSES
380 If WF(II)=0 then Goto 440
390 For I=1 to 3;For J=1 to 12;TEMP(I,J)=0;For K=1 to 3
400 TEMP(I,J)=TEMP(I,J)+D(I,K)*B(K,J);Next;Next;Next
410 SMUL=WF(II)*A
420 For I=1 to 12;For J=1 to 12;For K=1 to 3
430 S(I,J)=S(I,J)+B(K,I)*TEMP(K,J)*SMUL;Next;Next;Next
440 Next II         ;Rem END INTEGRATION LOOP ££££££££££££££££££££££££££££££££££££
450 T(1,2)=1;T(2,3)=1;T(3,5)=1;T(4,6)=1;T(5,8)=1;T(6,9)=1
455 Rem ***** NOW FILL BASIS TRANSFORMATION MATRIX OF EQN 8.58
460 T(7,1)=-1.5*X21/S21^2;T(7,4)=-T(7,1)
470 T(8,1)=-1.5*Y21/S21^2;T(8,4)=-T(8,1)
480 T(7,2)=0.5*CT(1,2)^2-0.25*CT(1,1)^2;T(7,5)=T(7,2)
490 T(7,3)=-0.75*CT(1,1)*CT(1,2);T(7,6)=T(7,3);T(8,2)=T(7,3)
500 T(8,5)=T(7,3);T(8,3)=0.5*CT(1,1)^2-0.25*CT(1,2)^2;T(8,6)=T(8,3)
510 T(9,4)=-1.5*X32/S32^2;T(9,7)=-T(9,4)
520 T(10,4)=-1.5*Y32/S32^2;T(10,7)=-T(10,4)
530 T(9,5)=0.5*CT(2,2)^2-0.25*CT(2,1)^2;T(9,8)=T(9,5)
540 T(9,6)=-0.75*CT(2,1)*CT(2,2);T(9,9)=T(9,6);T(10,5)=T(9,6)
550 T(10,8)=T(9,6);T(10,6)=0.5*CT(2,1)^2-0.25*CT(2,2)^2;T(10,9)=T(10,6)
560 T(11,1)=1.5*X13/S13^2;T(11,7)=-T(11,1)
570 T(12,1)=1.5*Y13/S13^2;T(12,7)=-T(12,1)
580 T(11,2)=0.5*CT(3,2)^2-0.25*CT(3,1)^2;T(11,8)=T(11,2)
590 T(11,3)=-0.75*CT(3,1)*CT(3,2);T(11,9)=T(11,3);T(12,2)=T(11,3)
600 T(12,8)=T(11,3);T(12,3)=0.5*CT(3,1)^2-0.25*CT(3,2)^2;T(12,9)=T(12,3)
610 For I=1 to 12;For J=1 to 9;TEMP(I,J)=0;For K=1 to 12
620 TEMP(I,J)=TEMP(I,J)+S(I,K)*T(K,J);Next;Next;Next
630 For I=1 to 9;For J=1 to 9;SS(I,J)=0;For K=1 to 12
640 SS(I,J)=SS(I,J)+T(K,I)*TEMP(K,J);Next;Next;Next ;Rem FINAL STIFFNESS MATRIX
650 For I=1 to 9;For J=1 to 9;CC(I,J)=0;For K=1 to 12
660 CC(I,J)=CC(I,J)+C(I,K)*T(K,J);Next;Next;Next    ;Rem FINAL STRESS MATRIX
670 For I=1 to 9;For J=1 to 9
680 Write £8,SS(I,J);Next;Next                       ;Rem FILE THE ESMs
```

```
690 For I=1 to 9;For J=1 to 9
700 Write £9,CC(I,J);Next;Next                  ;Rem FILE STRESS MATRICES
710 ! N                                         ;Rem PROGRESS REPORT ONLY
720 Return;Proc end
722 Rem Solution routine as in Sec. 15.7 with only necessary changes being
723 Rem dimensioning loops in lines 80 and 250 to 9.  One should also
724 Rem change line 50 to, say, Dim ESM(9,9),SK(90,60),SKP(60) to handle
725 Rem larger problems more sensibly - the small dimensions of SK,SKP used
726 Rem in Sec. 15.7 are to test the 'buffering' of the solution routine.
727 Rem Note that line 60 of SOLVE must then have SIZ=90 and LB=60.
730 End


10 Rem STRESS
20 Def shared proc STRUB
30 Dim F(9),B(9,9),R(9),FORCE(100,3)
40 For N=1 to NE
50 For I=1 to 9;For J=1 to 9
60 Read £9,B(I,J);Next;Next                     ;Rem READ ELEMENT STRESS MATRIX
70 For I=1 to NCN;M=NOP(N,I)
80 If M=0 then Goto 120
90 K=(I-1)*NDF
100 For J=1 to NDF;IJ=J+K
110 R(IJ)=DIS(J,M);Next J                       ;Rem COLLECT ELEMENT DISPLACEMENTS
120 Next I
130 IA=K+NDF
140 For I=1 to 9;F(I)=0;For J=1 to IA
150 F(I)=F(I)+B(I,J)*R(J)                        ;Rem CALCULATE ELEMENT STRESSES
160 Next;Next
170 For II=1 to 3;NI=NOP(N,II)
180 For I=1 to 3;IA=3*(II-1)+I
190 FORCE(NI,I)=FORCE(NI,I)+F(IA)/FNE(NI)   ;Rem AVERAGE NODAL STRESSES
200 Next;Next
210 Next N
215 !;! "NODAL AVERAGE DIRECT (X&Y) AND TWIST MOMENTS";!
220 For N=1 to NP
230 ! %"I5",N,%"15E6",FORCE(N,1),FORCE(N,2),FORCE(N,3)
240 Next
250 Return;Proc end
260 End
```

The data given in lines 1000–1120 uses eight elements to analyse a quadrant of a centrally loaded simply supported square plate. For this the boundary conditions are those shown in Fig. 8.9 with the curvature conditions omitted. Hence the data lines give the following informaton:

Line 1000   9 nodes, 8 elements, 8 b.c. nodes, 1 property set, half bandwidth $= 12$ (for example, the node numbers given by the program to the first element will be 1, 4, 2).

Line 1010 $E = 12$, $v = 0$, $\rho = 0$, $t = 1$, $p = 0$ (that is, zero UDL).

Line 1020   NX $= 3$, NY $= 3$, XLIM $= 0.5$, YLIM $= 0.5$ (that is 1×1 plate).

Line 1030–1100   For example ϕ_x and $\phi_y = 0$ at the plate centre.

Line 1110   Quarter of unit load applied to the quadrant of the plate.

The result for the deflection under the load is $w_9 = 0.01272824$ (the exact solution is 0.0116, and the result in reference 48 of Chapter 8 of 0.01273). The reader should find the results for this and other problems in agreement with those of Figs 8.11–8.14 and reference 48 of Chapter 8.

15.9 On the choice of element, extrapolation, and singularities

In choosing a particular element many considerations must be taken into account but a balance between economy and accuracy seems the best course. In earlier chapters the complexity of some higher-order elements such as the quintic was apparent and even higher-order elements exist, for example a C_2 triangular element which requires a ninth-order interpolation and fifty-five freedoms.[39]

In practice, however, a trade-off between accuracy and economy is desirable. Compare, for example, the quadratic shell element used in Section 15.7 (30 df) with a full quintic thin shell element (63 df) and assume these respectively require eighty-one and twenty-five nodes to model a problem with reasonable accuracy. If the quintic element is exactly integrated (6×6 Gaussian quadrature) we require 96 and 1152 integration points, respectively, in the quintic case with perhaps four times as much computation at each!

The foregoing argument is slightly artificial as reduced integration could be used in the quintic but it does suggest that more compact elements will remain more popular, particularly if their accuracy is able to be improved by means such as basis transformation. Moreover, with the use of extrapolation, which the results of Section 3.9 suggest will generally be possible with displacement elements, the results obtained with crude finite element models can be improved.

As an example of extrapolation let us consider singularity problems. Suppose we analyse a domain containing a singularity with two meshes with nodal spacings h_i and h_j ($h_j < h_i$) and obtain, respectively, stresses σ_{1i} and σ_{1j} at distance r_1 and stresses σ_{2i} and σ_{2j} at distance r_2 from the singularity. Then using eqn (3.60) the extrapolated solutions for these two radii are

$$\sigma_1 = \sigma_{1j} + \frac{\sigma_{1j} - \sigma_{1i}}{(h_i/h_j)^N - 1}$$

$$\sigma_2 = \sigma_{2j} + \frac{\sigma_{2j} - \sigma_{2i}}{(h_i/h_j)^N - 1}.$$

If the singularity is of the r^n type (that is, the stress distribution near the crack is given by $\sigma = \lambda r^n$ where λ is a *stress concentration factor*) and $r_1 < r_2$ one can write

$$\frac{\sigma_1}{\sigma_2} = \left(\frac{r_1}{r_2}\right)^n. \tag{15.25}$$

Taking the logarithm of both sides of eqn (15.25) the nature of the power law singularity is revealed:

$$n = \frac{\log(\sigma_1/\sigma_2)}{\log(r_1/r_2)}. \tag{15.26}$$

Once the nature of the singularity has been determined from eqn (15.26) the stress concentration factor can be determined from

$$\lambda = \tfrac{1}{2}\left(\frac{\sigma_1}{r_1^n} + \frac{\sigma_2}{r_2^n}\right). \tag{15.27}$$

An example of such a singularity occurs at the corner of a square hole at the centre of a square plate carrying a uniformly distributed load. Here there is a moment singularity of the $r^{-1/3}$ type. Using constant moment triangular elements around the hole (and rectangular elements elsewhere) and the foregoing extrapolation technique, Mohr and Medland obtain satisfactory results. The stress solutions obtained near the hole corner in this work are shown in Fig. 7.12.[40]

Note also that one can verify that there is indeed a singularity in the domain by applying eqn (3.62) to the stress solutions at the singularity (that is, at $r = 0$) and determining that $N = 0$ is the only value that yields $\alpha = 1$ (and therefore we must have r_1 and r_2 non-zero in the above procedure).

15.10 Minimum integrations, reduced integration

Observing eqn (5.112) the accuracy of a Gauss quadrature is easily estimated and compared to the required accuracy, which for an element with an interpolation of order p and requiring mth-order derivatives is $2(p - m)$, to integrate the stiffness matrix exactly. Table 15.2 shows examples of such exact integration for membrane and thin plate elements which bear this out and indeed higher-order integrations will not lead to improvement in results. On the contrary they may increase round-off errors.

The question of the minimum allowable integration is of greater interest because of the success of reduced integration in multivariate elements and in

Table 15.2 Exact numerical integrations with n points

Shape	df	m	p	$2(p - m)$	n	Accuracy	Reference
Rectangle	8	1	2	2	4	3	Section 6.4
Triangle	12	1	2	2	3*	2	Section 6.4
Rectangle	12	2	3	2	4	3	Section 8.2
Triangle	9	2	3	2	3*	2	Section 8.4
Triangle	12	2	3	2	3*	2	Razzaque[41]
Triangle	15	2	4	4	7*	5	Section 8.8
Triangle	18	2	5	6	16	7	Section 8.9

* Triangular integration

coupled problems.[42, 43, 44] Irons[42] proposed that for isoparametric elements the minimum integration is that which computes $|J|$ exactly. As Strang and Fix point out, this bears out for the CST element (where a one-point integration is used),[45] but the bilinear plane stress element is a better example. Although selective, one-point integration for the shear is helpful, but at least a 2×2 rule must be used for the extensional energy or the rank of the stiffness matrix will be less than eight. Irons' proposal holds here as, observing eqns (7.5) and (7.12), $|J|$ will involve quadratic terms.[27] In general, though, in $m = 1$ and $m = 2$ elements, examination of $2(p - m)$, using for p the length of the complete part of the polynomial, will suffice. The biquadratic plane stress element is an example. In this, four-point Gauss integration gives no significant loss in accuracy[44] and the same holds true in the bicubic plate element.

Table 15.3 summarizes some successful examples of reduced integration in multivariate, shell, and thick plate elements. Most of these cases are shell and thick plate elements in which mixed orders of differentiaton are involved (for example, $\varepsilon = \partial u/\partial x + w/R$, $\gamma = \partial w/\partial x - \phi$) and these are shown as M/N. Generally it is safe policy to use the integration of lowest accuracy predicted by the calculation $2(p - m)$ (for example, exact integration of the flexure in thick plates). However, in case 3 reduced integration of both flexure and shear is successful because the element is biquadratic, that is, the product interpolation involves 'trailing' high-order terms which need not be exactly integrated.

The fourth and fifth cases are thick shell elements in which penalty factors suppress the shear so that primarily reduced integration is approximating the strain energy of the radial strains (the w/R strain components). These cases are discussed in Section 11.1, 11.6, and 15.11.

The sixth case is an axisymmetric element in which all terms but those arising from the radial w/r strain are exactly integrated, and here only the reciprocal is approximated (whereas in case 4 terms arising from both w and r are approximated).

The seventh case is the triquadratic brick, in which reduced integration leads to a pronounced improvement in results,[27] and the last case is the classical case of the bilinear rectangle. As with other 'product interpolation' elements, multiplying the $1 - D$ interpolations causes a problem, leaving 'trailing' terms (for example eqn (5.39)). This is solved by not integrating these or some terms in the complete polynomial part exactly. In part, though, the phenomenon of *parasitic shear* is one of coupling (between the u and v freedoms through the shear stiffness) and does not occur in biharmonic multivariate elements. Thus, in the latter, reduced integration of trailing terms does not impede convergence but neither does it accelerate it, whereas in some extensional elements reduced integrations accelerate convergence by 'softening' the coupling. This is further discussed in the following section.

Table 15.3 Reduced integrations in 'mixed mode' finite elements. Integrations quoted are those found most successful

Type	Element df	Mode	m	p	$2(p-m)$	Integration Accuracy	Points	Reference
1. Plate	18	flexure	1	2	2	2	3	Section 9.3
		shear	1/0	2/2	2/4	2	3	
2. Plate	12	flexure	1	2	2	2	4	Pugh et al.[43]
		shear	1/0	2/2	2/4	0	1	
3. Plate	27	flexure	1	3	4	3	4	Pugh et al.[43]
		shear	1/0	3/3	4/6	3	4	
4. Shell	9	flexure	1	2	2	3	2	Section 11.3
		shear	1/0	2/2	2/4	3	2	
		membrane	1/0	2/2	2/4	3	2	
5. Shell	30	flexure	1	2	2	2	3	Section 11.6
		shear	1/0	2/2	2/4	2	3	
		membrane	1/0	2/2	2/4	2	3	
6, Shell	6	flexure	2	3	2	7	4	Section 10.1
		membrane	1/0	1/3	0/6	7	4	
7. Brick	60	3-D.	1	3	4	3	8	Hellen[44]
8. Membrane	8	direct	1	2	2	3	4	Section 6.4
		shear	1	1	0	0	1	

15.11 Integration in coupled systems

Reduced integration is a puzzling phenomenon: to this point it has been concluded (above) that a softening of the coupling occurs, and the effect of this must outweigh the effects of eliminating part of the stiffness matrix if the rate of convergence is to increase. Until now we have observed reduced integration in three situations:

(1) Extensional elements with

$$\varepsilon = \partial u/\partial y + \partial v/\partial x \qquad (15.28)$$

(2) Thick plate elements with

$$\gamma = \partial w/\partial x - \phi \qquad (15.29)$$

(3) Curved shell elements with

$$\varepsilon = \partial u/\partial x + w/R \qquad (15.30)$$

and in these one observes softening mechanisms respectively in eqns (6.48), (9.29) and (11.12). In each case diagonal entries in the stiffness matrix associated with the coupling effects of shear or 'radial' strains (of the form w/R) are reduced and the same phenomenon has been observed in three-dimensional elements.[25]

That reduction in trace(k) might accelerate convergence can be imagined by those familiar with the pronounced effect of lumped elastic boundary conditions applied by adding stiffnesses to the diagonal of the structure matrix. Whilst this reduced trace(k) phenomenon provides a useful empirical guide, however, a more formal explanation for it is obtained by a penalty factor argument. Considering the bilinear plane stress element, for example, we can write the fully integrated stress matrix as $k = k_e + k_s$ where k_e is for the extensional strains and k_s for the shear strains.

With one-point integration for k_s this reduces to k_s^* and the equivalent penalty factor statement is

$$k = k_e + k_s^* + \beta(k_s - k_s^*) \qquad (15.31)$$

that is, the penalty factor constrains out part of the shearing stiffness. Noting that k_s is obtained by using $\varepsilon_{xy} = \partial u/\partial y + \partial v/\partial x$ we may interpret this mathematically as a *constraint* coupling the freedoms u and v. In eqn (15.31), therefore, we see that this constraint is relaxed, thereby accelerating convergence.

Indeed we meet similar phenomenon in Section 19.5 where a continuity constraint $\partial u/\partial x + \partial v/\partial y = 0$ is relaxed by using 'mixed interpolation' for viscous flow elements. Here the objective is to eliminate spurious pressure solutions associated with the constraint but observing the shear stress solutions given in Table 6.1 (particularly those for the 12 df triangle) we see that similar difficulties occur in solid mechanics.

Finally it should be noted that successful reduced integration requires the use of suitable approximating points. At present the Gauss points appear to be the best points in line[36] and rectangular elements but the 'triangular integration points', for which there were two choices in Section 9.3, provide an excellent example of the need to choose the best approximating points. In the thick plate element case (Section 9.3) and also in the case of the quadratic shell element (Section 11.6), however, the choice was obvious, though only in retrospect, that is, integration at internal points as in the Gauss rules is preferable to integration at the element boundary by, for example, the Lobatto rules.[20]

References

1. Walsh, P. F. (1978). Intensive finite element grading for stress concentrations. *Engng Fract. Mech.*, **10**, 211.
2. Pedersen, P. (1973). Some properties of linear strain triangles and optimal finite element models. *Int. J. Num. Meth. Engng*, **7**, 415.
3. Meek, J. L. (1971). *Matrix Structural Analysis*. McGraw-Hill, New York.
4. Mohr, G. A. (1979). Displacement of reinforcement in perforated slabs. *Trans. I. E. Aust.*, **CE21** (1). 21.
5. Ingraffea, A. R. (1977). Nodal grafting for crack propagation studies. *Int. J. Num. Meth. Engng*, **11**, 1185.
6. Cheng, R. T. and Li, C. (1973). On the solution of transient free-surface flow problems in porous media by the finite element method. *J. Hydrol.*, **20**, 49.
7. Mohr, G. A. (1979). Design of shell shape using finite elements. *Comput. Struct.*, **10**, 745.
8. Parks, D. M. (1974). A stiffness derivative finite element method for determination of elastic crack tip stress intensity factors. *Int. J. Fract.*, **10**, 487.
9. Przemieniecki, J. S. (1968). *Theory of Matrix Structural Analysis*. McGraw-Hill, New York.
10. Rust, B. W. and Burrus, J. R. (1972). *Mathematical Programming and the Numerical Solution of Linear Equations*. Elsevier, New York.
11. Argyris, J. H. (1965). Continua and Discontinua. Proceedings of the First Conference on Matrix Methods in Structural Mechanics, Wright-Patterson AFB, Ohio.
12. Cheung, Y. K. and King, I. P. (1971). Computer programs and computer methods, In *The Finite Element Method in Engineering Science*, (2nd edn) (ed. O. C. Zienkiewicz). McGraw-Hill, London.
13. Jennings, A. (1977). *Matrix Computation for Engineers and Scientists*. Wiley, Chichester.
14. Martin, H. and Carey, G. F. (1973). *Finite Element Analysis – Theory and Application*. McGraw-Hill, New York.

15 Young, D. M. (1975). Iterative solution of linear systems arising from finite element techniques. In *MAFELAP II*, (ed. J. R. Whiteman). Academic Press, London.
16. Parekh, C. J. (1969). *Finite Element Solution System*. Ph. D. thesis, University of Wales, Swansea.
17. Sabir, A. B. (1976). The nodal solution routine. In *Finite Elements for Thin Shells and Curved Members*, (ed. D. G. Ashwell and R. H. Gallagher). Wiley, London.
18. Irons, B. M. (1970). A frontal solution program for finite element analysis. *Int. J. Num. Meth. Engng*, **2**, 5.
19. Atkin, J. E. and Pardue, R. M. (1975). Element ordering for frontal solution techniques. In *MAFELAP II*, (ed. J. R. Whiteman). Academic Press, London.
20. Irons, B. M. and Ahmad, S. (1980). *Techniques of Finite Elements*. Ellis-Horwood, Chichester.
21. Norrie, D. H. and de Vries, G. (1978). *An Introduction to Finite Element Analysis*. Academic Press, New York.
22. Argyris, J. H. and Bronlund, O. E. (1975). The natural factor formulation of the stiffness for the matrix displacement method. *Comp. Meth. Appl. Mech. Engng*, **5**, 97.
23. Johnsen, T. L. and Roy, J. R. (1974). On systems of linear equations of the form $A^t Ax = b$, error analysis and certain consequences for structural applications. *Comp. Meth. Appl. Mech. Engng*, **5**, 357.
24. Martin, C. W. and Harrold, A. J. (1975). Removal of truncation error in finite element analysis. In *MAFELAP II*, (ed. J. R. Whiteman). Academic Press, London.
25. Gallagher, R. H. (1975). *Finite Element Analysis Fundamentals*. Prentice-Hall, Englewood Cliffs, NJ.
26. Gordon, W. J. (1971). Blending function methods of bivariate and multivariate interpolation and approximation. *SIAM J. Num. Anal.*, **8**, 158.
27. Mitchell, A. R. and Wait, R. (1977). *The Finite Element Method in Partial Differential Equations*. Wiley, London.
28. Gordon, W. J. and Hall, C. A. (1973). Construction of curvilinear coordinate systems and application to mesh generation. *Int. J. Num. Meth. Engng*, **7**, 461.
29. Muspratt, M. A. (1972). Stochastic plastic analysis of a shell. *Int. J. Num. Meth. Engng*, **4**, 345.
30. Muspratt, M. A. (1972). Shakedown of steel plates. *ASME J. Appl. Mech.* **38**, 1088.
31. Chio, F. (1853). *Mémoire sur les fonctions connues sous le nom de résultantes ou de determinants*. Turin.
32. Gallagher, R. H. (1977). Accuracy of data input in stress calculations. In *Finite Elements in Geomechanics*, (ed. G. Güdehus). Wiley, London.
33. Ashwell, D. G. and Sabir, A. B. (1972). A new cylindrical shell finite element based in simple independent strain functions. *Int. J. Mech. Sci.*, **14**, 171.
34. Mohr, G. A. (1980). Numerically integrated triangular element for doubly curved thin shells. *Comput. Struct.*, **11**, 565.
35. Zienkiewicz, O. C. (1977). *The Finite Element Method*, (3rd edn). McGraw-Hill, London.
36. Mohr, G. A. and Garner, R. (1983). Reduced integration and penalty factors in an arch element. *Int. J. Struct. (INDIA)*, **3** (1), 9.

37. Oden, J. T. and Reddy, J. N. (1976). *An Introduction to the Mathematical Theory of Finite Elements.* Wiley, New York.
38. Mohr, G. A. (1982). On displacement finite elements for the analysis of shells. *Proceedings of the Fourth Australian International Conference on Finite Element Methods,* University of Melbourne.
39. Argyris, J. H. and Dunne, P. C. (1976). The finite element method applied to fluid mechanics. *Proceedings of a Conference on Computer Methods and Problems in Aeronautical Fluid Dynamics,* University of Manchester, 1974. Academic Press, London.
40. Mohr, G. A. and Medland, I. C. (1983). On convergence of displacement finite elements with an application to singularity problems. *Engng Fract. Mech.,* **17,** 481.
41. Razzaque, A. (1973). Program for triangular bending elements with derivative smoothing. *Int. J. Num. Meth. Engng,* **6,** 333.
42. Irons, B. M. (1971). Quadrature rules for brick based finite elements. *J. Amer. Inst. Aeron. Astron.,* **9,** 293.
43. Pugh, E. D. L., Hinton, E., and Zienkiewicz, O. C. (1978). A study of quadrilateral plate bending elements with reduced integration. *Int. J. Num. Meth. Engng,* **12,** 1059.
44. Hellen, T. K. (1976). Numerical integration considerations in two and three dimensional finite elements. In *MAFELAP II,* Vol. 2. Academic Press, London.
45. Strang, G. and Fix, G. J. (1973). *An Analysis of the Finite Element Method.* Prentice-Hall, Englewood Cliffs, NJ.

Part IV
Finite elements in fluid flow and other field problems

16
Basic differential equations
in continua

Previous chapters concentrated upon formulation of finite element models based upon energy or equivalent virtual work arguments and applied these to a wide range of problems in the mechanics of solids. In the mechanics of fluids and many other field problems, however, energy arguments are more difficult to construct; instead a governing differential equation is usually used as the starting point for the analysis. The present chapter introduces some of the basic differential equations that govern the behaviour of physical systems as a prelude to Chapter 17 in which techniques for attacking these are introduced. Later chapters apply these techniques to a wide range of fluid and field problems.

16.1 Introduction

In Section 1.1 we noted that physical systems must satisfy constitutive conservation, and continuity requirements. In our applications of finite elements to the mechanics of solids in previous chapters we satisfied these requirements at element level by:

(1) in plane stress, for example, using relationships of the form $\sigma_x = E(\varepsilon_x + v\varepsilon_y)/(1 - v^2)$;

(2) using the theorem of minimum potential energy or, equivalently, setting the change in energy cased by a virtual perturbation in $\{d\}$ to zero, thus energy was conserved and corresponding equilibrium equations were obtained;

(3) in plane stress, for example, using assumed displacement fields and strain–displacement relationships which satisfied the strain compatibility condition (see eqn (16.3)).

We also ensured that elements had some minimum level of interelement displacement continuity and assumed at least partial cancellation of the interelement reactions at the nodes when the element equations were summed.

In the mechanics of fluids we again have to satisfy these three require-ments but the equations involved generally take different forms, such as the following.

1. In viscous flow we base the constitutive relationships upon Newton's law of viscosity, that is, in a laminar shear flow with velocities $u =$ (const.)y, $v = 0$ the shear stresses are given by $\sigma_{xy} = \mu(du/dy)$ where μ is the viscosity.

2. For viscous flows we form equilibrium equations for an infinitesimal element and integrate these equations over each finite element.

3. For two-dimensional incompressible flows we ensure that the continuity condition $\partial u/\partial x + \partial v/\partial y = 0$ is satisfied.

In the following sections we discuss these and other basic differential equa-tions governing the behaviour of physical systems. We begin with the equa-tions for the behaviour of solids because, although most solid mechanics problems are solved by energy methods which do not require provision of a differential equation, these provide a useful introduction to the equations for fluids which are in many respects similar.

Moreover some special problems, such as that of plane torsion, in the mechanics of solids are governed by very simple differential equations and are thus easily solved by direct attack upon them. In Chapter 17, therefore, we use such simple problems as an introduction to the use of weighted residual method arguments much used in the remainder of the text.

16.2 Basic differential equations in solids

16.2.1 Plane stress

Figure 16.1 shows the stresses acting upon an element of dimensions δx, δy, and unity. Construction of such systems of stresses is the basic procedure of developing the equilibrium equations for continua and the case of plane stress provides a useful introduction to this procedure.

Multiplying the horizontal stresses by δy and the vertical stresses by δx and summing the resulting forces in each direction the equilibrium equations are

$$\frac{\partial \sigma_x}{\partial x} + \frac{\partial \sigma_{xy}}{\partial y} = 0 \qquad (16.1)$$

$$\frac{\partial \sigma_y}{\partial y} + \frac{\partial \sigma_{xy}}{\partial x} = 0 \qquad (16.2)$$

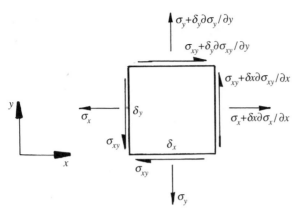

Fig. 16.1 Stresses acting upon a 'differential element'

after cancelling a common factor $\delta x \, \delta y$. If body forces X, Y are present we simply add these to the left-hand sides of eqns (16.1) and (16.2).

The strain–displacement relationships are $\varepsilon_x = \partial u / \partial x$, $\varepsilon_y = \partial v / \partial y$, and $\varepsilon_{xy} = \partial u / \partial y + \partial v / \partial x$ and it is easily verified that these satisfy the strain compatibility condition

$$\frac{\partial^2 \varepsilon_x}{\partial y^2} + \frac{\partial^2 \varepsilon_y}{\partial x^2} = \frac{\partial^2 \varepsilon_{xy}}{\partial x \partial y}. \tag{16.3}$$

Substituting the constitutive equations $E\varepsilon_x = \sigma_x - v\sigma_y$, $E\varepsilon_y = \sigma_y - v\sigma_x$ and $E\varepsilon_{xy} = 2(1 + v)\sigma_{xy}$ into eqn (16.3) and also substituting the derivatives of eqns (16.1) and (16.2) with respect to x and y, respectively into this result we obtain the stress compatibility condition

$$\frac{\partial^2 \sigma_x}{\partial y^2} - \frac{2 \partial^2 \sigma_{xy}}{\partial x \partial y} + \frac{\partial^2 \sigma_y}{\partial x^2} = 0. \tag{16.4}$$

Then solutions to plane stress problems must satisfy eqns (16.1), (16.2), and (16.4) subject to the boundary conditions

$$\sigma_x c_x + \sigma_{xy} c_y = T_x, \qquad \sigma_{xy} c_x + \sigma_y c_y = T_y \tag{16.5}$$

where c_x and c_y are the direction cosines of the outwardly directed surface normal.

It is possible to combine eqns (16.1), (16.2) and (16.4) to form a single governing differential equation by using the *Airy stress function* which is defined by

$$\sigma_x = \frac{\partial^2 \phi}{\partial y^2}, \qquad \sigma_y = \frac{\partial^2 \phi}{\partial x^2}, \qquad \sigma_{xy} = -\frac{\partial^2 \phi}{\partial x \partial y}. \tag{16.6}$$

Equations (16.6) identically satisfy eqns (16.1) and (16.2) and substituting eqns (16.6) into eqn (16.4) yields the *biharmonic equation*[†]

$$\nabla^4 \phi = \frac{\partial^4 \phi}{\partial x^4} + \frac{2\partial^4 \phi}{\partial x^2 \partial y^2} + \frac{\partial^4 \phi}{\partial y^4} = 0. \tag{16.7}$$

This reduces the problem to one of determining the distribution of a single variable ϕ and the Airy stress function is thus much used in finite difference analysis,[1] for example. Substituting eqns (16.6) into eqs (16.5) we see that the boundary conditions are

$$\frac{\partial^2 \phi}{\partial y^2} c_x - \frac{\partial^2 \phi}{\partial x \partial y} c_y = T_x \tag{16.8}$$

$$-\frac{\partial^2 \phi}{\partial x \partial y} c_x + \frac{\partial^2 \phi}{\partial x^2} c_y = T_y \tag{16.9}$$

and these are integrated to determine the boundary values of ϕ. This is straightforward when distributed forces are present on the boundary but it is difficult to model concentrated forces accurately.[1]

16.2.2 Three-dimensional elasticity

Equations (16.1) and (16.2) are readily generalized to three dimensions and if we include body forces X, Y, and Z we obtain

$$X + \frac{\partial \sigma_x}{\partial x} + \frac{\partial \sigma_{xy}}{\partial y} + \frac{\partial \sigma_{xz}}{\partial z} = 0 \tag{16.10}$$

$$Y + \frac{\partial \sigma_{yx}}{\partial x} + \frac{\partial \sigma_y}{\partial y} + \frac{\partial \sigma_{yz}}{\partial z} = 0 \tag{16.11}$$

$$Z + \frac{\partial \sigma_{zx}}{\partial x} + \frac{\partial \sigma_{zy}}{\partial y} + \frac{\partial \sigma_z}{\partial z} = 0. \tag{16.12}$$

Substituting the stress–strain relationships of eqn (2.15) and then the strain–displacement relationships of eqns (2.7) into eqns (16.10)–(16.12) leads, after a little manipulation, to Navier's equations

$$X + G \left\{ \nabla^2 u + \frac{\partial e}{\partial x} (1 - 2v) \right\} = 0 \tag{16.13}$$

$$Y + G \left\{ \nabla^2 v + \frac{\partial e}{\partial y} (1 - 2v) \right\} = 0 \tag{16.14}$$

$$Z + G \left\{ \nabla^2 w + \frac{\partial e}{\partial z} (1 - 2v) \right\} = 0 \tag{16.15}$$

[†] cf. variables ϕ that satisfy Laplace's equation $\nabla^2 \phi = \partial^2 \phi / \partial x^2 + \partial^2 \phi / \partial y^2 = 0$ are said to be *harmonic*.

where

$$\nabla^2( ) = \frac{\partial^2( )}{\partial x^2} + \frac{\partial^2( )}{\partial y^2} + \frac{\partial^2( )}{\partial z^2} \qquad (16.16)$$

$$e = \frac{\partial u}{\partial x} + \frac{\partial v}{\partial y} + \frac{\partial w}{\partial z}. \qquad (16.17)$$

The boundary conditions are a straightforward extension of eqns (16.5). For example for the x-direction we obtain

$$\sigma_x c_x + \sigma_{xy} c_y + \sigma_{xz} c_z = T_x \qquad (16.18)$$

and these are readily expressed in terms of displacements in order to correspond to eqns (16.13)–(16.15).

Equations (16.10)–(16.12) apply equally well to fluids and, with the substitution of appropriate constitutive relationships, the Navier–Stokes equations, which are similar in form to eqns (16.13)–(16.15), are obtained.

16.2.3 Bending of thin plates

If we consider the shear forces per unit width Q_x and Q_y and unit moment m_x, m_y, and m_{xy} in an element of area $\delta x\,\delta y$ of a thin plate carrying a transverse load per unit area q we can show that for moment equilibrium in the x- and y-directions we require

$$Q_x - \frac{\partial m_x}{\partial x} + \frac{\partial m_{xy}}{\partial y} = 0, \qquad Q_y - \frac{\partial m_y}{\partial y} + \frac{\partial m_{xy}}{\partial x} = 0 \qquad (16.19)$$

using the same differential approach used in Fig. 16.1. For vertical equilibrium of the element we can also show that we require

$$\frac{\partial Q_x}{\partial x} + \frac{\partial Q_y}{\partial y} + q = 0. \qquad (16.20)$$

Substituting eqns (16.19) into eqn (16.20) yields

$$\frac{\partial^2 m_x}{\partial x^2} - \frac{2\partial^2 m_{xy}}{\partial x\,\partial y} + \frac{\partial^2 m_y}{\partial y^2} = q. \qquad (16.21)$$

Substituting the moment–curvature relationships of eqn (2.29) and then the curvature–displacement relationships of eqns (2.30) into eqn (16.21) we obtain the governing differential equation of the plate as

$$\nabla^4 w = \frac{12(1 - v^2)q}{Et^3} \qquad (16.22)$$

where the biharmonic operator $\nabla^4( )$ is defined in eqn (16.7).

Again, as only a single variable is involved, eqn (16.22) is easily expressed in finite difference form.[2] Unlike the case of the Airy stress function the boundary conditions are now very simple. For example, for simply supported plates one sets $w = 0$ on the boundary and requires a row of external nodes adjacent to the plate edge whose displacements are set equal and opposite to those of the row of internal nodes adjacent to the boundary.[2] Very large numbers of iterations are sometimes required in this type of finite difference problem, however, and finite elements have proved much more successful in the analysis of plates.

16.2.4 Plane torsion

In plane torsion we consider a cross-section in the xy-plane of a member subjected to a torsional moment T applied about the z-axis. In this case the section will carry only shear stresses and the problem can be solved using *Prandtl's stress function*, which is defined by

$$\sigma_{zx} = \frac{\partial \psi}{\partial y}, \qquad \sigma_{zy} = -\frac{\partial \psi}{\partial x}. \tag{16.23}$$

Recalling that in circular cross-sections the shear stress at radius r is given by $G\theta r$, where θ is the angle of twist per unit length, we apply this to both sides of a small element $\delta x \delta y$. Then taking moments about one corner of the element the torsion carried by the element is $\delta T = 2G\theta \delta x \delta y$. Applying a similar differential argument to that in Fig. 16.1 we require, for moment equilibrium of the element,

$$\left(\frac{\partial \sigma_{zx}}{\partial y} - \frac{\partial \sigma_{zy}}{\partial x} + 2G\theta \right) \delta x \delta y = 0 \tag{16.24}$$

and substituting eqns (16.23) and deleting the common factor $\delta x \delta y$ the governing differential equation is *Poisson's equation* which is

$$\nabla^2 \psi + 2G\theta = \frac{\partial^2 \psi}{\partial x^2} + \frac{\partial^2 \psi}{\partial y^2} + 2G\theta = 0 \tag{16.25}$$

where ∇^2 is the Laplacian operator.

At the boundary it follows from eqns (16.23) that the magnitude of the normal shear stress is given by $\partial \psi / \partial S$. As this must be zero, ψ must be constant around the boundary. Because only the gradients of ψ determine the shear stresses we need only specify an arbitrary datum value, most simply zero, so that the boundary conditions are simply $\psi = 0$ on the boundary.

Returning to our small element we can also use eqns (16.23) to write an alternative condition for moment equilibrium which is

$$\delta T = \left(\frac{\partial \psi}{\partial y} \delta y + \frac{\partial \psi}{\partial x} \delta x \right) \delta x \delta y. \tag{16.26}$$

Integrating over the whole cross-section we obtain

$$T = 2G \iint \psi \, dx \, dy \tag{16.27}$$

including G, as it is omitted in eqns (16.23) (that is, $\partial\psi/\partial y$, for example, gives the corresponding shear strain and G times this gives the shear stress).

Plane torsion can also be solved in terms of *St Venant's warping function* which is related to the lateral (warping) displacement, w, of the cross-section and is therefore defined by

$$\sigma_{zx} = \frac{\partial\phi}{\partial x}, \qquad \sigma_{zy} = \frac{\partial\phi}{\partial y}. \tag{16.28}$$

The governing differential equation for this function is Laplace's equation: simply $\nabla^2\phi = 0$, but the boundary conditions are much more complicated than the simple homogeneous boundary conditions of Prandtl's stress function[1], the governing equation for which can in any case be reduced to Laplace's equation in the manner shown in Section 21.6.

Finally it is worth noting that the Prandtl and St Venant torsion functions are, analogous to the use of *stream* and *potential* functions respectively, in the mechanics of fluids. Thus plane torsion, although one of the more specialized problems in the mechanics of solids, provides a useful introduction to techniques for solving problems in the mechanics of fluids.

16.3 Basic equations for general field problems

In Section 1.1 we noted that physical problems can be classified as equilibrium, creep or diffusion, inertial or vibration, and eigenvalue problems. In earlier chapters we have dealt with each type of problem in the mechanics of solids. In the following section we examine some of the basic differential equations governing these problem types in 'general field problems.'

A field is either a scalar or vector point function.[3] By a scalar function we mean that if, at each point P of a region R, a scalar $u = u(P)$ is defined, then $u(P)$ is a scalar point function. Then in three-dimensional Cartesian coordinates, for example, we require three functions $u(P)$, $v(P)$, and $w(P)$ to define the components of the vector $V = u\hat{i} + v\hat{j} + w\hat{k}$. In solids u, v, w are typically displacements whilst in fluids u, v, w are typically velocity components. 'General field problems', therefore, encompass such physical phenomena as stress distributions in solids, fluid flow, heat flow, and electrical and magnetic fields and we shall encounter all of these in the following chapters.

Before considering the categories of equilibrium, diffusion, and so on, we shall consider continuity conditions which must be met by problems involving flows of material or energy. In fluid flow, for example, continuity conditions ensure conservation of mass. We arrive at the continuity condition by considering the mass contained in a volume V which is given by

$$M = \int \rho \, dV \qquad (16.29)$$

where ρ is the mass density. The rate of mass change is given by

$$\frac{\partial M}{\partial t} = \int \frac{\partial \rho}{\partial t} \, dV \qquad (16.30)$$

and if mass is to be conserved this must equal the rate of mass inflow through a surface S given by integrating the normal velocities V_n over the surface,

$$\frac{\partial M}{\partial t} = \int (-\rho |V_n|) \, dS$$

$$= \int - \left[\frac{\partial(\rho u)}{\partial x} + \frac{\partial(\rho v)}{\partial y} + \frac{\partial(\rho w)}{\partial z} \right] dV \qquad (16.31)$$

by Green's theorem, which as we shall see in Section 17.4, can be arrived at by integration by parts.

Equating the terms under the integral signs in eqns (16.30) and (16.31) the continuity condition is

$$\frac{\partial \rho}{\partial t} + \frac{\partial(\rho u)}{\partial x} + \frac{\partial(\rho v)}{\partial x} + \frac{\partial(\rho w)}{\partial z} = 0. \qquad (16.32)$$

If mass is generated within the region then eqn (16.32) becomes

$$\frac{\partial \rho}{\partial t} + \frac{\partial(\rho u)}{\partial x} + \frac{\partial(\rho v)}{\partial y} + \frac{\partial(\rho w)}{\partial z} + k\rho = 0 \qquad (16.33)$$

where k is called the growth factor. If $k = 0$ eqn (16.33) can also be written as

$$\frac{D\rho}{Dt} + \rho \left(\frac{\partial u}{\partial x} + \frac{\partial v}{\partial y} + \frac{\partial w}{\partial x} \right) = 0 \qquad (16.34)$$

where $D\rho/Dt$ is the total derivative which is given by writing the total increment in $\rho(x, y, z, t)$ for small increments in x, y, z, and t as

$$\delta\rho = \frac{\partial \rho}{\partial t} \delta t + \frac{\partial \rho}{\partial x} \delta x + \frac{\partial \rho}{\partial y} \delta y + \frac{\partial \rho}{\partial z} \delta z. \qquad (16.35)$$

Dividing by δt, taking the limit, and noting that $\mathrm{Lim}_{\delta x \to 0}(\delta x/\delta t) = dx/dt = u$ we obtain

$$\frac{D\rho}{Dt} = \frac{\partial \rho}{\partial t} + u\frac{\partial \rho}{\partial x} + v\frac{\partial \rho}{\partial y} + w\frac{\partial \rho}{\partial z} \qquad (16.36)$$

so that eqns (16.33) and (16.34) are indeed equivalent.

Under steady state conditions $\partial \rho/\partial t = 0$ and if the material is incompressible then ρ cannot vary with position and eqn (16.34) reduces to

$$\frac{\partial u}{\partial x} + \frac{\partial v}{\partial y} + \frac{\partial w}{\partial z} = 0. \tag{16.37}$$

In the remainder of this section we shall see that such continuity conditions govern many general field problems.

16.3.1 Equilibrium or quasi-static problems

Equation (16.37) governs many steady state incompressible fluid flow problems and a number of other analogous physical problems which differ only in terms of the associated 'flow law.' As an example consider steady state plane seepage for which the velocities of flow are given by *Darcy's laws*

$$u = -k_x \frac{\partial H}{\partial x}, \qquad v = -k_y \frac{\partial H}{\partial y} \tag{16.38}$$

where k_x and k_y are the permeability coefficients and H is the static pressure head. Note that the negative signs in eqns (16.38) ensure that flow is in the opposite directon to the gradient in the 'potential' H.

Substituting eqns (16.38) into eqn (16.37) (neglecting the term $\partial w/\partial z$) the governing differential equation for incompressible plane seepage is obtained as[†]

$$\frac{\partial}{\partial x}\left(k_x \frac{\partial H}{\partial x}\right) + \frac{\partial}{\partial y}\left(k_y \frac{\partial H}{\partial y}\right) = 0. \tag{16.39}$$

This reduces to Laplace's equation $\nabla^2 H = 0$ if the medium is isotropic, that is, $k_x = k_y$, and as shown in Section 16.3.2 below Laplace's equation also applies to steady state isotropic heat flow.

16.3.2 Diffusion problems

Transient heat flow provides perhaps the most useful introduction to diffusion problems. To obtain the governing differential equation we replace ρ in eqn (16.33) by the heat per unit volume, q, which is given by

$$q = c\rho T \tag{16.40}$$

where c is the specific heat of the medium (in J/kg°C), ρ is the mass density, and T is the temperature. In solids the specific heat is single valued but note that in gases we use the specific heat at constant volume, c_v, for incompressible conditions and the specific heat at constant pressure, c_p, under isobaric

---

[†] Because of the resemblance to Laplace's equation we refer to such problems as *pseudoharmonic problems*.

conditions or, in general, some combination thereof based upon the specific heat ratio $\gamma = c_v/c_p$.

The rates of heat conduction are given by applying *Fourier's law of heat conduction* in each of the Cartesian directions,

$$u = -\kappa_x \partial T/\partial x, \qquad v = -\kappa_y \partial T/\partial x, \qquad w = -\kappa_z \partial T/\partial z. \qquad (16.41)$$

Then replacing ρ by $q = c\rho T$ in the first term of eqn (16.32) and replacing ρu, ρv, and ρw by u, v, and w as obtained from eqns (16.41) we obtain

$$c\rho \frac{\partial T}{\partial t} = \frac{\partial}{\partial x}\left(\frac{\kappa_x \partial T}{\partial x}\right) + \frac{\partial}{\partial y}\left(\frac{\kappa_y \partial T}{\partial y}\right) + \frac{\partial}{\partial z}\left(\frac{\kappa_z \partial T}{\partial z}\right) = 0 \qquad (16.42)$$

where κ_x, κ_y, and κ_z are the coefficients of thermal conductivity.

If the medium is isotropic with conductivity κ then eqn (16.42) reduces to

$$c\rho \left(\frac{\partial T}{\partial t}\right) = \kappa \nabla^2 T = \kappa \left(\frac{\partial^2 T}{\partial x^2} + \frac{\partial^2 T}{\partial y^2} + \frac{\partial^2 T}{\partial z^2}\right) \qquad (16.43)$$

which is called the *diffusion equation*. Under steady state conditions we see that heat flow is also governed by Laplace's equation, $\nabla^2 T = 0$, if the medium is isotropic.

If some internal heat source is present then eqn (16.43) is generalized to

$$c\rho \frac{\partial T}{\partial t} = \kappa \nabla^2 T + G_v \qquad (16.44)$$

where G_v is the heat generated per unit volume. Heat can also be generated at the boundary by incident radiation or lost by convection, and an example of this type of problem is solved in Section 18.2.

16.3.3 Inertial effects

Equation 16.43 can be generalized to include inertial effects[3,4] when, for the distribution of a field variable ϕ, the governing differential equation is the wave equation

$$c \frac{\partial \phi}{\partial t} + m \frac{\partial^2 \phi}{\partial t^2} = \nabla^2 \phi \qquad (16.45)$$

the inertial term being derived in Section 14.1 for the one-dimensional case.

As shown in Section 13.1 we can use separation of variables and assume a solution in the form $\phi = \phi(x, y, z, t) = \psi(x, y, z)T(t)$ and thus obtain two governing equations

$$m \frac{\partial^2 \phi}{\partial t^2} + c \frac{\partial \phi}{\partial t} + k\phi = 0 \qquad (16.46)$$

$$\nabla^2 \psi + k\psi = 0 \qquad (16.47)$$

which are called the *vibration* and *Helmholtz* equations respectively.

In Chapter 13 we generalized eqn (16.46) to matrix form and deleted the damping term $c\partial\phi/\partial t$ to obtain a linear eigenvalue problem (that is, for the natural frequencies of free vibration), whilst in Chapter 14 the time derivatives in this generalized matrix form were replaced by finite difference approximations and time stepping used to obtain a solution.

The Helmholtz equation governs a wide range of eigenvalue problems, including those of incipient stability, acoustic vibration, seiche motion, and electromagnetic radiation, and an example of solution of the last type of problem is given in Chapter 18.

In most fluids mechanics problems, however, the inertia terms arise in connection with total derivatives of the velocity components. Considering a single particle, for example, the acceleration can be written as d^2x/dt^2 or equivalently as $d(dx/dt)/dt = (du/dx)(dx/dt) = udu/dx$ and we shall see that the inertia terms involve this form in Section 16.4.

16.4 Equations for the thermohydrodynamic behaviour of fluids

The general equations for the thermohydrodynamic behaviour of fluids are a good deal more complicated than those for the simple 'field' problems discussed in Section 16.3. The constitutive equations must take into account viscosity and the associated 'dynamic stresses', the fluid may be compressible so that the continuity condition must take the general form of eqn (16.34), equations of equilibrium between the body forces and stresses must be developed, and conditions for conservation of thermal energy are required. These various considerations are discussed in the following subsections.

16.4.1 Viscosity in fluids

The shear stresses in an ideal *Newtonian fluid* with velocities given by $u = f(y)$, $v = 0$ are given by[5]

$$\sigma_{xy} = \mu \frac{\partial u}{\partial y} \tag{16.48}$$

where μ is the *dynamic viscosity*. From this the *kinematic viscosity* is defined as $\nu = \mu/\rho$ and this is used to non-dimensionalize some problems.

Equation (16.48) corresponds to Hooke's law in elasticity and laws for non-linear fluids have also been proposed which are similar to those used in the mechanics of solids. An example is *Bingham's flow rule* in which $\sigma_{xy} = \sigma^* + (\text{const.})\partial u/\partial y$ where σ^* is a threshold value; this is equivalent to the assumption of ideal plasticity in solids. Another is the power law flow rule $\sigma_{xy} = (\partial u/\partial y)^n$; this is equivalent to the power law of creep used in solids. Such rules are sometimes used in combination to model biomechanical flows[6] but in most fluids Newtonian viscosity may be assumed.

Generalizing eqn (16.48) the result can be written in index notation as

$$\sigma_{ij} = \mu \left(\frac{\partial u_i}{\partial x_j} + \frac{\partial v_j}{\partial x_i} \right)$$ (16.49)

that is, $\sigma_x = 2\mu\partial u/\partial x$, $\sigma_{xy} = \mu(\partial u/\partial y + \partial v/\partial x)$, and so on. That $\sigma_x = 2\mu\partial u/\partial x$ is easily imagined if we realize that viscosity will resist the flow of a small element on two sides. Equation (16.49) can also be deduced from the generalized Hooke's laws of elasticity if we delete Poisson's ratio and replace G by μ and $E = 2(1 + v)G = 2G$ by 2μ. The second substitution $(E = 2\mu)$ is not physically correct (but the 'two sides' argument above is) but it is a useful mnemonic.

Now in compressible fluids bulk strains occur which do not cause stresses. In the theory of elasticity we calculate this as the fractional change in volume of a unit cube, which is

$$\varepsilon_x + \varepsilon_y + \varepsilon_z = \frac{\partial u}{\partial x} + \frac{\partial v}{\partial y} + \frac{\partial w}{\partial z} = \frac{3\sigma(1 - 2v)}{E}$$ (16.50)

for an isotropic stress σ and $K = E/3(1 - 2v)$ is the bulk modulus. Then for a viscous fluid we delete Poisson's ratio from eqn (16.50), replace E by 2μ so that the *fictitious* stress σ^* corresponding to the bulk 'dynamic strain' is given by

$$\sigma^* = \frac{2\mu}{3} \left(\frac{\partial u}{\partial x} + \frac{\partial v}{\partial y} + \frac{\partial w}{\partial z} \right).$$ (16.51)

Then subtracting σ^* and the ambient pressure p (which of course does not contribute to the dynamic stresses) from the direct stresses given by eqns (16.49) we obtain *Stokes' friction laws.*[7]

$$\sigma_x = -p + 2\mu \left(\frac{\partial u}{\partial x} - \frac{\nabla \cdot V}{3} \right), \qquad \sigma_{xy} = \mu \left(\frac{\partial u}{\partial y} + \frac{\partial v}{\partial x} \right)$$ (16.52a)

$$\sigma_y = -p + 2\mu \left(\frac{\partial v}{\partial y} - \frac{\nabla \cdot V}{3} \right), \qquad \sigma_{yz} = \mu \left(\frac{\partial v}{\partial z} + \frac{\partial w}{\partial y} \right)$$ (16.52b)

$$\sigma_z = -p + 2\mu \left(\frac{\partial w}{\partial z} - \frac{\nabla \cdot V}{3} \right), \qquad \sigma_{zx} = \mu \left(\frac{\partial w}{\partial x} + \frac{\partial u}{\partial z} \right)$$ (16.52c)

where

$$\nabla \cdot V = \left(\frac{\partial}{\partial x} \hat{\mathbf{i}} + \frac{\partial}{\partial y} \hat{\mathbf{j}} + \frac{\partial}{\partial z} \hat{\mathbf{k}} \right) \cdot (u\hat{\mathbf{i}} + v\hat{\mathbf{j}} + w\hat{\mathbf{k}})$$

$$= \frac{\partial u}{\partial x} + \frac{\partial v}{\partial y} + \frac{\partial w}{\partial z} = \text{div}(V).$$

Note that in eqns (16.52) the 'ambient' pressure in the fluid is a static pressure head or a thermodynamic pressure given by $p = \rho R T$ for an ideal gas, where R is the universal gas constant.

The mean pressure $\bar{p}$ is defined as

$$\bar{p} = \tfrac{1}{3}(\sigma_x + \sigma_y + \sigma_z) \tag{16.53}$$

and this is related to the ambient pressure by

$$\bar{p} = p - \sigma^* = \tfrac{2}{3}\mu(\nabla . V). \tag{16.54}$$

Note that the term $\nabla . V = \partial u/\partial x + \partial v/\partial y + \partial w/\partial z$ in eqns (16.52) will be zero for incompressible fluids (see eqn (16.37)) but subtracts the non-stress inducing bulk strain in compressible fluids. In the special case of fluids in which compression of the material does give rise to viscous stresses the term $\nabla . V$ is multiplied by $K - 2\mu/3$ (rather than $2\mu/3$) where K is the *bulk viscosity*.

16.4.2 Continuity condition

The continuity condition for viscous and other fluid flows was derived in Section 16.3 as eqn (16.34):

$$\frac{D\rho}{Dt} + \rho\left(\frac{\partial u}{\partial x} + \frac{\partial v}{\partial y} + \frac{\partial w}{\partial z}\right) = 0 \tag{16.55}$$

and this reduces to $\nabla . V = 0$ for incompressible fluids.

16.4.3 Momentum and Navier–Stokes equations

The momentum equations for a fluid particle can be obtained by using d'Alembert's principle to generalize eqns (16.10)–(16.12) to obtain

$$\rho\frac{Du}{Dt} = X + \frac{\partial\sigma_x}{\partial x} + \frac{\partial\sigma_{xy}}{\partial y} + \frac{\partial\sigma_{xz}}{\partial z} \tag{16.56a}$$

$$\rho\frac{Dv}{Dt} = Y + \frac{\partial\sigma_{yx}}{\partial x} + \frac{\partial\sigma_y}{\partial y} + \frac{\partial\sigma_{yz}}{\partial z} \tag{16.56b}$$

$$\rho\frac{Dw}{Dt} = Z + \frac{\partial\sigma_{zx}}{\partial x} + \frac{\partial\sigma_{zy}}{\partial y} + \frac{\partial\sigma_z}{\partial z} \tag{16.56c}$$

where X, Y, and Z are body forces per unit volume, by including the inertia forces. Note again that Du/Dt is the total derivative and is given by

$$\frac{Du}{Dt} = \frac{\partial u}{\partial t} + u\frac{\partial u}{\partial x} + v\frac{\partial u}{\partial y} + w\frac{\partial u}{\partial z} \tag{16.57}$$

and the last three terms of eqn (16.57), when multiplied by ρ, are called the *convective inertia* terms.

Substituting eqns (16.52) into eqns (16.56) we obtain the *Navier–Stokes equations*, the first of which is

$$\rho \frac{Du}{Dt} = X - \frac{\partial p}{\partial x} + \frac{\partial}{\partial x}\left\{2\mu\left(\frac{\partial u}{\partial x} - \frac{\nabla \cdot V}{3}\right)\right\}$$

$$+ \frac{\partial}{\partial y}\left\{\mu\left(\frac{\partial u}{\partial y} + \frac{\partial v}{\partial x}\right)\right\} + \frac{\partial}{\partial z}\left\{\mu\left(\frac{\partial w}{\partial x} + \frac{\partial u}{\partial z}\right)\right\} \qquad (16.58)$$

and the other two follow from cyclic progression of x, y, z and u, v, w.

16.4.4 The thermal energy equation

Fluid flow problems may also require a thermal energy equation to ensure conservation of energy. When a gas is compressed, for example, heat may be generated. This is obtained by generalizing eqn (16.43) to[7]

$$\rho c_v \frac{DT}{Dt} + \nabla \cdot \{g_r\} - \frac{\partial G}{\partial t} = \kappa \nabla^2 T + \mu \Phi(x, y, z, t) \qquad (16.59)$$

for isotropic conditions, where c_v is the specific heat at constant volume, g_r is a radiation heat flux vector, $\partial G/\partial t$ is the rate of internal heat generation and Φ is the *viscous dissipation function*.

The viscous dissipation function completes an equation of the total energy by including the energy per unit time consumed by the viscous stresses. This is calculated from the sum of the stresses multiplied by their 'strain rates':

$$\mu \Phi = \sigma_x \frac{\partial u}{\partial x} + \sigma_y \frac{\partial v}{\partial y} + \sigma_z \frac{\partial w}{\partial z} + \sigma_{xy}\left(\frac{\partial u}{\partial y} + \frac{\partial v}{\partial x}\right)$$

$$+ \sigma_{yz}\left(\frac{\partial v}{\partial z} + \frac{\partial w}{\partial y}\right) + \sigma_{zx}\left(\frac{\partial w}{\partial x} + \frac{\partial u}{\partial z}\right) \qquad (16.60)$$

where, as we are considering 'long-established' motion, we do not require a factor of one half as in elasticity. Substituting eqns (16.52) and dividing by μ we obtain

$$\Phi = u_z^2 + u_y^2 + v_z^2 + v_x^2 + w_x^2 + w_y^2 + 2(u_y v_x + v_z w_y + w_x u_z)$$

$$+ \tfrac{4}{3}\{u_x(u_x - v_y) + v_y(v_y - w_z) + w_z(w_z - u_x)\} \qquad (16.61)$$

where u_x denotes $\partial u/\partial x$ and so on.

Thus the term 'thermal energy equation' is a slight misnomer. We do begin with eqn (16.43), a heat flow equation, but with the addition of Φ we have a equation balancing viscous dissipation losses with thermal energy gains.

16.4.5 Constitutive equations

We have already used one constitutive equation, that for Newtonian viscosity, but in general others are needed to relate viscosities, densities, temperatures, and pressures. For gases, the *equation of state* is

$$p = \rho R T \tag{16.62}$$

for an ideal gas, where R is the universal gas constant, whilst for real gases *van der Waal's* equation[5]

$$p = \frac{\rho R T}{1 - \beta \rho} + \alpha \rho^2 \tag{16.63}$$

where α and β are constants, is sometimes used.

The viscosity of some fluids, particularly lubricating oils, for example, may be temperature dependent and Heubner assumes for lubricating oils that[7]

$$\mu = \mu_0 e^{-\gamma(T - T_0)} \tag{16.64}$$

where γ is a constant and T_0 is the reference temperature of measurement of μ_0. Here $d\mu/dT < 0$ but for gases the reverse applies, that is, the viscosity increases with temperature.

In ordinary fluids, however, it is generally satisfactory to assume that the viscosities and the thermal conductivities are related to the temperature by the same ratio law:

$$\frac{\mu}{\mu_0} = \frac{\kappa}{\kappa_0} = \left(\frac{T}{T_0}\right)^m \tag{16.65}$$

and m is often assumed to be unity.[5]

Use of equations such as (16.64) and (16.65) implies that μ and κ are pressure independent and this is in most cases a workable assumption, as any effects from pressure feed through from eqn (16.62).

In summary, many equations are needed to govern the thermohydrodynamic behaviour of fluids and these are:

(1) the continuity condition;

(2) three momentum equations;

(3) the thermal energy equation;

(4) several constitutive equations, such as eqns (16.48) and (16.62)–(16.65).

The large majority of fluid analyses, however, are governed by simplified forms of the equations presented in this section. In steady state, isothermal, inviscid, incompressible flow eqn (16.58) reduces to $X = \partial p/\partial x$ and the problem is governed only by the continuity condition. This simple case is called *potential flow* and can be analyzed using either stream or potential functions in terms of which the governing equation is Laplace's equation and is discussed in detail in Section 19.2.

16.5 Classification of differential equations

Most partial differential equations of the type $L(\phi) - f = 0$, where $L(\ )$ is a differential operator, can be classified as *elliptic, parabolic,* or *hyperbolic*. For example, the second-order equation

$$a\frac{\partial^2 \phi}{\partial x^2} + b\frac{\partial^2 \phi}{\partial x \partial y} + c\frac{\partial^2 \phi}{\partial y^2} = f \tag{16.66}$$

is elliptic, parabolic, or hyperbolic when $b^2 - 4ac$ is respectively negative, zero, or positive.[8]

For the purposes of applying this test we assume the actual coefficients are either -1, 1, or 0, with corresponding values of a, b, and c in eqn (16.66). If the problem is time dependent we simply replace y by t. Then we can conclude that most equilibrium problems are elliptic, perhaps the best example being the harmonic and pseudoharmonic problems involving the Laplacian operator for which $a = c = 1$ and $b = 0$.

Most diffusion problems are parabolic, for example those governed by the diffusion equation $\partial^2 \phi / x^2 = \partial \phi / \partial t$ for which we take $b = c = 0$ in eqn (16.67). Most vibration problems, on the other hand, are hyperbolic and the wave equation $\partial^2 \phi / \partial x^2 = \partial^2 \phi / \partial t^2$, for which we take $a = 1$, $b = 0$ and $c = -1$, is an example.

This terminology is of interest in the *method of characteristics*, for example, where diffusion problems have only a single set of characteristic curves (that is, diffusion occurs over only one front) whereas vibration problems have two sets of characteristics because propagation of material varies in pattern with time.[8]

It is also possible for equations to have a dual nature. An example is *Tricomi's equation*

$$\frac{\partial^2 \phi}{\partial x^2} + \frac{y\partial^2 \phi}{\partial y^2} = f. \tag{16.67}$$

When $y > 0$ eqn (16.67) is elliptic and when $y < 0$ it is hyperbolic, and this property makes it useful in boundary layer studies.[9]

Another useful application of this classification is to plane strain plasticity. Although intuitively this might be thought to be an elliptic problem it is in fact a diffusion problem, that is, it is parabolic and diffusion occurs over a single expanding front as yield spreads through the material.[8]

16.6 Concluding remarks

In the solid mechanics applications of earlier chapters, in all but the simplest cases, appeal to an energy principle is more usual and less laborious than solution through the use of a governing differential equation. Nevertheless the equations developed for solids in Section 16.2 provide a useful introduction to the equations for field and fluid dynamics problems which follow. The case of plane torsion, in particular, provides a useful example of the application of weighted residual techniques in the following chapter and the 'MWR', or equivalent virtual work arguments, provide the most direct means of attacking the heat flow, electromagnetic field, and fluid flow problems in the chapters that follow.

References

1. Timoshenko, S. P. and Goodier, J. N. (1951). *Theory of Elasticity*, (3rd edn). McGraw-Hill, New York.
2. Timoshenko, S. P. and Woinowsky-Krieger, W. K. (1959). *Theory of Plates and Shells* (2nd edn). Mcraw-Hill, New York.
3. Rainville, E. D. and Beafait, W. P. (1974). *Elementary Differential Equations*, (5th edn). Macmillan, New York.
4. Riley, K. F. (1974). *Mathematical Methods for the Physical Sciences.* Cambridge University Press.
5. S. W. Yuan, (1967). *Foundations of Fluid Mechanics.* Prentice-Hall, Englewood Cliffs, NJ.
6. Brown, J. H. U., Jacobs, J. E., and Stark, L. (1971). *Biomedical Engineering.* Davis, Philadelphia.
7. Heubner, K. H. (1975). *The Finite Element Method for Engineers.* Wiley, New York.
8. Crandall, S. H. (1956). *Engineering Analysis.* McGraw-Hill, New York.
9. Carey, G. F., Cheung, Y. K., and Lau, S. K. (1980). Mixed operator problems using least squares finite element collocation. *Comp. Meth. Appl. Mech. Engng*, **22**, 121.

17
Variational and weighted residual methods

In preceding chapters a wide range of solid mechanics problems was solved using the principle of minium potential energy or an equivalent virtual work argument. In subsequent chapters attention turns to the field and fluid flow problems introduced in Chapter 16. Minimum principles for some of the simplest of these can be constructed by energy or virtual work arguments but they can be solved more expeditiously by direct attack upon the governing differential equation using the *method of weighted residuals*, most commonly the *Galerkin method*. These methods and an equivalent virtual work approach are discussed in this chpater.

17.1 Introduction

At the outset it is perhaps worth attempting to draw together some basic ideas concerning energy methods and the method of weighted residual (MWR) and the role that virtual work arguments can play in both.

17.1.1 Energy methods

Energy methods played an important role in the development of the finite element method[1] and until the early 1970s were used almost exclusively to develop finite element equations. In the minimum potential energy method, for example, we begin by writing the total potential energy of a finite element as

$$\Pi_e = \tfrac{1}{2} \int \{\varepsilon\}^t \{\sigma\} \, dV - \{\bar{u}\}^t \{T\} \, dS - \{d\}^t \{q_r\}. \tag{17.1}$$

We introduce constitutive relationships $\{\sigma\} = D\{\varepsilon\}$ and displacement interpolations $\{\bar{u}\} = F^t\{d\}$ and differentiate F according to the strain–displacement relationships to form a strain interpolation matrix B such that $\{\sigma\} = B\{d\}$. Substituting these into eqn (17.1) and performing $\partial \Pi_e/\partial \{d\} = \{0\}$ we obtain the equilibrium equations for the element, $k\{d\} = \{q_t\} + \{q_r\}$ where $k = \int B^t DB \, dV$, $\{q_t\} = \int F\{T\} \, dS$ and $\{q_r\}$ are the nodal interelement reactions. Summing the element equations we assume

$\Sigma\{q_r\} = \{0\}$ and include external nodal loads $\{Q_n\}$ to form the system equations which are then solved simultaneously to determine the unknown displacements in $\{D\}$.

As we saw in Chapter 3 many additional terms can be included in eqn (17.1), for example leading to the contact stiffness matrix k_b, and many other energy principles exist, for example the principle of minimum complementary energy which leads to the development of equilibrium elements with stress of force freedoms and this approach is the *dual* of the primal problem with displacement freedoms.

Methods in which a minimum principle is used belong to a class of methods known generically as *variational methods*. In these we seek to minimize a *functional*, that is, the integral of a function of scalar or vector quantities which yields a scalar result. In eqn (17.1), for example, if we substitute the matrices D, B and F we obtain the functional $F(\{d\}) = \Pi_e$.

The term variational methods is not to be confused with the classical *calculus of variations* in which the *Euler–Lagrange equations* are applied to a functional to discover the corresponding differential equation. We have seen an example of this procedure in Section 14.1 where the integral of the potential and kinetic energy of a stretched string was operated upon to obtain the wave equation.

When a functional is already available, however, it is more expeditious to seek a solution by minimization or virtual work and the governing differential equation is not needed (in thick shells, for example, it would be rather intractable if it could be discovered at all). Indeed, given a governing differential equation, we should be more interested in seeking the corresponding functional, and in Section 17.3 we shall see that virtual work again proves useful by providing a means of achieving this aim.

17.1.2 Weighted residual methods

The weighted residual methods are those in which *weighting functions* are applied to the *residual errors* in a differential equation and these residual errors may be evaluated at discrete points or in an integral sense over a number of subdomains or the complete domain. There are a number of alternative weighted residual methods, and some of these are discussed in Section 17.5, but in the finite element method we evaluate the residual errors in a differential equation $L(\{\bar{u}\}) = 0$ by substituting the interpolations $\{\bar{u}\} = F^t\{d\}$, giving the residual error $R = L(F^t\{d\})$. Multiplying the residual error by weighting functions w_i and integrating over the volume of the element we equate the result to zero:

$$\int w_i R \, dV = \int w_i L(F^t\{d\}) = 0 \qquad i = 1 \rightarrow N \qquad (17.2)$$

where N is the number of freedoms of the element.

If we choose these weighting functions as virtual perturbations $\{\delta\bar{u}\} = F^t\{\delta d\}$ then eqn (17.2) becomes

$$\{\delta d\}^t\,(\textstyle\int F F^{*t}\,\mathrm{d}V)\,\{d\} = \{\delta d\}^t\{0\} \tag{17.3}$$

where F^* denotes the matrix F differentiated in accordance with the terms appearing in $L(\{\bar{u}\}) = 0$ and the common factor $\{\delta d\}^t$ is deleted.

As we shall see in Section 17.5 eqn (17.3) is completely equivalent to the *Galerkin method*, in which the weighting functions are chosen as the interpolation functions F. The Galerkin method is the MWR most widely used for finite elements, so that the equivalence of virtual work arguments provides yet another demonstration of their usefulness.

In practice eqn (17.3) is integrated by parts and this results in forcing terms on the right-hand side. Moreover, when the differential equation is *self-adjoint*, it is always possible to obtain element matrices identical to those obtained by applying minimization or virtual work to the corresponding functional. We discuss self adjointness of differential equations in the following section.

17.2 Self-adjoint differential equations

A differential equation $L(\phi) = g$ of order $2m$ may be subject to *essential boundary conditions* involving derivatives of order m or less, and *natural boundary conditions* of higher order. The differential operator $L(\ )$ is self-adjoint if, for two arbitrary functions u and v which satisfy the same essential and natural boundary conditions,[2]

$$\textstyle\int u L(v)\,\mathrm{d}V = \int v L(u)\,\mathrm{d}V \tag{17.4}$$

and is positive if $\int u L(u)\,\mathrm{d}V \geq 0$ and positive definite if $\int u L(u)\,\mathrm{d}V = 0$ only when u is identically zero everywhere.

Considering a second-order differential operator $L''(\ )$, for example, integrating by parts twice we obtain

$$\textstyle\int u L''(v)\,\mathrm{d}V = u L'(v)| - \int u'L'(v)\,\mathrm{d}V$$
$$= u L'(v)| - u'L^0(v)| + \int u''L^0(v)\,\mathrm{d}V \tag{17.5}$$

where $|$ denotes boundary terms, $L'(\ )$ is a first-order operator, and $L^0(v) = v$. Putting $u'' = L^*(u)$, $u' = S^*(u)$, $L'(v) = S'(v)$ in eqn (17.5) it becomes

$$\textstyle\int u L''(v)\,\mathrm{d}V = \{u S'(v) - S^*(u)v\}| + \int L^*(u)v\,\mathrm{d}V. \tag{17.6}$$

Then the operator $L''(\ )$ is self adjoint if $L''(\ ) = L^*(\ )$, that is, if $\int u L''(v)\,\mathrm{d}V = \int v L''(u)\,\mathrm{d}V$, as required by eqn (17.4). If this is the case then $S(\ )$ will also be self-adjoint, that is, $S^*(\ ) = S'(\ )$.

Taking a beam on an elastic foundation as an actual example, the differential equation for this can be expressed in the form[2].

$$L(\phi) = \frac{d^4\phi}{dx^4} + \phi = g \qquad (17.7)$$

where $\phi = EI\,y/qL^4$ and x are dimensionless, and g is a function of the foundation stiffness. Integrating $\int uL(v)\,dx$ by parts twice gives

$$\int_0^1 u(v_{xxxx} + v)\,dx = (uv_{xxx} - u_x v_{xx})\big|_0^1 + \int_0^1 (u_{xx}v_{xx} + uv)\,dx \qquad (17.8)$$

where $v_{xx} = d^2v/dx^2$, and so on.

If the boundary conditions are $\phi = \phi_{xx} = 0$ (simple support) then $u = v_{xx} = 0$ and the boundary term vanishes. If they are $\phi = \phi_x = 0$ (clamped support) then $u = u_x = 0$ and the boundary term also vanishes. Finally, if they are $\phi_{xx} = \phi_{xxx} = 0$ (free) then $v_{xx} = v_{xxx} = 0$ and again the boundary term in eqn (17.8) vanishes. Then as the remaining term on the right-hand side is symmetrical, $L(\;)$ is self-adjoint and replacing v by u we obtain

$$\int_0^1 u(u_{xxxx} + u)\,dx = \int_0^1 ((u_{xx})^2 + u^2)\,dx. \qquad (17.9)$$

As the right-hand side of eqn (17.9) is a sum of squares it can vanish only when u is identically zero and hence $L(\;)$ is positive definite.

Now integrating the right side of eqn (17.8) by parts twice more we obtain

$$\int_0^1 u(v_{xxxx} + v)\,dx = (uv_{xxx} - u_x v_{xx} + u_{xx}v_x - u_{xxx}v)\big|_0^1 + \int_0^1 v(u_{xxxx} + u)\,dx$$

that is

$$\int uL(v)\,dx = (uS(v) - vS(u) - u_xB(v) + v_xB(u))| + \int vL(u)\,dx. \quad (17.10)$$

Again it is clear that $L(\;)$ is self-adjoint and that $S(\;)$ and $B(\;)$ are also self-adjoint (replacing u by u_x in eqn (17.4) is of little consequence as u is an arbitrary function.)

Note that if on the right-hand side of eqn (17.8) we put $v = \{f\}^t\{d\}$ and $u = \{f\}^t\{\delta d\}$ the last term corresponds to the contact stiffness matrix and the penultimate term to the usual beam stiffness matrix. Here $\{\delta d\}$ is a perturbation which is cancelled as a common factor, leaving $\{f\}$ to act as weights on the left-hand side of eqn (17.8), which in accordance with eqn (17.2) is then set to zero.

Thus, when the differential equation is self-adjoint (e.g. it involves even-order derivatives) we have already discovered a virtual work argument that leads to the same results as those from variational methods. This means that through virtual work we have discovered, in effect, a means of forming functionals and we discuss this further in the next section.

17.3 Formation of functionals by virtual work

In the previous section we concluded with the realization that virtual work arguments could be applied directly to differential equations to form a functional and that if the differential equation is self-adjoint we obtain the same symmetrical result as that obtained by energy arguments.

To amplify this point we return to the beam but begin at the outset with a virtual work argument. The shape $v = f(x)$ of a transversely loaded beam is governed by $EI\,d^4v/dx^4 = q$ and if we allow an arbitrary displacement $\delta v = g(x)$ along the beam and integrate this along the length to determine the associated virtual work δF, we obtain

$$\delta F = \int_0^L \left(EI \frac{d^4v}{dx} - q \right) g(x)\,dx.$$

Integrating the fourth derivative by parts twice

$$\delta F = EI\left[\delta v v_{xxx} - \delta v_x v_{xx} \right]_0^L + \int_0^L (EI\,\delta v_{xx} v_{xx} - \delta v q)\,dx. \qquad (17.11)$$

As $\delta v = g(x)$ is arbitrary it can be replaced by $v = f(x)$ when the required functional is obtained as

$$F(v) = \int \left[\tfrac{1}{2} EI(v_{xx})^2 - qv \right] dx. \qquad (17.12)$$

The factor $\tfrac{1}{2}$ is required as can be demonstrated by taking a perturbation $\delta(\tfrac{1}{2} v_{xx}^2) = \delta v_{xx} v_{xx}$ and the boundary terms in eqn (17.11) have been neglected. These terms are the interelement reactions which are clearly unknown but are assumed to cancel upon element summation.

Equation (17.12) can be verfied using the calculus of variations. To this end we apply the second-order Euler–Lagrange equation for line integrals which is[3]

$$\frac{\partial F}{\partial v} - \frac{d}{dx}\left(\frac{\partial F}{\partial v_x} \right) + \frac{d^2}{dx^2}\left(\frac{\partial F}{\partial v_{xx}} \right) = 0. \qquad (17.13)$$

This minimizes the line integral $\int F(x, v, v_x, v_{xx})\,ds$ where $ds = \sqrt{(dx^2 + dv^2)}$, of a functional $F(\ )$ along the deflected beam. From eqn (17.12) we have

$$\frac{\partial F}{\partial v} = -q, \qquad \frac{\partial F}{\partial v_x} = 0, \qquad \frac{\partial F}{\partial v_{xx}} = EIv_{xx} \qquad (17.14)$$

and substituting eqns (17.14) into eqn (17.13) we obtain

$$\frac{d^2(EIv_{xx})}{dx^2} - q = 0 \qquad (17.15)$$

which is the expected result.

If we note that the Euler–Lagrange equations are themselves derived by a perturbation argument,[3] however, then we can circumvent their use by applying a perturbation directly to the functional. Then allowing a perturbation δv in eqn (17.12) and integrating by parts twice we obtain

$$\delta F(v) = \int [\tfrac{1}{2} EI\delta(v_{xx})^2 - q\delta v]\,dx = \int [EI\delta v_{xx}v_{xx} - q\delta v]\,dx$$

$$= EI(\delta v_x v_{xx} - \delta v v_{xxx})| + \int [EI\delta v v_{xxxx} - q\delta v]\,dx. \qquad (17.16)$$

Seeking to minimize the integral we put $\lim_{v\to 0}(\delta F/\delta v) = 0$ when, neglecting the boundary terms, we obtain the required differential equation. Note, however, that the boundary terms in eqn (17.16) are opposite in sign to those of eqn (17.11) that is, had we included these reaction terms in the original function then our integration by parts would lead to their cancellation.

In summary, we can use virtual work to carry out the following operations:

(1) attack functionals directly, as shown originally in Section 3.1;

(2) attack differential equations directly, as we discovered at the close of Section 17.1 (by a virtual weighting of the residuals);

(3) form functionals from a given differential equation:

(4) obtain the differential equation corresponding to a given functional.

The last two applications also require the use of integration by parts. The second application does not strictly require integration by parts; we could leave things as they were but usually integration by parts is used to obtain more convenient (usually symmetric) matrices for finite elements. In the following section we use the functional for plane torsion to obtain finite element matrices by minimization of the functional but in Section 17.6 we shall see that (2) above leads to the same results.

17.4 Variational formulation for plane torsion

The governing differential equation for plane torsion is Poisson's equation

$$\nabla^2\psi + 2G\theta = \frac{\partial^2\psi}{\partial x^2} + \frac{\partial^2\psi}{\partial y^2} + 2G\theta = 0 \qquad (17.17)$$

where ψ is Prandtl's stress function. To obtain a functional from this equation we allow a virtual increment $\delta\psi$ and integrate by parts to obtain

$$\iint \delta\psi\,\psi_{xx}\,dx\,dy = \int \delta\psi\,\psi_x\,dy - \iint \delta\psi_x\psi_x\,dx\,dy \qquad (17.18a)$$

$$\iint \delta\psi\,\psi_{yy}\,dx\,dy = \int \delta\psi\,\psi_y\,dx - \iint \delta\psi_y\psi_y\,dx\,dy. \qquad (17.18b)$$

Defining $c_x = \partial x/\partial n$, $c_y = \partial y/\partial n$ as the direction cosines of the outwardly

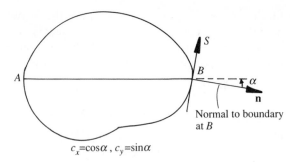

$c_x = \cos\alpha$, $c_y = \sin\alpha$

Fig. 17.1 Boundary integration

directed normal at the boundary shown in Fig. 17.1 and correspondingly writing $c_x = \partial y/\partial S$, $c_y = \partial x/\partial S$[†] as the anticlockwise directed tangent at the boundary, the boundary term in eqns (17.18) are combined to give

$$\int \delta\psi(\psi_x c_x + \psi_y c_y)\,\mathrm{d}S = \int \delta\psi \frac{\partial\psi}{\partial n}\,\mathrm{d}S \qquad (17.19)$$

as $\partial\psi/\partial n = (\partial\psi/\partial x)(\partial x/\partial n) + (\partial\psi/\partial y)(\partial y/\partial n) = \psi_x c_x + \psi_y c_y$.

As noted in Section 16.2.4 $\partial\psi/\partial n$ vanishes on the boundary of the domain and at element boundaries it is an 'interelement flux' which we assume to be self-cancelling between elements, that is, it is an interelement reaction. Hence neglecting the boundary terms in eqns (17.18) and noting that $\delta(\frac{1}{2}\psi_x)^2 = \delta\psi_x\psi_x$ and $\delta(\frac{1}{2}\psi_y)^2 = \delta\psi_y\psi_y$ we finally obtain the functional as

$$F(\psi) = \iint \left\{ \frac{1}{2}\left(\frac{\partial\psi}{\partial x}\right)^2 + \frac{1}{2}\left(\frac{\partial\psi}{\partial y}\right)^2 - 2G\theta\psi \right\} \mathrm{d}x\,\mathrm{d}y. \qquad (17.20)$$

In fact this functional gives the potential energy of the element multiplied by G which can be inferred by noting that $\sigma_{zx} = \partial\psi/\partial y$ and $\sigma_{zy} = -\partial\psi/\partial x$ so that the strain energy corresponding to these stresses is

$$U = \iint \left[\frac{\sigma_{zx}^2}{2G} + \frac{\sigma_{zy}^2}{2G} \right] \mathrm{d}x\,\mathrm{d}y = \frac{1}{2G}\iint \left[\left(\frac{\partial\psi}{\partial x}\right)^2 + \left(\frac{\partial\psi}{\partial y}\right)^2 \right] \mathrm{d}x\,\mathrm{d}y. \quad (17.21)$$

Note, however, that the right-hand side of eqn (17.21) can also be derived from first principles by the equilibrium arguments of Section 16.2.4.

Introducing finite element interpolation functions such that $\psi = \{f\}^{\mathrm{t}}\{\psi\}$ eqn (17.20) yields

$$F(\psi) = \iint [\frac{1}{2}\{\psi\}^{\mathrm{t}}(\{f_x\}\{f_x\}^{\mathrm{t}} + \{f_y\}\{f_y\}^{\mathrm{t}})\{\psi\} - 2G\theta\{\psi\}^{\mathrm{t}}\{f\}]\,\mathrm{d}x\,\mathrm{d}y$$

$$(17.22)$$

[†] To ensure that the arc length is integrated positively in all quadrants this term must be taken as positive.

where $\{f_x\} = \{\partial f/\partial x\}$ and $\{f_y\} = \{\partial f/\partial y\}$. Minimizing $F(\psi)$ with respect to $\{\psi\}$ yields the finite element equations

$$\iint [\{f_x\}\{f_x\}^{\mathrm{t}} + \{f_y\}\{f_y\}^{\mathrm{t}}]\,\mathrm{d}x\,\mathrm{d}y = 2G\theta \iint \{f\}\,\mathrm{d}x\,\mathrm{d}y \qquad (17.23)$$

or

$$k\{\psi\} = \{q\} \qquad (17.24)$$

where

$$k = \iint [\{f_x\}\{f_x\}^{\mathrm{t}} + \{f_y\}\{f_y\}^{\mathrm{t}}]\,\mathrm{d}x\,\mathrm{d}y \qquad (17.25a)$$

$$\{q\} = 2G\theta \iint \{f\}\,\mathrm{d}x\,\mathrm{d}y. \qquad (17.25b)$$

Equation (17.25b) yields forces comparable to body forces and the stiffness matrix of eqn (17.24) ca be written in the form

$$k = \iint B^{\mathrm{t}}DB\,\mathrm{d}x\,\mathrm{d}y \qquad (17.26)$$

where

$$B = \begin{bmatrix} \partial f_1/\partial x & \partial f_2/\partial x & \cdots & \partial f_n/\partial x \\ \partial f_1/\partial y & \partial f_2/\partial y & \cdots & \partial f_n/\partial y \end{bmatrix} \qquad (17.27a)$$

$$D = \begin{bmatrix} 1 & 0 \\ 0 & 1 \end{bmatrix} \qquad (17.27b)$$

which is reminiscent of the results of previous chapters. In the following chapters, however, we shall operate directly upon governing differential equations and use the notation of eqns (17.25) rather than introduce an artificial modulus matrix like that of eqn (17.27b).

Assembly of the element equations follows the usual procedure and after imposing boundary conditions $\psi = 0$ for all boundary nodes the system equations are solved for the remaining freedoms at the internal nodes. Finally the element stresses are calculated from

$$\sigma_{zx} = \frac{\partial \psi}{\partial y}, \qquad \sigma_{zy} = -\frac{\partial \psi}{\partial x}. \qquad (17.28)$$

As an example consider the square section subjected to plane torsion shown in Fig. 17.2. Taking advantage of symmetry a quadrant of the section is analysed using a single bilinear finite element. The interpolation functions are

$$\{f\} = \tfrac{1}{4}\{(1-a)(1-b), (1+a)(1-b), (1+a)(1+b), (1-a)(1+b)\} \qquad (17.29)$$

where $a = -1 \to 1$ and $b = -1 \to 1$.

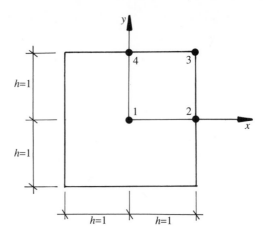

Fig. 17.2 Quadrant of a square cross-section subjected to plane torsion with a single bilinear finite element

Eliminating the freedoms $\psi_2 = \psi_3 = \psi_4 = 0$ only a single freedom remains and we obtain only a single equation $k_{11}\psi_1 = q_1$ where

$$k_{11} = \frac{1}{4h^2} \int_{-1}^{1} \int_{-1}^{1} [(a-1)^2 + (b-1)]^2 \frac{h^2}{4} \, da \, db = \frac{2}{3} \qquad (17.30)$$

$$q_1 = \frac{2G\theta}{4} \int_{-1}^{1} \int_{-1}^{1} (1 - a - b + ab) \frac{h^2}{4} \, da \, db = \frac{1}{2} \qquad (17.31)$$

If $G = \theta = 1$. Then the approximate solution is $\psi_1 = q_1/k_{11} = 0.75$ whereas the exact solution is 0.5833.[2,4]

The corresponding torque acting upon the section is given by eqn (16.27). that is, $T = 2G \int\int \psi \, dx \, dy$, so that the torsional rigidity of the section is given by

$$GJ = \frac{T}{\theta} = \frac{2G}{\theta} \int\int \psi \, dx \, dy = \Sigma 2 \int\int \{f\}^t \{\psi\} \, dx \, dy$$

$$= 4(2) \int_{-1}^{1} \int_{-1}^{1} f_1 \psi_1 \frac{h^2}{4} \, da \, db = 1.5 \qquad (17.32)$$

if we sum the result for the four quadrants, whereas the exact solution is $GJ = 2.2495$.[2,4]

Clearly more elements are needed to obtain a better approximation, although as Crandall shows for the finite difference method,[2] a sufficiently accurate result can be obtained with only nine nodes by applying h^2 extrapolation to the two solutions then available.

As we shall see in Section 19.2 eqns (17.28) are identical to those used to define stream functions for 'potential flow', which is then governed by Laplace's equation, $\nabla^2 \psi = 0$. A computer program for such potential flow is given in Chapter 20 and this can therefore be used for the analysis of plane torsion and other pseudoharmonic problems. For those more familiar with the mechanics of solids, therefore, plane torsion provides a useful introduction to finite element analysis of fluid flows.

17.5 Weighted residual methods

As noted in Section 17.1 weighted residual methods involve weighting the residual errors in differential equations in which *trial functions* are used to approximate the distributions of the dependent variables. In the finite element method the trial functions are the element interpolation functions and as eqn (17.3) shows the use of weighting functions alllows us to obtain simultaneous equations in which the unknowns are the element freedoms.

Such weighting functions, however, have a relatively abstract character and the motivation for their use is clarified by virtual work arguments. If we have a differential equation of the form $L(\{\bar{u}\}) = c$, where c is constant, then we introduce the interpolation $\{\bar{u}\} = F^t\{d\}$ and multiply by a perturbation $\{\delta\bar{u}\} = F^t\{\delta d\}$, yielding

$$\{\delta d\}_i^t \int FL(F^t\{d\}) \, dV = \{\delta d\}^t \int cF \, dV. \qquad (17.33)$$

Cancelling the common factor $\{\delta d\}^t$ we can write eqn (17.33) in the form

$$k\{d\} = \int FF^{*t} \, dV\{d\} = c \int F \, dV = \{q\} \qquad (17.34)$$

where F^* is a matrix obtained by differentiating F in accordance with the derivatives of the operator $L(\ )$.

Equation (17.34) is the same result as that given by the *Galerkin method* which is one of several weighted residual methods.[5, 6] The basic MWRs are described below.

17.5.1 Point collocation

In point collocation the approximate solution is made to intersect the true solution at N sampling points, as shown in Fig. 17.3. In order to achieve this the weighting functions are *Dirac delta functions* which for a typical point $P(x_i)$ in one dimension, for example, are defined by

$$\delta(x) = 0 \quad x \neq x_i, \quad \delta(x) = \infty \quad x = x_i, \quad \int_{-\infty}^{\infty} \delta(x) = 1. \qquad (17.35)$$

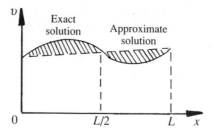

Fig. 17.3 Illustration of the use of MWR with three sampling points

Then replacing F in eqn (17.34) by a vector of Dirac functions we obtain

$$\{I\}F^{*\mathrm{t}}\{d\} = c\{I\} \qquad (17.36)$$

where $\{I\}$ is a unit vector.

Usually point collocation is applied to the complete domain so that eqn (17.36) becomes $F^{*\mathrm{t}}\{D\} = c\{I\}$ and F^* will be a fully populated matrix. If we work with a 'cell' of three equally spaced nodes and apply quadratic interpolation to the freedoms u_1, u_2, and u_3 we obtain

$$u = \tfrac{1}{2}(s^2 - s)u_1 + (1 - s^2)u_2 + \tfrac{1}{2}(s^2 + s)u_3 \qquad s = -1 \rightarrow 1. \quad (17.37)$$

If the differential equation is simply $\mathrm{d}^2 u/\mathrm{d}x^2 = 0$, for example, then differentiating eqn (17.37) twice gives

$$\frac{\mathrm{d}^2 u}{\mathrm{d}x^2} = \left\{\frac{\mathrm{d}^2 f}{\mathrm{d}x^2}\right\}^{\mathrm{t}} \{u\} = \frac{1}{h^2}\left\{\frac{\mathrm{d}^2 f}{\mathrm{d}s^2}\right\}^{\mathrm{t}} \{u\} = \frac{1}{h^2}(u_1 - 2u_2 + u_3). \quad (17.38)$$

If we apply this to each of N nodes, changing the subscripts accordingly, then the matrix $F^{*\mathrm{t}}$ of eqn (17.36) is a tridiagonal matrix with diagonal entries of 2 and off-diagonal entires of unity. In fact this speical case is the finite difference method which, because of its simplicity, is more widely used than point collocation, but as the trial function is implicitly quadratic in second-order problems (rather than of order N) it is much less accurate.

17.5.2 Subdomain collocation

In subdomain collocation the domain is divided into N subdomains and the integral of the residual error is equated to zero in each. Thus the weighting functions take the form of step functions which are zero outside the domain and unity within it. Then F in eqn (17.34) is deleted and we write for each subdomain

$$\int F^{*\mathrm{t}}\,\mathrm{d}V\{d\} = c\int \mathrm{d}V \qquad (17.39)$$

where, as in the finite element method, $\int \mathrm{d}V$ now denotes integration over the subdomain. In the one-dimensional example illustrated in Fig. 17.3 subdomain collocation is equivalent to balancing the hatched error areas. Once equation (17.39) has been evaluated for each subdomain, the subdomain equations are assembled in the same manner as in the finite element method. The method appears therefore to have much in common with the finite element method but there are major differences in practice, principal amongst these being that it is not customary to alter eqn (17.39) (say by integration by parts as in FEM) and the functions used to form $'F^*$ are not necessarily polynomials as is generally the case in FEM; transcendental functions, for example, are often used.

17.5.3 Method of least squares

In the method of least squares the weighting function is half the residual and the resulting square of the residual error is minimized with respect of each of the unknowns. Thus we have

$$\tfrac{1}{2} \frac{\partial}{\partial\{d\}} \int (F^{*\mathrm{t}}\{d\} - c)^2 \, \mathrm{d}V = 0. \tag{17.40}$$

Expanding the square term of eqn (17.40) we can easily show that after minimization the equations obtained are

$$\int F^* F^{*\mathrm{t}} \, \mathrm{d}V \{d\} = c \int F^* \, \mathrm{d}V. \tag{17.41}$$

In the one-dimensional example illustrated in Fig. 17.3 the method of least squares can be interpreted as minimizing the hatched area shown.

Observing eqn (17.41) it is clear that symmetric equations are obtained and if the method is applied to subdomains then it has much in common with the finite element method. There are fundamental differences, however, because these is no energy or equilibrium equation basis attached to the squaring of the residual error; it is simply a computational device which is more frequently applied in statistics than to physical problems.

17.5.4 The Galerkin method

The Galerkin method is the most wiedely used weighted residual method for finite elements. In the Galerkin method the weighting functions are the same functions as those used to form F^* so that we obtain the same result as eqn (17.34) which resulted from a virtual work agrument:

$$\int F F^{*\mathrm{t}} \, \mathrm{d}V \{d\} = c \int F \, \mathrm{d}V. \tag{17.42}$$

As with the collocation methods (except for the finite difference method if we include this as a special case) eqn (17.42) does not yield a symmetric matrix.

Hence it is the usual practice when the Galerkin method is applied to finite elements to use integration by parts to alter the governing differential equation. Then, as we saw in Section 17.2, so long as the differential equation is self-adjoint (that is, in general, it has even-order derivatives) we can obtain half the order of the derivatives in the original differential equation and obtain terms that are squared. These squared terms then lead to the formation of symmetric matrices when the finite element interpolation functions are introduced.

When symmetric matrices are obtained in this way they will be identical to those obtained by variational methods. This is because, as shown in Section 17.3, variational functionals can be obtained by a 'parallel' integration by parts procedure allied with a virtual work argument.

We shall further demonstrate this equivalence by applying the Galerkin method to the plane torsion problem in Section 17.6 to obtain finite element matrices identical to those obtained in Section 17.4 by variational methods.

If we are using polynomial bases for our trial functions, as is customary in the finite element method, then a special case of the Galerkin method arises. This is the *method of moments* in which the weighting function vector is, for example, $\{1, x, x^2 \ldots, x^n\}$, in one-dimensional problems. Here n is the order of the polynomial used to form the strain interpolation matrix F.[7] Except in certain special one-dimensional problems, however, this method has limited application and the Galerkin method is more appropriate in the finite element context.

We now consider a one-dimensional example of the point collocation, subdomain collocation, least squares, and Galerkin methods to demonstrate these. This is the equilibrium equation for an eccentrically loaded column which takes the form

$$\frac{\mathrm{d}^2 v}{\mathrm{d}x^2} + c^2(v - 10) = 0 \tag{17.43}$$

with boundary conditions $v = 0$ at $x = 0$ and $x = L$; $L = 2000$; and $c = 7.5 \times 10^{-4}$. Equation (17.43) has the exact solution

$$v = \frac{-10[\sin cx + \sin c(L - x) - \sin cL]}{\sin cL}. \tag{17.44}$$

We choose the fourth-order interpolation

$$v = \{x^4 - L^3 x, x^3 - L^2 x, x^2 - Lx\}^{\mathrm{t}}(a_1, a_2, a_3) = \{f\}^{\mathrm{t}}\{a\} \tag{17.45}$$

as this satisfies the boundary conditions, and to simplify calculations we solve for the coefficients a, b, and c rather than for nodal displacements.

Differentiating eqn (17.45) twice and substituting the result into eqn (17.43) lead to

$$L(\{a\}) = [12x^2 + c^2 x^4 - c^2 L^3 x, 6x + c^2 x^3 - c^2 L^2 x, 2 + c^2 x^2 - c^2 Lx]$$

$$\times \{a\} - 10c^2 = F^*\{a\} - 10c^2 = 0. \tag{17.46}$$

Using point collocation at $x = 0$, $L/2$, and L we simply substitute these values in eqn (17.46) to obtain

$$\begin{bmatrix} 0 & 0 & 2c \\ 3L^2 - 0.4375c^2L^4 & 3L - 0.375c^2L^3 & 1 - 0.25c^2L^2 \\ 12L^2 & 6L & 1 \end{bmatrix} \begin{Bmatrix} a_1 \\ a_2 \\ a_3 \end{Bmatrix} = \begin{Bmatrix} 10c^2 \\ 10c^2 \\ 10c^2 \end{Bmatrix}$$

(17.47)

After solving for $\{a\}$ the result is substituted back into eqn (17.45) yielding the results shown in Table 17.1

Using subdomain collocation with the subdomains $0 < x < L/3$, $L/3 < x < 2L/3$, and $2L/3 < x < L$ we obtain $k\{a\} = 10c^2L/3\{I\}$ where

$$k = \begin{vmatrix} 4L^3 - 133c^2L^5/2430 & L^2/3 - 17c^2L^4/324 & 2L/3 - 7c^2L^3/162 \\ 28L^3 - 343c^2L^5/2430 & L^2 - 39c^2L^4/324 & 2L/3 - 13c^2L^3/162 \\ 76L^3 - 253c^2L^5/2430 & L^2 - 25c^2L^4/324 & 2L/3 - 7c^2L^3/162 \end{vmatrix}$$

(17.48)

yielding the solution shown in Table 17.1.

Using the method of least squares we substitute the matrix F^* of eqn (17.46) into eqn (17.41), yielding $k\{a\} = \{q\}$ where, denoting c^2L^2 as β,

$$k = \begin{bmatrix} L^2(288/5 - 36\beta/7 + 2\beta^2/9) & L^4(36 - 4\beta + 11\beta^2/60) & L^3(16 - 12\beta/5 + 5\beta^2/42) \\ & 8L^3(3 - 2\beta/5 + 2\beta^2/105) & L^2(12 - 2\beta + \beta^2/10) \\ \text{Symmetric} & & L(8 - 4\beta/3 + \beta^2/15) \end{bmatrix}$$

(17.49)

Table 17.1 Weighted residual solutions of eqn (17.43)

x	Exact	Point collocation	Subdomain collocation	Least squares	Galerkin
0	0	0	0	0	0
200	−1.2799	−1.2828	−1.2804	−1.2801	−1.3061
400	−2.3064	−2.3113	−2.3076	−2.3067	−2.3222
600	−3.0566	−3.0626	−3.0580	−3.0566	−3.0480
800	−3.5135	−3.5199	−3.5150	−3.5132	−3.4837
1000	−3.6670	−3.6735	−3.6685	−3.6665	−3.6290

and

$$\{q\} = \{80c^2 L^3 - 6c^4 L^5, 60c^2 L^2 - 5c^4 L^4, 40c^2 L - 10c^4 L^3/3\}. \quad (17.50)$$

Solving for $\{a\}$ and substituting the result into eqn (17.45) yields the solutions shown in Table 17.1

Finally using the Galerkin method we substitute the matrix F^* of eqn (17.46) and the matrix $F = \{f\}$ given by eqn (17.45), that is,

$$F = \{x^4 - L^3 x, x^3 - L^2 x, x^2 - Lx\} \quad (17.51)$$

into eqn (17.42). The resulting equations are

$$\begin{bmatrix} L^7(\beta/9 + 9L^2/7) & L^6(11\beta/120 - 1) & L^5(5\beta/84 - 3/5) \\ & L^5(8\beta/105 - 4/5) & L^4(\beta/20 - 1/2) \\ \text{Symmetric} & & L^3(\beta/30 - 1/3) \end{bmatrix} \begin{Bmatrix} a_1 \\ a_2 \\ a_3 \end{Bmatrix}$$

$$= \begin{Bmatrix} -3c^2 L^5 \\ -5c^2 L^4/2 \\ -5c^2 L/3 \end{Bmatrix}. \quad (17.52)$$

Solving for $\{a\}$ and substituting the result in eqn (17.45) yields the solutions shown in Table 17.1. Note that the symmetry which arises in eqn (17.52) is due to the special nature of the interpolation used in eqn (17.45) and will not in general occur unless the governing equation is self-adjoint and Galerkin weighting is followed by integration by parts.

In Table 17.1 the least squares method gives the most accurate results but this method involves a good deal more computation than the others. Point collocation and the Galerkin method perform fairly well but are less accurate than subdomain collocation. This is because the use of three intervals in the suddomain collocation is comparable to the use of three finite elements whereas we have, in effect, only one element with the Galerkin method. Moreover the Galerkin method will yield better results if integration by parts is used to reduce the order of the govering differential equation and this is the usual procedure in the finite element method and leads to results of similar accuracy to those with the least squares method.

17.6 Application of the Galerkin method to plane torsion

Returning to the plane torsion problem of Fig. 17.2 the governing differential equation is Poisson's equation:

$$\iint \left(\frac{\partial^2 \psi}{\partial x^2} + \frac{\partial^2 \phi}{\partial y^2} + 2G\theta \right) dx\,dy = 0. \quad (17.53)$$

Introducing the finite element interpolation $\psi = \{f\}^t \{\psi\}$ and weighting eqn (17.53) by the vector $\{f\}$ (note again that this follows from the virtual work argument used in eqn (17.33)) we obtain

$$\iint \{f\}\,(\{f_{xx}\}^t + \{f_{yy}\}^t + 2G\theta)\,\mathrm{d}x\,\mathrm{d}y\{\psi\} = \{0\} \qquad (17.54)$$

where $\{f_{xx}\} = \{\partial^2 f/\partial x^2\}$ and $\{f_{yy}\} = \{\partial^2 f/\partial y^2\}$.

Applying integration by parts to the terms involving second derivatives in eqn (17.54) we obtain

$$\iint \{f\}\,\{f_{xx}\}^t\,\mathrm{d}x\,\mathrm{d}y = \int \{f\}\,\{f_x\}^t\,\mathrm{d}y - \iint \{f_x\}\,\{f_x\}^t\,\mathrm{d}x\,\mathrm{d}y \qquad (17.55a)$$

$$\iint \{f\}\,\{f_{yy}\}^t\,\mathrm{d}x\,\mathrm{d}y = \int \{f\}\,\{f_y\}^t\,\mathrm{d}x - \iint \{f_y\}\,\{f_y\}^t\,\mathrm{d}x\,\mathrm{d}y. \qquad (17.55b)$$

Defining the direction cosines of the normal and tangent vector at the boundary (see Fig. 17.1) as

$$\frac{\mathrm{d}x}{\mathrm{d}n} = c_x, \qquad \frac{\mathrm{d}y}{\mathrm{d}n} = c_y \qquad (17.56a)$$

$$\frac{\mathrm{d}x}{\mathrm{d}S} = c_y, \qquad \frac{\mathrm{d}y}{\mathrm{d}S} = c_x \qquad (17.56b)$$

the boundary terms in eqns (17.55) can be combined to give

$$\int \{f\}\,(\{f_x\}^t c_x + \{f_y\}^t c_y\,\mathrm{d}S. \qquad (17.57)$$

Noting that

$$\left\{\frac{\partial f}{\partial n}\right\} = \left\{\frac{\partial f}{\partial x}\right\}\frac{\partial x}{\partial n} + \left\{\frac{\partial f}{\partial y}\right\}\frac{\partial y}{\partial n}$$

$$= \{f_x\}^t c_x + \{f_y\}^t c_y \qquad (17.58)$$

the boundary term of eqn (17.57) reduces to $\int \{f\}\,\{\partial f/\partial n\}\,\mathrm{d}S$ and substituting this result and the last terms in eqns (17.55) into eqn (17.54) we obtain the element eqnations in the form

$$k\{\psi\} = \{q\} + \{q_r\} \qquad (17.59)$$

where

$$k = \iint (\{f_x\}\,\{f_x\}^t + \{f_y\}\,\{f_y\}^t)\,\mathrm{d}x\,\mathrm{d}y \qquad (17.60)$$

$$\{q\} = 2G\theta \iint \{f\}\,\mathrm{d}x\,\mathrm{d}y \qquad (17.61)$$

$$\{q_r\} = \int \{f\}\,\frac{\partial\psi}{\partial n}\,\mathrm{d}S. \qquad (17.62)$$

Equation (17.62) gives the interelement reactions and we assume these cancel between elements in the assembly process. Then eqn (17.62) need only be evaluated at the domain boundary but in plane torsion, as shown in Section 16.2.4, $\partial\psi/\partial n$ is zero at the domain boundary.

Equations (17.60) and (17.61) are identical to eqns (17.25) obtained by minimization of the functional for plane torsion, so that numerical solution of the problem proceeds proceeds in the manner demonstrated in Section 17.4. Solution of other pseudoharmonic problems (that is, with a differential equation of the form $\nabla^2 \phi = c$) proceeds in a similar fashion and other examples are discussed in Chapter 18.

17.7 Application of the Galerkin method to fourth-order problems

To demonstrate application of the Galerkin method to fourth-order problems we return to the familiar beam. Applying Galerkin weighting to the differential equation we have for a beam with uniform rigidity

$$EI \int \{f\} \frac{d^4_v}{dx^4} dx = \int q \{f\} dx. \tag{17.63}$$

Substituting for v using the finite element interpolation, $v = \{f\}^t\{d\}$, and reducing the order of the differential equation by applying integration by parts to the left-hand side we obtain

$$\int \{f\} \{f_{xxxx}\}^t dx = \{f\} \{f_{xxx}\}^t | - \int \{f_x\} \{f_{xxx}\}^t dx \tag{17.64}$$

where | denotes evaluation at the ends of the element, and $\{f_x\} = \{df/dx\}$, and so on. The last term in eqn (17.64) clearly does not result in a symmetric matrix so that we integrate this term by parts again, yielding

$$- \int \{f_x\} \{f_{xxx}\}^t dx = - \{f_x\} \{f_{xx}\}^t | + \int \{f_{xx}\} \{f_{xx}\}^t dx. \tag{17.65}$$

Then substituting the boundary terms of eqns (17.64) and (17.65) and the last term of eqn (17.65) for the left-hand side of eqn (17.63) we obtain

$$EI \int \{f_{xx}\} \{f_{xx}\}^t dx \{d\} = EI (\{f\} \{f_{xxx}\}^t - \{f_x\} \{f_{xx}\}^t) \{d\} | + \int q \{f\} dx$$

or

$$k\{d\} = \{q_r\} + \{q_t\}. \tag{17.66}$$

Now the stiffness matrix is symmetric and is identical with that obtained by variational methods. The consistent loads for the surface traction, $\{q_t\}$, are also identical with those obtained variationally. The reason for this lies in the virtual work arguments discussed at the close of Section 17.2 and also in Section 17.3, that is, we have in effect obtained $\{q_t\}$ by cancellation of perturbations $\{\delta d\}^t$ from the expession $\{\delta d\}^t \int \{f\} q\, dx$. Finally the loads $\{q_r\}$ in eqn (17.66) are clearly unknown, for example we cannot calculate the term

$$F_{v1} = EI(\{f\} \{f_{xxx}\}^t \{d\})_{x=0} = EI \left(\frac{d^3 v}{dx^3} \right)_{x=0} \tag{17.67}$$

until $\{d\}$ is known. These terms arise out of the virtual work done by the interelement reactions and we must assume that these cancel when the element equations are summed.

If we use the freedoms $\{d\} = \{v_1, \phi_1, v_2, \phi_2\}$ and cubic interpolation then there will be slope continuity between elements, that is, they are C_1 continuous, so that the essential boundary conditions of the fourth-order governing diferential equation (eqn (17.63)) are satisfied at element level. This emphasizes an advantage of using integration by parts, namely that it reduces continuity requirements and, moreover, if it yields quadratic forms such as $\{f_{xx}\}\{f_{xx}\}^t$ then we obtain a formulation of a least squares character which, as we saw in Table 17.1, will generally be more accurate than other forms.

To see why satisfaction of the essential boundary conditions does ensure cancellation of the interelement reactions we transpose the second boundary term of eqn (17.66) and introduce perturbations $\{\delta d\}$ to obtain

$$(EI\{\delta d\}^t \{f_{xx}\})\{f_x\}^t\{d\} = \delta M_1 \phi_1 \qquad (17.68)$$

at $x = 0$. Then, as another element abutting at node 1 will have the same slope ϕ_1, a loading δM_1 is associated with a uniquely defined quantity of virtual work.

Many further applications of the Galerkin method appear in the following chapters and in Section 19.45 we shall see that it is possible to obtain symmetric element matrices even when the governing differential equations involve derivatives of odd order. As we have seen above, integration by parts clearly reveals the interelement reactions and virtual work arguments in fact provide the soundest basis for the Galerkin and other weighted residual methods in physical problems because these always involve conservation and continuity conditions. For brevity, in our descriptions, however, we shall retain the term Galerkin method, but the reader would do well to recall the underlying virtual work argument in the present text.

17.8 Concluding remarks

Previous chapters concentrated upon structural applications in which stiffness matrices with an energy basis were developed and direct appeal to differential equations was not required. In many field and fluid flow problems, however, direct formation of an energy functional is very difficult and it is more expeditious to form instead a govering differential equation; many examples of this process were given in Chapter 16.

Then, as we have seen, virtual work arguments can be applied directly to the governing differential equation and integration by parts used to seek a solution of the same form as would be obtained by energy arguments. In the finite element method we are most likely to use the same interpolations for the virtual perturbations as for the element freedoms for reasons of consistency. This corresponds to Galerkin weighting of the residuals and the following integration by parts, as shown in Section 17.3, corresponds to formation of a variational functional.

Following this process we then obtain a solution with the same least squares character of energy arguments which also reduces the order of the essential boundary conditions and thence the interelement continuity requirements.

Extensions of the weighted residual methods to incorporate additional interelement continuity constraints, for example, can be implemented using the techniques discussed in Chapter 4 and extensions of these are discussed in Chapter 23.

Finally it should be remarked that we have not considered incompressible solids (that is, $v = 0.5$ which 'collapses' D in eqn (2.15)) for which special variational principles[8] or, perhaps more directly, a fluid analogy may be used.

Reference

1. Argyris, J. H. (1960). *Energy Theorems and Structural Analysis*. Butterworths, London. (1956).
2. Crandall, S. H. (1956). *Engineering Analysis*. McGraw-Hill, New York.
3. Irving, I. and Mollineaux, N. (1959). *Mathematics in Physics and Engineering*. Academic Press, London.
4. Timoshenko, S. P. and Goodier, J. N. (1951). *Theory of Elasticity*, (3rd edn). McGraw-Hill, New York.
5. Finlayson, B. A. and Scriven, L. E. (1966). The method of weighted residuals; a review. *Appl. Mech. Rev.* **19**, 9.
6. Finlayson, B. A. (1972). *The Method of Weighted Residuals and Variational Principles*. Academic Press, New York.
7. Connor, J. J. and Brebbia, C. A. (1977). *Finite Element Techniques for Fluid Flow*. Butterworths, London.
8. Herrmann, L. R. (1965). Elastic equations for incompressible and nearly incompressible materials by a variational theorem. *J. Amer. Inst. Aeron. Astron.*, **3**, 1896.

18
Pseudoharmonic field problems

Chapter 16 introduced the basic differential equations for many fluid flow and other field problems and Chapter 17 discussed methods for solving them. The following chapter discusses the solution of heat and fluid flow problems governed by the diffusion equation. The object of finite element heat flow analysis is to determine the distribution of nodal temperatures in a discretized domain. For heat flow in solids such an analysis may be followed by an analysis of the thermal stresses following the procedure outlined in Section 3.6 and in the case of transient heat flow this may take place after each time step. Two further diffusion problems, those of groundwater seepage and foundation consolidation, are then discussed and the chapter closes with an electromagnetic wave problem governed by the Helmholtz equation.

18.1 Introduction

In Chapter 16 we introduced the concept of continuity (or more strictly conservation) equations for a variety of flow problems for heat, fluids and so on. This chapter deals principally with diffusion problems the governing equations for which are continuity equations. In steady state one-dimensional diffusion problems the velocity of diffusion u per unit area per unit time of a substance of concentration c per unit volume is given by Fick's first law of diffusion,

$$u = -D\frac{\partial c}{\partial x}$$

that is, the rate of diffusive flow is proportional to the concentration gradient and D is a diffusion coefficient with units of area/time. The same result is known as Darcy's law in the context of seepage problems or Fourier's law of heat conduction in heat flow problems.

If the diffusion problem is time dependent then we require the sum of the time rate of increase of the diffusing substance in an infinitesimal control volume $\delta x\,\delta A$ (that is, $(\partial c/\partial t)\delta x\,\delta A$) and the rate of outflow of the substance

$(\partial u/\partial x)\delta x\delta A$, to be zero. This leads to Fick's second law of diffusion which, for an isotropic material, is

$$\frac{\partial c}{\partial t} = D\frac{\partial^2 c}{\partial x^2}.$$

Generalizing to three dimensions one obtains

$$\frac{\partial c}{\partial t} = D\nabla^2 c = D\left(\frac{\partial^2 c}{\partial x^2} + \frac{\partial^2 c}{\partial y^2} + \frac{\partial^2 c}{\partial z^2}\right).$$

With appropriate changes in variables this equation governs many other problems, including those of heat flow and seepage, both of which are discussed in the following sections. The basic diffusion equations can also be generalized to include internal source terms and boundary flux terms and this is demonstrated in the following section.

18.2 Steady state heat flow

Steady state heat flow problems were one of the first non-structural field problems to which the finite element method was applied.[1] Steady state heat flow in an infinitesimal two-dimensional anisotropic element with temperature-independent thermal conductivities κ_x and κ_y is governed by the equation

$$\frac{\partial}{\partial x}\left(\kappa_x\frac{\partial T}{\partial x}\right) + \frac{\partial}{\partial y}\left(\kappa_y\frac{\partial T}{\partial y}\right) + G_v = 0 \tag{18.1a}$$

where T is the temperature and G_v is a known rate of internal heat generation. In addition, the heat flux across an infinitesimal element dS at the boundary is given by

$$\alpha(T - T_a) + G_s = 0 \tag{18.1b}$$

where T_a is the ambient temperature and G_s is a known heat flux. The first term corresponds to Newton's law of cooling and α is a conduction coefficient. Then for elements at the boundary we weight the sum of the residuals of eqns (18.1). Applying Galerkin weighting to eqns (18.1) and integrating the second derivatives by parts one obtains for a homogeneous (that is, κ_x and κ_y are constant) finite element of constant thickness, equations of the form

$$(k + k_b)\{T\}_e = \{q\} \tag{18.2}$$

where

$$k = t\kappa_x \int \{f_x\}\{f_x\}^t \, \mathrm{d}x\,\mathrm{d}y + t\kappa_y \int \{f_y\}\{f_y\}^t \, \mathrm{d}x\,\mathrm{d}y \tag{18.3}$$

$$k_b = \alpha \int \{f\}\{f\}^t \, dS \qquad (18.4)$$

$$\{q\} = t \int \{f\} G_v \, dx \, dy + \int \{f\} G_s \, ds + \alpha \int \{f\} T_a \, dS$$

$$+ t\kappa_x \int \{f\} \frac{\partial T}{\partial x} dy + t\kappa_y \int \{f\} \frac{\partial T}{\partial y} dx. \qquad (18.5)$$

Here $\{f_x\} = \partial\{f\}/\partial x$ and $\{f_y\} = \partial\{f\}/\partial y$, and k_b corresponds to the contact stiffness matrix of solid mechanics introduced in Chapter 3. The loading terms involving G_v, G_s, and α are comparable to body forces and surface tractions, and those involving κ_x and κ_y are self-cancelling *interelement fluxes*.

Considering a one-dimensional element of circular cross-section with two freedoms and linear interpolation with $\{f\} = \{1 - s, s\}$ one obtains the conduction and boundary convection matrices of eqns (18.3) and (18.4) as

$$k = \frac{A\kappa}{L} \begin{bmatrix} 1 & -1 \\ -1 & 1 \end{bmatrix}, \qquad k_b = \frac{2\pi r \alpha L}{6} \begin{bmatrix} 2 & 1 \\ 1 & 2 \end{bmatrix} \qquad (18.6)$$

where $A = \pi r^2$. If G_v and G_s are zero the element 'loads' are given by

$$\{q\} = 2\pi r \alpha T_a L \{\tfrac{1}{2}, \tfrac{1}{2}\}. \qquad (18.7)$$

As an example, consider the problem shown in Fig. 18.1, where a rod of circular cross-section has specified temperature $T = 150°C$ at the left end and the ambient temperature is $T_a = 40°C$. Using three elements one obtains for the shorter element

$$k = \pi \begin{bmatrix} 48 & -48 \\ -48 & 48 \end{bmatrix}, \qquad k_b = \pi \begin{bmatrix} 10 & 5 \\ 5 & 10 \end{bmatrix}, \qquad \{q\} = \pi\{600, 600\}. \qquad (18.8)$$

k is halved, and k_b and $\{q\}$ are doubled for the longer elements. At the right-hand end a convection term

$$\alpha A(T_4 - T_a) = \pi\alpha(T_4 - T_a) \qquad (18.9)$$

must be included, $\pi\alpha$ being added to the fourth pivot of the conductance matrix. (This corresponds to a lumped boundary stiffness in stress analysis; in the present context it is termed a *point sink*.) and $\pi\alpha T_a = 400\pi$ is added to the

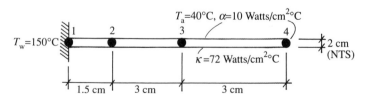

Fig. 18.1 A spar of circular cross-section

'loads' on the right-hand side. Thus assembling all element contributions one obtains

$$
\begin{bmatrix}
58 & -43 & 0 & 0 \\
-43 & 102 & -14 & 0 \\
0 & -14 & 88 & -14 \\
0 & 0 & -14 & 54
\end{bmatrix}
\begin{Bmatrix} T_1 \\ T_2 \\ T_3 \\ T_4 \end{Bmatrix}
=
\begin{Bmatrix} 600 \\ 1800 \\ 2400 \\ 1600 \end{Bmatrix}
\tag{18.10}
$$

and, as $T_1 = 150$ is specified, this reduces to

$$
\begin{bmatrix}
102 & -14 & 0 \\
-14 & 88 & -14 \\
0 & -14 & 54
\end{bmatrix}
\begin{Bmatrix} T_2 \\ T_3 \\ T_4 \end{Bmatrix}
=
\begin{Bmatrix} 8250 \\ 2400 \\ 1600 \end{Bmatrix}
\tag{18.11}
$$

adding 43×150 to q_2 as an equivalent boundary loading (that is, $\delta\{q\} = -K_{bf}^t\{T\}_b$ as given by eqn (15.15). The solution is then $\{T\} = \{150, 87.4, 47.9, 42.0\}$, a reasonable approximation to the exact solution which is[2] $\{T\} = \{150, 89.9, 50.6, 43.3\}$ considering that only three elements are used. Repeating the exercise with five equal elements one obtains $\{T\} = \{150, 86.3, 61.7, 49.8, 44.8, 43.0\}$ which is a good approximation to the exact solution $\{T\} = \{150, 89.9, 62.8, 50.6, 45.2, 43.3\}$ obtained from

$$
\frac{T - T_a}{T_w - T_a} = \frac{\cosh\beta(L - x) + (\beta/2)\sinh\beta(L - x)}{\cosh\beta L + (\beta/2)\sinh\beta L} \qquad \text{where } \beta = \sqrt{\left(\frac{2\alpha}{\kappa}\right)}.
$$

18.3 Transient heat flow

For transient heat flow in one dimension the governing differential equation, introduced in Section 18.1, is Fick's second law of diffusion which, with appropriate changes in parameters, is

$$
\frac{\kappa \partial^2 T}{\partial x^2} = \frac{\rho c_v \partial T}{\partial t}.
\tag{18.12}
$$

Using the Galerkin method and integration by parts for the left-hand side, first-order time-dependent equations are obtained of the form[3,4]

$$
k_D\{\dot{T}\}_e + k\{T\}_e = \{q\}
\tag{18.13}
$$

where

$$
k_D = \rho c_v \int \{f\}\{f\}^t \, dV = \frac{\rho c_v AL}{6}\begin{bmatrix} 2 & 1 \\ 1 & 2 \end{bmatrix}
\tag{18.14}
$$

is the 'damping' matrix, and the other contributions are as in Section 18.2.

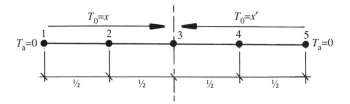

Fig. 18.2 Initial conditions in a bar of unit cross-sectional area

As an example consider the problem shown in Fig. 18.2, where a bar initially has a linear variation of temperature up to a maximum at the middle and is perfectly insulated except at the ends where $T_a = 0$. Using forward differences (see eqn (14.5)) for time stepping the system equations are

$$K_D\{T\}_{n+1} = (K_D - \delta K)\{T\}_n. \tag{18.15}$$

Using two elements to analyse the left half only, but including a third element and node 4 temporarily, one obtains

$$\frac{\rho c_v}{4} \begin{Bmatrix} T_1 \\ 2T_2 \\ 2T_3 \\ T_4 \end{Bmatrix}_{n+1} = \begin{bmatrix} \frac{1}{4}\rho c_v - 2\delta\kappa & 2\delta\kappa & 0 & 0 \\ 2\delta\kappa & \frac{1}{2}\rho c_v - 4\delta\kappa & 2\delta\kappa & 0 \\ 0 & 2\delta\kappa & \frac{1}{2}\rho c_v - 4\delta\kappa & 2\delta\kappa \\ 0 & 0 & 2\delta\kappa & \frac{1}{4}\rho c_v - 2\delta\kappa \end{bmatrix} \begin{Bmatrix} T_1 \\ T_2 \\ T_3 \\ T_4 \end{Bmatrix}_n$$

$$\tag{18.16}$$

for simplicity using lumping in the K_D matrix. With $\rho, c_v, \kappa = 1$, $T_1 = 0$, and $T_2 = T_4$ (symmetry) one obtains, first multiplying through by the factor 4, the two recurrence relations

$$2T_3 = 8\delta T_2 + (2 - 16\delta)T_3 + 8\delta(T_4 = T_2)$$

or

$$T_3 = 8\delta T_2 + (1 - 8\delta)T_3 \tag{18.17}$$

$$2T_2 = (2 - 16\delta)T_2 + 8\delta T_3. \tag{18.18}$$

With the initial conditions $(T_2)_0 = 0.5$, $(T_3)_0 = 1$ and using time steps $\delta = 0.02$ one obtains at $t = 0.10$ (after five steps), $T_2 = 0.45$, $T_3 = 0.70$, which compares well with the exact series solution $T_2 = 0.45$, $T_3 = 0.64$ from Kreyszig.[5]

The problem is highly idealized but demonstrates the simplicity of the recurrence relations provided by the finite element method for time-dependent problems when forward differences and lumping are used. Nevertheless it should be noted that the Crank–Nicolson formula, obtained by using eqns (14.2), is more stable[6] and involves a smaller truncation error, as was shown in Section 14.2.

18.4 Plane seepage

For steady state plane seepage the governing equation is of the same form as eqn (18.1a) that is, neglecting the source term we have

$$\frac{\partial}{\partial x}\left(k_x \frac{\partial H}{\partial x}\right) + \frac{\partial}{\partial y}\left(k_y \frac{\partial H}{\partial y}\right) = 0 \qquad (18.19)$$

where H is the static pressure head and k_x, k_y are the permeability coefficients. The velocities of flow are given by Darcy's laws which are

$$u = -k_x \partial H/\partial x, \qquad v = -k_y \partial H/\partial y \qquad (18.20)$$

so that flow is in the reverse direction to the pressure gradient and the governing equation (eqn 18.19) can be obtained by substituting eqns (18.20) into the continuity equation $\partial u/\partial x + \partial v/\partial y = 0$. The finite element formulation will closely resemble that given for steady state heat flow in Section 18.2 except that the forcing terms are now simply specified fluid fluxes at the boundary.

In some cases permeabilities k'_x, k'_y will be known for axes inclined at some angle α to the global axes, in which case the matrix D in eqn (17.27b) becomes[7]

$$D = \begin{bmatrix} c & s \\ -s & c \end{bmatrix} \begin{bmatrix} k'_x & 0 \\ 0 & k'_y \end{bmatrix} \begin{bmatrix} c & -s \\ s & c \end{bmatrix} \qquad (18.21)$$

where $c = \cos\alpha$, $s = \sin\alpha$. This follows from eqn (2.50) if we note that in elasticity the anisotropic modulus matrix relating to axes at angle α to the xy-axes is obtained using the same transformation as for stresses. Thus, E_x and E_y have the same dimensions as the stresses and are transformed in the same way. A similar argument justifies the permeability transformation of eqn (18.21).

In free surface flows when the location of the *phreatic surface* (where $H = 0$) is not known one begins with an initial guess for this and calculates the nodal pressures, adjusting the mesh at the surface upwards wherever $H_{node}/\rho g > y_{node}$, repeating the process until convergence is obtained (generally with about five iterations). The location of the free surface may also be time dependent (for example following rapid drawdown in an earth dam) when a time stepping analysis similar to that in Section 18.3, with the first analysis at $t = 0$ giving the initial conditions, is needed.[8]

18.5 One-dimensional consolidation

The differential equation governing one-dimensional consolidation is again the diffusion equation,[9]

$$c_v \frac{\partial^2 p}{\partial x^2} = \frac{\partial p}{\partial t} \tag{18.22}$$

where p is the pressure of the diffusing pore water and the *coefficient of vertical consolidation* c_v is given by

$$c_v = \frac{k_x(1 + e)}{\gamma_w a_v} \tag{18.23}$$

where k_x is the permeability, e is the void ratio, γ_w is the specific weight of the water, and a_v is the coefficient of compressibility of the soil. Once again the element equations are of the form of eqns (18.13):

$$k_D\{\dot{p}\} + k\{p\} = \{q\}. \tag{18.24}$$

Using linear elements we obtain the element matrices

$$k = \frac{c_v}{L}\begin{bmatrix} 1 & -1 \\ -1 & 1 \end{bmatrix}, \quad k_D = \frac{L}{6}\begin{bmatrix} 2 & 1 \\ 1 & 2 \end{bmatrix}, \quad \{q\} = c_v\begin{bmatrix} (\mathrm{d}p/\mathrm{d}x)_i \\ (\mathrm{d}p/\mathrm{d}x)_j \end{bmatrix} \tag{18.25}$$

using eqns (18.6), (18.14), and the fourth term of (18.5) with altered parameters. Here $\{q\}$ are self-cancelling interelement fluxes. Assembling the element equations and using backward differences (eqn (14.4)) for $\{\dot{p}\}$ the recurrence relation obtained is

$$\left(K + \frac{1}{\delta}K_D\right)\{p\}_{n+1} = \{Q\}_n + \frac{1}{\delta}K_D\{p\}_n. \tag{18.26}$$

Figure 18.3 shows an example problem and the results obtained by Desai and Johnson[10] using the twenty linear elements, which these authors find to give satisfactory results for this type of problem. Here loading at the surface gives a vertical stress σ which remains constant and is initially equilibrated by the pore pressure throughout the layer.[11]

With time the pore pressure gradually dissipates, transferring the stress to the soil skeleton, so that the effective stress $\sigma' = \sigma - p$ gradually increases giving rise to long-term settlements (to which the initial elastic settlement should be added) governed by Hooke's law,

$$\mathrm{d}\sigma = -\frac{\mathrm{d}e}{a_v}. \tag{18.27}$$

The settlement at the surface is given by integrating the strain $\mathrm{d}e$ through the depth. The problem is of the same type as that in Section 18.3 and assembly of the system equations follows the same lines but here good results cannot be obtained with as few elements, because the initial 'square' pressure distribution changes very rapidly in the initial time steps near $x = 0, H$.

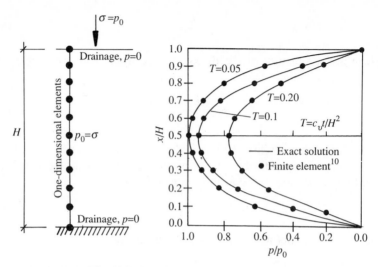

Fig. 18.3 One-dimensional consolidation

Once again the problem is highly idealized but such finite element analyses are easily extended to deal with layered media and two-dimensional consolidation.

18.6 Eigenvalues of electromagnetic vibration

Figure 18.4 shows an electromagnetic waveguide of rectangular cross-section in which both the transverse electric wave $\psi(x, y)$ and the transverse magnetic wave $\phi(x, y)$ satisfy the Helmholtz equation:

$$\nabla^2\psi + \lambda\psi = 0 \qquad \text{and} \qquad \nabla^2\phi + \lambda\phi = 0 \qquad (18.28)$$

subject to the boundary conditions $\partial\psi/\partial n = 0$, $\phi = 0$ around the edges.[12] Thus we have a *Neumann problem* for ψ and a *Dirichlet problem* for ϕ. If only

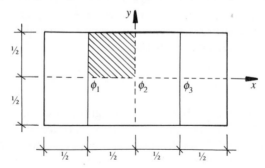

Fig. 18.4 Rectangular waveguide

the magnetic wave is analysed, eight bilinear elements are used and, after applying the boundary conditions, only three freedoms remain. Applying Galerkin weighting to the second of eqns (18.28) and integrating the second derivatives by parts one obtains element equations of the form

$$k\{\phi\} - \lambda m\{\phi\} = \{0\} \tag{18.29}$$

where k is similar to that obtained in the heat flow problem of Section 18.2, that is, it is obtained by omitting κ_x and κ_y in eqn (18.3), and $m = \int \{f\}\{f\}^t \, dv$. For the element hatched in Fig. 18.4 one obtains, using the $\{f\}$ of eqns (5.37) and assuming $h = 1/2$, $t = 1$,

$$k_{11} = \frac{h^2}{4} \frac{1}{4h^2} \int_{-1}^{1} \int_{-1}^{1} \left[\left(\frac{\partial f_1}{\partial a} \right)^2 + \left(\frac{\partial f_1}{\partial b} \right)^2 \right] da \, db = \frac{2}{3} \tag{18.30}$$

$$k_{12} = \frac{h^2}{4} \frac{1}{4h^2} \int_{-1}^{1} \int_{-1}^{1} \left[\frac{\partial f_1}{\partial a} \frac{\partial f_2}{\partial a} + \frac{\partial f_1}{\partial b} \frac{\partial f_2}{\partial b} \right] da \, db = -\frac{1}{6} \tag{18.31}$$

$$m = \frac{h^2}{4} \int_{-1}^{1} \int_{-1}^{1} \{f\}\{f\}^t \, da \, db = \frac{h^2}{36} \begin{bmatrix} 4 & 2 & 1 & 2 \\ 2 & 4 & 2 & 1 \\ 1 & 2 & 4 & 2 \\ 2 & 1 & 2 & 4 \end{bmatrix}. \tag{18.32}$$

Summing all element contributions the system equations are obtained as

$$\left[\begin{bmatrix} 8/3 & -1/3 & 0 \\ -1/3 & 8/3 & -1/3 \\ 0 & -1/3 & 8/3 \end{bmatrix} - \frac{\lambda}{36} \begin{bmatrix} 4 & 1 & 0 \\ 1 & 4 & 1 \\ 0 & 1 & 4 \end{bmatrix} \right] \begin{Bmatrix} \phi_1 \\ \phi_2 \\ \phi_3 \end{Bmatrix} = \{0\} \tag{18.33}$$

giving the eigenvalues 14.597, 24, and 43.689. (Note in passing that if one uses symmetry to analyse only half the system one obtains only the first and third values, corresponding to the symmetric eigenmodes.) On the other hand, if we use lumping in m that is, with diagonal entries $h^2/4$, we obtain the eigenvalues 8.781, 10.667, and 12.552.

These results may be compared to the exact eigenvalues which are 12.337, 32.076, and 71.555.[12] Clearly lumping in m produces much poorer results, and more nodes in the y-direction are needed to improve the solution.

18.7 A finite difference method comparison

The electromagnetic wave problem of the preceding section provides a good opportunity to consider the finite difference method. In this, second derivatives are approximated by the central difference result

$$\frac{\partial^2 \phi}{\partial x^2} \simeq \frac{\phi_1 - 2\phi_2 + \phi_3}{h^2} \tag{18.34}$$

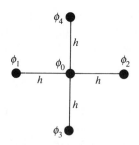

Fig. 18.5 Regular finite different cell for the Laplacian operator

where ϕ_1, ϕ_2, and ϕ_3 are three equally spaced values at distance h apart. Applying eqn (18.34) in both directions in the regular finite difference cell shown in Fig. 18.5 one obtains the central difference approximation for the Laplacian operator as

$$\nabla^2\phi \simeq \frac{\phi_1 + \phi_2 - 4\phi_0 + \phi_3 + \phi_4}{h^2}. \tag{18.35}$$

Applying this result to the equation for the magnetic wave, $\nabla^2\phi + \lambda\phi = 0$, at the three nodes indicated in Fig. 18.4 one obtains the three simultaneous equations

$$\left(\frac{1}{h}\right)^2\left[\begin{bmatrix} -4 & 1 & 0 \\ 1 & -4 & 1 \\ 0 & 1 & -4 \end{bmatrix} + \lambda\begin{bmatrix} 1 & 0 & 0 \\ 0 & 1 & 0 \\ 0 & 0 & 1 \end{bmatrix}\right]\begin{Bmatrix} \phi_1 \\ \phi_2 \\ \phi_3 \end{Bmatrix} = \{0\} \tag{18.36}$$

where $h = 1/2$, yielding the eigenvalues as 10.343, 16, and 21.657. Using a second mesh with $h = 1/4$ and applying h^2 extrapolation to the two sets of results Crandall obtains close approximations to the exact solutions[12] (which are noted in Section 18.6).

The finite difference results are a little better than the finite element results obtained with a lumped m matrix but not as good as those when a consistent m matrix is used. This highlights an advantage of the finite element method in that the consistent loads, contact stiffness matrices, geometric matrices, large displacement matrices, and tangent stiffness matrices for non-linear problems provide us with a powerful general method.

18.8 Discussion

The resemblance of the terms in eqns (18.2)–(18.5) to those outlined in Section 3.4 is striking and indeed such diffusion problems can also be formulated on a variational basis, for instance using convolution products.[11]

It will be clear to the reader that the steady state heat flow problem is indeed an equilibrium problem whilst the transient case is a 'creep' problem in which the initial conditions transition exponentially into the equilibrium conditions. Seepage and consolidation problems are very similar to those of heat conduction as the example of Section 18.5 shows and the closing electromagnetic wave problem provides a good opportunity to compare the finite element and finite difference methods, reminding us that more sophisticated elements than the bilinear one are worthwhile.

References

1. Wilson, E. L. and Nickell, R. E. (1966). Applications of finite element method to heat conduction analysis. *Nucl. Engng Des.*, **23**, 1.
2. Huebner, K. H. (1975). *The Finite Element Method for Engineers*. Wiley, New York.
3. Wilson, E. L., Bathe, K. J., and Pederson, F. E. (1974). Finite element analysis of linear and nonlinear heat transfer. *Nucl. Engng Des.*, **29**, 110.
4. Siauw, T. M. (1979). Heat flow analysis in metal casting processes using the finite element method. Proceedings of the Third Australian International Conference on Finite Element Methods, University of NSW, Sydney.
5. Kreyszig, E. (1979). *Advanced Engineering Mathematics*, (4th edn). Wiley, New York.
6. Irons, B. M. and Ahmad, S. (1980). *Techniques of Finite Elements*. Wiley, Chichester.
7. Connor, J. J. and Brebbia, C. A. (1976). *Finite Element Techniques for Fluid Flow*. Butterworths, London.
8. Cheng, R. T. and Li, C. (1973). On the solution of transient free surface flow problems in porous media by the finite element method. *J. Hydrol.* **20**, 49.
9. Terzaghi, K. and Peck, R. B. (1955). *Soil Mechanics in Engineering Practice*. Wiley, New York.
10. Desai, C. S. and Johnson, L. D. (1972). Evaluation of two finite element formulations for one dimensional consolidation. *Comput. Struct.*, **2**, 469.
11. Sandhu, R. S. and Pister, K. S. (1971). Variational principles for boundary and initial value problems in continuum mechanics. *Int. J. Solids Struct.*, **7**, 5.
12. Crandall, S. H. (1956). *Engineering Analysis*. McGraw-Hill, New York.

19
Potential flow, viscous flow
and compressible flow

At the close of Chapter 16 the Navier–Stokes, continuity, and thermal energy equations for fluid flow were introduced. The present chapter presents finite element formulations for a wide range of fluid flow problems governed by these equations. Of particular interest are the *uvp* or *primitive variable* formulations for Stokes flow. These have resulted in many new insights into the finite element method and can be extended to deal with convective, transient, and compressible flows. Chapter 20 then gives several worked examples and two programs for the analysis of fluid flows.

19.1 Introduction

Although the basic governing differential equations for the thermohydro-dynamic behaviour of fluids were introduced in Chapter 16 it is worth stating some simplified versions and introducing some specialized concepts which find wide application in the present chapter.

The first of these is the continuity equation for steady state conditions[1]

$$\operatorname{div}(\rho V) = \nabla \cdot (\rho V) = \frac{\partial(\rho u)}{\partial x} + \frac{\partial(\rho v)}{\partial x} + \frac{\partial(\rho w)}{\partial z} = 0 \tag{19.1}$$

which, for incompressible flow, reduces to

$$\nabla \cdot V = \nabla \cdot (u\hat{\mathbf{i}} + v\hat{\mathbf{j}} + w\hat{\mathbf{k}}) = \frac{\partial u}{\partial x} + \frac{\partial v}{\partial y} + \frac{\partial w}{\partial z} = 0. \tag{19.2}$$

A skew symmetric *vorticity tensor* for three dimensions can be defined as[2]

$$\frac{1}{2}\left(\frac{\partial u_j}{\partial x_i} - \frac{\partial v_i}{\partial x_j}\right) \qquad i \neq j; \quad i = 1, 2, 3; \quad j = 1, 2, 3. \tag{19.3}$$

More usefully at present, a vorticity vector can be defined by

$$\{\omega_x\hat{\mathbf{i}}, \omega_y\hat{\mathbf{j}}, \omega_z\hat{\mathbf{k}}\} = \operatorname{curl}(\{V\}) = \nabla \times \{u\hat{\mathbf{i}}, v\hat{\mathbf{j}}, w\hat{\mathbf{k}}\} \tag{19.4}$$

giving

$$\{\omega\} = \begin{vmatrix} \hat{\mathbf{i}} & \hat{\mathbf{j}} & \hat{\mathbf{k}} \\ \partial/\partial x & \partial/\partial y & \partial/\partial z \\ u & v & w \end{vmatrix}.$$ (19.5)

In two-dimensional flow we have only a single vorticity given by

$$\omega_z = \left(\frac{\partial v}{\partial x} - \frac{\partial u}{\partial y} \right) \hat{\mathbf{k}}.$$ (19.6)

Equations (19.2) and (19.6) are used in the following section to develop finite element models for *potential flow*. For two-dimensional potential flow analysis we also define a *potential function* ϕ such that

$$V = -\operatorname{grad}(\phi) = -\nabla(\phi) = -\frac{\partial \phi}{\partial x}\hat{\mathbf{i}} - \frac{\partial \phi}{\partial y}\hat{\mathbf{j}}$$ (19.7)

and the contours of ϕ are perpendicular to the direction of flow.

We can also define a *stream function* ψ, the contours of which (the *streamlines*) are a set of curves orthogonal to the contours of ϕ. Then to define ψ we note that V^*, the velocity at an angle α to V, is given by

$$V^* = (cu + sv)\hat{\mathbf{i}} - (su - cv)\hat{\mathbf{j}}$$ (19.8)

where α is measured anticlockwise from the direction of V and $c = \cos \alpha$, $s = \sin \alpha$. If V^* is perpendicular to V ($\alpha = 90$ degrees) eqn (19.8) reduces to

$$V^* = V_n = v\hat{\mathbf{i}} - u\hat{\mathbf{j}}.$$ (19.9)

Noting the form of eqn (19.7) one now defines the stream function such that

$$V_n = -\operatorname{grad}(\psi) = -\frac{\partial \psi}{\partial x}\hat{\mathbf{i}} - \frac{\partial \psi}{\partial y}\hat{\mathbf{j}}$$ (19.10)

and, comparing eqns (19.9) and (19.10) we obtain

$$u = \frac{\partial \psi}{\partial y}, \qquad v = -\frac{\partial \psi}{\partial x}.$$ (19.11)

For viscous flows the key equations are the Navier–Stokes equations. In two dimensions one replaces the 1/3 factor in eqn (16.58) by 1/2 so that for an isotropic medium eqn (16.58) and its couterpart in the y-direction become

$$\rho \frac{Du}{Dt} = X - \frac{\partial p}{\partial x} + \mu \frac{\partial^2 u}{\partial x^2} + \mu \frac{\partial^2 u}{\partial y^2}$$ (19.12)

$$\rho \frac{Dv}{Dt} = Y - \frac{\partial p}{\partial y} + \mu \frac{\partial^2 v}{\partial x^2} + \mu \frac{\partial^2 v}{\partial y^2}$$ (19.13)

and these simplified results are particularly useful in later sections.

For compressible flows one must also include thermal energy and constitutive equations, such as $p = \rho RT$. Sufficient details of these equations were given in Chapter 16 and they are used at the close of the present chapter.

19.2 Inviscid incompressible flow

Potential flow refers to inviscid incompressible irrotational flow characterized by a potential function ϕ such that the velocities are given by $\{u\} = -\nabla\phi$ or

$$u = -\frac{\partial\phi}{\partial x}, \qquad v = -\frac{\partial\phi}{\partial y}. \tag{19.14}$$

Equations (19.14) identically satisfy the zero vorticity or irrotationality condition[3]

$$\omega_z = \left(\frac{\partial v}{\partial x} - \frac{\partial u}{\partial y}\right)\hat{\mathbf{k}} = 0. \tag{19.15}$$

Alternatively one defines an orthogonal stream function ψ such that

$$u = \frac{\partial\psi}{\partial y}, \qquad v = -\frac{\partial\psi}{\partial x} \tag{19.16}$$

Equations (19.16) identically satisfy the two-dimensional continuity condition

$$\nabla.\{u\} = \frac{\partial u}{\partial x} + \frac{\partial v}{\partial y} = 0 \tag{19.17}$$

and lines of constant ψ are streamlines. Both ϕ and ψ identically satisfy Laplace's equation which is seen by substituting eqns (19.14) into eqn (19.17), giving

$$\nabla.\{u\} = \nabla.\nabla\phi = \nabla^2\phi = 0 \tag{19.18}$$

and substituting eqns (19.16) into eqn (19.15) gives $\nabla^2\psi = 0$ as the governing differential equation.

Indeed we have already seen examples of both types of problem, ϕ corresponding to T in heat conduction (see Section 18.2) and to H in plane seepage (see Section 18.4) and ψ corresponding to Prandtl's stress function for plane torsion (see Section 16.2). In fact the problems of heat flow or seepage are potential flow problems whilst, like plane torsion problems, potential flow problems may be solved by the membrane analogy.[4]

First directing attention at potential function formulations the general problem is governed by eqn (19.18) subject to Neumann boundary conditions $\partial\phi/\partial n = 0$ on impermeable boundaries and $\partial\phi/\partial n = V_n$ elsewhere, as shown in Fig. 19.1. Using Galerkin weighting and integrating the second derivatives in eqn (19.18) by parts one obtains the element equations as

$$k\{\phi\} = \{q\} \tag{19.19}$$

where

$$k = \int [\{f_x\}\{f_x\}^t + \{f_y\}\{f_y\}^t]\,dV \tag{19.20}$$

$$\{q\} = \int \{f\}(c_x\phi_x + c_y\phi_y)\,dS \tag{19.21}$$

$$= -\int \{f\}(uc_x + vc_y)\,dS. \tag{19.22}$$

Where $\{f_x\} = \{\partial f/\partial x\}$ and so on, and eqns (19.14) are used to yield the result for the velocity flux forcing terms at the boundaries. Thus with Lagrangian elements $\partial\phi/\partial x$ and $\partial\phi/\partial y$ provide the forcing terms at inlet and outlet but with Hermitian elements, on the other hand, skew boundary conditions will be required to enforce $\partial\phi/\partial n$ on the aerofoil in Fig. 19.1.

Without any Dirichlet boundary conditions on ϕ, however, the system equations are singular and the remedy is to specify $\phi = 0$ as a datum value for at least one node.[5]

With stream function formulations one simply replaces ϕ with ψ in eqns (19.19)–(19.21) but eqn (19.22) becomes

$$\{q\} = \int \{f\}(uc_y - vc_x)\,dS \tag{19.23}$$

and this vanishes when the resultant velocity flux V is perpendicular to the inlet or outlet boundary, that is, $u = Vc_x$, $v = Vc_y$, as in Fig. 19.1. This simplification is seen to advantage in Sections 20.1 and 20.6 where the simpler Dirichlet boundary conditions of the stream function formulation also considerably reduce the number of equations to be solved.

Typical stream and potential function boundary conditions are shown in Fig. 19.1, setting datum values on one boundary and using eqns (19.14) and

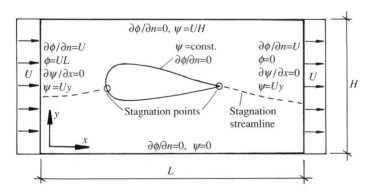

Fig. 19.1 Potential flow past an aerofoil

(19.16) to obtain other boundary values:

$$\psi = \int \frac{\partial \psi}{\partial y} \, dy = U y \tag{19.24}$$

is used to find values at the inlet, outlet, and upper boundaries whilst

$$\phi = \int \frac{\partial \phi}{\partial x} \, dx = - U x \tag{19.25}$$

is used to obtain values at the inlet boundary. (However, note that this result only applies if the inlet boundary is relatively remote from the aerofoil, otherwise one leaves the values at the inlet unspecified, that is the only specified values will be the datum values at outlet, and generally this is the safer course.)

In the example of Section 20.1 symmetry is used to advantage, allowing $\psi = 0$ to be imposed on both the boundary of the immersed body and the horizontal centreline of the control volume. Note too that the example of Section 17.4, is of the stream function type, again possessing simple boundary conditions.

When the distribution of ϕ (or ψ) has been found, the pressure distribution is found by Bernoulli's equation which is the sum of the velocity, elevation, and pressure heads, giving

$$\tfrac{1}{2}[\{u, v\}]^2 + q + \frac{p}{\rho} = \text{const.} \tag{19.26}$$

where

$$[\{u, v\}]^2 = u^2 + v^2 = \left(\frac{\partial \phi}{\partial x}\right)^2 + \left(\frac{\partial \phi}{\partial y}\right)^2 = \left(\frac{\partial \psi}{\partial y}\right)^2 + \left(\frac{\partial \psi}{\partial x}\right)^2.$$

q is the body force potential (for example, gravity, when $q = - gy$) and the Navier–Stokes equations of Section 19.1, with $\mu = 0$ followed by integration, yield the same result for steady two-dimensional inviscid incompressible flow.[6]

The aerofoil in Fig. 19.1 is unsymmetrical with respect to the flow and will experience lift. To obtain correct results, it is found necessary to enforce the Kutta–Joukowsky condition that the downstream stagnation point, where $u = v = 0$, must occur at the sharp trailing edge. This requires a discontinuity in ϕ to be allowed along the stagnation streamline which can only be achieved by trial and error[4] and the Kutta–Joukowsky condition is more easily enforced with the stream function approach, the boundary conditions for which are also shown in Fig. 19.1, by analysing with two different values ψ_1 and ψ_2 specified on the aerofoil. Then the required solution follows from an appropriate superposition of these results.[4]

The lift on the aerofoil is perpendicular to the free stream velocity U and is given by calculating $\int pc_y \, dS$ around the aerofoil or by the Kutta–Joukowsky theorem,[1] which applies to aerofoils of any shape,

$$P_{\text{lift}} = -\rho U \Gamma. \qquad (19.27)$$

Γ is the *circulation* which is evaluated as the integral of the tangential velocity around the aerofoil:

$$\Gamma = \int V_t \, dS = -\int \frac{\partial \psi}{\partial n} \, dS \qquad (19.28)$$

and is positive if anticlockwise. Applying Gauss' theorem to the circulation, or using eqns (17.56),

$$\Gamma = -\int \frac{\partial \psi}{\partial n} \, dS = -\int \left(c_x \frac{\partial \psi}{\partial x} + c_y \frac{\partial \psi}{\partial y} \right) dS \qquad (19.29)$$

$$= \int \left(\frac{\partial^2 \psi}{\partial x^2} + \frac{\partial^2 \psi}{\partial y^2} \right) dA = -\int \left(\frac{\partial v}{\partial x} - \frac{\partial u}{\partial y} \right) dA = -\int 2\omega \, dA \qquad (19.30)$$

that is, if the flow is irrotational the circulation is zero and this will be the case if the immersed object is symmetrical relative to U. Thus, although our analysis assumes irrotationality, we are able to deal with *vortex flows* in which $\Gamma \neq 0$ by the aforementioned artifices of:

(1) allowing a discontinuity in ϕ across the stagnation streamline (the magnitude of which equals the circulation[4]);
(2) using superposition of two stream function solutions;[4] this is generally the preferred approach.

An important type of potential flow problem is that of free surface flows, for example outlet flow from a sluicegate, and, as described in Section 18.4, these require an iterative procedure beginning with an initial estimate of the free surface position.[6, 7]

19.3 Harmonic response to seiches

Seiches are relatively small oscillations in depth in the ocean caused by sudden changes in barometric pressure. In certain harbours with narrow openings to the sea, like that shown in Fig. 19.2, seiche motions may be considerably amplified in the harbour itself. The problem is therefore one of *harmonic forcing* and if the frequency of this is close to one of the natural frequencies for the harbour (which depend upon its dimensions) resonance occurs and causes large depth fluctuations in the harbour.

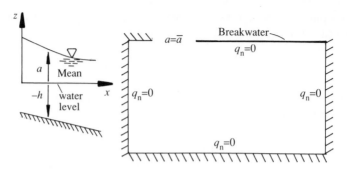

Fig. 19.2 Tidal basin subject to harmonic forcing

The equilibrium equations are obtained by neglecting viscosity, body force, and convective inertia terms like $u(\partial u/\partial x)$ in the Navier–Stokes equations (eqns (19.12) and (19.13)), leading to the results

$$\rho\frac{\partial u}{\partial t} = -\frac{\partial p}{\partial x}, \qquad \rho\frac{\partial v}{\partial t} = -\frac{\partial p}{\partial y} \tag{19.31}$$

and, as

$$p = g\rho(a + h) \tag{19.32}$$

in this case we have

$$\frac{\partial u}{\partial t} = -g\frac{\partial a}{\partial x}, \qquad \frac{\partial v}{\partial t} = -g\frac{\partial a}{\partial y}. \tag{19.33}$$

The continuity equation is

$$\frac{\partial u}{\partial x} + \frac{\partial v}{\partial y} + \frac{\partial w}{\partial z} = 0. \tag{19.34}$$

Integrating over the depth to define flow rates q_x and q_y as

$$q_x = \int_{-h}^{a} u\,dz, \qquad q_y = \int_{-h}^{a} v\,dz \tag{19.35}$$

and noting that the vertical velocity is simply

$$w = \frac{\partial a}{\partial t} \tag{19.36}$$

it can be shown using the Leibniz integration rule (see eqn (19.88)) that eqn (19.34) becomes[6]

$$\frac{\partial q_x}{\partial x} + \frac{\partial q_y}{\partial y} + \frac{\partial a}{\partial t} = 0 \tag{19.37}$$

a result which is obvious on comparing eqns (19.34), (19.35), and (19.36). Equations (19.33) become, assuming $h \gg a$,

$$\frac{\partial q_x}{\partial t} = -gh\frac{\partial a}{\partial x}, \qquad \frac{\partial q_y}{\partial t} = -gh\frac{\partial a}{\partial y}. \qquad (19.38)$$

The differentiating eqns (19.38) with respect to x and y, respectively, and substituting in eqn (19.37) differentiated with respect to t gives a single governing equation

$$\frac{\partial}{\partial x}\left(h\frac{\partial a}{\partial x}\right) + \frac{\partial}{\partial y}\left(h\frac{\partial a}{\partial y}\right) - \frac{1}{g}\frac{\partial^2 a}{\partial t^2} = 0 \qquad (19.39)$$

subject to

$$a = \bar{a} \qquad \text{on } S_1 \text{ (ocean boundary)} \qquad (19.40)$$

and

$$\frac{h\partial a}{\partial n} = -\frac{\partial}{\partial t}\left(\frac{\bar{q}_n}{g}\right) \qquad \text{on } S_2 \text{ (land boundary)}. \qquad (19.41)$$

The problem can be solved by time stepping techniques or as a harmonic forcing problem in which the forcing and response are in phase (or have a constant phase difference). Such a dynamic steady state occurs after sufficient time has elapsed since the start of the excitation, when the problem reduces to an eigenvalue one.

Using separation of variables (that is, let $a(x, y, t) = \phi(x, y)T(t)$), virtual weighting with $\delta\phi$, and integration by parts, including the boundary condition of eqn (19.41), gives

$$\iint\left\{h\phi_x\delta\phi_x + h\phi_y\delta\phi_y - \frac{\omega^2}{g}\phi\delta\phi\right\}dx\,dy = \int\bar{q}_n\delta\phi\,dS \qquad (19.42)$$

where ϕ_x denotes $\partial\phi/\partial x$, and so on. The element equations are then obtained in the form

$$k\{\phi\} - \omega^2 m\{\phi\} = \{q\} \qquad (19.43)$$

where

$$k = \iint h(\{f_x\}\{f_x\}^t + \{f_y\}\{f_y\}^t)\,dx\,dy \qquad (19.44)$$

$$m = \frac{1}{g}\iint\{f\}\{f\}^t\,dx\,dy \qquad (19.45)$$

$$\{q\} = \int\phi\bar{q}_n\,dS. \qquad (19.46)$$

In many cases the flux $\bar{q}_n = 0$ and $\bar{a} = \bar{\phi}$ is prescribed at inlet (as shown in Fig. 19.2). Then eqn (19.43) is solved for the eigenvalues and eigenmodes.

19.4 Shallow water circulation

For the analysis of the motion of water in harbours and lakes viscosity is neglected and bed and wind shear stresses are included directly so that one begins with eqns (19.12) and (19.13) together with the continuity condition. Defining flow rates per unit width by integrating over the depth

$$q_x = \int_{-h}^{a} u \, dz, \qquad q_y = \int_{-h}^{a} v \, dz$$

where a, h are as shown in Fig. 19.2, eqns (19.12) and (19.13) become, using Leibnitz's theorem (eqn (19.88)) and incorporating the continuity condition,

$$\frac{\partial q_x}{\partial t} + \frac{\partial}{\partial x}\left(\frac{q_x^2}{H}\right) + \frac{\partial}{\partial y}\left(\frac{q_x q_y}{H}\right) = \frac{\partial(N_x - N_p)}{\partial x} + \frac{\partial N_{xy}}{\partial y} + T_x \quad (19.47)$$

$$\frac{\partial q_y}{\partial t} + \frac{\partial}{\partial x}\left(\frac{q_x q_y}{H}\right) + \frac{\partial}{\partial y}\left(\frac{q_y^2}{H}\right) = \frac{\partial(N_y - N_p)}{\partial y} + \frac{\partial N_{xy}}{\partial x} + T_y \quad (19.48)$$

where $H = a + h$. The stress resultants are given by

$$N_x = 2\varepsilon_x \frac{\partial q_x}{\partial x}, \qquad N_y = 2\varepsilon_y \frac{\partial q_y}{\partial y}, \qquad N_p = \frac{\rho g H^2}{2} + H p_a$$

$$N_{xy} = \varepsilon_{xy}\left(\frac{\partial q_x}{\partial y} + \frac{\partial q_y}{\partial x}\right) . \quad (19.49)$$

where ε_x, ε_y, ε_{xy} are the *eddy viscosity coefficients* which are equal in the case of isotropic behaviour. The driving forces T_x and T_y are

$$T_x = fq_y + \gamma^2 p_a W^2 \cos\theta - gq_x \bar{q}/\rho c^2 H^2 + p_a \frac{\partial H}{\partial x} + \rho g H \frac{\partial a}{\partial x} \quad (19.50)$$

$$T_y = -fq_x + \gamma^2 p_a W^2 \sin\theta - gq_y \bar{q}/\rho c^2 H^2 + p_a \frac{\partial H}{\partial y} + \rho g H \frac{\partial a}{\partial y} \quad (19.51)$$

where f is the Coriolis parameter, equal to twice the angular velocity of the earth and signs are here for the Northern hemisphere. p_a is the atmospheric pressure at the water surface, W is the wind velocity, θ is the wind direction from the x-axis, $\bar{q} = \sqrt{(q_x^2 + q_y^2)}$, γ^2 is the wind stress coefficient and is equal to 0.0026 from Reference 6, and c is the *Chezy* coefficient or friction factor.

The boundary conditions are $q_n = 0$ on land boundaries (that is, no flow normal to the boundary), $q_n = \bar{q}_n$ at river entry points (that is, $\bar{q}_n$ is a specified inflow) and $N_n = -N_p$ at ocean boundaries (boundary tractions $N_x = -c_x N_p$ and $N_y = -s_x N_p$). Equations (19.47) and (19.48) are highly non-linear but some simplifying assumptions can be made, including

(1) $H \simeq h$;

(2) eddy viscosities are negligible;

(3) bed shear stresses are βq_x and βq_y;

(4) non-linear convective terms such as $\partial q_x^2/\partial x$ are neglected;

(5) P_a is neglected.

With these assumptions, and assuming that steady state conditions exist, eqns (19.47) and (19.48) reduce to $T_x = 0$ and $T_y = 0$ or

$$fq_y + \rho gh\frac{\partial a}{\partial x} + \tau_{xs} - \beta q_x = 0 \qquad \text{where } \tau_{xs} = \gamma^2 p_a W^2 \cos \theta \quad (19.52)$$

$$-fq_x + \rho gh\frac{\partial a}{\partial y} + \tau_{ys} - \beta q_y = 0 \qquad \text{where } \tau_{ys} = \gamma^2 p_a W^2 \sin \theta \quad (19.53)$$

subject to the continuity requirement

$$\frac{\partial q_x}{\partial x} + \frac{\partial q_y}{\partial y} = 0. \qquad (19.54)$$

Differentiating eqn (19.52) with respect to y and eqn (19.53) with respect to x, subtracting the results, and taking into account eqn (19.54) one obtains

$$Q = \frac{\partial \tau_{xs}}{\partial y} - \frac{\partial \tau_{ys}}{\partial x} = \beta \left\{ \frac{\partial q_x}{\partial y} - \frac{\partial q_y}{\partial x} \right\} \qquad (19.55)$$

and the Coriolis effects have cancelled. Defining a stream function $q_x = \partial \psi/\partial y$, $q_y = -\partial \psi/\partial x$ eqn (19.55) reduces to

$$Q = \beta \nabla^2 \psi \qquad (19.56)$$

and the boundary conditions are formed in the same manner as those for stream function potential flow analysis. Thus we obtain a pseudoharmonic steady state problem soluble in the same way as those in Chapter 18.

For the transient case one can include $\partial q_x/\partial t$ and $\partial q_y/\partial t$ in eqns (19.52) and (19.53) and use eqns (19.52)–(19.54) to obtain a *primitive variable* formulation with q_x, q_y, and a as nodal variables. Indeed the *uvp* formulations of the following sections can also be used to model lake circulation problems.

19.5 Incompressible viscous flow without inertia

Slow incompressible viscous flow in which the inertia terms are neglected is generally referred to as *Stokes flow*. This occurs at very low values of the dimensionless *Reynolds number* Re (usually Re < 1), which appears when the Navier–Stokes equations are non-dimensionalized, practical examples of

Stokes flow being in lubrication, visco-elasticity, and plasticity problems. The Reynolds number is defined as the ratio of the dynamic pressure ρu^2 to the magnitude of a typical shearing stress $\mu u/L$, so that $\mathrm{Re} = \rho u L/\mu$.

Assuming constant viscosity and deleting the time-dependent and convective inertia terms in eqns (19.12) and (19.13) the Navier–Stokes equations for two-dimensional isothermal flow reduce to

$$0 = X - \frac{\partial p}{\partial x} + \mu\left(\frac{\partial^2 u}{\partial x^2} + \frac{\partial^2 u}{\partial y^2}\right) \tag{19.57}$$

$$0 = Y - \frac{\partial p}{\partial y} + \mu\left(\frac{\partial^2 v}{\partial x^2} + \frac{\partial^2 v}{\partial y^2}\right) \tag{19.58}$$

subject to the continuity condition $\partial u/\partial x + \partial v/\partial y = 0$. Introducing a stream function (eqns (19.16)) and differentiating eqn (19.57) with respect to y and eqn (19.58) with respect to x and subtracting the results gives

$$\mu\left(\frac{\partial^4 \psi}{\partial x^4} + \frac{2\partial^4 \psi}{\partial x^2 \partial y^2} + \frac{\partial^4 \psi}{\partial y^4}\right) + \rho\left(\frac{\partial X}{\partial y} - \frac{\partial Y}{\partial x}\right) = 0 \tag{19.59}$$

which is a pseudobiharmonic problem. Noting that the governing equation for thin plate bending is $\nabla^4 w = 12(1 - v^2)q/Et^3$ we see that a standard plate bending program can be used to solve eqn (19.59) (In most fluid problems body forces are constant or can be taken so within an element so that the terms involving these in eqn (19.59) will vanish.) Specification of the boundary conditions[4] requires a little more care than with the *uvp* formulations described below and the latter are now more widely used.

An alternative formulation is had by integrating the velocity and pressure terms in eqns (19.57) and (19.58) by parts, giving, with Galerkin weighting,

$$\mu\iint \{f_x^i u_x + f_y^i u_y - f_x^i p\}\,\mathrm{d}x\,\mathrm{d}y = \iint f^i X\,\mathrm{d}x\,\mathrm{d}y + \int \mu f^i \frac{\partial u}{\partial n}\,\mathrm{d}S - \int f^i p\,\mathrm{d}S \tag{19.60}$$

$$\mu\iint \{f_x^i v_x + f_y^i v_y - f_y^i p\}\,\mathrm{d}x\,\mathrm{d}y = \iint f^i Y\,\mathrm{d}x\,\mathrm{d}y + \int \mu f^i \frac{\partial v}{\partial n}\,\mathrm{d}S - \int f^i p\,\mathrm{d}S \tag{19.61}$$

where f_x^i denotes $\partial f_i/\partial x$ for the ith interpolation function, and $u_x = \partial u/\partial x$, and so on. One must also use Galerkin weighting for the continuity equation, giving

$$\iint \{f^i u_x + f^i v_y\}\,\mathrm{d}x\,\mathrm{d}y = 0 \tag{19.62}$$

so that in matrix form the problem is obtained as

$$k\{u, v, p\} = \{q\} \tag{19.63}$$

where

$$
k = \mu \iint \begin{bmatrix}
\overset{u \text{ columns}}{\{f_x\}\{f_x\}^t + \{f_y\}\{f_y\}^t} & \overset{v \text{ columns}}{O} & \overset{p \text{ columns}}{-\{f_x\}\{f\}^t/\mu} \\
O & \{f_x\}\{f_x\}^t + \{f_y\}\{f_y\}^t & -\{f_y\}\{f\}^t/\mu \\
-\{f\}\{f_x\}^t/\mu & -\{f\}\{f_y\}^t/\mu & O
\end{bmatrix} dx\,dy.
$$

(19.64)

A symmetric result is achieved by reversing the signs of the continuity constraint equations.[3] Note that eqn (19.64) has the same form as that obtained with Lagrange multiplier constraints (see Section 4.3). Pivoting difficulties can be avoided using small penalty numbers (see eqn (19.73)). The forcing terms are given by

$$
\{q\} = \int \begin{Bmatrix} \{f\}X \\ \{f\}Y \\ \{0\} \end{Bmatrix} dx\,dy + \mu \int \begin{Bmatrix} \{f\}\partial u/\partial n \\ \{f\}\partial v/\partial n \\ \{0\} \end{Bmatrix} dS - \int \begin{Bmatrix} \{f\}c_x p \\ \{f\}c_y p \\ \{0\} \end{Bmatrix} dS
$$

(19.65)

and the second term on the right-hand side will apply only to elements at the boundary. The usual advantage of direct formulation with primitive variables rather than a stream or potential function prevails, namely that boundary conditions are physically more obvious. Further, the governing equation is second order rather than fourth order so that the essential boundary conditions are the nodal freedoms and not their first derivatives.[8]

Remarkably the six-node quadratic triangle was chosen in all the first published *uvp* solutions for viscous flow[9-13] but doubts existed on whether linear or quadratic interpolation should be used for the pressure. The problem here is that quadratic pressure interpolation leads to six incompressibility constraint equations but, as these involve first derivatives of the velocities, the six velocity sampling points are insufficient to provide six independent equations. The result is sometimes a spurious chequerboard mode in the pressure solutions, that is, these exhibit sometimes large oscillations.[14-19] To overcome this difficulty, Hood and Taylor,[10] Kawahara *et al.*[11] and Yamada *et al.*[12] use quadratic interpolation for the velocities and linear interpolation for the pressures.

Another approach is to use manipultion of the momentum equations to relegate the pressure freedoms to a separate Poisson equation. Then, once the velocity solution has been obtained, the pressures can be calculated at the same sampling points. Using such a *segregated formulation* and a Hermitian cubic triangle, with *consistent interpolation*, Olson and Tuann[15] are able to avoid chequerboard problems.

Segregated formulations can also be obtained with the use of penalty factors[16-19] and mixed interpolation,[16,17] or selectively reduced integration[18] of the penalty contributions. Most of these formulations still fail to eliminate the chequerboard effect[16-18] but Engelman *et al.*[19] obtain penalty factor formulations with consistent integration which eliminate it.

Another aproach is to apply 'smoothing' and 'filtering' techniques to extract satisfactory pressures[14,20] from numerical results obtained with consistent interpolation, but polluted by the chequerboard effect. Sani *et al.*[20] conclude that penalty methods themselves have an 'automatic' filtering effect. Moreover the wide applicability of penalty techniques is worth noting. We have already seen a number of such applications (such as in Chapters 9 and 11) whilst others include contact problems[21] and optimization (Chapter 22). In the present context we shall describe two penalty formulations for Stokes flow, the second a modification of eqn (19.64). Comparing eqns (4.10) and (19.64) one observes that the $\{p\}$ freedoms play the role of Lagrange multipliers to enforce the continuity condition; these freedoms are the 'stresses' necessary to enforce the continuity condition. Now comparing eqns (4.1) and (4.2) we deduce that in general $\lambda = \beta g(\ )$, or in the present case†

$$p = \beta \left(\frac{\partial u}{\partial x} + \frac{\partial v}{\partial y} \right) \tag{19.66}$$

where β is the penalty factor and $g(\ )$ is the constraint equation (in this case a differential one) and as $\beta \to \infty$, $g(\ ) \to 0$.

This result is established more formally by Reddy[17] who compared the Euler equations for the variational Lagrange multiplier and penalty factor formulations of the problem which, referring to eqns (4.3a) and (4.7), are

$$\min \{ f(u, v, \lambda) = F(u, v) + \iint \lambda g(u, v) \, dx \, dy \} \tag{19.67}$$

and

$$\min \{ f(u, v) = F(u, v) + \tfrac{1}{2} \beta \iint [g(u, v)]^2 \, dx \, dy \} \tag{19.68}$$

where $g(u, v) = \partial u/\partial x + \partial v/\partial y$ and $F(u, v)$ is the functional corresponding to eqns (19.60) and (19.61). This is obtained in the manner described in Section 17.3.

But once eqn (19.66) is established penalty formulation is straightforward: one simply substitutes it into eqns (19.57) and (19.58) and integrates all terms by parts to obtain, in matrix form,

$$\begin{bmatrix} \mu(A + B) + \beta A & \beta C \\ \beta C^t & \mu(A + B) + \beta B \end{bmatrix} \begin{Bmatrix} \{u\} \\ \{v\} \end{Bmatrix} = \iint \begin{Bmatrix} \{f\} X \\ \{f\} Y \end{Bmatrix} dx \, dy + \int \begin{Bmatrix} F_1 \\ F_2 \end{Bmatrix} dS \tag{19.69}$$

---

† See also Chapter 23 where conversion from Lagrange mutiplier to penalty formulations is discussed.

where
$$A = \iint \{f_x\}\{f_x\}^t \, dx \, dy, \qquad B = \iint \{f_y\}\{f_y\}^t \, dx \, dy$$

$$C = \iint \{f_x\}\{f_y\}^t \, dx \, dy$$

$$F_1 = \{f\} \left[\mu \left(\frac{\partial u}{\partial n}\right) + \beta \left(\frac{\partial u}{\partial x}\right) c_x + \beta \left(\frac{\partial v}{\partial y}\right) c_x \right]$$

$$F_2 = \{f\} \left[\mu \left(\frac{\partial v}{\partial n}\right) + \beta \left(\frac{\partial u}{\partial x}\right) c_y + \beta \left(\frac{\partial v}{\partial y}\right) c_y \right]$$

so that the assembled equations are solved for u, v throughout the domain and the pressures are then calculated from eqn (19.66). With the pressures thus eliminated from the matrix solution only two thirds as many equations must be solved and this is the advantage of penalty formulations compared to Lagrange multipliers.

An even more direct penalty approach is to rearrange equation (19.66) into

$$\frac{\partial u}{\partial x} + \frac{\partial v}{\partial y} = -\frac{p}{\beta} \tag{19.70}$$

and, using this as a modified continuity constraint, one obtains with Galerkin weighting an entry

$$-\frac{1}{\beta} \iint \{f\}\{f\}^t \, dx \, dy \tag{19.71}$$

replacing the null partition in eqn (19.64). Values of $\beta \geq 10^4$ are found to provide satisfactory results with both these formulations[17] but unfortunately neither eliminates the chequerboard syndrome.[17] As already remarked this is easily overcome by reducing the order of the pressure interpolation but alternative modifications of the continuity constraints are worth noting. Kawahara and Hirano,[22] for example, obtain a modified continuity condition by including a small artificial compressibility and combining the equation of state, $p = \rho R T$, with $\partial u/\partial x + \partial v/\partial y = 0$.

An alternative way of modifying the continuity equation is to integrate it by parts, integrating the $\partial u/\partial x$ term with respect to dx and the $\partial v/\partial y$ term with respect to dy. (Compare this with the momentum equations which are integrated by parts with respect to only one variable.) Then in order to retain symmetry the pressure entries in the momentum equations (eqns (19.57) and (19.58)) are not partially integrated, so that one obtains[23]

$$\iint \begin{bmatrix} C & O & -A \\ O & C & -B \\ -A^t & -B^t & \alpha I \end{bmatrix} dx \, dy \begin{Bmatrix} \{u\} \\ \{v\} \\ \{-p\} \end{Bmatrix} = \iint \begin{Bmatrix} \{f\}X \\ \{f\}Y \\ \{0\} \end{Bmatrix} dx \, dy$$

$$+ \int \begin{Bmatrix} \mu\{f\}(\partial u/\partial n) \\ \mu\{f\}(\partial v/\partial n) \\ \{f\}V_n \end{Bmatrix} dS \tag{19.72}$$

where

$$A = \{f\}\{f_x\}^t, \qquad B = \{f\}\{f_y\}^t$$

$$C = \mu[\{f_x\}\{f_x\}^t + \{f_y\}\{f_y\}^t]$$

and V_n is the normal outward velocity at the domain boundary. Negative pressure freedoms are used to retain symmetry and the diagonal entries for the pressure columns αI use an arbitrary 'penalty number' given by

$$\alpha = \left(\sum_i \sum_j C_{ij}^2\right)^{1/2} \times 10^{-5}/36. \qquad (19.73)$$

This value was used for single precision computation (nominally seven digits). With higher-precision computation, of course, smaller values could be used.

The formulations of eqns (19.69), and (19.64) with (19.71) are equivalent to two 'penalty' formulations derived by Reddy[17] (the second is more strictly a Lagrange multiplier formulation) which he shows to be numerically equivalent. Equation (19.72) then constitutes a modification of the second of these in which:

(1) the pressure terms in the momentum equations are not integrated by parts, simplifying the loading terms for these equations;

(2) the continuity equation is integrated by parts in an exceptional fashion, that is, the first term is reduced with respect to x and the second with respect to y; this results in loading terms $\int\{f\} V_n \, \mathrm{d}S$ for the continuity constraints;

(3) small errors are associated with the continuity equations via the entries $\alpha I \{\lambda\}$.

The first two results are of practical interest: any term in any governing differential equation or side constraint can be integrated by parts (but need not be), regardless of physical implications.

That the second result leads to right-hand sides for the constraint equations is of importance. As we shall see in Chapter 23, this is the more general form of the Lagrange multiplier problem, namely that for inequality constraints.

The third result is of considerable importance as it constitutes a new numerical procedure, the entries $\alpha I \{\lambda\}$ being *small slack variables*. This question is taken up again in Chapter 23 after discussion of slack variables, penalty factors, and Lagrange multipliers in the context of optimization in Chapter 22. It will be shown that any Lagrange multiplier formulation can be converted into an equivalent penalty factor one. Moreover, by extending the small slack variable procedure to deal with inequality constraints and then using 'Lagrange multiplier to penalty' condensation we see, in Chapter 23, that a useful procedure for optimization can be arrived at.

The formulation of eqn (19.72) was implemented with the six-node triangle in which $\{f\}$ is given by eqns (5.60) and $\{f_x\}$ and $\{f_y\}$ are obtained as in Section 7.3. Figure 19.3 shows the mesh and boundary conditions used for the Couette flow problem, that is, flow between two parallel infinite plates, the upper moving with constant velocity $U = 1$, under a constant pressure gradient.

Here full quadratic interpolation for the pressures and three-point integration (either eqn (5.135) or (5.136) which give similar results), which is a reduced integration for the terms A and B in eqn (19.72), is used.

Table 19.1 compares the transverse velocity and longitudinal pressure gradients obtained with the exact solution (see eqn (20.8)) indicating good agreement. For the special case where $U = 0$ on the upper boundary (plane Poiseille flow) one obtains equally good results. The formulation is easily generalized to deal with axisymmetric flows in pipes but in the numerical integration for the viscosity and other matrices, $\partial(\ )/\partial y$ is replaced by $\partial(\ )/\partial r$ and $\omega_i \Delta$ by $2\pi r \omega_i \Delta$ where ω_i are the integration point weights.

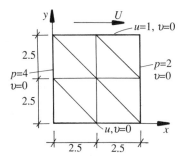

Fig. 19.3 Couette flow

Table 19.1 Finite element solutions for the analysis of Fig. 19.1

x, y	Horizontal velocity, u FEM	Exact	x, y	Pressure, p FEM	Exact
2, 0	0.0000	0.0000	0, 2	4.0000	4.0000
2, 1	1.1875	1.1875	1, 2	3.5003	3.5000
2, 2	1.7500	1.7500	2, 2	3.0025	3.0000
2, 3	1.6875	1.6875	3, 2	2.5004	2.5000
2, 4	1.0000	1.0000	4, 2	2.0000	2.0000

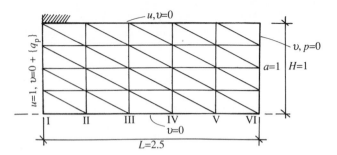

Fig. 19.4 Entry flow problem

Figure 19.4 shows the mesh used for analysis of entry flow between parallel plates, using symmetry to study half the domain. Here integration at the midside nodes was used and the velocity loading terms required in eqn (19.72) were evaluated using

$$\{q_p\} = \int \{f\} V_n \, dS = \int \{f\} \{f\}^t \{u\} \, dy = \frac{(H/4)\{U, 4U, U\}}{6} \quad (19.74)$$

for the nodes at inlet.

The developing velocity profiles, which transition from uniform velocity at inlet to parabolic Poiseille flow at section VI in Fig. 19.4,[23] were in good agreement with those obtained by Atkinson *et al.*[24] and Kawahara *et al.*[11] At section II the velocity profile shows a 'hump' near the upper boundary not predicted by the idealized analytical solution[25] but the weight of evidence now supports the finite element result.

The pressure profiles, which indicate a singularity in pressure at the right-angled corner of entry,[26] were in good agreement with those observed by Yamada *et. al.*,[12] the singularity indicating that rounded corners should in practice be provided at inlet. Clearly better results were obtained using consistent quadratic interpolation, although a slight unevenesss was observed in the pressure profiles, a very mild incidence of the chequerboard syndrome.

In relation to boundary conditions, note that generally one must specify either velocity or pressure at inlet and outlet to force the flow, but not both. In the problem of Fig. 19.4, unlike that of Fig. 19.3, velocity and pressure are respectively specified at inlet and outlet.

In conclusion the *uvp* finite element formulations for Stokes flow have not only provided another useful method of analysis, but also remarkable insights into the use of penalty factors and Lagrange multipliers in the finite element method, whilst eqn (19.72) is able to be extended in later sections.

19.6 Incompressible viscous flow with inertia

Inclusion of the non-linear convective inertia terms of eqns (19.12) and (19.13) in FEM flow formulations has been accomplished by three approaches[3,6] using

(1) stream functions;

(2) stream and vorticity functions (the latter defined by eqn (19.5));

(3) pressure and velocity freedoms (that is, primitive variables).

Primitive variables have several advantages, not the least of which is greater economy,[27] and their use is a straightforward extension of the formulations discussed in Section 19.5.

Including these convective inertia terms in eqns (19.57) and (19.58) and neglecting body forces one obtains

$$\rho u \frac{\partial u}{\partial x} + \rho v \frac{\partial u}{\partial y} = -\frac{\partial p}{\partial x} + \mu \nabla^2 u \tag{19.75}$$

$$\rho u \frac{\partial v}{\partial x} + \rho v \frac{\partial v}{\partial y} = -\frac{\partial p}{\partial y} + \mu \nabla^2 v \tag{19.76}$$

subject to the continuity condition $\partial u/\partial x + \partial v/\partial y = 0$. Here the equations are linearized by using values of u_k, v_k from a previous trial solution. Applying Galerkin weighting to the inertia terms the element equations are now obtained as

$$k\{u, v, -p\} + k^*\{u, v\} = \{q\} \tag{19.77}$$

where k, $\{q\}$ are as in eqn (19.72) and the non-linear matrix k^* is

$$k^* = \rho \iint \begin{bmatrix} \{f\}u_k\{f_x\}^t + \{f\}v_k\{f_y\}^t & O \\ O & \{f\}u_k\{f_x\}^t + \{f\}v_k\{f_y\}^t \end{bmatrix} dx\,dy \tag{19.78}$$

in which $u_k = \{f\}^t\{u_k\}$ and $v_k = \{f\}^t\{v_k\}$ are interpolated values at each integration point, following the same procedure as in Section 12.7 with $\dot{v}$. Thus the initial solution is the linear Stokes solution and iteration then proceeds with u_k, v_k used to form the tangent matrix, convergence generally being achieved in ten or less iterations.[12]

The foregoing result is unsymmetric, doubling the computation requirements. Alternatively one can juse the derivatives $\partial u/\partial x$ and $\partial v/\partial y$ as the scalar multipliers, that is, the virtual work of the first convective term of eqn (19.75) is written as

$$\{\delta u\}^t \iint \{f\} \rho u \frac{\partial u}{\partial x} dx\,dy = \{\delta u\}^t \rho \iint \{f\} \{f\}^t\{u\} \left(\frac{\partial u}{\partial x}\right)_k dx\,dy \tag{19.79}$$

so that the convection matrix k^* is

$$\begin{bmatrix} D(\partial u/\partial x)_k & D(\partial u/\partial y)_k \\ D(\partial v/\partial x)_k & D(\partial v/\partial y)_k \end{bmatrix} \qquad (19.80)$$

where $D = \iint \rho \{f\} \{f\}^t \, dx \, dy$. This result is still unsymmetric but this can be remedied by shifting the offending terms to the right-hand side, so that the non-linear correction to be added to eqn (19.72) is now[26]

$$\begin{bmatrix} D(\partial u/\partial x)_k & O & O \\ O & D(\partial v/\partial y)_k & O \\ O & O & O \end{bmatrix} \begin{Bmatrix} \{u\} \\ \{v\} \\ \{-p\} \end{Bmatrix} = \begin{bmatrix} -(\partial u/\partial y)_k D\{u\}_k \\ -(\partial v/\partial x)_k D\{v\}_k \\ \{0\} \end{bmatrix}. \quad (19.81)$$

This arrangement is motivated by the Crank–Nicolson scheme (see Section 14.2), which also distributes damping matrix contributions to both sides of the recurrence relation and one hopes here to achieve similar stability advantages. There is also another precedent, namely in Sections 12.7 and 12.8, where large strains and curvatures involve non-linear stiffness and residual load effects on both sides of the recurrence relation, these being calculated using derivatives $(\partial v/\partial x)_k$ and $(\partial^2 v/\partial x^2)_k$ as scalar multipliers in forming the non-linear terms.

Equation (19.81) also has an interesting physical interpretation. The left-hand side has the form of a mass damping matrix, the damping effect depending upon the velocity gradients, whilst the right-hand side can be viewed as a vorticity loading, that is, when $\partial u/\partial y = \partial v/\partial x$, or the vorticity is zero, the forcing effect is the same upon both the horizontal and vertical velocities.

As an example of convective flow the cavity flow problem of Fig. 19.5 is solved using a regular mesh of thirty two elements (9×9 nodes), this being relatively coarse in view of the singularities at the upper corners. The velocity boundary conditions shown are those of a sliding lid with the pressure set to zero at the central basal node to establish a datum.

Here reduction to linear pressure interpolation gave slightly better velocity solutions. This is illustrated in Table 19.2 where the velocity profiles at the vertical centreline for $Re = 0$, obtained using both quadratic and linear pressure interpolations, are compared to Burggraf's finite difference results.[28] Also included is the profile obtained using the continuation method (that is the density is increased in steps, comparable to the use of 'load stepping' in Section 12.2) for $Re = 100$, again indicating reasonable agreement with Burggraf's results. The agreement is as good as that obtained by Hughes *et al.*[27] for $Re = 400$.

With quadratic pressure interpolation, however, spurious chequerboard pressures were obtained, apparently owing to the scarcity of pressure boundary conditions (at only one point rather than several as in Figs 19.3 and 19.4).

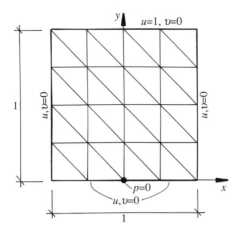

Fig. 19.5 Shear-induced cavity flow

Table 19.2 Numerical results for the horizontal velocities at the vertical centreline of the cavity problem of Fig. 19.5

	Re = 0			Re = 100	
y	quadratic p	linear p	Reference 28	linear p	Reference 28
1.000	1.00	1.00	1.00	1.00	1.00
0.875	0.35	0.37	0.35	0.27	0.28
0.750	0.01	− 0.02	− 0.04	− 0.07	0.01
0.625	− 0.12	− 0.16	− 0.17	− 0.25	− 0.11
0.500	− 0.14	− 0.20	− 0.20	− 0.21	− 0.20
0.375	− 0.15	− 0.18	− 0.16	− 0.13	− 0.19
0.250	− 0.07	− 0.13	− 0.12	− 0.06	− 0.12
0.125	− 0.05	− 0.07	− 0.05	− 0.03	− 0.05
0.000	0.00	0.00	0.00	0.00	0.00

But with reduction to linear pressure interpolation satisfactory results were obtained. These are illustrated by comparison of the minimum nodal pressure obtained for Re = 100 (indicating the vortex centre) with values obtained by Olson and Tuann[15] and Burggraf[28] in Table 19.3. The complete pressure contours were also in satisfactory agreement with those obtained by these authors.

Thus the formulation of eqn (19.72) must be restricted to linear pressure interpolation in some cases to avoid chequerboard effects. An alternative

Table 19.3 Numerical results for pressure at the vortex centre in Fig. 19.5

| | Location | | |
Method	x	y	Pressure
uvp formulation[15]	− 0.13	0.73	− 9.557
Segregated formulation[15]	− 0.13	0.75	− 7.028
Stream function[15]	− 0.11	0.72	− 9.447
Present method[26]	0.00	0.75	− 8.440
Finite difference[28]	− 0.13	0.74	− 9.1

remedy is the use of reduced integration. In the case of the bilinear quadrilateral element, for example, reduced integration of the penalty terms is equivalent to piecewise constant pressure interpolation,[19] but in the six-node triangle used for eqn (19.72) the integration is already reduced (it is quadratic, whereas the pressure terms *A*, *B* are cubic).

Hence, apart from linear pressure interpolation, the only other remedy is the use of filtering and smoothing techniques.[20] This is a difficult exercise so that the author recommends three-point integration and linear pressure interpolation with the six-node triangle and the formulation of eqn (19.72) together with eqn (19.81) when required.

19.7 Transient incompressible viscous flow

For transient flows the terms $\rho \partial u/\partial t$ and $\rho \partial v/\partial t$ of eqns (19.12) (19.13) must also be included, when the element equations are now obtained as

$$k\{u, v, -p\} + k^*\{u, v\} + k_D\{\dot{u}, \dot{v}\} = \{q\} \qquad (19.82)$$

where, with Galerkin weighting, the mass damping matrix for the accelerations is given by

$$k_D = \iint \rho \begin{bmatrix} \{f\}\{f\}^t & O \\ O & \{f\}\{f\}^t \end{bmatrix} dx\,dy. \qquad (19.83)$$

A recurrence relation can then be formed in the usual way, by using the finite difference approximations

$$\frac{\partial u}{\partial t} = \frac{u_{n+1} - u_n}{\delta t}, \qquad \frac{\partial v}{\partial t} = \frac{v_{n+1} - v_n}{\delta t}$$

but note that if forward differences are used in an attempt to obtain an explicit solution of the form of eqn (18.15) then eqn (19.83) provides no entries associated with the pressures. Remedies are to form an implicit scheme, at least for the pressures,[29] or to use the segregated penalty formulation of eqn (19.69).

An alternative approach proposed by Oden[30] is to use a *Lagrangian coordinate system* (that is, a moving frame of reference, allowing the mesh to distort after the fashion described in Section 12.6). In this case the non-linear terms in eqns (19.12) and (19.13) are eliminated so that the matrix k^* in eqn (19.82) is eliminated. This approach is particularly useful in the analysis of free surface flows.[31]

19.8 Incompressible lubrication problems

There have been comparatively few applications of finite element models to hydrodynamic lubrication,[3, 32] where a thin wedge-shaped space is formed between surfaces, the funnelling effect of the lubricant trapped between increasing the downstream pressure and keeping the surfaces apart. However, such models have wide application and are particularly economical because integration acorss the joint thickness reduces the analysis to two dimensions at most. Additional simplifications are the following.

1. Velocity derivàtives across the lubricating film are assumed dominant so that velocity derivatives parallel to the plane of the film are neglected. This is the thin film approximation.

2. Convective inertia effects are neglected.

3. Bearing curvature may be assumed to have only second-order effects.

Then with repeated integrations across the film most terms in the governing momentum equations are forced to the right-hand side of the matrix formulation, resulting in an economical analysis in which only pressure freedoms are required.

Neglecting convective inertia and assuming steady state, incompressible, and isothermal conditions (and thus constant viscosity) the Navier–Stokes equations for three-dimensional flow reduce to

$$\frac{\partial p}{\partial x} = \mu \nabla^2 u + X, \qquad \frac{\partial p}{\partial y} = \mu \nabla^2 v + Y, \qquad \frac{\partial p}{\partial z} = \mu \nabla^2 w + Z$$

where $\nabla^2 = \partial^2(\ )/\partial x^2 + \partial^2(\ )/\partial y^2 + \partial^2(\ )/\partial z^2$, subject to the continuity constraint $\partial u/\partial x + \partial v/\partial y + \partial w/\partial z = 0$.

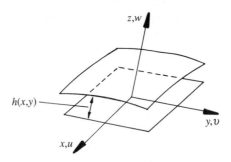

Fig. 19.6 Flow between two surfaces

Considering flow between two surfaces (Fig. 19.6) one assumes $w = 0$ and, invoking the thin film approximation, the governing equations reduce to

$$\frac{\partial p}{\partial x} = \mu \frac{\partial^2 u}{\partial z^2} + X \tag{19.84}$$

$$\frac{\partial p}{\partial y} = \mu \frac{\partial^2 v}{\partial z^2} + Y \tag{19.85}$$

$$\frac{\partial u}{\partial x} + \frac{\partial v}{\partial y} = 0. \tag{19.86}$$

Defining average velocities $\bar{U}$, $\bar{V}$ as

$$\bar{U} = \frac{1}{h} \int_0^h u \, dz, \qquad \bar{V} = \frac{1}{h} \int_0^h v \, dz \tag{19.87}$$

and using Liebnitz's integration theorem, that is,

$$\frac{\partial}{\partial x_1} \int_{h_1}^{h_2} F \, dx_3 = \int_{h_1}^{h_2} \frac{\partial F}{\partial x_1} dx_3 + F \bigg|_{h_2} \frac{\partial h_2}{\partial x_1} - F \bigg|_{h_1} \frac{\partial h_1}{\partial x_1} \tag{19.88}$$

where $F = F(x_1, x_2, x_3)$, $h_1 = h_1(x_1, x_2)$, $h_2 = h_2(x_1, x_2)$, one obtains

$$\frac{\partial(h\bar{U})}{\partial x} = \int_0^h \frac{\partial u}{\partial x} dz + U \frac{\partial h}{\partial x} \tag{19.89}$$

$$\frac{\partial(h\bar{V})}{\partial y} = \int_0^h \frac{\partial v}{\partial y} dz + V \frac{\partial h}{\partial y} \tag{19.90}$$

where U, V are the velocities of the upper surface ($z = h$) relative to the lower ($z = 0$). Integrating eqn (19.86) across the thickness and substituting eqns (19.89) and (19.90) yields

$$\frac{\partial(h\bar{U})}{\partial x} + \frac{\partial(h\bar{V})}{\partial y} - U \frac{\partial h}{\partial x} - V \frac{\partial h}{\partial y} = 0. \tag{19.91}$$

Now integrating eqns (19.84) and (19.85) three times across the film thickness and using eqns (19.89) and (19.90) one obtains

$$\frac{h^3}{6\mu}\frac{\partial p}{\partial x} = h\bar{U} + \frac{h^3}{6\mu}X \tag{19.92}$$

$$\frac{h^3}{6\mu}\frac{\partial p}{\partial y} = h\bar{V} + \frac{h^3}{6\mu}Y. \tag{19.93}$$

Using eqns (19.92) and (19.93) to substitute for $\bar{U}$ and $\bar{V}$ in eqn (19.91) gives the governing differential equation

$$\frac{\bar{h}^3}{6\mu}\left(\frac{\partial^2 p}{\partial x^2} + \frac{\partial^2 p}{\partial y^2}\right) = U\frac{\partial h}{\partial x} + V\frac{\partial h}{\partial y} + \frac{\partial h}{\partial t} + v_\mathrm{d}$$

$$+ \frac{\bar{h}^3}{6\mu}\left(\frac{\partial X}{\partial x} + \frac{\partial Y}{\partial y}\right) \tag{19.94}$$

where $\partial h/\partial t$ maintains mass conservation when the joint thickness changes with time, and v_d is the diffusion flow if the boundary surfaces at $z = 0, h$ are porous. Note that $\bar{h}$ denotes the average thickness because the terms involving this are obtained by assuming h constant and it is the first two terms on the right-hand side that introduce the hydrodynamic funnelling effect.

Applying Galerkin weighting and partially integrating the second derivatives in eqn (19.94) one obtains the element equations as[33]

$$k\{p\} = \{q_u\} + \{q_v\} + \{q_X\} + \{q_Y\} + k^*\left(\left\{\frac{\partial h}{\partial t}\right\} + \{v_\mathrm{d}\}\right) \tag{19.95}$$

where

$$k = \frac{\bar{h}^3}{6\mu}\int[\{f_x\}\{f_x\}^\mathrm{t} + \{f_y\}\{f_y\}^\mathrm{t}]\,\mathrm{d}A \tag{19.96}$$

$$k^* = \int\{f\}\{f\}^\mathrm{t}\,\mathrm{d}A \tag{19.97}$$

$$\{q_u\} = U\left(\frac{\delta h}{\delta x}\right)\int\{f\}\,\mathrm{d}A = U\int\{f\}\{f\}^\mathrm{t}\,\mathrm{d}A\left\{\frac{\partial h}{\partial x}\right\} \tag{19.98}$$

$$\{q_v\} = V\left(\frac{\delta h}{\delta y}\right)\int\{f\}\,\mathrm{d}A = V\int\{f\}\{f\}^\mathrm{t}\,\mathrm{d}A\left\{\frac{\partial h}{\partial y}\right\} \tag{19.99}$$

$$\{q_X\} = \frac{\bar{h}^3}{6\mu}\int\{f\}\{f_x\}^\mathrm{t}\,\mathrm{d}A\,\{X\} \tag{19.100}$$

$$\{q_Y\} = \frac{\bar{h}^3}{6\mu}\int\{f\}\{f_y\}^\mathrm{t}\,\mathrm{d}A\,\{Y\} \tag{19.101}$$

and $\{f_x\} = \{\partial f/\partial x\}$, $\{f_y\} = \{\partial f/\partial y\}$.

Two alternative forms are given for the 'velocity loadings' in eqns (19.98) and (19.99), the first using average values for $\partial h/\partial x$ and $\partial h/\partial y$, and these give identical results when these slopes are constant. Note that in these equations U and V are relative velocities and, of course, if there is no relative velocity between the surfaces, hydrodynamic lubrication will not occur.

Once the nodal pressures have been determined by assembling and solving eqns (19.95) these can be used to calculate nodal loadings as $\{q\} = \int \{f\} p \, dS$ on the bearing surfaces and these in turn used to carry out a stress analysis of the bearing components.

Calculation of the overall efficiency of the bearing, however, is generally of greatest interest. Considering one-dimensional elements for example, to this end one calculates the total thrust on each surface as

$$P = \int p \, dx = \sum \int_0^{L_e} \{f\}\{p\} \, dx = \sum \bar{p} L_e \qquad (19.102)$$

where $\bar{p}$ is the average pressure relative to the ambient pressure in each element, and L_e is the horizontal length of the element. The total frictional force on each bearing is given by

$$F = \int \tau \, dx = \int \mu \frac{\partial u}{\partial y} \, dx = \sum \frac{\mu U L_e}{\bar{h}} \qquad (19.103)$$

and the overall coefficient of friction for the bearing is given by the ratio F/P. Noting from eqns (19.95), (19.96), and (19.98) that the pressures used in eqn (19.102) will be proportional to μ and U one sees that F/P will be independent of viscosity and velocity, one of the key features of hydrodynamic lubrication.[33]

19.9 Inviscid compressible flow

Compressible flows allow variation in density throughout the domain and are classified as subsonic, transonic, and supersonic according to the *Mach number*, which is the ratio of the local velocity to that of sound in air. For subsonic flows the governing equation is elliptic, whereas transonic and supersonic flows are hyperbolic and involve shock fronts (lines of pressure discontinuity) and are thus more difficult to analyse.

For steady two-dimensional inviscid compressible flow the continuity and momentum equations are obtained by putting $\mu = 0$, $\partial(\ )/\partial t = 0$ in eqns (19.1), (19.12), and (19.13), giving, in index notation,

$$\frac{\partial(\rho u_i)}{\partial x_i} = 0 \qquad i = 1, 2 \qquad (19.104)$$

$$\rho u_j \frac{\partial u_i}{\partial x_j} + \frac{\partial p}{\partial x_i} = 0 \qquad i = 1, 2 \quad j = 1, 2. \qquad (19.105)$$

if body forces are neglected. For isentropic flow (that is, the process is reversible) one uses the constitutive equation[8]

$$c^2 = \frac{\partial p}{\partial \rho} \qquad (19.106)$$

where c is the velocity of sound at the given conditions. Multiplying eqn (19.105) by u_i, replacing $\partial p/\partial x_i$ by $c^2 (\partial \rho/\partial x_i)$, which follows from eqn (19.106), and using the chain rule of differentiation one obtains

$$u_i u_j \frac{\partial u_i}{\partial x_j} + \frac{u_i c^2}{\rho} \frac{\partial \rho}{\partial x_i} = 0. \qquad (19.107)$$

Applying the product rule to eqn (19.104), so that

$$u_i \frac{\partial \rho}{\partial x_i} + \rho \frac{\partial u_i}{\partial x_i} = 0,$$

eqn (19.107) can be written as

$$u_i u_j \frac{\partial u_i}{\partial x_j} - c^2 \frac{\partial u_i}{\partial x_i} = 0. \qquad (19.108)$$

Expanding the indices in eqn (19.108) one obtains

$$\left(1 - \frac{u^2}{c^2}\right)\frac{\partial u}{\partial x} + \left(1 - \frac{v^2}{c^2}\right)\frac{\partial v}{\partial y} - \frac{uv}{c^2}(\partial u/\partial y + \partial v/\partial x) = 0 \qquad (19.109)$$

and substituting $uv(\partial v/\partial x - \partial u/\partial y)/c^2 = 0$ (for irrotational flow) we now obtain

$$\left(1 - \frac{u^2}{c^2}\right)\frac{\partial u}{\partial x} + \left(1 - \frac{v^2}{c^2}\right)\frac{\partial v}{\partial y} - \frac{2uv}{c^2}\frac{\partial u}{\partial y} = 0. \qquad (19.110)$$

Using a velocity potential (defined by eqns (19.14) yields

$$(c^2 - \phi_x^2)\phi_{xx} + (c^2 - \phi_y^2)\phi_{yy} - 2\phi_x \phi_y \phi_{xy} = 0 \qquad (19.111)$$

where $\phi_x = \partial\phi/\partial x$, $\phi_{xx} = \partial^2\phi/\partial x^2$, and so on. Using Galerkin weighting, partially integrating the second derivatives, and writing the non-linear terms of eqn (19.111) on the right-hand side, one obtains the element equations[34]

$$k\{\phi\}^{n+1} = \{F\}_n + \{R\}_n + \{S\}_n \qquad (19.112)$$

where

$$k = c^2 \int (\{f_x\}\{f_x\}^t + \{f_y\}\{f_y\}^t) dV \qquad (19.113)$$

$$\{F\}_n = c^2 \int \{f\}[c_x \phi_x + c_y \phi_y] dS \qquad (19.114)$$

$$\{R\}_n = \int [\phi_x^3\{f_x\} + \phi_y^3\{f_y\} + \phi_x \phi_y^2\{f_x\} + \phi_x^2 \phi_y\{f_y\}] dV \qquad (19.115)$$

$$\{S\}_n = -\int \{f\}[c_x^* \phi_x + c_y^* \phi_y][\phi_x^3 + \phi_y^3 + \phi_x \phi_y^2 + \phi_x^2 \phi_y] dS. \qquad (19.116)$$

The last two terms deal approximately with the non-linearities as forcing functions by integrating by parts with first derivatives of ϕ held constant. In $\{F\}_n$, c_x and c_y are direction cosines at the boundaries and in $\{S\}_n$, c_x^* and c_y^* are the direction cosines at the shock front. Solving eqn (19.112) iteratively the first trial solution is that for incompressible flow with $\{R\}_n = \{S\}_n = \{0\}$.

An alternative approach is to neglect the non-linear functions $\{R\}_n$ and $\{S\}_n$ and iteratively solve eqn (19.112) using

$$k_n \{\phi\}_{n+1} = \{F\}_n \tag{19.117}$$

where

$$k_n = \int \rho_n [\{f_x\}\{f_x\}^t + \{f_y\}\{f_y\}^t] \, dV$$

$$\{F\}_n = \int \{f\} [c_x \phi_x + c_y \phi_y] \, dS.$$

ρ_n is updated at each integration point using the isentropic flow relation

$$\rho/\rho_0 = \left[1 - \frac{(\gamma - 1)\nabla\phi \cdot \nabla\phi}{2V_0^2} \right]^{-1/(\gamma - 1)} \tag{19.118}$$

where $\gamma = c_v/c_p$ is the specific heat ratio (which is 1.4 for air) and V_0 is the velocity of sound in the fluid at rest (the *stagnation velocity*). $M_0 = \sqrt{\nabla\phi \cdot \nabla\phi}/v_0$ gives the Mach number corresponding to stagnation conditions, which can be compared with the local Mach number given by $M = \sqrt{\nabla\phi \cdot \nabla\phi}/V$, V being the local velocity of sound, and the *free stream Mach number* given by $M_\infty = \sqrt{\nabla\phi \cdot \nabla\phi}/V_\infty$, V_∞ being the free stream velocity. Note that the V here are resultant velocities.

Such iteration of eqn (19.117) satisfies the continuity condition (eqn (19.104)) but does not conserve momentum (eqns (19.105)), so that the solutions are termed *non-conservative*, but note that eqn (19.117) is similar to that for simple potential flow (eqn (19.19)).

Alternative potential function formulations have been obtained by de Vries et al.[35] whilst Periaux[36] has also obtained a stream function formulation of the problem. To deal effectively with shock fronts, across which a change in entropy occurs,[1] Chung uses interpolations of both continuous and discontinuous velocity potential functions (the latter are Lagrange multipliers) to form a special shock front element.[34]

The most direct approach, which we have seen in several preceding sections, is through the use of primitive variables, allowing a small artifical viscosity. Equation (19.77) is solved iteratively using eqn (19.118) to evaluate ρ in eqn (19.78). Indeed, viscous compressible flow is dealt with in the following section using such a formulation but now temperature freedoms must also be included.

19.10 Viscous compressible flow

The influence of the viscosity of a fluid on its flow is generally restricted to a thin layer over rigid boundary surfaces and the general flow pattern is influenced very little by viscosity. Hence inviscid flow models, which frequently utilize stream or potential functions, prove useful in determining flow patterns but cannot deal with such phenomena as skin friction, form drag on a body, and separation of flow, which are of considerable importance.

If the flow is compressible, heat caused by friction, as well as temperature change caused by compression, must be taken into account. In addition, it is necessary to consider the effects of the variation of viscosity with temperature. Thus, to deal with the complete thermohydrodynamic state all the equations of Section 16.4 must be taken into account.

Under steady state conditions eqns (19.77) may be used to determine the velocity and pressure distributions and the temperature distribution is governed by eqn (16.59). When direct heating from chemical reaction and radiation heating is absent this equation reduces in two dimensions to [1]

$$\rho u c_v \frac{\partial T}{\partial x} + \rho v c_v \frac{\partial T}{\partial y} - \kappa \frac{\partial^2 T}{\partial x^2} - \kappa \frac{\partial^2 T}{\partial y^2}$$

$$- u \frac{\partial p}{\partial x} - v \frac{\partial p}{\partial y} - \mu \Phi = 0 \tag{19.119}$$

where

$$\Phi = \left(\frac{\partial u}{\partial y}\right)^2 + \left(\frac{\partial v}{\partial x}\right)^2 + 2\frac{\partial u}{\partial y}\frac{\partial v}{\partial x} + \frac{4}{3}\left[\left(\frac{\partial u}{\partial x}\right)^2 - \frac{\partial u}{\partial x}\frac{\partial v}{\partial y} + \left(\frac{\partial v}{\partial y}\right)^2\right]$$

is the viscous dissipation function and c_v and κ are assumed constant. Using Galerkin weighting and integration by parts to reduce the second-order derivatives, the element thermal energy equations are obtained as

$$k_T\{T\} + k_p\{p\} = \{q_T\} \tag{19.120}$$

where

$$k_T = \int \rho c_v [u_k\{f\}\{f_x\}^t + v_k\{f\}\{f_y\}^t] \, dV$$

$$\quad + \int \kappa [\{f_x\}\{f_x\}^t + \{f_y\}\{f_y\}^t] \, dV \tag{19.121}$$

$$k_p = -\int [u_k\{f\}\{f_x\}^t + v_k\{f\}\{f_y\}^t] \, dV \tag{19.122}$$

$$\{q_T\} = \mu \int \{f\}\Phi \, dV + \int \kappa \{f\}\left[c_x\frac{\partial T}{\partial x} + c_y\frac{\partial T}{\partial y}\right] dS. \tag{19.123}$$

The convective inertia terms involving u_k and v_k are dealt with iteratively in the manner described in Section 19.6, the complicated viscous dissipation function is used as a loading term, and the boundary integral term in eqn (19.123) arises in the same way as those for plane torsion and potential flow. Finally eqns (19.120) are superimposed over eqns (19.77).

Thus the complete formulation involves velocity, pressure, and temperature freedoms, the k_p matrix replacing the submatrix αI in eqn (19.72). Note that μ, ρ and κ are now functions of temperature and must be recalculated during each iteration using[1]

$$\frac{\mu}{\mu_0} = \frac{\kappa}{\kappa_0} = \left(\frac{T}{T_0}\right)^m, \qquad p = \rho RT \qquad (19.124)$$

where μ_0 and κ_0 are the values measured at a reference temperature T_0.

Extension of the complete formulation to deal with transient problems is straightforward[34] but a more economical approach is achieved by following the procedure described by Heubner for compressible lubrication problems,[3] where the incompressible uvp formulation of eqn (19.72) is iterated using recalculated viscosities and densities (from eqns (19.124)), including solution of eqn (19.120) with k_p moved to the right-hand side to determine the temperatures as a second step in each iteration.[23, 26] This *tandem iterative* procedure is illustrated by a simple example in Section 20.5 and, with the assumption of a small artificial viscosity, has also been applied to the analysis of convection problems.[37]

Note that eqns (19.121) and (19.122) involve unsymmetric matrices but symmetry can be restored in the manner described in Section 19.6, that is, the scalar multipliers u_k, v_k are replaced by $(\partial u/\partial x)_k$, $(\partial v/\partial y)_k$ but now there is no need to shift some terms to the right-hand side as in eqn (19.81).

19.11 Concluding remarks

Clearly, applications of finite elements to a wide range of fluid flow problems are well established, of particular note being the use of primitive or natural variables (those variables appearing in the classical governing equations) rather than stream or potential functions. These have many advantages but coupling of different types of freedoms is not without its hazards, particularly the chequerboard syndrome discussed in Section 19.5 which also appears elsewhere, for example in shallow water circulation problems.[38]

Similar difficulties also occur in solid mechanics applications where measures such as reduced integration (see Section 15.10) or reduced interpolation (see Section 11.5) have also proved effective. Moreover the chequerboard syndrome may also occur in the finite difference method when two or more variables are associated with each grid point but here remedies such as reduced integration are not available. It appears, therefore, that formulations such as those of Section 19.5 and 19.10 are as firmly established as some of the finite element formulations for plane stress, plate bending, and shell analysis discussed in earlier chapters.[39]

References

1. Yuan, S. W. (1967). *Foundations of Fluid Mechanics*. Prentice-Hall, Englewood Cliffs, N.J.
2. Fung, Y. C. (1965). *Foundations of Continuum Mechanics*. Prentice-Hall, Englewood Cliffs, NJ.
3. Heubner, K. H. (1975). *The Finite Element Method for Engineers*. Wiley, New York.
4. Argyris, J. H. and Dunne, P. C. (1976). The finite element method applied to fluid mechanics. *Proceedings of Conference on Computer Methods and Problems in Aeronautical Fluid Dynamics*, University of Manchester, 1974. Academic Press, London.
5. de Vries, G. and Norrie, D. H. (1971). The applicaions of the finite element technique to potential flow problems. *Trans ASME*, Series E, **38**, 978.
6. Connor, J. J. and Brebbia, C. A. (1977). *Finite Element Techniques for Fluid Flow*. Butterworths, London.
7. McCorquodale, J. A. and Li Y. (1971). Finite element analysis of sluice gate flow. *Trans. Inst. Eng. Canada*, **14** No. C-2.
8. Crandall, S. (1956). *Engineering Analysis*. McGraw-Hill, New York.
9. Oden, J. T. and Wellford, L. C. (1972). Analysis of flow of viscous fluids by the finite element method. *J. Amer. Inst. Aeron. Astron.*, **10**, 1590.
10. Hood, P. and Taylor, C. (1974). Navier–Stokes equations using mixed interpolation. *Proceedings of the Symposium on Finite Element Methods in Flow Problems*, University College, Wales, Swansea.
11. Kawahara, M., Yoshimura, N., Nakagawa, K., and Ohsaka, H. (1976). Steady and unsteady finite element analysis of incompressible viscous fluid. *Int. J. Num. Meth. Engng*, **10**, 437.
12. Yamada, Y., Ito, K., Yokouchi, Y., Tamano T., and Ohtsubo, T. (1974). Finite element anlysis of steady fluid and metal flow. In *Finite Element Methods in Flow Problems*, (ed. J. T. Oden, O. C. Zienkiewicz, K. H. Gallagher, and C. Taylor). UAH Press, Huntsville.
13. Zienkiewicz, O. C. (1977). *The Finite Element Method*, (3rd edn). McGraw-Hill, London.
14. Sani, R. L., Gresho, P. M., Lee, R. L., and Griffiths, D. F. (1981). The cause and cure (?) of the spurious pressures generated by certain FEM solutions of the incompressible Navier-Stokes equations: part 1. *Int. J. Num. Meth. Fluids*, **1**, 17.
15. Olson, M. D. and Tuann, S. Y. (1978). Primitive variables versus stream function finite element solutions of the Navier–Stokes equations. In *Finite Elements in Fluids*, Vol. 3, (ed. R. H. Gallagher *et al.*). Wiley, New York.
16. Carey, G. F. and Krishnan, R. (1982) Penalty approximation of Stokes flow. *Comp. Meth. Appl. Mech. Engng*, **35**, 169.

17. Reddy, J. N. (1982). On penalty function methods in finite element analysis of flow problems. *Int. J. Num. Meth. Fluids,* **2**, 151.
18. Oden, J. T., Krikuchi, N., and Song, Y. J. (1982). Penalty-finite element methods for the anaysis of Stokesian flows. *Comp. Meth. Appl. Mech. Engng,* **31**, 297.
19. Engelman, M. S., Sani, R. L. Gresho, P. M., and Bercovier, M. (1982). Consistent vs reduced integration penalty methods for incompressible media using several old and new elements. *Int. J. Num. Meth. Fluids,* **2**, 25.
20. Sani, R. L., Gresho, P. M., Leww, R. L., Griffiths, D. F., and Engelman, M. (1981). The cause and cure (!) of the spurious pressures generated by certain FEM solutions of the incompressible Navier-Stokes equations: part 2. *Int. J. Num. Meth. Fluids,* **1**, 171.
21. Oden, J. T., Kikuchi, N., and Song, Y. J. (1979). An analysis of exterior penalty methods and reduced integration for finite element approximations of contact problems in incompressible elasticity. TICOM Report 79–10, University of Texas, Austin.
22. Kawahara, M. and Hirano, H. (1983). A finite element method for high Reynolds number viscous fluid flow using two step explicit scheme. *Int. J. Num. Meth. Fluids,* **3**, 137.
23. Mohr, G. A. (1983). Finite element analysis of viscous flows. *Conference on Computer Techniques and Applications,* University of Sydney.
24. Atkinson, B., Brocklebank, M. P., Card, C. C. H., and Smith, J. M. (1969). Low Reynolds number developing flows. *J. Amer. Inst. Chem. Engng,* **15**, 548.
25. Schlicting, H. (1979) *Boundary-Layer Theory,* (7th edn). McGraw-Hill, New York.
26. Mohr, G. A. (1984). Finite element analysis of viscous flow. *Comput. Fluids,* **12**, 217.
27. Hughes, T. J. R., Taylor, R. L., and Levy, J. F. (1978). High Reynolds number, steady incompressible flows by a finite element method. In *Finite Elements in Fluids,* Vol. 3, (ed. R. H. Gallagher, O. C. Zienkiewicz, J. T. Oden, M. Morandi Cecchi, and C. Taylor). Wiley, New York.
28. Burggraf, O. C. (1966). Analytical and numerical studies of the structure of steady separated flows. *J. Fluid Mech.,* **24**, Part I, 113.
29. Gresho, P. M., Lee, R. L., and Sani, R. L. (1980). On the time-dependent solution of incompressible Navier-Stokes equations in two and three dimensions. In *Recent Advances in Numerical Methods in Fluids* Vol. 1, (ed. C. Taylor and K. Morgan). Pineridge Press, Swansea.
30. Oden, J. T. (1972). *Finite Elements of Nonlinear Continua.* McGraw-Hill, New York.
31. Hirt, C. W., Cook, J. L., and Butler, T. D. (1970). A Lagrangian method for calculating the dynamics of an incompressible fluid with free surface. *J. Comput. Phys.* **5** (1).
32. Argyris, J. H. and Scharpf, D. W. (1969). The incompressible lubrication problem. *Aero. Jour.* **73**, 1044.
33. Mohr, G. A. and Broom, N. D. (1982). A finite element lubrication model. *Fourth Australian International Conference on Finite Element Methods in Engineering,* University of Melbourne.
34. Chung, T. J., (1978) *Finite Element Analysis in Fluid Dynamics.* McGraw-Hill, New York.

35. de Vries, G., Berard, G. P., and Norrie, D. H. (1971). Application of the finite element technique to compressible flow problems. *Trans ASME*, Series E, **38**.
36. Periaux, J. (1975). Three-dimensional anaysis of compressible potential flows with the finite element method. *Int. J. Num. Meth. Engng*, **9**, 775.
37. Gatling, D. K. and Nickell, R. E. (1978). Finite element analysis of free and forced convection. In *Finite Elements in Fluids*, Vol. 3, (ed. R. H. Gallagher, O. C. Zienkiewicz, J. T. Oden, M. Morandi Cecchi, and C. Taylor). Wiley, New York.
38. Walters, R. A. and Carey, G. F. (1983). Analysis of spurious oscillation modes for the Navier–Stokes and shallow water equations. *Comput. Fluids*, **11**, 51.
39. Hughes, T. J. R. (1987). *The Finite Element Method: Linear Static and Dynamic Finite Element Analysis*. Prentice-Hall, Englewood Cliffs, NJ.

20
Fluid flow
examples and programs

The preceding chapter gave an account of a wide variety of fluid flow problems. Here we follow with several worked examples. The first is simple potential flow but those which follow are of special interest as they are *pseudo-one-dimensional* and thus are able to be solved very economically. Finally, to complement the discussion of programming in Chapter 15, a miniature program for potential flow and a FEMSS program for Stokes flow is given. The latter is based on eqn (19.72) and introduces the reader to the use of *small slack variables* (given by eqn (19.73)). The role of these as inverse penalty factors is discussed further in Chapter 23.

20.1 Two-dimensional potential flow

As an example of two-dimensional potential flow consider the problem of ducted flow with unit free stream velocity around an infinitely long cylinder of unit radius. Figure 20.1(a) shows the boundary conditions for analysis of a quadrant of the domain using five bilinear elements and the stream function formulation described in Section 19.2.

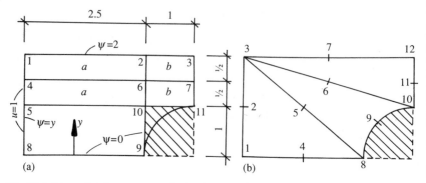

Fig. 20.1 Potential flow over a cylinder (one quadrant analysed): (a) bilinear elements and (b) quadratic elements

Using a rectangular cut-out to approximate the cylinder crudely there are only two unspecified ψ freedoms (ψ_6 and ψ_7) whereas if a potential function is used there·are eight unknown nodal values. Further, choosing a stream function formulation the forcing terms of eqns (19.23) vanish as the velocity flux is perpendicular to the inlet and outlet boundaries ($c_y u = 0$). The element matrices are obtained from eqn (19.20) by the same calculations used in Sections 17.4 and 18.6 or by putting $v = 0$, neglecting the shear contributions, and combining the u, v contributions in eqns (6.47), giving

$$
k = \frac{1}{12}
\begin{bmatrix}
4\lambda + 4/\lambda & -4\lambda + 2/\lambda & -2\lambda - 2/\lambda & 2\lambda - 4/\lambda \\
 & 4\lambda + 4/\lambda & 2\lambda - 4/\lambda & -2\lambda - 2/\lambda \\
\text{Symmetric} & & 4\lambda + 4/\lambda & -4\lambda + 2/\lambda \\
 & & & 4\lambda + 4/\lambda
\end{bmatrix}
\tag{20.1}
$$

where $\lambda = L_b/L_a$ and the local node numbers are as shown in Fig. 6.1(b).

Assembling the element equations, deleting the specified freedoms, and calculating their equivalent loadings in the manner of eqn (15.15) gives

$$
\begin{bmatrix}
2k_{66}^a + 2k_{66}^b & 2k_{67}^b \\
2k_{67}^b & 2k_{77}^b
\end{bmatrix}
\begin{Bmatrix} \psi_6 \\ \psi_7 \end{Bmatrix} =
\begin{Bmatrix} q_6 \\ q_7 \end{Bmatrix}
\tag{20.2}
$$

where

$$
q_6 = -(k_{61}\psi_1 + k_{62}\psi_2 - 2k_{64}\psi_4 - k_{65}\psi_5)^a - (k_{62}\psi_2 - k_{63}\psi_3)^b \tag{20.3}
$$

$$
q_7 = -(k_{73}\psi_3 + k_{72}\psi_2)^b \tag{20.4}
$$

with the superscripts a, b denoting the element shapes thus indicated in Fig. 20.1. Taking care to map from local (that is, 1–2–3–4 anticlockwise) to global node numbers and putting $\psi_1 = \psi_2 = \psi_3 = 2$, $\psi_4 = 1.5$, $\psi_5 = 1$, $\lambda^a = 0.2$, and $\lambda^b = 0.5$ one obtains

$$
\begin{bmatrix}
61.6 & 2 \\
2 & 20
\end{bmatrix}
\begin{Bmatrix} \psi_6 \\ \psi_7 \end{Bmatrix} =
\begin{Bmatrix} 66.8 \\ 24 \end{Bmatrix}
\tag{20.5}
$$

giving $\psi_7 = 1.0951$, which is in good agreement with the analytical result, $\psi_7 = 1.1025$, obtained by the method of images.[1]

Once the nodal stream function solution has been obtained the velocities follow from eqns (19.16). For example, approximate velocities above the crest of the cylinder may be calculated as

$$
u_{11-7} \doteq \frac{\psi_{11} - \psi_7}{y_{11} - y_7} = \frac{0 - 1.0951}{1.0 - 1.5} = 2.1902 \tag{20.6}
$$

$$
u_{7-3} = \frac{\psi_7 - \psi_3}{y_7 - y_3} = \frac{1.0951 - 2.0}{1.5 - 2.0} = 1.8098. \tag{20.7}
$$

Figure 20.1(b) shows the problem analysed using three six-node triangular elements and the program of Section 20.6, yielding the result $\psi_{11} = 1.0296$ which is an excellent result with so coarse a mesh.

20.2 One-dimensional solution for plane Couette flow

Figure 20.2 shows the velocity profile obtained in Couette flow, that is, steady flow between parallel stationary and moving plates under a constant pressure gradient, the exact solution for which is[2]

$$\frac{u}{U} = \frac{y}{h} - \frac{\partial p}{\partial x}\frac{hy}{2\mu U}\frac{1-y}{h}.$$ (20.8)

From eqn. (19.72) the element equations for the vertical elements reduce to

$$k\{u\} + k_D\left\{\frac{\partial p}{\partial x}\right\} = \{0\}$$ (20.9)

where

$$k = \mu \int \left\{\frac{\partial f}{\partial y}\right\}\left\{\frac{\partial f}{\partial y}\right\}^t dV = \frac{\mu}{L}\begin{bmatrix} 1 & -1 \\ -1 & 1 \end{bmatrix}$$ (20.10)

$$k_D = \int \{f\}\{f\}^t dV = \frac{L}{6}\begin{bmatrix} 2 & 1 \\ 1 & 2 \end{bmatrix}$$ (20.11)

taking direct advantage of the constant value of $\partial p/\partial x$ in eqn (20.9) to allow k_D to be symmetric. Now using two elements (see Fig. 20.2) and lumping in

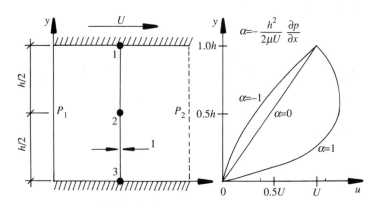

Fig. 20.2 One-dimensional FEM solution for plane Couette flow

k_D, the assembled equations are

$$\frac{2\mu}{h}\begin{bmatrix} 1 & -1 & 0 \\ -1 & 2 & -1 \\ 0 & -1 & 1 \end{bmatrix}\begin{Bmatrix} u_1 \\ u_2 \\ u_3 \end{Bmatrix} = -\frac{h}{4}\begin{bmatrix} 1 & 0 & 0 \\ 0 & 2 & 0 \\ 0 & 0 & 1 \end{bmatrix}\begin{Bmatrix} \partial p/\partial x \\ \partial p/\partial x \\ \partial p/\partial x \end{Bmatrix} \quad (20.12)$$

and, with $u_3 = Q$, $u_1 = U$, one obtains

$$\left(\frac{2\mu}{h}\right)(2)u_2 = \left(\frac{-h}{4}\right)(2)\frac{\partial p}{\partial x} - \left(\frac{2\mu}{h}\right)(-1)(u_1 = U)$$

or

$$\frac{u_2}{U} = 0.5 - \frac{h^2}{8\mu U}\partial p/\partial x \quad (20.13)$$

which is exactly the result given by eqn (20.8).

In the special case when $U = 0$ (plane Poiseille flow)[2] one also sets $u_1 = 0$ in eqn (20.12), giving the solution $u_2 = -(h^2/8\mu)(\partial p/\partial x)$ which is again the exact result. Clearly the philosophy of using lumping with low-order elements (see Section 13.5) is being justified here as in some of the examples of Chapter 18. The problem is admittedly a very simple one, principally because it is essentially one dimensional so that the continuity condition need not be enforced, but is frequently used to test two-dimensional elements.

20.3 Hydrodynamic lubrication

Figure 20.3 shows relative motion between a slider bearing and the bearing guide, the gap being filled with lubricant which we idealize with four linear elements in the horizontal direction. The problem is to determine the pressure distribution between the bearing and the guide relative to the ambient

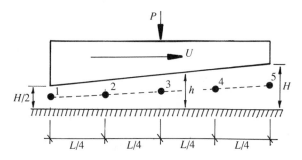

Fig. 20.3 Slider bearing

pressure which we shall assume to be zero.[3] Using eqns (19.96) and (19.98) the element equations are

$$k\{p\} = \{q_u\} \tag{20.14}$$

where

$$k = \int \frac{h^3}{6\mu} \{f_x\} \{f_x\}^t \, dx = \frac{\bar{h}^3}{6\mu L_e} \begin{bmatrix} 1 & -1 \\ -1 & 1 \end{bmatrix} \cdot \tag{20.15}$$

$$\{q_u\} = U \frac{\delta h}{\delta x} \int \{f\} \, dx = U L_e \frac{\delta h}{\delta x} \{\tfrac{1}{2}, \tfrac{1}{2}\} \tag{20.16}$$

using averaged thicknesses for the tapered elements. The assembled equations are, after imposing $p_1 = p_5 = 0$,

$$\frac{(H/16)^3}{6\mu L/4} \begin{bmatrix} 2060 & -1331 & 0 \\ -1331 & 3528 & -2197 \\ 0 & -2197 & 5572 \end{bmatrix} \begin{Bmatrix} p_2 \\ p_3 \\ p_4 \end{Bmatrix} = H \begin{Bmatrix} U/8 \\ U/8 \\ U/8 \end{Bmatrix} \tag{20.17}$$

giving the solution

$$\{p_2, p_3, p_4\} = \frac{\mu U L}{H^2} \{0.94, 0.87, 0.48\} \tag{20.18}$$

whereas the exact solution is[2]

$$\{p_2, p_3, p_4\} = \frac{\mu U L}{H^2} \{0.96, 0.89, 0.49\} \tag{20.19}$$

to which the ambient pressure is added if significant. Now using eqns (19.102) and (19.103) one obtains the overall coefficient of friction as $F/P = 2.415H/L$ (cf. the exact result which is $2.431\ H/L$).[2] Typical values of H/L are approximately 1/100, so that the coefficient of friction will be about 0.025 or 5 per cent of the value without lubrication.[4]

20.4 Blasius flow

Figure 20.4 shows a flat plate, initially at rest, suddenly given a constant horizontal velocity U. The pressure is constant and the streamlines will be parallel to the plate so that $v = 0$. Combining eqns (19.72) and (19.83), and using forward difference time stepping (eqn (14.5)), one obtains the element recurrence relation as

$$k_D \{u\}_{n+1} = (k_D - \delta k) \{u_n\} \tag{20.20}$$

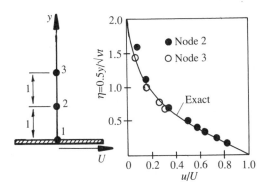

Fig. 20.4 Motion of a plate in an infinite medium

where

$$k = \frac{\mu A}{L} \begin{bmatrix} 1 & -1 \\ -1 & 1 \end{bmatrix}, \qquad k_\mathrm{D} = \frac{AL}{6} \begin{bmatrix} 2 & 1 \\ 1 & 2 \end{bmatrix} \qquad (20.21)$$

using linear elements. Using two elements of unit width and thickness (that is, $A = 1$) and lumping in k_D one obtains the system equations as

$$\begin{Bmatrix} u_1/2 \\ u_2 \\ u_3/2 \end{Bmatrix}_{n+1} = \begin{bmatrix} \frac{1}{2} - \mu\delta & \mu\delta & 0 \\ \mu\delta & 1 - 2\mu\delta & \mu\delta \\ 0 & \mu\delta & \frac{1}{2} - \mu\delta \end{bmatrix} \begin{Bmatrix} u_1 \\ u_2 \\ u_3 \end{Bmatrix}_n. \qquad (20.22)$$

The boundary conditions are $u_1 = U(t > 0)$ and $u = 0$ at $y = \infty$ (for all t). To obtain an approximate solution we can solve the problem using (a) $u_3 = 0$ and (b) u_3 unspecified or free, and average the two results obtained to approximate the 'infinity' condition. Thus the recurrence relations become

(a) $$(u_2)_{n+1} = (1 - 2\mu\delta)(u_2)_n + \mu\delta U \qquad (20.23)$$

(b) $$(u_2)_{n+1} = (1 - 2\mu\delta)(u_2)_n + \mu\delta U + \mu\delta(u_3)_n \qquad (20.24\mathrm{a})$$

$$(u_3)_{n+1} = 2\mu\delta(u_2)_n + (1 - 2\mu\delta)(u_3)_n. \qquad (20.24\mathrm{b})$$

Using $\delta = 0.1$ and averaging the results from eqns (20.23) and (20.24) one obtains the results shown in Fig. 20.4 expressed in terms of the similarity parameter[2]

$$\eta = \frac{y}{2\sqrt{vt}} \qquad (20.25)$$

where $v = \mu/\rho$ is the kinematic viscosity. The thickness of the boundary layer (in which the viscous stresses are negligible) can be estimated by the limits[2]

$$0.564 < \eta < 1.253 \qquad (20.26)$$

and, taking the lower limit, this defines the region in which η/u is approximately constant, and the range in eqn (20.26) can be interpreted as a transition zone.

In the preceding examples simplifications were possible so that the use of linear elements proved sufficient, whilst in the last case and those of Section 18.3 and 18.5 this gave a consistent order of approximation for the spatial and time interpolations. Indeed, in one-dimensional consolidation for example, linear elements may perform better than cubic ones.[5]

20.5 Compressible Couette flow

The problem of adiabatic compressible Couette flow under constant pressure between parallel plates with relative velocity U possesses an analytical solution[2] and is thus a useful example. This is similar to Fig. 20.2 but there is no pressure gradient and the fluid is compressible. Using eqns (19.72) and (19.121) and noting that the domain is isobaric (that is, at constant pressure so that nodal pressure freedoms are not required), $\partial(\ )/\partial x = 0$, and $v = 0$, one obtains the element equations for the velocities and temperatures as

$$k\{u\} = \{0\} \tag{20.27}$$

$$k_\mathrm{T}\{T\} = \{q_\mathrm{T}\} \tag{20.28}$$

where

$$k = \int \mu \{f_y\} \{f_y\}^\mathrm{t}\,\mathrm{d}V \tag{20.29}$$

$$k_\mathrm{T} = \int \kappa \{f_y\} \{f_y\}^\mathrm{t}\,\mathrm{d}V \tag{20.30}$$

$$\{q_\mathrm{T}\} = \mu \int \{f\} \left(\frac{\partial u}{\partial y}\right)^2 \mathrm{d}V \tag{20.31}$$

and all convective terms (those involving u_k, v_k) have vanished.

As the boundary surfaces are adiabatic these equations do not include surface heat convection terms (which, were they required, would be given by eqns (18.4) and (18.5)).

The solution is obtained by assuming an initial temperature distribution and solving equations (20.27) and (20.28) iteratively using updated element values of μ and κ at the last temperature distribution. Denoting by T, μ, and κ, the values at the moving plate, and using two linear elements across the field (as in Fig. 20.2) eqn (20.27) yields, after assembly,

$$\frac{2\mu}{h} \begin{bmatrix} 1 & -1 & 0 \\ -1 & 2 & -1 \\ 0 & -1 & 1 \end{bmatrix} \begin{Bmatrix} u_1 \\ u_2 \\ u_3 \end{Bmatrix} = \begin{Bmatrix} 0 \\ 0 \\ 0 \end{Bmatrix} \tag{20.32}$$

initially, assuming $T_1 = T_2 = T_3 = T$ and thus $\mu_i = \mu$ in each element.

Imposing $u_1 = U$ and $u_3 = 0$ one obtains, using eqn (15.15),

$$\left(\frac{2\mu}{h}\right)(2)u_2 = -\left(\frac{2\mu}{h}\right)(-1)U \qquad (20.33)$$

or $u_2 = U/2$ as the first approximation.

Now eqn (20.28) yields, after assembly,

$$\frac{2\kappa}{h}\begin{bmatrix} 1 & -1 & 0 \\ -1 & 2 & -1 \\ 0 & -1 & 1 \end{bmatrix}\begin{Bmatrix} T_1 \\ T_2 \\ T_3 \end{Bmatrix} = \mu\left(\frac{U}{h}\right)^2\left(\frac{h}{2}\right)\begin{Bmatrix} 1/2 \\ 1 \\ 1/2 \end{Bmatrix} \qquad (20.34)$$

and imposing $T_1 = T$ one obtains

$$T_2 = \frac{3\mu U^2}{8\kappa} + T \qquad (20.35)$$

$$T_3 = \frac{\mu U^2}{2\kappa} + T. \qquad (20.36)$$

We will assume that, at the plate, $\mathrm{Pr}\,M^2 = 2$, where Pr is the *Prandtl number* for the fluid, defined by $\mathrm{Pr} = \mu c_p/\kappa$. Also assuming $m = 1$ in eqn (19.124) and that $M = U/c$ is the Mach number of the plate, we have

$$\frac{\mu U^2}{\kappa} = \frac{2c^2}{c_p} \qquad (20.37)$$

where the velocity of sound at the plate is given by[6]

$$c^2 = \gamma\frac{p}{\rho} = \gamma RT. \qquad (20.38)$$

The *Universal gas constant* R is given by[2]

$$R = c_p - c_v = c_p\frac{\gamma - 1}{\gamma} \qquad (20.39)$$

where the specific heat ratio $\gamma = 1.405$ for air.

Combining eqns (20.37)–(20.39) we have $\mu U^2/\kappa = 0.81\,T$ and thus $T_2 = 1.30\,T$ and $T_3 = 1.405\,T$. Using average temperatures in each element and eqn (19.124) (with $m = 1$), the element viscosities are obtained as 1.152μ and 1.355μ, the same ratios applying to the thermal conductivities (see eqn (19.124)). Using these values to modify eqn (20.32) and solving, one obtains $u_2 = 0.460\,U$. In turn, using this result to modify the velocity gradients on the right-hand side of eqn (20.34) and adjusting the values of κ and μ in each element, we obtain

$$T_2 = 1.319\,T, \qquad T_3 = 1.405\,T. \qquad (20.40)$$

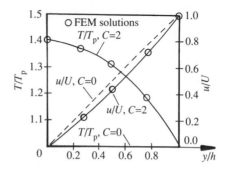

Fig. 20.5 Exact and FEM solutions for compressible Couette flow, $C = \Pr M^2$

The analytical solutions[2] for u_2 and T_2 are $0.46U$ and $1.32T$ whilst the analytical result for the temperature T_3 at the adiabatic fixed plate (the *recovery temperature*) is[2]

$$T_3/T = 1 + \frac{\Pr M^2 \gamma - 1}{2} \doteq 1.405. \qquad (20.41)$$

Therefore, as shown in Fig. 20.5, the finite element model yields satisfactory results with two iterations.

More general two-dimensional problems will require iterative treatment of the convective terms involving u_k and v_k and inclusion of the pressure freedoms. Observing eqn (19.78), these will also require iterative recalculation of ρ in each element (using the gas law $p = \rho R T$) in the same way as μ and κ were dealt with in the foregoing example.

20.6 A simple program for potential flow

The following program for potential flow, like that given in Section 1.11, is suitable for introductory purposes. It uses the stream function approach described in Section 19.2 and the six-node isoparametric triangle. Thus the local derivative calculations are those detailed in Section 7.3. The program uses the same miniscule Gauss–Jordan reduction solution routine as the program of Section 1.11.

The program does not include forcing or 'load' terms at the nodes. When these are required they must be calculated using eqn (19.23) (in many cases leading to simple results like those of eqn (19.74)) and read in as the last data lines by inserting the line

375 For I = 1 to NL; Read N, F; Q(N) = Q(N) + F; Next

where NL is the number of loaded nodes (which should be read in line 40).

```
5   Rem; POTENTIAL FLOW PROGRAM
10  Dim CORD(50,2),EM(6,6),S(50,50),Q(25),XY(6,2),NEL(6)
20  Dim CI(6,2),DL(2,6),TJ(2,2),T(2,6)
25  Rem ***** INTEGRATION AT MIDSIDE NODES
30  CI(1,1)=0.5;CI(1,2)=0.5;CI(2,1)=0;CI(2,2)=0.5;CI(3,1)=0.5;CI(3,2)=0
40  Read NP,NE,NB                       ;Rem PROBLEM SIZE
50  For N=1 to NP;Read CORD(N,1),CORD(N,2);Next   ;Rem NODAL COORDINATES
60  For L=1 to NE       ;Rem LOOP ON ELEMENTS £££££££££££££££££££££££££££££££££££
70  For I=1 to 6;Read NEL(I);Next I     ;Rem NODE NUMBERS FOR ELEMENT
80  For I=1 to 6;K=NEL(I)
90  XY(I,1)=CORD(K,1);XY(I,2)=CORD(K,2)
100 For J=1 to 6;EM(I,J)=0;Next;Next
110 For IP=1 to 3       ;Rem COMMENCE NUMERICAL INTEGRATION LOOP *************
120 F1=4*CI(IP,1);F2=4*CI(IP,2)
130 DL(1,1)=F1-1;DL(1,2)=0;DL(1,3)=F1+F2-3
140 DL(1,4)=F2;DL(1,5)=-F2;DL(1,6)=4-2*F1-F2  ;Rem MATRIX OF EQN 7.15
150 DL(2,1)=0;DL(2,2)=F2-1;DL(2,3)=F1+F2-3
160 DL(2,4)=F1;DL(2,5)=4-F1-2*F2;DL(2,6)=-F1
170 For I=1 to 2;For J=1 to 2;TJ(I,J)=0;For K=1 to 6
180 TJ(I,J)=TJ(I,J)+DL(I,K)*XY(K,J);Next;Next;Next  ;Rem J AS PER EQN 7.16
190 DJ=TJ(1,1)*TJ(2,2)-TJ(1,2)*TJ(2,1);DD=TJ(1,1)
200 TJ(1,1)=TJ(2,2)/DJ;TJ(2,2)=DD/DJ            ;Rem INVERT JACOBIAN
210 TJ(1,2)=-TJ(1,2)/DJ;TJ(2,1)=-TJ(2,1)/DJ
220 For I=1 to 2;For J=1 to 6;T(I,J)=0;For K=1 to 2
230 T(I,J)=T(I,J)+TJ(I,K)*DL(K,J);Next;Next;Next   ;Rem EQN 7.17
240 F=abs(DJ)/6
250 For I=1 to 6;For J=1 to 6
260 EM(I,J)=EM(I,J)+F*(T(1,I)*T(1,J)+T(2,I)*T(2,J))  ;Rem EQN 7.18
270 Next;Next
280 Next IP       ;Rem END NUMERICAL INTEGRATION LOOP ************************
290 For I=1 to 6;NR=NEL(I)
300 For J=1 to 6;NC=NEL(J)
310 S(NR,NC)=S(NR,NC)+EM(I,J);Next;Next   ;Rem ADD k TO SYSTEM MATRIX
320 Next L       ;Rem END LOOP ON ELEMENTS ££££££££££££££££££££££££££££££££££££
330 For L=1 to NB                        ;Rem LOOP BOUNDARY CONDITION NODES
340 Read N,F                             ;Rem N=NODE NO., F=SPECIFIED VALUE
350 For I=1 to NP;Q(I)=Q(I)-F*S(I,N)     ;Rem F*COLUMN TO RHS = 'EFFECTIVE LOAD'
360 S(I,N)=0;S(N,I)=0;Next I             ;Rem ZERO ROW AND COLUMN
370 S(N,N)=1;Q(N)=F;Next L               ;Rem WITH UNIT PIVOT AND F ON RHS
380 For I=1 to NP
390 X=S(I,I);Q(I)=Q(I)/X                 ;Rem X=PIVOT
400 For J=I+1 to NP
410 S(I,J)=S(I,J)/X;Next J               ;Rem ROW/PIVOT
420 For K=1 to NP;If K=I then Goto 460
430 X=S(K,I);S(K,I)=0;Q(K)=Q(K)-X*Q(I)   ;Rem X='ROW MULTIPLIER'
440 For J=I+1 to NP
450 S(K,J)=S(K,J)-X*S(I,J);Next J         ;Rem ROW SUBTRACTION OPERATION
460 Next K
465 Next I
470 !;! "NODAL STREAM FUNCTION VALUES";!
480 For I=1 to NP
490 ! %"I5",I,%"15F6",Q(I);Next
500 End
```

```
1000 Data 12,3,9                    ;Rem NP,NE,NB
1010 Data 0,0                       ;Rem FIRST NODE COORDINATES
1020 Data 0,1
1030 Data 0,2
1040 Data 1.25,0
1050 Data 1.25,1
1060 Data 1.75,1.5
1070 Data 1.75,2
1080 Data 2.5,0
1090 Data 2.79289,0.70711
1100 Data 3.5,1
1110 Data 3.5,1.5
1120 Data 3.5,2                     ;Rem LAST NODE COORDINATES
1130 Data 1,8,3,4,5,2
1140 Data 3,8,10,5,9,6              ;Rem ELEMENT NODE NUMBER SETS
1150 Data 3,10,12,6,11,7
1160 Data 1,0                       ;Rem NODE NUMBERS AND SPECIFIED STREAM
1170 Data 2,1                       ;Rem FUNCTION VALUES FOR B.C.s
1180 Data 3,2
1190 Data 4,0
1200 Data 7,2
1210 Data 8,0
1220 Data 9,0
1230 Data 10,0
1240 Data 12,2
```

Data input requirements are

Lines	Data
1	number of nodes (NP); number of elements (NE); number of boundary condition nodes (NB).
NP lines	nodal x- and y-coordinates (CORD(N, 1) and CORD(N, 2))
NE lines	node number sets for each element (NEL(I), I = 1 → 6).
NB lines	node number (N); value of ψ at this node (F).

The data for the problem of Fig. 20.1(b) is included in lines 1000–1240, the boundary conditions for which are shown in Fig. 20.1(a). Hence the data lines give the following information:

Line 1000	12 nodes, 3 elements, and 9 b.c. nodes.
Line 1010–1120	nodal coordinates.
Line 1130–1150	node number sets for each element.
Line 1160–1240	boundary stream function values.

The solution for ψ_{11} should be close to the value given in Section 20.1. Generally, finer meshes are required, though for the present problem the reader should be able to obtain a solution for ψ half way between the crest of the cylinder and the upper boundary within 1 per cent of the analytical solution given in Section 20.1.

20.7 'FEMSS' program for Stokes flow

The FEMSS (Finite Element Microcomputer Solution System) of Appendix 1 can, of course, be used for any type of finite element problem by substituting an appropriate element routine. The program given in this section uses a six-node triangular element based on eqn (19.72) for the analysis of Stokes flow.

As in the programs of Section 15.8 automatic mesh generation is used following the scheme illustrated in Fig. 15.3. For primitive variable (u, v, p) viscous flow formulations we require non-zero boundary values for velocity and pressure. Hence the solution routine given in Section 15.7 must be modified to deal with non-zero boundary conditions (which are read in line 330 of subroutine MAIN). The reader will find, however, that the solution routine given here differs very little from that given in Section 15.7.

The data input requirements are:

Lines	Data
1	number of nodes (NP); number of elements (NE); number of b.c. nodes (NB); number of property sets (NT); half bandwidth (NBW).
1	PENF, POW, giving the multiplier in eqn (19.73) as $PENF \times 10^{POW}$.
NT lines	μ (PROP(N, 1)); element thickness t (PROP(N, 2)); ρ(PROP(N, 3)); body forces X and Y (PROP(N, 4) and PROP(N, 5)). Once again data input is simplified by using property set 1 for each element (line 290).
1	number of nodes in X-direction (NX), number of nodes in Y-direction (NY), domain size in X-direction (XLIM), domain size in Y-direction (YLIM); see Fig. 15.3.
NB lines	b.c. node number (N); zero/one (one for a b.c.) flags for freedoms u, v, p (IBC(N, I), I = 1 → 3); specified values of u, v, p (SPEC(N, I), I = 1 → 3).
NL lines	NL = number of loaded nodes; node number (NQ); loads corresponding to freedoms u, v, p (R(1), R(2), R(3)).

The data required for the problem of Fig. 20.2 using twenty-five nodes and eight elements is given in lines 500–700. Here we set $h = 5$, $U = 1$, $P_1 = 4$, and $P_2 = 2$. P_1 and P_2 are the pressures at inlet and outlet and, as indicated in eqn (19.72), these must be input as negative values. Note that the nodal load data (none in the example problem) must be terminated by a data line with four zeros.

```
 10 Include 'ELMAT','SOLVE'
 20 Access * from *                        ;Rem INCLUDES SAME MESH GENERATION
 30 Open £8,"B:EMATS"                      ;Rem CODE AS THE LINEAR STRAIN
 50 Open £10,"B:STIFM"                     ;Rem TRIANGLE PROGRAM OF SEC. 15.8
 60 Dim FNE(100),IBC(100,3),Q(300),SPEC(100,3),IMAT(100),CORD(100,2),PROP(20,5)
 70 Dim NOP(100,6),CI(6,2),WF(6),R(5)
 80 Def shared data NP,NE,NB,NT,NBW,NCN,NDF,PENF,FNE( )
 90 Def shared data IBC( ),SPEC( ),Q( ),CI( ),WF( )
100 Def shared data IMAT( ),CORD( ),PROP( ),NOP( )
110 NDF=3;NCN=6
120 CI(1,1)=0.5;CI(1,2)=0.5;CI(2,1)=0;CI(2,2)=0.5;CI(3,1)=0.5;CI(3,2)=0
130 WF(1)=1/6;WF(2)=1/6;WF(3)=1/6           ;Rem INTEGRATION AT MIDSIDE NODES
150 Read NP,NE,NB,NT,NBW;NBW=NBW-NDF
155 Read PENF,POW;PENF=PENF*10^POW          ;Rem PENALTY FACTOR - SEE EQN 19.73
160 For N=1 to NT;For I=1 to 5;Read PROP(N,I);Next;Next
170 Read NX,NY,XLIM,YLIM                    ;Rem NX,NY ARE NO. NODES IN X,Y
180 NEX=NX-1;NEY=NY-1;RNX=NEX;RNY=NEY       ;Rem DIRECTIONS AND XLIM,YLIM ARE THE
190 DX=XLIM/RNX;DY=YLIM/RNY                 ;Rem DOMAIN SIZES IN THOSE DIRECTIONS
200 For I=1 to NX;For J=1 to NY
210 RNDX=I-1;RNDY=J-1;NN=NY*(I-1)+J
220 CORD(NN,1)=RNDX*DX;CORD(NN,2)=RNDY*DY;Next;Next   ;Rem NODAL COORDINATES
230 NEX=(NX-1)/2;NEY=(NY-1)/2
240 For I=1 to NEX;For J=1 to NEY           ;Rem ELEMENT NODE NUMBERS ARE SET
250 NI=(I-1)*2*NY+(J-1)*2+1;NJ=NI+2*NY      ;Rem UP AS SHOWN IN FIG. 15.3
260 NS=NEY*(I-1)+J;NN=2*NS-1
270 NOP(NN,1)=NI;NOP(NN,2)=NJ;NOP(NN,3)=NI+2
280 NOP(NN,4)=NI+NY;NOP(NN,5)=NOP(NN,4)+1;NOP(NN,6)=NI+1
290 IMAT(NN)=1;NN=2*NS
300 NOP(NN,1)=NI+2;NOP(NN,2)=NJ;NOP(NN,3)=NJ+2
310 NOP(NN,4)=NOP(NN-1,5);NOP(NN,5)=NJ+1;NOP(NN,6)=NOP(NN-1,4)+2
320 IMAT(NN)=1;Next;Next                    ;Rem ONLY ONE PROPERTY SET USED
325 For I=1 to NB
330 Read N,IBC(N,1),IBC(N,2),IBC(N,3),SPEC(N,1),SPEC(N,2),SPEC(N,3)
335 Next       ;Rem IBC ARE B.C. FLAGS AND SPEC ARE SPECIFIED BOUNDARY VALUES
340 Read NQ,R(1),R(2),R(3)                  ;Rem READ NODAL FORCING TERMS
350 If NQ=0 then Goto 380
360 For K=1 to NDF;IC=(NQ-1)*NDF+K;Q(IC)=Q(IC)+R(K);Next
370 Goto 340
380 Rem
385 For N=1 to NE;For II=1 to NCN;NN=NOP(N,II);FNE(NN)=FNE(NN)+1
387 Next;Next                              ;Rem 'NODAL VALENCY'
390 For N=1 to NE
400 ELSUB N;Next
410 Filepos(8)=0
415 SOLSUB
430 End
500 Data 25,8,16,1,33                       ;Rem DATA READ AS ABOVE
510 Data 1,-5
520 Data 1,1,0,0,0
530 Data 5,5,5,5
540 Data 1,1,1,1,0,0,-4
550 Data 2,0,1,1,0,0,-4
560 Data 3,0,1,1,0,0,-4
570 Data 4,0,1,1,0,0,-4
580 Data 5,1,1,1,1,0,-4
590 Data 6,1,1,0,0,0,0
600 Data 10,1,1,0,1,0,0
610 Data 11,1,1,0,0,1,0,0
620 Data 15,1,1,0,1,0,0
630 Data 16,1,1,0,0,0,0
640 Data 20,1,1,0,1,0,0
650 Data 21,1,1,1,0,0,-2
660 Data 22,0,1,1,0,0,-2
670 Data 23,0,1,1,0,0,-2
680 Data 24,0,1,1,0,0,-2
690 Data 25,1,1,1,1,0,-2
700 Data 0,0,0,0                            ;Rem ZEROES TO TERMINATE LOAD DATA
```

```
10 Rem ELMAT
20 Def shared proc ELSUB @N                    ;Rem ONLY ELEMENT NO. AS ARGUMENT
30 Dim S(18,18),F(6),TEMP(3,6),Z(2,3),V(3,6),T(18,18),TT(18,18)
40 L=IMAT(N);VISC=PROP(L,1);TH=PROP(L,2);DENS=PROP(L,3)
50 FX=PROP(L,4);FY=PROP(L,5)                    ;Rem COLLECT ELEMENT PROPERTIES
60 I=NOP(N,1);J=NOP(N,2);K=NOP(N,3)
70 X21=CORD(J,1)-CORD(I,1);X32=CORD(K,1)-CORD(J,1);X13=CORD(I,1)-CORD(K,1)
80 Y21=CORD(J,2)-CORD(I,2);Y32=CORD(K,2)-CORD(J,2);Y13=CORD(I,2)-CORD(K,2)
90 A=X21*Y32-X32*Y21                            ;Rem A=TWICE ELEMENT AREA
100 For I=1 to 3;NI=I+3;NF=NOP(N,NI)*NDF-2
110 Q(NF+1)=Q(NF+1)+A*FY/6                       ;Rem ADD BODY FORCES TO L D VECTOR
120 Q(NF)=Q(NF)+A*FX/6;Next
130 Z(1,1)=-Y32/A;Z(1,2)=-Y13/A;Z(1,3)=-Y21/A   ;Rem EXTENDED FORM OF EQN 6.28
140 Z(2,1)=X32/A;Z(2,2)=X13/A;Z(2,3)=X21/A       ;Rem WHEN L3 NOT ELIMINATED
150 For II=1 to 3        ;Rem COMMENCE INTEGRATION LOOP ££££££££££££££££££££££££££
160 F1=4*CI(II,1);F2=4*CI(II,2);F3=4-F1-F2
170 C1=CI(II,1);C2=CI(II,2);CC=C1+C2
180 F(1)=2*C1*C1-C1;F(2)=2*C2*C2-C2;F(3)=1-3*CC+2*CC*CC
190 F(4)=4*C1*C2;F(5)=4*C2*(1-CC);F(6)=4*C1*(1-CC)  ;Rem INTERPOLATION FUNCTIONS
200 V(1,1)=F1-1;V(1,4)=F2;V(1,6)=F3
210 V(2,2)=F2-1;V(2,4)=F1;V(2,5)=F3                 ;Rem EXTENDED FORM OF EQN 6.26
220 V(3,3)=F3-1;V(3,5)=F2;V(3,6)=F1                 ;Rem WHEN L3 NOT ELIMINATED
230 For I=1 to 2;For J=1 to 6;TEMP(I,J)=0;For K=1 to 3
240 TEMP(I,J)=TEMP(I,J)+Z(I,K)*V(K,J);Next;Next;Next
250 VOL=TH*A*WF(II);SUM=0
260 For I=1 to 6;For J=1 to 6
270 G=VOL*VISC*(TEMP(1,I)*TEMP(1,J)+TEMP(2,I)*TEMP(2,J))
280 S(3*I-2,3*J-2)=S(3*I-2,3*J-2)+G              ;Rem FILL ELEMENT MATRIX AS PER
290 S(3*I-1,3*J-1)=S(3*I-1,3*J-1)+G              ;Rem EQN 19.72
300 S(3*I-2,3*J)=S(3*I-2,3*J)-VOL*F(I)*TEMP(1,J)
310 S(3*I-1,3*J)=S(3*I-1,3*J)-VOL*F(I)*TEMP(2,J)
320 S(3*I,3*J-2)=S(3*I,3*J-2)-VOL*TEMP(1,I)*F(J)
330 S(3*I,3*J-1)=S(3*I,3*J-1)-VOL*TEMP(2,I)*F(J)
340 SUM=SUM+G*G
350 Next;Next
360 For K=1 to 6
370 S(3*K,3*K)=S(3*K,3*K)+sqrt(SUM)*PENF/36;Next  ;Rem ADD FACTOR OF EQN 19.73
380 Next II            ;Rem END INTEGRATION LOOP ££££££££££££££££££££££££££££££££
385 Goto 495                                      ;Rem SKIP SUPPRESSION OF MIDSIDE Ps
390 For I=1 to 18;T(I,I)=1;Next                   ;Rem FOR THE PRESENT PROBLEM
400 T(12,3)=0.5;T(12,6)=0.5;T(12,12)=0
410 T(15,6)=0.5;T(15,9)=0.5;T(15,15)=0
420 T(18,9)=0.5;T(18,3)=0.5;T(18,18)=0            ;Rem SUPPRESSION OF MIDSIDE NODE
430 For I=1 to 18;For J=1 to 18                   ;Rem PRESSURES AS PER EQN 4.29
440 TT(I,J)=0;For K=1 to 18
450 TT(I,J)=TT(I,J)+S(I,K)*T(K,J);Next;Next;Next
460 For I=1 to 18;For J=1 to 18
470 S(I,J)=0;For K=1 to 18
480 S(I,J)=S(I,J)+T(K,I)*TT(K,J);Next;Next;Next
490 S(12,12)=1;S(15,15)=1;S(18,18)=1
495 Rem
500 For I=1 to 18;For J=1 to 18
510 Write £8,S(I,J);Next;Next                     ;Rem FILE THE ELEMENT MATRICES
520 ! N                                           ;Rem PROGRESS REPORT ONLY
530 Return;Proc end
600 End
```

Note: the above coding is for straight-sided elements. The program of Section 23.9 gives the coding for isoparametric elements which may have curved sides.

```
10 Rem SOLVE                     Rem SAME AS SEC. 15.7 BUT SK,SKP (AND THUS SIZ,LB)
20 Def shared proc SOLSUB       ;Rem ARE LARGER, LOOPS OF LINES 90,260 SMALLER AND
45 Dim DIS(3,100)               ;Rem LINES 320-380,780-790 CHANGE TO HANDLE THE
46 Rem                           Rem NONZERO BOUNDARY CONDITIONS IN SPEC( , ).
50 Dim ESM(18,18),SK(120,90),SKP(90)
55 Def shared data DIS( )       ;Rem SHARE DISPLACEMENT SOLUTION WITH STRESS
60 SIZ=120;LB=90;BUF=SIZ-LB;NRW=0;NTW=NBW+NDF;NLOAD=NP*NDF;NBN=1
70 L=0;N=1                      ;Rem SIZ,LB ARE STIFFNESS BLOCK DEPTH & WIDTH
80 For I=1 to 18;For J=1 to 18
90 Read £8,ESM(I,J);Next;Next ;Rem READ FIRST k (ELEMENT MATRIX)
100 L=L+1        ;Rem COMMENCE LOOP FOR NODE BY NODE FORWARD REDUCTION ££££££££££££
110 For M=1 to 8
120 If N=(NE+1) then Goto 280
130 For I=1 to NCN
140 If NOP(N,I)=L then Goto 170           ;Rem CHECK IF NEXT k NEEDED YET
150 Next
160 Goto 280
170 Rem
180 For I=1 to NCN;For J=1 to NCN
190 For IL=1 to NDF;IE=(I-1)*NDF+IL;NR=(NOP(N,I)-1)*NDF+IL
200 NRE=NR-NRW                            ;Rem NRW = NO. ROWS OF K FILED
210 For JL=1 to NDF;JE=(J-1)*NDF+JL;NC=(NOP(N,J)-1)*NDF+JL
220 NCB=NC-NR+1;If NR>NC then Goto 240
230 SK(NRE,NCB)=SK(NRE,NCB)+ESM(IE,JE)    ;Rem ASSEMBLY OF K (SEE FIG. 15.1)
240 Next;Next;Next;Next
245 If N=NE then Goto 265
250 For I=1 to 18;For J=1 to 18
260 Read £8,ESM(I,J);Next;Next            ;Rem READ NEXT k
265 N=N+1
270 Next M
280 Rem
290 NDIF=(NP-L+1)*NDF;If NDIF>NBW then LIM=NBW+NDF
320 For ID=1 to NDF
330 LIM=LIM-1;IP=ID+NDF*(L-1);IPE=IP-NRW;R=Q(IP);NOB=0
340 If IBC(L,ID)<>0 then NOB=1;If NOB=1 then RS=-R
350 If NOB=1 then R=SPEC(L,ID)
360 If NOB=1 then Goto 380
370 XK=1/SK(IPE,1);Q(IP)=XK*R;Goto 430
380 Rem
385 ! L,ID                                ;Rem REPORT SOLUTION PROGRESS
420 Q(IP)=RS+SK(IPE,1)*R;XK=1;R=-R        ;Rem Q(IP) = BOUNDARY 'REACTION'
430 Rem
440 For J=1 to LIM;JA=J+1;SKP(J)=SK(IPE,JA);Next ;Rem STORE 'ROW MULTIPLIERS'
450 NC=LIM+1
460 For J=1 to NC;SK(IPE,J)=SK(IPE,J)*XK        ;Rem DIVIDE ROW BY PIVOT
470 If NOB=1 then SK(IPE,J)=-SK(IPE,J);Next     ;Rem NEGATE BOUNDARY ROW
480 If (L+ID-NP-NDF)=0 then Goto 660            ;Rem END TEST
490 For I=1 to LIM;NR=IP+I;NRE=IPE+I
500 If SKP(I)=0 then Goto 550;If NOB=1 then Goto 530;NC=LIM-I+1
510 For J=1 to NC;JP=J+I
520 SK(NRE,J)=SK(NRE,J)-SK(IPE,JP)*SKP(I);Next  ;Rem FORWARD REDUCTION
530 JP=I+1
540 Q(NR)=Q(NR)-SK(IPE,JP)*R                    ;Rem REDUCTION IN LOAD VECTOR
550 Next I
560 If (IPE+NTW)<SIZ then Goto 630              ;Rem TEST IF STIFFNESS BLOCK FULL
570 If (NLOAD-NRW)<=SIZ then Goto 630
580 For I=1 to BUF;For J=1 to LB
590 Write £10,SK(I,J);Next;Next                 ;Rem FILE PART OF STIFFNESS BLOCK
600 NRW=NRW+BUF                                 ;Rem NRW = NO. ROWS OF K FILED
610 For I=1 to LB;For J=1 to LB;IA=I+BUF
620 SK(I,J)=SK(IA,J);SK(IA,J)=0;Next;Next ;Rem SHIFT REMAINING ROWS UP
630 Rem
640 Next ID
650 Goto 100       ;Rem END NODE BY NODE FORWARD REDUCTION LOOP ££££££££££££££
660 Rem
670 NR=NDF*NP;NRE=NR-NRW;DIS(NDF,NP)=Q(NR);Rem LAST DISPLACEMENT NOW KNOWN
680 Q(NR)=0;I=NDF;L=NP
690 Goto 780
700 L=L-1                       ;Rem LOOP ON NODES FOR BACK SUBSTITUTION
```

```
710 I=I-1                           ;Rem LOOP ON D.F./NODE FOR BACK SUBSTITUTION
720 NR=NDF*(L-1)+I;NRE=NR-NRW
730 DIS(I,L)=Q(NR);Q(NR)=0
740 If LIM<(NBW+NDF-1) then LIM=LIM+1
750 For J=1 to LIM;JA=J+1
760 LJ=L+(J+I-1)/NDF;LJ=trunc(LJ);K=I+J-(LJ-L)*NDF
770 DIS(I,L)=DIS(I,L)-SK(NRE,JA)*DIS(K,LJ);Next    ;Rem BACK SUBSTITUTION
780 If IBC(L,I)=0 then Goto 800
790 Q(NR)=DIS(I,L);DIS(I,L)=SPEC(L,I)
800 Rem
810 If (NRE-NTW)>0 or NRW=0 then Goto 880
820 For II=1 to LB;For J=1 to LB
830 IA=SIZ-II+1;IB=LB-II+1
840 SK(IA,J)=SK(IB,J);Next;Next
850 NRW=NRW-BUF
855 Filepos(10)=filepos(10)-BUF*LB*5    ;Rem BACKSPACE K FILE
860 For II=1 to BUF;For J=1 to LB        ;Rem READ BACK FILED PARTS OF
870 Read £10,SK(II,J);Next;Next          ;Rem REDUCED K AS NEEDED
875 Filepos(10)=filepos(10)-BUF*LB*5    ;Rem BACKSPACE AGAIN - NOTE 5 IS
880 Rem                                  ;REM BYTES/NUMBER IN MEGABASIC
890 If (I+L-2)=0 then Goto 930           ;Rem END TEST
900 If I<>1 then Goto 710                ;Rem END LOOP ON FREEDOMS/NODE
910 I=NDF+1
920 Goto 700                             ;Rem END BACKSUB LOOP ON NODES
930 Rem
935 !;! "SOLUTIONS FOR U,V,P AT EACH NODE";!
940 For N=1 to NP
950 ! %"I5",N,%"15E6",DIS(1,N),DIS(2,N),DIS(3,N)
960 Next
970 Return;Proc end
980 End
```

Hence the data given in lines 500–700 is as follows:

Line 500	25 nodes, 8 elements, 16 b.c. nodes, 1 property set, half bandwidth = 33 (e.g. the mesh generation will assign the node numbers 1, 11, 3, 6, 7, 2 to the first element).
Line 510	the factor 10^{-5} in eqn (19.73).
Line 520	$\mu = 1$, $t = 1$, $\rho = 0$, $X = 0$, $Y = 0$.
Line 530	NX = 5, NY = 5, XLIM = 5, YLIM = 5.
Line 540–690	$u = 0$ on lower boundary, $u = 1$ on upper boundary, $v = 0$ on upper and lower boundaries, $p = 4$ at inlet boundary, $p = 2$ at outlet boundary.
Line 700	terminates reading of data.

Results should agree with the exact solution of eqn (20.8) (for the horizontal velocity) whilst the pressure solutions should indicate a linear pressure gradient in the horizontal direction. Note that the program can suppress the midside pressure freedoms (in lines 390–490 of ELMAT using eqn (4.29)).As noted in Section 19.5 this is not necessary in the present case but is generally advisable. To suppress them, line 385 of of ELMAT becomes

385 Rem Goto 495

Note also that potential flow problems can be solved by using a small *artificial viscosity*[7] in the foregoing program and this artifice is used in the program of Section 23.9.

References

1. Chung, T. J. (1978). *Finite Element Analysis in Fluid Dynamics.* McGraw-Hill, New York.
2. Yuan, S. W. (1967). *Foundations of Fluid Mechanics.* Prentice-Hall, Englewood Cliffs, NJ.
3. Heubner, K. H. (1975). *The Finite Element Method for Engineers.* Wiley, New York.
4. Bowden, F. P. and Tabor, D. (1954). *The Friction and Lubrication of Solids.* Oxford University Press.
5. Desai, C. S. and Johnson, L. D. (1972). Evaluation of two finite element formulations for one-dimensional consolidation. *Comput. Struct.* **2**, 469.
6. de Laplace, P. S. (1816). Sur la vitesse du son dans l'air et dans l'eau. *Annales de Chimie et de Physique*, Series 2, **3**, 238.
7. Argyris, J. H., St Doltsnis, I., and Fritz, H. (1989). Hermes Space Shuttle: exploration of reentry aerodynamics. *Comp. Meth. Appl. Mech. Engng*, **73**, 1.

Part V
Further applications of the finite element method

21
Boundary solution techniques

Problems involving infinite domains were encountered in Sections 7.3 and 20.4. In the first instance use of a known *far field solution* provided approximate elastic boundary conditions. The *boundary element method* (BEM) provides an elegant approach to such problems and, whilst the present text is intended to concentrate upon finite elements, the reader may find some of the boundary element concepts both useful and interesting. Of particular note, for example, is the extension of the integration by parts processes used with the Galerkin or virtual work methods and the weighted residual character of the boundary element method. Weighting is now with the *free-space Green's function* which, as noted at the close of the chapter, can also be used to model far field behaviour in the finite element method.

21.1 Introduction

The boundary element method is based upon the use of *integral equations*. To obtain an integral equation solution to the problem of determining the distribution of a function $y = y(s)$ in a domain one begins by expressing the solution in the form

$$y = \int F(x, s) \, dx$$

and integration by parts is used to force the solution to the boundary. In practice $F(x, s)$ consists of two parts: a forcing function $b(x)$ describing the variation in intensity of some 'loading' effect, and a response or *influence function* $G(x, s)$ which gives the response at s resulting from a point loading at x.

The latter is called *Green's function* (or the *free-space Green's function* if it applies to an infinite domain) or the *fundamental solution*. For a unit point source potential in three-dimensional space the freespace Green's function is simply $1/4\pi r$, that is, it is given by dividing the unit magnitude of the source by the spherical surface area of radius r. In two dimensions also Green's function often has an inverse radius character. However, in one-dimensional problems it is often expressed in terms of a *step function* but an alternative approach is to use a modulus function. In a beam simply supported at $s = -L$ and $s = L(L \to \infty)$ and loaded by a point load at $x = 0$, for example, the bending moment in the beam can be expressed as

$$M = \tfrac{1}{2} P(L - |x - s|).$$

By differentiating or integrating this result expressions for the shear, slope, and deflection of the beam can be found. However, before this can be done we need to develop a differentiation rule for $|x - s|$. This is given by

$$\frac{\partial |x - s|^n}{\partial x} = \frac{\partial |x - s|^n}{\partial |x - s|} \frac{\partial |x - s|}{\partial x}$$

$$= n|x - s|^{n-1} \frac{\partial |x - s|}{\partial (x - s)} \frac{\partial (x - s)}{\partial x}$$

$$= n|x - s|^{n-1} \operatorname{sgn}(x - s) \quad (1)$$

and the integration rule follows. An example of the use of this result in the boundary element method occurs in Section 21.4 whilst an example of the use of integral equations is given in the following section.

21.2 Integral equation solution for a beam

As an introductory example of the use of integral equations consider the beam on an elastic foundation shown in Fig. 21.1. The governing differential equation is eqn (17.7) and an integral equation solution to this can be obtained by using superposition to write[1]

$$y = y(s) = \int_0^L \{b(x) - \mu y(x)\} G(x, s)\,\mathrm{d}x \qquad (21.1)$$

where $b(x)$ describes the distribution of loading and $G(x, s)$ is the appropriate *Green's function*, that is, an influence function which gives the deflection at s caused by a unit load at x. Here this is the solution of the differential equation

$$\frac{EI\,\mathrm{d}^4 y}{\mathrm{d}s^4} = \delta(x, s) \qquad (21.2)$$

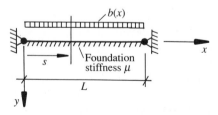

Fig. 21.1 Beam on an elastic foundation

which satisfies the boundary conditions y, $d^2 y/ds^2 = 0$ at $s = 0$, L. $\delta(x, s)$ is the *Dirac delta function* which is defined by

$$\delta(x, s) = \begin{cases} 0 & \text{for } x \neq s \\ \infty & \text{for } x = s, \end{cases} \qquad (21.3)$$

$$\int_{-\infty}^{\infty} \delta(x, s) \, ds = 1(x) \qquad (21.4)$$

and this last property enables it to represent the unit load at x. Integrating eqn (21.2) four times and applying the boundary conditions one obtains the results

$$G(x, s) = \frac{s(L - x)(2Lx - s^2 - x^2)}{6EIL} \qquad s \leq x \qquad (21.5)$$

$$G(x, s) = \frac{x(L - s)(2Ls - x^2 - s^2)}{6EIL} \qquad s \geq x \qquad (21.6)$$

and if $b(x) = q$ and μ are constant, the solution can be obtained by expressing eqn (21.1) as

$$y(s) = \frac{qL^4}{24EI} \left[\frac{s}{L} - 2\left(\frac{s}{L}\right)^3 + \left(\frac{s}{L}\right)^4 \right] - \mu \int_0^L y(x) G(x, s) \, dx. \qquad (21.7)$$

The first term on the right-hand side is the solution of $EI(d^4 y/ds^4) = q$ which is obtained as

$$q \int_s^L G_1(x, s) \, dx + q \int_0^s G_2(x, s) \, dx$$

where $G_1(x, s)$ and $G_2(x, s)$ denote the results in eqns (21.5) and (21.6), respectively.

Noting that $y(s)$ and $y(x)$ in eqn (21.7) must be identical, eqn (21.7) can be solved for a given trial function after some rearrangement. The foregoing approach has much in common with the boundary element method but in the latter integration by parts is used to obtain a matrix formulation.

21.3 Boundary element method for one-dimensional problems

In this section we shall again use the beam on an elastic foundation problem as an example of the application of the *direct boundary element method* (DBEM). In this method the complete governing equation is weighted by a Green's function (we use the term 'weighted' as the process may be likened to the weighted residual methods[2]). We shall now denote the Green's function as y^* for brevity:

$$\int_0^L \frac{EI\,d^4 y}{dx^4}\, y^*\, dx = \int_0^L (b(x) - \mu y)\, y^*\, dx \tag{21.8}$$

and note that one integrates the 'load coordinate' x on the right-hand side and follows suit on the left-hand side.

Integrating the left-hand side of eqn (21.8) four times, dropping the limits 0, L, and using $dy/dx = y_x$ to simplify notation, one obtains

$$\int EI y_{xxxx}\, y^*\, dx = \{EI y_{xxx}\, y^*\} - \int EI y_{xxx}\, y_x^*\, dx \tag{21.9}$$

$$-\int EI y_{xxx}\, y_x^*\, dx = -\{EI y_{xx}\, y_x^*\} + \int EI y_{xx}\, y_{xx}^*\, dx \tag{21.10}$$

$$\int EI y_{xx}\, y_{xx}^*\, dx = \{EI y_x\, y_{xx}^*\} - \int EI y_x\, y_{xxx}^*\, dx \tag{21.11}$$

$$-\int EI y_x\, y_{xxx}^*\, dx = -\{EI y y_{xxx}^*\} + \int EI y y_{xxxx}^*\, dx \tag{21.12}$$

Collecting these results and using the notations

$$v = -EI\frac{d^3 y}{dx^3} \qquad v^* = -EI\frac{\partial^3 y^*}{\partial x^3}$$

$$m = -EI\frac{d^2 y}{dx^2} \qquad m^* = -EI\frac{\partial^2 y^*}{\partial x^2} \tag{21.13}$$

$$\phi = \frac{dy}{dx} \qquad \phi^* = \frac{\partial y^*}{\partial x}$$

here noting that $y^* = G(x, s)$ in connection with the partial derivatives above, one obtains

$$[-y^*v + \phi^*m - m^*\phi + v^*y]_0^L + \int_0^L EI y y_{xxxx}^*\, dx = \int_0^L (b(x) - y)\, y^*\, dx. \tag{21.14}$$

Noting that

$$\int_0^L EI y_{xxxx}^*\, dx = \int_0^L (\delta(x, s) - \mu y^*)\, dx \tag{21.15}$$

is the integration of the equation governing a beam with a point load at x replacing the distributed load $b(x)$, eqn (21.14) becomes

$$[-y^*v + \phi^*m - m^*\phi + v^*y]_0^L + \int_0^L y(\delta(x, s) - \mu y^*)\, dx = \int_0^L (b(x) - \mu y)\, y^*\, dx \tag{21.16}$$

so that the terms involving μ cancel. Using the property of eqn (21.4),

$$\int_0^L \delta(x, s)\, y\, dx = y(s) \tag{21.17}$$

one obtains

$$y(s) = [y^*v - \phi^*m + m^*\phi - v^*y]_0^L + \int_0^L b(x) y^* \, dx$$

$$= [y^*, -\phi^*, m^*, -v^*]_0^L \{v, m, \phi, y\} + \int_0^L b(x) y^* \, dx \qquad (21.18)$$

where v, m, ϕ, y are boundary values at each end of the beam and ϕ^*, m^*, v^* are obtained by using eqns (21.13) to differentiate the influence function y^*. Now differentiating eqn (21.18) with respect to s one obtains an equation for the slope of the beam,

$$\phi(s) = [y_s^*, -\phi_s^*, m_s^*, -v_s^*]_0^L \{v, m, \phi, y\} + \int_0^L b(x) y_s^* \, dx \qquad (21.19)$$

where $y_s^* = \partial y^*/\partial s$, and so on (recall again that $y^* = G(x, s)$).

Now as shown in Fig. 21.2 one chooses an arbitrary origin O and considers the beam end-points to be *boundary elements* of size $\varepsilon(\varepsilon \to 0)$ at $s_1 = X + \varepsilon, s_2 = X + L - \varepsilon$ (that is, they are *inside* the domain) with freedoms y_1, ϕ_1, y_2, ϕ_2. Then, allowing $x, s \to \infty$, one chooses for y^* the *free-space Green's function*, that is, the solution of eqn (21.15) with limits $-\infty, \infty$. This permits the final formulation to be used for various geometries and a variety of boundary conditions.

Applying eqns (21.18) and (21.19) to each boundary element at $s_1 = X + \varepsilon$, $s_2 = X + L - \varepsilon$ one obtains four equations which may be written in the matrix form

$$\{y_2, y_1, \phi_2, \phi_1\} = A_2\{d_2\} - A_1\{d_1\} + \{f\} \qquad (21.20)$$

where

$$A_1 = \begin{bmatrix} \{g\}^t, s = s_2 \\ \{g\}^t, s = s_1 \\ \{\partial g/\partial s\}^t, s = s_2 \\ \{\partial g/\partial s\}^t, s = s_1 \end{bmatrix}_{x = s_1 - \varepsilon} \qquad A_2 = \begin{bmatrix} \{g\}^t, s = s_2 \\ \{g\}^t, s = s_1 \\ \{\partial g/\partial s\}^t, s = s_2 \\ \{\partial g/\partial s\}^t, s = s_1 \end{bmatrix}_{x = s_2 + \varepsilon}$$

$$\{d_1\} = \{v_1, m_1, \phi_1, y_1\}, \qquad \{d_2\} = \{v_2, m_2, \phi_2, y_2\}$$

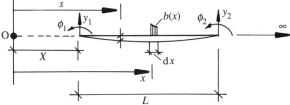

Fig. 21.2 Coordinate system for beam represented by two boundary elements

are the vectors of nodal freedoms,

$$\{f\} = \int\limits_0^L \begin{bmatrix} b(x)\,y^*(x,s_2) \\ b(x)\,y^*(x,s_1) \\ b(x)\,y_s^*(x,s_2) \\ b(x)\,y_s^*(x,s_1) \end{bmatrix} dx$$

is the load vector, and

$$\{g\} = \{y^*,\ -\phi^*, m^*,\ -v^*\}$$

$$= \{y^*,\ -\partial y^*/\partial x,\ -EI\partial^2 y^*/\partial x^2, EI\partial^3 y^*/\partial x^3\}$$

is the vector of influence functions.

Hence eqn (21.20) matches the boundary element displacements (on the left) with those calculated from:

(1) integration of the governing differential equation;

(2) use of the free-space Green's function solution.

In the *indirect boundary element method* (IBEM) this coupling is achieved by a superposition argument similar to that used to obtain eqn (21.1).[3]

The direct boundary element method , to which we shall restrict attention in the present chapter, may also be viewed as a coupling of these solutions (1) and (2),[4] whilst schemes for coupling the method with finite element procedures are based on equating the boundary displacements of the finite element mesh with those on the left-hand side of eqn (21.20).[5]

In the elastically supported beam problem $y^* = G(x, s)$ will be the free-space solution of eqn (21.15) and this involves relatively complicated transcendental function expressions less amenable to numerical solution than the polynomial functions so much used in the finite element method. To overcome this difficulty one can use for y^* the free-space Green's function for a beam without elastic support, in which case the last term in eqn (21.15) is deleted so that there is no cancellation of the terms involving μ in eqn (21.16). In this case, therefore, an iterative solution is required and Butterfield[6] shows that such iterative solutions can also deal with inhomogeneous systems.

21.4 Application of the boundary element method to beams

As a worked example consider the cantilever beam shown in Fig. 21.3, choosing the origin at node 1 for simplicity. The beam is not elastically supported by a foundation so that eqn (21.20) applies directly so long as the influence functions are those for such a beam.

The first influence function can be written by inspection, taking $EI = 1$, as

$$v^* = -\tfrac{1}{2}\,\mathrm{sgn}(x - s) \tag{21.21}$$

where $\mathrm{sgn}(x - s) = 1$, $x > s$, $\mathrm{sgn}(x - s) = -1$, $x < s$ and $\mathrm{sgn}(0)$ is undefined.

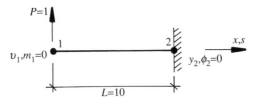

Fig. 21.3 End loaded cantilever

Thus, in an infinite beam with a positively directed (upwards) unit load the shear is -0.5 to the left ($s < x$) and $+0.5$ to the right ($s > x$). Introducing a modulus function $|x - s|$ and noting that

$$\frac{\partial(|x - s|^n)}{\partial x} = n|x - s|^{n-1} \operatorname{sgn}(x - s) \tag{21.22a}$$

$$\frac{\partial(|x - s|^n)}{\partial s} = -n|x - s|^{n-1} \operatorname{sgn}(x - s) \tag{21.22b}$$

and that $\operatorname{sgn}^2(x - s) = +1$, one can use integration to obtain the four influence functions, and partial differentiation with respect to s to obtain their derivatives with respect to s. Hence one obtains

$$y^* = \tfrac{1}{12}(2L^3 + |x - s|^3 - 3L|x - s|^2)$$

$$y_s^* = -\tfrac{1}{4}|x - s|(|x - s| - 2L)\operatorname{sgn}(x - s)$$

$$\phi^* = \tfrac{1}{4}|x - s|(|x - s| - 2L)\operatorname{sgn}(x - s) \qquad \phi_s^* = \tfrac{1}{2}(L - |x - s|)$$

$$m^* = -\tfrac{1}{2}(|x - s| - L) \qquad\qquad m_s^* = \tfrac{1}{2}\operatorname{sgn}(x - s)$$

$$v^* = -\tfrac{1}{2}\operatorname{sgn}(x - s) \qquad\qquad v_s^* = 0. \tag{21.23}$$

Note that integration to obtain y^* and m^* has taken into account that we require $y^* = 0$ and $m^* = 0$ at $s = L$ and that m_s^*, for example, can be obtained as $\partial m^*/\partial s$ (by definition) or as $-\partial \phi_s^*/\partial x$ (using eqns (21.13)).

Evaluating and substituting these influence functions into eqn (21.20) one obtains

$$\begin{bmatrix} y_2 \\ y_1 \\ \phi_2 \\ \phi_1 \end{bmatrix} = \begin{bmatrix} 500/3 & 0 & 5 & -1/2 \\ 0 & -25 & 0 & -1/2 \\ 0 & -5 & -1/2 & 0 \\ 25 & 0 & 1/2 & 0 \end{bmatrix} \begin{bmatrix} v_2 \\ m_2 \\ \phi_2 \\ y_2 \end{bmatrix}$$

$$- \begin{bmatrix} 0 & 25 & 0 & -1/2 \\ 500/3 & 0 & 5 & -1/2 \\ -25 & 0 & -1/2 & 0 \\ 0 & -5 & 1/2 & 0 \end{bmatrix} \begin{bmatrix} v_1 \\ m_1 \\ \phi_1 \\ y_1 \end{bmatrix} + \begin{bmatrix} 0 \\ 500/3 \\ 25 \\ 0 \end{bmatrix}$$

$$\tag{21.24}$$

where the non-zero loading terms on the right are obtained as

$$\int_0^L \delta(x, s_1) y^*(x, s_1)\, dx = 1(x = 0) y^*(x = 0, s = \varepsilon) \qquad \varepsilon \to 0$$

$$\int_0^L \delta(x, s_2) y_s^*(x, s_2)\, dx = 1(x = 0) y_s^*(x = 0, s = L - \varepsilon) \quad \varepsilon \to 0. \qquad (21.25)$$

Taking account of the boundary conditions $y_2, \phi_2, v_1, m_1 = 0$ these columns are deleted from the right-hand side matrices (note that one puts $y_2 = \phi_2 = 0$ on the left-hand side in eqn (21.24) but unit entries remain for y_1 and ϕ_1). The reduced equations are then

$$\begin{bmatrix} 500/3 & 0 & 0 & 1/2 \\ 0 & -25 & -5 & -1/2 \\ 0 & -5 & -1/2 & 0 \\ 25 & 0 & -1/2 & 0 \end{bmatrix} \begin{bmatrix} v_2 \\ m_2 \\ \phi_1 \\ y_1 \end{bmatrix} = \begin{bmatrix} 0 \\ -500/3 \\ -25 \\ 0 \end{bmatrix}. \qquad (21.26)$$

Solving eqn (21.26) one obtains $\phi_1 = -50$, $y_1 = 1000/3$, $v_2 = -1$, and $m_2 = 10$. These are in agreement with the well-known beam theory solutions $y_{end} = PL^3/3EI$, and so on, which are readily obtained by solving eqn (1.10) with v_2 and $\phi_2 = 0$.

Physically the foregoing problem is a trivial one more easily handled by the finite element method but mathematically the procedure is relatively complex. However, through the use of the free-space Green's function to represent the point source (here a point load) it provides a rigorous means of dealing with infinite domains.

21.5 Application of BEM to pseudoharmonic problems

Most applications of the boundary element method to date have centred around pseudoharmonic problems and are simpler in at least one respect than the foregoing example, namely that integration by parts need only be carried out twice. Taking as an example the Poisson equation

$$\nabla^2 u + b = \frac{\partial^2 u}{\partial x^2} + \frac{\partial^2 u}{\partial y^2} + b = 0, \qquad (21.27)$$

weighting by a source function u^*, and integrating over the area, we have

$$\iint u^* \nabla^2 u \, dx \, dy + \iint b u^* \, dx \, dy = 0 \qquad (21.28)$$

(cf. eqn (21.8)). The source function (sometimes termed a fundamental solution) obeys

$$\iint \nabla^2 u^* \, dx \, dy + \delta = 0 \qquad (21.29)$$

where δ is the Dirac delta function (cf. eqn (21.15)). Integrating the first integral of eqn (21.28) by parts twice one obtains

$$\iint u^* \frac{\partial^2 u}{\partial x^2} \, dx \, dy = \int u^* \frac{\partial u}{\partial x} \, dy - \iint \frac{\partial u^*}{\partial x} \frac{\partial u}{\partial x} \, dx \, dy \qquad (21.30)$$

where

$$-\iint \frac{\partial u^*}{\partial x} \frac{\partial u}{\partial x} \, dx \, dy = -\int \frac{\partial u^*}{\partial x} u \, dy + \iint \frac{\partial^2 u^*}{\partial x^2} u \, dx \, dy;$$

and

$$\iint u^* \frac{\partial^2 u}{\partial y^2} \, dx \, dy = \int u^* \frac{\partial u}{\partial y} \, dx - \iint \frac{\partial u^*}{\partial y} \frac{\partial u}{\partial y} \, dx \, dy \qquad (21.31)$$

where

$$-\iint \frac{\partial u^*}{\partial y} \frac{\partial u}{\partial y} \, dx \, dy = -\int \frac{\partial u^*}{\partial y} u \, dx + \iint \frac{\partial^2 u^*}{\partial y^2} u \, dx \, dy.$$

Noting that

$$\frac{\partial x}{\partial n} = c_x, \qquad \frac{\partial y}{\partial n} = c_y \qquad (21.32a)$$

$$\frac{\partial x}{\partial S} = c_y, \qquad \frac{\partial y}{\partial S} = c_x \qquad (21.32b)$$

where c_x and c_y are the direction cosines of the outwardly directed normal to the boundary, and S is the anticlockwise directed tangent at the boundary, and that

$$\frac{\partial u}{\partial n} = \frac{\partial u}{\partial x} \frac{\partial x}{\partial n} + \frac{\partial u}{\partial y} \frac{\partial y}{\partial n}$$

$$= c_x \frac{\partial u}{\partial x} + c_y \frac{\partial u}{\partial y} \qquad (21.33)$$

the integration variable in the single integrals in eqns (21.30) and (21.31) may be changed to dS and the resulting integrals combined in pairs so that we obtain

$$\iint u^* \nabla^2 u \, dx \, dy = \int u^* \frac{\partial u}{\partial n} \, dS - \int \frac{\partial u^*}{\partial n} u \, dS + \iint u \nabla^2 u^* \, dx \, dy. \qquad (21.34)$$

In view of eqn (21.29) the last term in eqn (21.34) can be written

$$\iint (-\delta) u \, dx \, dy = -c_i u \qquad (21.35)$$

where $c_i = 1$ if point i is inside the domain (as is usually the case), $c_i = 1/2$ on a smooth boundary, and $c_i = 0$ outside the boundary (this occurs in tunnel problems like that of Fig. 7.4).[7] Thus using eqns (21.34) and (21.35) in eqn (21.28) one obtains the governing integral equation as

$$\int u^* \frac{\partial u}{\partial n} \, dS - \int \frac{\partial u^*}{\partial n} \, u \, dS - c_i u + \iint b u^* \, dx \, dy = 0. \tag{21.36}$$

Forming interpolations

$$u = \{f\}^t \{u\}, \qquad \frac{\partial u}{\partial n} = \{g\}^t \left\{ \frac{\partial u}{\partial n} \right\} = \{g\}^t \{u_n\}$$

for each boundary element one obtains

$$\left[\int \{f\}^t \frac{\partial u^*}{\partial n} \, dS \right] \{u\} - \left[\int \{g\}^t u^* \, dS \right] \{u_n\} + c_i u = \iint b u^* \, dx \, dy \tag{21.37}$$

where $\{u\}$ and $\{u_n\}$ are the unknown boundary values at the nodes of the boundary element (cf. eqn (21.20)).

Figure 21.4 shows a domain approximated by two-node boundary elements which will thus have linear functions $\{f\}$ and $\{g\}$. To evaluate the integral on the right-hand side of eqn (21.37) the domain must be divided into two-dimensional 'cells' as shown in Fig. 21.4 and numerical integration can be used to evaluate the area integrals for each cell when one obtains scalar loading values b_i for each boundary node. The same procedure is used to deal with body forces; b in eqn (21.27) could in some cases be a body force.

If 'constant' boundary elements are used with a single node at the centre of each then the c_i are known ($= 1/2$), and applying eqn (21.37) for each element and assembling the results one obtains a system of equations of the form

$$A\{u\} - B\{u_n\} + \{c_i\}^t \{u\} = \{b_i\}. \tag{21.38}$$

The c_i terms can be included in matrix A as diagonal entries (a similar exercise was required for y_1 and ϕ_1 in forming eqn (21.26)).

If linear elements are used, on the other hand, the boundary is not smooth at their intersection, as shown in Fig. 21.4, and the c_i are not known explicitly

Fig. 21.4 Domain enclosed by straight boundary elements

but this difficulty can be overcome in some situations, an example of which follows.

Neglecting body forces and considering the c_i absorbed into matrix A of eqn (21.38) we obtain

$$A\{u\} = B\{u_n\}. \tag{21.39}$$

Noting that when a uniform 'potential' u is applied to a domain the values of the boundary fluxes $(\partial u/\partial n)$ will be zero[2] so that one can write

$$A\{I\} = \{0\} \tag{21.40}$$

that is, the sums of entries in each row of A must equal zero so that the diagonal entries can be calculated from the sum of the off-diagonal entries.

Finally eqn (21.39) can be solved for the unknown values in $\{u\}$ and $\{u_n\}$, requiring similar rearrangements to those needed to obtain eqn (21.26). In the special case of a Dirichlet problem, of course, all the $\{u\}$ are known (for example, plane torsion with Prandtl's stress function in Section 17.4) whereas in Neumann problems all the $\{u_n\}$ are known (for example, the transverse electric wave in the electromagnetic waveguide problem of Section 18.6).

Recalling now that body forces were omitted in eqn (21.39), one means of including these is by superposing a known analytic solution for the body force case.[8]

21.6 Solution of Poisson's equation with boundary elements

As an example of the solution of two-dimensional problems using the boundary element method we consider the solution of the plane torsion problem. If we again use Prandtl's stress function the governing equation is Poisson's equation:

$$\nabla^2 \psi - b = 0 \tag{21.41}$$

where $b = -2G\theta$. The boundary condition is $\psi = 0$ around the boundary of the section, posing an unsatisfactory BEM problem, so that we substitute

$$\psi = b(x^2 + y^2) + u \tag{21.42}$$

in eqn (21.41). The governing equation is then Laplace's equation

$$\nabla^2 u = 0 \tag{21.43}$$

with

$$u = -b(x^2 + y^2) \tag{21.44}$$

on the boundaries.

Analysing a quadrant using four constant boundary elements the boundary conditions are those shown in Fig. 21.5 if $b = 2$, that is $G\theta = -1$. The integral equation for formation of the boundary element equations is eqn (21.37) with $b = 0$ and $\{f\}^t$ and $\{g\}^t$ deleted (because the elements are constant):

$$\sum_{j=1}^{N} \left(\int \frac{\partial u}{\partial n} \, dS \right) u - \sum_{j=1}^{N} (\int u^* \, dS) \frac{\partial u}{\partial n} + c_i u_i = 0 \qquad (21.45)$$

where the subscripts j indicate that for each boundary node i we evaluate the integrals all around the boundary, that is, over each boundary element. Thus we imagine each boundary node as a point source u_i and eqn (21.45) relates this freedom to the other nodal freedoms.

Then applying eqn (21.45) at each node we obtain the simultaneous equations

$$\mathbf{H}\{u\} - \mathbf{G} \left\{ \frac{\partial u}{\partial n} \right\} + \langle c_i \rangle \{u\} = \{0\} \qquad (21.46)$$

where $\langle c_i \rangle$ denotes a matrix with diagonal entries c_i.

The fundamental solution for an isotropic two-dimensional continuum is

$$u^* = \frac{1}{2\pi} \ln \left(\frac{1}{r} \right). \qquad (21.47)$$

The entries in matrix H of eqn (21.46) are thus given by

$$H_{ij} = \frac{1}{2\pi} \int_j \frac{\partial u^*}{\partial n} \, dS = \frac{1}{2\pi} \int_j \frac{\partial r}{\partial n} \frac{\partial u^*}{\partial r} \, dS$$

$$= \int_j \frac{r_n}{r_i} \left(\frac{-1}{2\pi r_i} \right) dS = - \int_j \frac{r_n}{2\pi r_i^2} \, dS \qquad (21.48)$$

where r_i is measured from node i, $\int_j$ denotes integration over element j, and r_n is the 'offset' distance shown in Fig. 21.6.

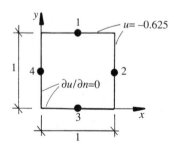

Fig. 21.5 Quadrant of a square section in plane torsion with four constant boundary elements

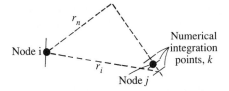

Fig. 21.6 Illustration of offset distance r_n

Evaluating eqn (21.48) using N-point numerical integration we write

$$H_{ij} = - \sum_{k=1}^{N} \frac{r_n}{2\pi r_i^2} (\omega_k S_j) \tag{21.49}$$

where S_j is the length of element j.

For the diagonal entries H_{ii}, however, we use the artifice mentioned in connection with the evaluation of the coefficients c_i in Section 21.5, that is, $H_{ii} = -\Sigma H_{ij}(ij \neq ii)$ and we term this the 'uniform potential condition.' Again using numerical integration the entries in matrix G are given by

$$G_{ij} = \sum_{k=1}^{N} \frac{1}{2\pi} \ln\left(\frac{1}{r_i}\right)(\omega_k S_j) \tag{21.50}$$

but for the diagonal entries this result is singular, that is, $1/r_i = \infty$ at node i. The remedy is to evaluate G_{ii} explicitly using[3]

$$
\begin{aligned}
G_{ii} &= \frac{1}{2\pi} \int_{-S_i/2}^{S_i/2} \ln\left(\frac{1}{r_i}\right) \mathrm{d}S = \frac{1}{\pi} \int_{0}^{S_i/2} \ln\left(\frac{1}{r_i}\right) \mathrm{d}S \\
&= \frac{1}{\pi} \left[r_i \ln\left(\frac{1}{r_i}\right) + \int r_i \left(\frac{1}{r_i}\right) \mathrm{d}S \right]_{0}^{S_i/2} \\
&= \frac{1}{\pi} \left(\frac{S_i}{2}\right) \left[1 - \ln\left(\frac{S_i}{2}\right) \right].
\end{aligned}
\tag{21.51}
$$

Using three-point Gauss quadrature and eqns (21.49)–(21.51) the matrices H and G in eqn (21.46) for the problem shown in Fig. 21.5 are obtained as

$$H = \frac{1}{2\pi} \begin{bmatrix} 3.15 & -1.11 & -0.93 & -1.11 \\ -1.11 & 3.15 & -1.11 & -0.93 \\ -0.93 & -1.11 & 3.15 & -1.11 \\ -1.11 & -0.93 & -1.11 & 3.15 \end{bmatrix} \tag{21.52}$$

$$G = \frac{1}{2\pi} \begin{bmatrix} 1.90 & 0.34 & 0.04 & 0.34 \\ 0.34 & 1.90 & 0.34 & 0.04 \\ 0.04 & 0.34 & 1.90 & 0.34 \\ 0.34 & 0.04 & 0.34 & 1.90 \end{bmatrix} \tag{21.53}$$

where note again that the diagonal entries in H have been calculated from the sum of the other entries in each row, that is, the coefficients c_i have been absorbed into H.

Then substituting these matrices into $H\{u\} - G\{\partial u/\partial n\} = \{0\}$ and introducing the boundary conditions shown in Fig. 21.5 (note that for $u_1 = u_2 = -0.625$ we multiply the corresponding columns by this figure and transpose the results to the right-hand side) we obtain

$$
\begin{bmatrix}
-0.93 & -1.11 & 1.90 & 0.34 \\
-1.11 & -0.93 & 0.34 & 1.90 \\
3.15 & -1.11 & 0.04 & 0.34 \\
-1.11 & 3.15 & 0.34 & 0.04
\end{bmatrix}
\begin{Bmatrix}
u_3 \\
u_4 \\
(\partial u/\partial n)_1 \\
(\partial u/\partial n)_2
\end{Bmatrix}
=
\begin{Bmatrix}
1.275 \\
1.275 \\
-1.275 \\
-1.275
\end{Bmatrix}
\quad (21.54)
$$

which yields the solutions $(\partial u/\partial n)_1 = (\partial u/\partial n)_2 = 0$ and $u_3 = u_4 = -0.625$. Using eqn (21.42) with $b = 2$ we thus obtain $\psi_3 = \psi_4 = -0.438$ which may be compared with an accurate result of -0.461 obtained by the Rayleigh–Ritz method.[9]

The shear stress at node 2 is given by

$$
\sigma_{zy} = -\frac{\partial \psi}{\partial x} = -\frac{\partial u}{\partial n} - \frac{b}{4} \frac{\partial (x^2 + y^2)}{\partial x} = 0 - \frac{2b}{4} = -1 \quad (21.55)
$$

whereas with $G\theta = -1$ the series solution for the maximum shear stress (which occurs at $x = 0$, $y = 1$ and $x = 1$, $y = 0$) is -0.675. Clearly more than four constant boundary elements are needed to obtain accurate stress results but the foregoing exercise provides a useful demonstration of the method which may be usefully compared with the finite element solution calculated in Section 17.4.

Once the solution for the boundary values has been obtained the value of u (and thus ψ) at any internal point is obtained using eqn (21.45). As the point is internal, c_i is unity so that we have

$$
u_i = \sum_{j=1}^{N} G_{ij} \frac{\partial u}{\partial n_j} - \sum_{j=1}^{N} H_{ij} u_j \quad (21.56)
$$

where G_{ij} and H_{ij} are evaluated using eqns (21.49) and (21.50), with r measured from the internal point, for each boundary element.

As numerical integration is used, extension to higher-order boundary elements is straightforward. Now eqn (21.37) is applied at each node within each element to obtain an equation coupling each node to the others.

The process can also be extended to three-dimensional problems when the boundary elements are now surfaces, though when these are curved the geometrical calculations become complicated, as they do in shell analysis with doubly curved shell elements. Hence calculations similar to those described in Section 11.8 are required.[10]

Likewise the procedure used for beam problems demonstrated in Section 21.4 can be extended to deal with plate and shell problems. To date, however, there has been little application to plates and shells, and boundary element techniques have only been proposed for shallow shell problems. This is because shallow shells may be dealt with by analogy to plates on elastic foundations, the 'radial strain' components of the form w/R corresponding to the elastic foundation, for which a sufficiently simple governing differential equation exists.[11]

21.7 Elastic boundary conditions and infinite elements

The problem shown in Fig. 7.4 provides a useful example of two-dimensional finite element analysis in an infinite domain. The infinite domain solution is a simple modification of the theory of thick cylinders, in which the radial and circumferential stresses are given by[9]

$$\sigma_r = \frac{-A}{r^2} + 2C, \qquad \sigma_c = \frac{A}{r^2} + 2C. \tag{21.57}$$

Setting the boundary conditions $\sigma_r = p_i$, where p_i is the interfacial pressure between the pipe and the surrounding medium, at $r = b$ and $\sigma_r = 0$ at $r = \infty$, one obtains

$$\sigma_r = \frac{p_i b^2}{r^2}, \qquad \sigma_c = \frac{-p_i b^2}{r^2}. \tag{21.58}$$

For plane stress the radial displacement is calculated from $E_o w/r = -(\sigma_c - v\sigma_r)$ giving

$$w = \frac{p_i b^2 (1 + v)}{E_o r} \tag{21.59}$$

where E_o is the modulus for the outer zone. The same result applies for plane strain.

Then if each boundary node subtends an angle θ at the centre of the domain the first of eqns (21.58) gives the boundary force at the nodes as $F = p_i b^2 \theta t/r$, where t is the thickness of the domain measured parallel to the axis of the buried pipe. From eqn (21.59) the corresponding displacement is given by $w = p_i b^2 (1 + v)/E_o r$ so that the radial stiffnesses for the boundary nodes are calculated as

$$k = \frac{F}{w} = \frac{E_o h \theta}{1 + v}. \tag{21.60}$$

Resolving these stiffnesses to the Cartesian directions (in the fashion used for simple spar elements) and including these in the system matrix, much improved results are obtained, as shown in Fig. 7.5. Using a similar approach good results are obtained in the three-dimensional problem of Fig. 7.8, as shown in Fig. 7.9.[12]

Wood[13] also uses an inverse square far field solution in the analysis of a seepage problem. Such solutions can also be used to form 'infinite elements' in which special integration rules are used to integrate with an infinite limit, comparable to the special rules required for singular isoparametric elements developed in Section 7.7.

Bettess,[14] for example, incorporates exponential decay terms in the interpolation functions to formulate infinite elements. These considerably complicate the task of integration and simple elastic boundary stiffnesses such as those described above may in many cases yield results of comparable accuracy.

21.8 Concluding remarks

The boundary element method is of particular interest as a means of dealing with infinite domain problems. It can also be included amongst the considerable number of weighted residual techniques. As discussed in the preceding section inverse square and exponential far field solutions have been incorporated into the finite element method and in general any free-space Green's function could be incorporated, either by calculating appropriate stiffnesses at the boundary or by using special integrations to form a layer of infinite elements around the boundary. In either case substructuring using eqns (6.41) and (6.42) can then be used to separate the boundary and 'interior' problems.

Finally, infinite domain problems can, of course, be dealt with by combining finite and boundary[15] or infinite[16] elements. Programming of the BEM is in some respects simpler than for FEM, for banding of the matrices need not be considered. Brebbia[3] gives full listings of programs for BEM analysis of pseudoharmonic problems using constant and linear elements. These are in FORTRAN but the reader should have little difficulty converting them to the BASIC used in the present text for microcomputer applications (or directly mounting them as they stand if a FORTRAN compiler is available for their PC).

References

1. Crandall, S. H. (1956). *Engineering Analysis.* McGraw-Hill. New York.
2. Brebbia, C. A. and Walker, S. (1978). Introduction to boundary element methods.

   In *Recent Advances in Boundary Element Methods*, (ed. C. A. Brebbia). Pentech Press, Plymouth.
3. Brebbia, C. A. (1978). *The Boundary Element Method for Engineers*. Pentech Press, London.
4. Herrera, I. (1978). Theory of connectivity: a systematic formulation of boundary element methods. In *Recent Advances in Boundary Element Methods*, (ed. C. A. Brebbia). Pentech Press, Plymouth.
5. Kelly, D. W., Mustoe, G. G. W., and Zienkiewicz, O. C. (1979). Coupling boundary element methods with other numerical methods. In *Developments in Boundary Element Methods*, Vol. 1, (ed. P. K. Banerjee and R. Butterfield). Applied Science Publications, London.
6. Butterfield, R. (1979). New concepts illustrated by old problems. In *Developments in Boundary Element Methods*, Vol. 1, (ed. P. K. Banerjee and R. Butterfield). Applied Science Publications, London.
7. Courant, R. and Hilbert, D. (1953). *Methods of Mathematical Physics*. Interscience, New York.
8. Wardle, L. J. and Crotty, M. J. (1979). Two-dimensional boundary integral equation analysis for non homogeneous mining applications. In *Developments in Boundary Element Methods*, Vol. 1, (ed. P. K. Banerjee and R. Butterfield). Applied Science Publications, London.
9. Timoshenko, S. P. and Goodier, J. N. (1951). *Theory of Elasticity*, (3rd edn). McGraw-Hill, New York.
10. Alarcon, E., Martin, A., Paris, F. (1978). Improved boundary elements in torsion problems. In *Recent Advances in Boundary Element Methods*, (ed. C. A. Brebbia). Pentech Press, Plymouth.
11. Tottenham, H. (1979). Boundary element analysis of plates and shells. In *Developments in Boundary Element Methods*, Vol. 1, (ed. P. K. Banerjee and R. Butterfield). Applied Science Publications, London.
12. Mohr, G. A. and Power, A. S. (1978). Elastic boundary conditions for finite elements of infinite and semi-infinite media. *Proc. Inst. Civ. Eng.*, Part 2, **65**, 675.
13. Wood, W. L. (1976). On the finite element solution of an exterior boundary value problem. *Int. J. Num. Meth. Engng.* **10**, 885.
14. Bettess, P. (1977). Infinite elements. *Int. J. Num. Meth. Engng*, **11**, 53.
15. O'Donoghue, P. E. and Atluri, S. N. (1987). Field/boundary element approach to the large deflection of thin flat plates. *Comput. Struct.*, **27**, 427.
16. Godbole, P. N., Viladkar, M. N., and Noorzori, J. (1990). Nonlinear soil-structure interaction using coupled finite-infinite elements. *Comp. Struct.*, **36**, 1089.

22
Optimization techniques
and applications

The majority of the problems encountered in preceding chapters were linear whilst relatively economical solutions can be obtained for most of the non-linear problems encountered, except perhaps when time dependency is also included, resulting in a 'twice iterative' situation. One of the most beneficial applications of the finite element method, however, is to the optimization of engineering systems and much remains to be done in this area. This chapter briefly outlines some optimization fundamentals, the classical Michell 'constant strain' principle, applications of linear programming in matrix structural analysis, non-linear optimization techniques, and finally summarizes some applications of optimization techniques in the finite element method.

22.1 Classical theory of optimization

The necessary condition for a stationary point of a function $f(x_i)$ is

$$\frac{\partial f}{\partial x_i} = 0 \qquad i = 1 \to n$$

and a necessary condition for a point to be a *relative* minimum is that

$$|H| = \left| \frac{\partial^2 f}{\partial x_i \partial x_j} \right| > 0, \qquad \frac{\partial^2 f}{\partial x_i \partial x_j} \neq 0 \qquad i, j = 1 \to n \qquad (22.1)$$

whereas for a relative maximum $|H| < 0$ and for a saddle point $|H| = 0$. Analytical solutions to non-linear optimization problems are governed by these conditions. However, when more than a few variables are involved and $f(x)$ is not continuous iterative numerical methods must be used.

If the function is constrained by *inequality constraints*

$$c_j(x_i) + y_j = 0 \qquad i = 1 \to n, \quad j = 1 \to m \qquad (22.2)$$

where the c_j are in general non-linear operators and y_j are *slack variables*, the minimum is a turning point of the *Lagrangian function*

$$\phi(x, \lambda) = f(x) + \sum_{j=1}^{m} \lambda_j c_j(x_i) \qquad (22.3)$$

where $\{\lambda\}$ are *Lagrange multipliers*. The minimum is then an *unconstrained* solution of

$$\left\{\frac{\partial f}{\partial x_i} + \sum_{j=1}^{m} \lambda_j \frac{\partial c_j}{\partial x_i} = 0 \,\middle|\, c_j + y_j = 0, j = 1 \to m\right\} \qquad i = 1 \to n. \quad (22.4)$$

Thus one obtains a minimum of $\phi(x, \lambda)$ with respect to variations in $\{x\}$ and when some $y_i \neq 0$ the point is also a maximum with respect to variations in $\{\lambda\}$, that is, it is a saddle point of the Lagrangian function.

Note that any equality constraint $e(x) = 0$ can be used to eliminate one variable by writing $x_i = E(x_1, x_2, \ldots, x_{i-1}, x_{i+1}, \ldots, x_n)$ or can be split into two inequality constraints $e(x) < 0$ and $e(x) > 0$.

The Lagrange multiplier technique demonstrates that in constrained problems the location of the minimum depends as much upon the constraints, which block access to the 'natural' unconstrained minimum, as it does upon the nature of the *merit* or *objective* function $f(x)$.

Note that optimization problems are separable if they can be expressed in the form[1]

$$\text{Min}: \quad f(x) = \sum_{j=1}^{m} f_j(x_j) \,\middle|\, c_i(x) = \sum_{j=1}^{m} c_{ij}(x_j) \geq 0 \qquad i = 1 \to n \quad (22.5)$$

that is, the variables are coupled together only in an additive fashion so that

$$\frac{\partial^2 f}{\partial x_k \partial x_j} = \frac{\partial^2 c_i}{\partial x_k \partial x_j} = 0 \qquad k \neq j. \quad (22.6)$$

All linear programming problems are separable but quadratic programming problems are only separable if the Hessian matrix H (see eqn (22.1)) is a diagonal matrix or if transformations are applied to the variables to diagonalize H (for example by Jacobi diagonalization[2]).

22.2 Special applications of classical theory

A spectracular early application of the classical theory of Section 22.1 was made by A.G.M. Michell[3] ('the paper would have been impressive had it appeared fifty years later, and must rank as one of the foremost and most original contributions to the optimization literature. It also foreshadowed the modern theory of plastic solids'[4]). This led to the formation of optimum shapes of trusses (*Michell structures*) taking the form of *Hencky–Prandtl nets* comprising (two orthogonal systems of curves.[5]

The following proof is essentially equivalent to that given by Michell but more nearly follows the fundamentals of Section 22.1 and the plastic optimum theory of beams due to Rozvany.[6] The problem is to minimize

$$F = \Sigma |A_i| L_i, \qquad \text{sgn}(A_i) = \text{sgn}(T_i) \qquad i = 1 \to n \quad (22.7)$$

subject to the equilibrium constraints

$$T_i - EA_i\varepsilon_i = 0 \qquad i = 1 \to n \tag{22.8}$$

where A_i, L_i, T_i, ε_i are respectively the cross-sectional areas, lengths, thrusts, and strains in a truss with n members. If in the optimum truss $|A_i|L_i = f_i$ (so that $F_{\min} = \Sigma f_i$) then if the truss is not optimal

$$A_i L_i \geq f_i, \qquad A_i L_i \leq -f_i \tag{22.9}$$

or

$$A_i L_i + f_i + s_1 = 0 \tag{22.10}$$

$$- A_i L_i + f_i + s_2 = 0 \tag{22.11}$$

where s_1 and s_2 are slack variables for the compression and tension members, respectively.

Then using the Lagrange multiplier technique the constrained minimization problem is written as

$$\text{Min}: F^* = \Sigma\left[f_i - \lambda_1(A_i L_i + f_i + s_1) - \lambda_2(- A_i L_i + f_i + s_2) \right.$$

$$\left. + \lambda_3 \frac{L_i}{\sigma_{\lim}}(T_i - EA_i\varepsilon_i) \right] \tag{22.12}$$

where λ_1, λ_2, and λ_3 are Lagrange multipliers, the factor $(L/\sigma_{\lim})$ is added for dimensional consistency and $\sigma_{\lim}$ is the limiting stress magnitude for the material.

For a minimum of F^* we require

$$\frac{\partial F^*}{\partial s_i} = \lambda_i = 0 \qquad (\text{but } \lambda_i \neq 0 \text{ if } s_i = 0) \tag{22.13}$$

$$\frac{\partial F^*}{\partial f_i} = 1 - (\lambda_1 + \lambda_2) = 0 \qquad \text{or} \qquad \lambda_1 + \lambda_2 = 1 \tag{22.14}$$

$$\frac{\partial F^*}{\partial (A_i L_i)} = - \lambda_1 + \lambda_2 + \frac{E\varepsilon_i}{\sigma_{\lim}} = 0$$

or

$$\varepsilon_i = \frac{\sigma_{\lim}}{E}(\lambda_1 - \lambda_2) = C(\lambda_1 - \lambda_2) \qquad C = \text{const.} \tag{22.15}$$

$$\frac{\partial F^*}{\partial \lambda_3} = \frac{L_i}{\sigma_{\lim}}(T_i^* - EA_i\varepsilon_i) = 0 \tag{22.16}$$

where T_i^* are the optimum thrusts. Minimization with respect to λ_1 and λ_2 is not required here and reveals only eqns (22.10) and (22.11) again. Combining eqns (22.13)–(22.15) one obtains the three following results.

1. Tension members. If $T_i^*, \varepsilon_i > 0$ we require $s_2 > 0$ (in eqn (22.11)) and hence $\lambda_2 = 0$ (eqn 22.13), so that $\lambda_1 = 1$ (eqn (22.14)) and hence from eqn (22.15)

$$\varepsilon_i = C. \qquad (22.17)$$

2. Compression members. If $T_i^*, \varepsilon_i < 0$ then $s_1 > 0$ (in eqn (22.10)) and hence $\lambda_1 = 0$ (eqn 22.13)), so that $\lambda_2 = 1$ (eqn (22.14)) and hence from eqn (22.15),

$$\varepsilon_i = -C. \qquad (22.18)$$

3. Vanishing members. If $T_i^* = 0$ then $\lambda_1, \lambda_2 \neq 0$ and from eqn (22.15) $\varepsilon_i = C(\lambda_1 - \lambda_2) = C(2\lambda_1 - 1)$ (from eqn (22.14)) so that

$$-C \leq \varepsilon_i \leq C \qquad (22.19)$$

as $-1 \leq \lambda_1 \leq 1$.

The last result (eqn (22.19)) shows that, as in general $\varepsilon_i \neq 0$, we require $L_i = 0$ to satisfy eqn (22.16) when $T_i^* = 0$, that is, the member vanishes (but note that in the present argument we still define a strain between its end-points).

Thus we obtain Michell's conclusion that the optimum truss will have constant strain magnitude C.[3] It is for this reason that *fully stressed design* (FSD) is often used in structural analysis to obtain more economical structures.[7] In statically determinate trusses of fixed geometry this corresponds to the optimum result but condition 3 for vanishing members may not be satisfied in statically indeterminate structures (see for example Section 23.6).

The foregoing proof follows that given by Rozvany for plastic beams[6] except that we add eqn (22.16) (to deal with the possibility of redundant members of a truss vanishing) whilst in beams, conditions 1, 2, and 3 now relate to the generalized strains or curvatures.

In eqn (22.7), noting that vanishing members can be neglected, one can assume $A_i = A_0 \sigma_i / \sigma_{\mathrm{lim}}$ where A_0 is the initial area assumed for each member. Thus one seeks to minimize $F = (\text{const.}) \Sigma \sigma_i L_i$ and generalizing this result to beams and plates one seeks in these to minimize the *moment area* $\int |M| \, dx$ and the *moment volume*[6]

$$\iint (|m_1| + |m_2|) \, dx \, dy \qquad (22.20)$$

respectively, m_1 and m_2 being the principal unit moments.

An example of the application of the latter criterion is the 'fibre-optimum' method of Lowe and Melchers[8] which might, like Rozvany's method,[6] be classified as a subdomain method, the former being non-conforming[6] and the latter being conforming. As we shall see in Section 23.8 such criteria may be generalized to combine extensional and flexural stresses and used to optimize finite element models of more general problems.

22.3 Linear optimization problems

The linear programming (LP) optimization problem is

$$\text{Min}: \quad z = \{c\}^{t}\{x\} \qquad x_i \geq 0, \quad i = 1 \rightarrow n \tag{22.21}$$

subject to the m constraints

$$A\{x\} + I\{y\} = \{b\} \qquad y_i \geq 0, \quad j = 1 \rightarrow m. \tag{22.22}$$

A is an $m \times n$ matrix and $\{x\}$, $\{c\}$, $\{y\}$ are, respectively, the *design variables*, *unit costs*, and the *slack variables*. The slack variables deal with inequality constraints. Combining eqns (22.21) and (22.22) the problem can be written in the form

$$\begin{bmatrix} A & I \\ \{c\}^{t} & \{0\}^{t} \end{bmatrix} \begin{Bmatrix} \{x\} \\ \{y\} \end{Bmatrix} = \begin{Bmatrix} \{b\} \\ z \end{Bmatrix}. \tag{22.23}$$

As the problem is overdetermined (that is, the total number of variables is $m + n$ which exceeds the number of equations, m, available for solution) the problem is solved iteratively by beginning with the initial *infeasible* solution $\{x\} = \{0\}$, $\{y\} = \{b\}$, $z = 0$. In each iteration Gauss–Jordan reduction is applied to eqn (22.23), choosing the most 'attractive' pivots and eliminating the cost entries in their columns.

For maximization the usual procedure is to choose the pivot in the column with the largest positive cost entry and then in the row which, considering only the column already chosen, gives the smallest ratio b/a ($b > 0$, $a > 0$). Thus, in the n-dimensional problem, one moves along the most profitable axis of the x_i until the first constraint intersecting it is encountered.

An interesting alternative pivoting rule for minimization utilizes the concepts of search and steepest descent to be encountered in connection with non-linear problems in the following sections to minimize the number of Gauss-Jordan reductions required.[9] Such pivots must be chosen so that all entries in the cost row remain positive (otherwise one overshoots the minimum). Therefore the pivots a_{ij} are selected according to

$$\text{Max} \left\{ \frac{b_i c_j}{a_{ij}} \,|\, b_i < 0, a_{ij} < 0, c_j > 0, \quad \text{all} \quad c_j' > 0 \right\} \tag{22.24}$$

where c_j' are the new entries in the cost row after reduction with this pivot.

Note that for the initial problem to be well posed, so that a feasible non-trivial ($\{x\} \neq \{0\}$) minimum exists, some of the b_i must be negative. Hence there must be some $\geq$ constraints which, to form eqn (22.23), are reversed in sign. Then the minimum solution has been obtained when no legitimate pivots remain, that is, when all $b_i \geq 0$ which means that the solution is feasible.

We have not considered equality constraints which are most easily dealt with either as 'split constraints' (see Section 22.1) or by ensuring that the corresponding entries in $\{y\}$ are eliminated from the basis by selecting a pivot in the corresponding row, and disallowing their return by thereafter disallowing pivots in the columns for these artificial variables.

If the problem possesses a feasible maximum as well, one may then proceed to determine this using the pivot selection rule[10]

$$\text{Max}\left\{\frac{b_i c_j}{a_{ij}} \mid b_i > 0, c_j > 0, a_{ij} > 0, \quad \text{all} \quad b'_j \geq 0\right\} \qquad (22.25)$$

the last condition ensuring that the solution remains feasible. The optimum solution is obtained when no eligible pivots remain, that is, when all $c_j < 0$ so that no further increase in z is possible.

Rather than travelling as far as permitted along the axis with the largest current c_j, eqns (22.24) and (22.25) search for that axis along which permissible travel produces the largest change in the objective function, but the whole A matrix must be searched to achieve this. Consider as an example the problem

$$\text{Min. and Max.:} \quad z = 2x_1 + 7x_2 \qquad (22.26)$$

subject to

$$-x_1 + 2x_2 \leq 14 \qquad (22.27a)$$

$$5x_1 + 2x_2 \leq 50 \qquad (22.27b)$$

$$x_1 + 2x_2 \geq 18. \qquad (22.27c)$$

The graphical solution of this is shown in Fig. 22.1 and the tabular Gauss–Jordan reduction solution is given in Table 22.1.

The maximum solution is obtained in only one step after the minimum, which here provides an initial feasible solution prior to maximization, whereas three steps are required if one returns first to the initial solution, when the simple Max(c) and Min(b/a) rule no longer applies as this initial solution is not feasible. Observing Fig. 22.1 one sees that the solution steps proceed along the sides of a two-dimensional simplex or triangle, giving rise to the term simplex method.[11]

As observed in Section 22.1 the minimum of $z = f(\{x\}, \{y\})$ with respect to $\{x\}$ (the *primal problem*) is the maximum with respect to the supplementary variables $\{y\}$, the latter being the *dual problem*. The conjunction of the two extrema is called the *saddle point condition*. The dual problem to that of eqns (22.21) and (22.22) is

$$\text{Max. and Min.:} \quad z = \{b\}^t\{y\}$$

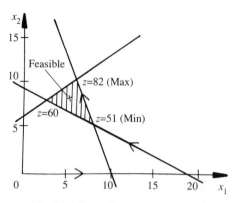

Fig. 22.1 Two-dimensional simplex

Table 22.1 Simplex table for eqns (22.26) and (22.27)

	− 1	2	1	0	0	14
Initial	5	2	0	1	0	50
solution	(− 1)	− 2	0	0	1	− 18
(pivot) ╱	2	7	0	0	0	0
	0	4	1	0	− 1	32	
First	0 (− 8)	0	1	5	− 40	
reduction	1	2	0	0	− 1	18
	0	3	0	0	2	− 36
	0	0	1	1/2	(3/2)	12	$(y_1 = 12)$
Second	0	1	0	− 1/8	− 5/8	5	$(x_2 = 5)$
reduction	1	0	0	1/4	1/4	8	$(x_1 = 8)$
	0	0	0	3/8	31/8	− 51	(Min)
	0	0	2/3	1/3	1	8	$(y_3 = 8)$
Third	0	1	5/12	1/12	0	10	$(x_2 = 10)$
reduction	1	0	− 1/6	1/6	0	6	$((x_1 = 6)$
	0	0	− 31/12	− 11/12	0	− 82	(Max)

subject to

$$- A^t\{y\} + I\{x\} = \{c\}. \tag{22.28}$$

Both the maximum and minimum solutions of this are obtained using eqns (22.25) and (22.24) in that order, noting that before the minimization steps signs must be reversed in the right-hand column and bottom row. Following this procedure exactly the same sequence of pivots as in the primal is obtained, that is eqns (22.24) and (22.25) are dual pivoting rules.

With the arrangements used in eqns (22.22) and (22.28) a dual pivoting rule is formed by interchanging b with c and reversing the sign restrictions for a and b. Comparison of eqns (22.24) and (22.25) on this basis verifies that these rules are indeed dual. Thus the dual of the widely used $\text{Max}(c)-\text{Min}(b/a)$ rule for maximization is in minimization to choose the pivot row as that with the least $b_i(<0)$ and then the row with $\text{Min}(|c/a|)$, $c > 0$, $a < 0$.

A linear programming problem of interest is that of plastic collapse of rigid jointed frames. In the *mechanism method*[12] one minimizes the plastic moment area $\Sigma M_{pi} L_i$ whilst a dual form of the problem is obtained by the *equilibrium method*.[12] Plastic analysis to determine the moment distribution at collapse is also a linear programming problem which possesses dual forms in which the design variables are either nodal displacements[13] or nodal forces.[13,14]

Such analyses bear some relationship to the finite element method and indeed, using the LP method, equivalent elastic analyses may be performed. Analysing a truss structure, for example, using displacement variables we assemble the element strain matrices to form a single system matrix B and thus obtain the LP problem

$$\text{Min}: \quad \{Q\}^t\{D\} | DB\{D\} - \{T\} = \{0\} \tag{22.29}$$

the constraints ensuring equilibrium (as in Section 22.2). In such cases, however, it is generally preferable to solve the problem in the quadratic form $B^t DB\{D\} = \{Q\}$ as this requires only a single step. However, the plastic frame problems cannot be posed in quadratic form because only 'critical' mechanisms operate in the determination of the minimum collapse load so that stepwise solution as an LP problem is necessary, the pivot selection rule of eqn (22.24) finding the most critical mechanisms.

22.4 Univariate non-linear problems

A number of univariate search techniques have been proposed, including *golden section search, Fibonacci search* and *bisection* all of which seek to bracket the optimum increasingly closely.[4]

A preferable approach is to use quadratic interpolation to estimate the location of the minimum once this has been crudely bracketed. This is done by taking steps that are successively doubled in size from a starting point and moving in the direction of decreasing $f(x)$ until an increase is first obtained. The last step is then halved giving four equally spaced points bracketing the minimum.[1] Discarding the worst point (that furthest from that which gives the lowest $f(x)$) and using quadratic interpolation with the remaining three,

$$f = \tfrac{1}{2}f_1(s^2 - s) + f_2(1 - s^2) + \tfrac{1}{2}f_3(s^2 + s) \qquad (22.30)$$

where f_1 is at $x_1 = -L/2$, f_2 is at $x_2 = 0$, and f_3 is at $x_3 = L/2$, the minimum point is given by

$$\frac{df}{dx} = \frac{df}{ds}\frac{ds}{dx} = (2x - 1)f_1 - 2xf_2 + (2x + 1)f_3 = 0$$

giving

$$x^* = \frac{0.5L(f_1 - f_3)}{f_1 - 2f_2 + f_3} \qquad (22.31)$$

as the approximate location of the minimum.

Alternatively, if the points are not equally spaced one uses[15]

$$Cx^* = \frac{(x_3^2 - x_2^2)f_1 + (x_1^2 - x_3^2)\,f_2 + (x_2^2 - x_1^2)f_3}{(x_3 - x_2)f_1 + (x_1 - x_3)f_2 + (x_2 - x_1)f_3} \qquad (22.32)$$

where $C = x_2 x_3 x_{32} + x_1 x_3 x_{13} + x_1 x_2 x_{21}$ and $x_{32} = x_3 - x_2$, and so on.

22.5 Search methods for non-linear problems

22.5.1 Relaxation methods

These 'relax' each variable in turn using univariate search. Such methods are excessively laborious; for example for twenty variables Fibonacci search, which is the most efficient, requires 10^{26} function evaluations.[1]

22.5.2 Simplex methods

More effective are the simplex methods which use $n + 1$ function evaluations to form a simplex in the n-dimensional design space, iteratively forming new simplexes by reflecting the worst vertex about the centroid of the last simplex. Ultimately the sequence of simplexes thus formed revolves in the region of the optimum and their size is then progressively reduced to more closely find the optimum.[1]

22.5.3 Vector search methods

In the crudest of these, due to Hooke and Jeeves,[16] exploratory moves are made with vectors formed by perturbations $\pm \delta x_i | \delta f < 0$ of each design variable in turn and these are followed by *pattern moves* with the same *move vector* or further exploratory moves.

With the use of such perturbations the gradient vector $g_i = \{\partial f/\partial x_i\}$ at each point i may be evaluated using finite difference approximations and the method of steepest descent used (see Section 22.6), that is, the search direction is $-g_i$. In the local sense this is the most efficient direction but in the global sense convergence is somewhat erratic (see Fig. 22.2).[1, 17]

Improved methods use orthonormal vectors (Rosenbrock)[18] in conjunction with quadratic search in the move directions.[1] Still better results are obtained with the *conjugate gradient methods* that guarantee to find the minimum of the quadratic function

$$f(x) = \tfrac{1}{2}\{x\}^t H\{x\} + \{b\}^t\{x\} + c \qquad (22.33)$$

in a specified number of iterations and posess quadratic convergence. The search directions $\{\delta x\}_i$ and $\{\delta x\}_j$ are conjugate with respect to H if

$$\{\delta x\}_i^t H\{\delta x\}_j = 0 \qquad i \neq j. \qquad (22.34)$$

Then the minimum point can be expressed in terms of n linearly independent and mutually conjugate directions $\{\delta x\}_i$, $i = 1 \rightarrow n$,

$$\{x^*\} = \{x_0\} + \sum_{i=1}^{n} \alpha_i \{\delta x\}_i. \qquad (22.35)$$

In Smith's method,[19] the orthogonal unit search vectors are formed by the Gram–Schmidt orthonormalization relationships[20] and the α_i are found by a search in each of these directions. This process must be repeated iteratively with new sets of search directions and is thus impractical for more than a few variables.[1]

A much simpler and more economical approach is that used by Hestenes and Steifel[21] for solving linear equations, which Fletcher and Reeves applied to non-linear optimization problems.[22] One begins with a 'minimizing' steepest descent step (found with quadratic search, for example) in the direction $-g_i$ and the new search direction is obtained as[17]

$$d_i = -g_i + \frac{g_i^t g_{i^*}}{g_{i-1}^t g_{i-1}} d_{i-1}. \qquad (22.36)$$

The procedure is restarted with a steepest descent step after a prespecified limiting number of iterations.[22]

Various modified conjugate gradient methods have since been proposed[1, 15, 23] and the second-order gradient methods (see Section 22.6) may also be regarded as search methods if derivatives are calculated by perturbation techniques (see eqns (22.44)).

22.6 Gradient methods for non-linear problems

22.6.1 First-order methods

The classical method is that of steepest descent, first applied by Cauchy,[24] which chooses the gradient of the objective function as the search direction. Considering a perturbation $|\delta x| = (\Sigma \delta x^2)^{1/2}$ and using the method of Lagrange multipliers we can write

$$\delta f = \sum \frac{\partial f}{\partial x_i} \delta x_i + \lambda(\Sigma \delta x^2 - \Delta^2) \qquad (22.37)$$

where Δ is the optimum or most effective perturbation which is obtained if

$$\frac{\partial(\delta f)}{\partial(\delta x_i)} = 0 = \frac{\partial f}{\partial x_i} + 2\lambda \delta x_i \qquad i = 1 \to n. \qquad (22.38)$$

Thus, $\delta x_i = -(\partial f/\partial x_i)/2\lambda$ so that the greatest change in f is obtained by proceeding in the direction of the gradient vector $g = \{\partial f/\partial x_i\}$. This direction is orthogonal to the contours $f = $ const. and if univariate search is used in this direction alternate steps in problems with $n = 2$ will be mutually orthogonal.[1] This gives somewhat erratic convergence in elliptic problems, which are perhaps the most common type, and convergence is sometimes improved by using an underconvergence factor of less than one.[25]

In finite element analysis the most widely used method for non-linear problems is Newton's method and when equilibrium iteration is possible the step lengths are directly given by the residuals.[26] As we saw in Section 12.2 improved search directions can be obtained by simple means (comparable to the use of eqn (22.36)),[27] and an equally simple improvement in the method of steepest descent is demonstrated in Fig. 22.2.

In general non-linear optimization problems the disadvantages of Newton's method are that it will only find the local optimum, the Jacobian matrix may become singular, search is required at each step, and convergence will have the same erratic 'criss-cross' character typical of the steepest descent method.

22.6.2 Second-order gradient methods

The basis of these is the 'generalized Newton method' which, like the conjugate gradient methods, possesses quadratic convergence and is based upon the quadratic prolem

$$f(x) = \tfrac{1}{2}\{x\}^t H\{x\} + \{g\}^t\{x\} + c \qquad (22.39)$$

where

$$H = \left[\frac{\partial^2 f}{\partial x_i \partial x_j}\right], \qquad \{g\} = \left\{\frac{\partial f}{\partial x_i}\right\}. \qquad (22.40)$$

To obtain the minimum we require

$$\frac{\partial f}{\partial \{x\}} = H\{x\} + \{g\} = \{0\} \tag{22.41}$$

giving the minimum solution $\{x^*\} = -H^{-1}\{g\}$. This gives the solution to the quadratic programming problem in a single step and in a general problem can be used as a search direction in the iterative scheme

$$\{x\}_{n+1} = \{x\}_n - \alpha_n H^{-1}\{g\} \tag{22.42}$$

using search to determine α_n.

Fletcher and Powell have shown that the search direction $-H^{-1}\{g\}$ always reduces $f(x)$.[28] Numerical computation of the Hessian matrix is relatively laborious on d Box *et al.*[1] suggest that it be recomputed every few iterations. But for this search direction to be 'profitable', (that is, in the descent direction, H must be positive definite. This is ensured in the 'quasi-Newton algorithms' by using an approximate inverse Hessian matrix A (usually beginning with $A = I$) which is iteratively improved using the algorithm

$$A_{n+1} = A_n + \frac{\{p\}\{p\}^t}{\{d\}^t\{\delta g\}} - \frac{A_n\{\delta g\}\{\delta g\}^t A_n}{\{\delta g\}^t A_n\{\delta g\}} + \beta\gamma\{v\}\{v\}^t \quad 0 \le \beta \le 1 \tag{22.43}$$

where

$$\{p\} = \{x_i\}_{n+1} - \{x_i\}_n = -\alpha_n A_n\{g\}$$

$$\{\delta g\} = \left\{\frac{\partial f}{\partial x_i}\right\}_{n+1} - \left\{\frac{\partial f}{\partial x_i}\right\}_n$$

$$\{v\} = \{p\} - \frac{1}{\gamma} A_n\{\delta g\}$$

$$\gamma = \frac{\{\delta g\}^t A_n\{\delta g\}}{\{p\}^t\{\delta g\}}.$$

The first adjustment term on the right-hand side of eqn (22.43) ensures that A converges towards $[\partial^2 f/\partial x_i \partial x_j]^{-1}$ and the second ensures positive definiteness provided that α_n yields a sufficiently accurate minimization in the line search.[27]

If $\beta = 0$ one obtains the Davidon–Fletcher–Powell (DFP) method (Davidon,[29] Fletcher and Powell[28]) which was the first method of this type. It is perhaps worth noting in an FEM text that Davidon used cubic interpolation with f and $\{g\}^t\{x\}$ as 'end values', rather than the quadratic interpolation of Section 22.4 in the line searches at each step.

If $\beta = 1$ one obtains the Broyden–Fletcher–Goldfarb–Shanno (BFGS) method[30-33] for which 'there is growing evidence that it is the best general-purpose quasi-Newton method currently available.'[27]

The quasi-Newton methods give a smooth transition from steepest descent behaviour to that of Newton's method (eqn (22.42)) as $A \to H^{-1}$. As the latter performs best near the optimum such a transition is desirable.

All the gradient methods require calculation of at least the gradient $\{g\} = \nabla\{f\}$ and possibly the Hessian matrix $\nabla^2\{f\}$ at each point $\{x\}_n$. When these derivatives are not explicitly available they can be evaluated by using the forward difference approximations

$$\frac{\partial f(x)}{\partial x_i} \simeq \frac{f(x + h_i) - f(x)}{h}$$

$$\frac{\partial^2 f(x)}{\partial x_i \partial x_j} \simeq \frac{1}{h}\left[\frac{\partial f(x + h_j)}{\partial x_i} - \frac{\partial f(x)}{\partial x_i}\right] \qquad (22.44a)$$

where h_i denotes a 'small' increase in x_i in $\{x\}$, but not so small as to cause computational difficulties. In poorly scaled problems this will vary with i. Equations (22.44a) involve a truncation error $O(h)$ whereas the central difference formulae

$$\frac{\partial f(x)}{\partial x_i} \simeq \frac{f(x + h_i) - f(x - h_i)}{2h}$$

$$\frac{\partial^2 f(x)}{\partial x_i \partial x_j} \simeq \frac{1}{2h}\left[\frac{\partial f(x + h_j)}{\partial x_i} - \frac{\partial f(x - h_j)}{\partial x_i}\right] \qquad (22.44b)$$

involve a truncation error of $O(h^2)$ but involve twice as much computation. Clearly the second derivative computations are laborious but forward differences generally suffice for these, whereas central differences are advisable for the first differences when these tend to vanish at local minima.[27] With the introduction of such finite difference approximations, however, the gradient methods can equally well be classified as search methods.

22.7 Constrained non-linear optimization techniques

22.7.1 Stepwise application of linear programming[11]

The stepwise application of LP to finite element models is a somewhat impractical exercise and, of more relevance to material in earlier chapters, we shall concentrate upon penalty and Lagrange multiplier techniques.

22.7.2 Constraint riding techniques[34]

These have a steepest descent basis but sustitute the constraint gradients $\{\partial c/\partial x_i\}$ for $\{\partial f/\partial x_i\}$ when a constraint is violated, first using interpolation to locate the constraint surface. When the constraints are non-linear, though, very small step lengths may be needed to 'ride' these.

22.7.3 Hemstitching techniques

These initially search in the gradient direction but take steps orthogonal to a violated constraint to return to the feasible region.[34] It is of interest to note the relationship of this procedure to that illustrated in Fig. 22.2.

22.7.4 Modified quasi-Newton methods

Davidon's method (eqn (22.43) with $\beta = 0$) can be extended to deal with linear constraints by reducing the rank of the matrix A using the formula[35]

$$A_{n+1} = A_n - \frac{A_n\{e\}\{e\}^t A_n}{\{e\}^t A_n \{e\}} \tag{22.45}$$

to successively incorporate each equality constraint $\{e\}^t\{x\} = q$. Inequality constraints are also incorporated when these become 'active.' Note that this is comparable to the 'deflation' method of reducing the rank of matrices with embedded eigenvalues; here $\{e\}$ are the eigenvectors successively determined by vector iteration (see eqn (13.103)).

22.7.5 Penalty function methods

These are amongst the simplest to program and are the most widely applicable optimization techniques. They are also of considerable interest in the finite element method. In non-linear optimization problems there are two basic approaches:[4]

(1) exterior point methods;

(2) interior point methods.

Both of these transform the constrained problem to a sequence of unconstrained minimizations. In the exterior point methods one minimizes

$$F(x) = f(x) + \beta \Sigma |c_i(x)|^2 \tag{22.46}$$

so that the merit function is not modified at feasible points and the *sequence of unconstrained minimization technique* (SUMT)[36] proceeds with $\beta \to \infty$. These methods have the disadvantage that at infeasible points program failure can be caused by computations such as $\sqrt{x}$ and $\ln(x)$ with $x < 0$, and the interior point methods which restrict attention to feasible points avoid this.

For the interior point methods one must first determine a feasible point, and Box *et al.*[1] suggest that this be accomplished by first applying SUMT to minimize

$$C = -\Sigma s_i + \Sigma c_i^2(x) \tag{22.47}$$

where s_i are the slack variables in the inequality constraints $c_i(x) + s_i = 0$. When $C = 0$ has been obtained the point is feasible and SUMT is applied to minimize

$$F(x) = f(x) + \beta \Sigma 1/c_i(x) + \beta^{-1/2} \Sigma |e_j(x)|^2 \qquad (22.48)$$

with $\beta \to 0$, although $\beta = 0$ is obviously disallowed with equality constraints.

Here the inequality constraint term creates a *response surface* or barrier that confines the search to the feasible region,[37] and the equality constraint term is of the usual form,[38] the least squares nature of which is worth note.

Alternatively we can combine eqns (22.47) and (22.48) and minimize[4]

$$F(x) = f(x) + \beta \Sigma 1/s_i + \beta^{-1/2} \Sigma |e_j(x)|^2 + \beta^{-1/2} \Sigma |c_i(x) + s_i|^2 \qquad (22.49)$$

which permits a commencement from an infeasible point.

22.7.6 Search techniques

Box[39] has extended the simplex search method (see Section 22.5) to deal with constrained problems and constraints of the type $L_i \le c(x_i) \le U_i$, beginning with an initial feasible point and using pseudorandom numbers α_i to generate k further trial points:

$$\{x\} = \{L\} + \alpha_i \{L - U\} \qquad 0 < \alpha < l. \qquad (22.50)$$

These are shifted towards the centroid of the 'complex' until they are feasible.

Once a feasible simplex has been constructed the method follows the reflection procedure described in Section 22.5, except that $k > n + 1$ trial points are found necessary. In fact $k = 2n$ are desirable but impractical when $n > 10$. 'Over-reflection' is used, so that rejected points are replaced by reflecting them θ times their distance from the centroid. Box *et al.*[1] recommend using $\theta = 1.3$.

22.7.7 Augmented Lagrangian methods

These are an extension of the penalty techniques that aim to avoid the ill-conditioning that will inevitably occur if too long a β sequence is used. In these the Lagrangian function (eqn (22.3)) rather than the objective function is penalized. Considering only equality constraints the augmented Lagrangian function is written

$$\phi(x, \lambda, \beta) = f(x) - \Sigma \lambda_j e_j(x) + \tfrac{1}{2} \beta \Sigma |e_j(x)|^2 \qquad (22.51a)$$

and if $\beta > \bar{\beta}$ (a threshold value[40]) the optimum solution $\{x\} = \{x^*\}$ minimizes $\phi(x, \lambda, \beta)$ rather than finding a saddle point in $\phi(x, \lambda)$.

If the optimum Lagrange multiplier values were known, only a single unconstrained minimization would be required. In practice one uses a sequence of vectors $\{\lambda\}_n$, generally beginning with $\{\lambda\}_0 = \{0\}$ so that the first step is a simple penalty one,[41] which may be computed from[42]

$$\{\lambda\}_{n+1} = \{\lambda\}_n - \beta\{e_j(x)\}_n. \tag{22.51b}$$

This update formula yields a linear rate of convergence and this can be improved by increasing β by a factor of 10 if[43]

$$\{e_j(x)\}_n^t\{e_j(x)\}_n < \tfrac{1}{4}\{e_j(x)\}_{n-1}^t\{e_j(x)\}_{n-1}.$$

Generalization of such methods to deal with inequality constraints is described by Fletcher.[44] Clearly they offer an attractive alternative to penalty methods but, like those, require numerous unconstrained minimizations. Recently some attention has been focused upon limiting the accuracy of each but the last minimization step to reduce computation times.[45,46] The following section demonstrates such an approach with an exterior penalty method.

22.8 Predictor–corrector penalty method for constrained optimization

In preceding sections difficulties such as the meandering convergence of the steepest descent methods, formation of Hessian matrices, and large total numbers of minimizations in penalty techniques have been noted. In finite element applications, where the number of design variables may be large, such difficulties increase. If we retain a first-order gradient approach and incorporate the 'adjust the aim' philosophy of the predictor–corrector methods we can reduce some of these difficulties.[47] As an example consider the problem

$$\text{Min.:} \quad f(x_1, x_2) = (x_1 - 2)^2 + (x_2 - 1)^2 \tag{22.52}$$

subject to the constraints

$$1 - \tfrac{1}{4}x_1^2 - x_2 \leq 0 \tag{22.53}$$

$$x_1 - 2x_2 + 1 = 0. \tag{22.53b}$$

Using eqn (22.46) one obtains the SUMT problem of minimizing

$$F = f(x_1, x_2) + \beta|1 - \tfrac{1}{4}x_1^2 - x_2|^2 + \beta(x_1 - 2x_2 + 1)^2 \qquad \beta \to \infty \tag{22.54}$$

where $|\;\;|$ here denotes a Macaulay step function: $|x| = x, x \geq 0, |x|$

= 0, $x < 0$. Using the steepest descent method the unconstrained minimizations are given by the iteration

$$\{x\}_{i+1} = \{x\}_i - \alpha_i \left\{ \frac{\partial F}{\partial x_1}, \frac{\partial F}{\partial x_2} \right\} \tag{22.55}$$

with α_i determined by quadratic search (see Section 22.4).

Beginning with $\{x_0\} = \{2, 2\}, \beta = 1$ the results obtained are given in
Table 22.2. As Fig. 22.2 shows (for $\beta < 1000$) convergence is somewhat
erratic and the steps becomes orthogonal as the minimum point is approached.

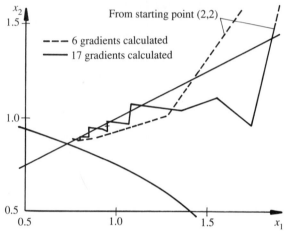

Fig. 22.2 Comparison of convergence of steepest descent (solid line) and modified
steepest descent (dashed line) methods

Table 22.2 N = number of gradients calculated

β	N	x_1	x_2	$f(x_1, x_2)$
Steepest descent				
1	4	1.360	1.040	0.410
10	7	0.932	0.937	1.417
100	6	0.765	0.877	1.567
1000	4	0.732	0.867	1.626
Predictor–corrector				
1	2	1.286	1.008	0.509
10	2	0.880	0.889	1.267
100	2	0.776	0.879	1.512
1000	4	0.732	0.867	1.626
Exact solution		0.732	0.867	1.626

In this problem the generalized Newton method gives similar search directions until β is large but involves much more computation, whilst Davidon's method requires many iterations to develop an accurate Hessian matrix. However, in SUMT problems the optimum will not be closely approached until a few values of β have been used so that first-order gradient methods are initially more appropriate.

Following the scheme suggested in connection with Newton's method in Section 12.3, therefore, one can use a predictor–corrector scheme[47] with the earlier values of β. Thus for the nth unconstrained minimization one uses as a predictor step

$$\{x^*\}_n = \{x\}_{n-1} - \tfrac{1}{2}\alpha\left[\frac{\beta_n}{\beta_{n-1}}\bar{g}_{n-1} + g_n\right] \tag{22.56}$$

and as a corrector step

$$\{x\}_n = \{x\}_{n-1} - \alpha\bar{g}_n = \{x\}_{n-1} - \tfrac{1}{2}\alpha[g_n + g_n^*] \tag{22.57}$$

where $g_n = g(\{x\}_{n-1}, \beta_n)$ is evaluated at the termination point of the last minimization (with β_{n-1}) and $\bar{g}_n$ averages this with $g_n^* = g(\{x^*\}_n, \beta_n)$ evaluated at the predictor point.

In the first step ($\beta = 1$ in Table 22.2) one uses $\bar{g}_0 = g_1 = g(\{x\}_0, \beta_1)$ and $\beta_0 = 1$ (actually, of course, there is no 'zeroth step') in eqn (22.56). In later steps use of the factor β_n/β_{n-1} assumes the use of substantial step sizes in β which quickly allow the penalty terms to dominate the gradient calculation. Then this factor ensures that the gradients to be averaged are of the same order of magnitude (cf. eqn (22.36) which involves a comparable process). Observing Table 22.2 and Fig. 2 much improved reslts are obtained (compared to steepest descent) and the method is also more economical than the conjugate gradient method in this problem.

22.9 Optimization with finite elements

The problem of optimization of finite element models adds a third type of non-linearity that can occur in minimization of the terms

$$\tfrac{1}{2}\int\{d\}^t B^t DB\{d\}\,\mathrm{d}V - \tfrac{1}{2}\int\{d\}^t B\{\sigma_0\}\,\mathrm{d}V.$$

The three non-linear problems resulting are:

(1) Non-linear constitutions

$$\{\sigma_0\} = DB\{d\}, \quad D = F(\{\varepsilon\}) \tag{22.58}$$

(2) Non-linear geometry

$$\{\sigma_0\} = DB\{d\}, \quad B = F(\{d\}) \tag{22.59}$$

(3) Non-linear synthesis

$$\{\sigma_0\} = \{\sigma_{\lim}\}, \quad \text{Volume} = F(\{t\}, \{x, y\}). \tag{22.60}$$

The third case can be written as a Newton process[48]

$$K_T\{\delta D\} = \{\Sigma \int B^t \{\sigma\} \, dV\} - \{\Sigma \int B^t \{\sigma_{\lim}\} \, dV\} \tag{22.61}$$

where $\{\sigma_{\lim}\}$ are the specified stress limits, giving the fully stressed design (FSD) problem,[7] some motivation for which was observed in Section 22.2.

Here equilibrium is not directly violated (that is, $\Sigma \int B^t \{\sigma\} \, dV = \{Q\}$) but the set stress limits are violated (either exceeded unsafely or there is a significant local understress and wastage of material). In the case of plate structures with fixed nodal geometry the analysis is iterated with updated element thicknesses using in each element

$$t_{\text{new}} = t\left(1 + \frac{(\sigma/\sigma_{\lim}) - 1}{\beta}\right), \quad \frac{1}{M} < \frac{t_{\text{new}}}{t} < M \qquad M > 1. \tag{22.62}$$

A convergence factor $\beta > 1$ and a move limit M and used and iteration continues until $\{\sigma\} \simeq \{\sigma_{\lim}\}$. This can be achieved without the use of the residual type calculation on the right-hand side of eqn (22.61).[49]

FSD does not guarantee a globally optimal result[7] but confidence in the result can be increased with the use of several analyses with different starting points.

Figure 22.3 shows the FSD result for the thickness variation in a clamped circular slab[49] compared with a result obtained by the SUMT method.[50] The agreement is good despite the fact that one analysis used triangular plate elements[49] and the other axisymmetric solid elements.[50]

FSD can be classified as an *optimality criterion* method. One satisfies the optimality criterion and finds a minimum, although in some cases this may only be a local one. Another example is the iterative design of shell shapes.[51]

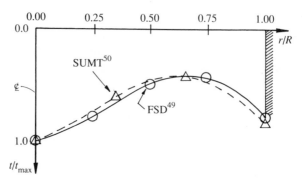

Fig. 22.3 Optimal thickness variation in a clamped homogeneous circular slab with blanket loading

If the shell is of uniform thickness the problem is simplified and one iterates the analysis seeking both to eliminate bending as nearly as possible and to minimize the integral of the membrane stresses over the surface. (cf. eqn (22.20)). The 'best' shape having been found iteratively, FSD can be used to distribute the shell material logically.[51]

To minimize bending in the shell the *general cable theorem*, $y = M_{eb}/T$ where M_{eb} is the bending couple in the 'equivalent' beam[52] and T is the tension in the cable, can be used as a basis. Thus a move vector of vertical nodal translations is based upon the nodal bending moments and this vector is augmented by a Newton type calculation based upon the membrane stresses. Iteration continues until a turning point in the shell weight (estimated with the use of a specified stress limit) is obtained.

Table 22.3 shows the result obtained using this optimality criterion approach for the 'optimal' nodal elevations in a simply supported wide arch modelled with sixteen flat triangular shell elements (for half the arch) and carrying horizontally or surface distributed loads. There is satisfactory agreement with the analytical optimum solution.[51]

Whilst the optimality criterion methods have less theoretical basis, they are very economical and mixed optimization methods have been devised which seek a compromise between this economy and the greater generality of such methods as the BFGS method.[53]

Other examples of application of optimization techniques to finite elements include:

(1) optimization of trusses using the simplex method of linear programming in a stepwise fashion;[54]

Table 22.3 Numerical results for the optimal shape of a simply supported arch of span L. An initial triangular shape was assumed with elevation $= H_0$ at $x = L/2$. The same problem is studied further in Section 23.8.

Case	Elevations			
	$x = L/8$	$x = L/4$	$x = 3L/8$	$x = L/2$
Horizontal load				
Theoretical	6.1	10.4	13.0	13.9
Numerical, $H_0 = 4$	5.3	9.0	11.1	13.6
Numerical, $H_0 = 8$	5.6	9.3	11.6	13.5
Surface load				
Theoretical	4.3	7.4	9.2	9.8
Numerical, $H_0 = 4$	4.7	7.6	9.2	10.2

(2) optimization of axisymmetric plates and shells using SUMT in tandem with Davidson's method[50] or Rosenbrock's conjugate gradient method;[55]

(3) optimization of axisymmetric solids using Box's 'complex' method;[56]

(4) optimization of frame structures using steepest descent techniques;[57]

(5) use of LP techniques to optimize the interpolation amplitudes in finite elements in which the number of amplitudes required to form a complete polynomial exceeds the desired number of degrees of freedom;[58]

(6) optimization of finite element meshes.[59, 60]

Clearly much further work remains to be done in combining optimization techniques with the finite element method and the following section gives a simple 'hands on' example of the application of penalty factors to the weight minimization of trusses. Extension of such processes to other finite element applications is possible and a number of programs are given in Chapter 23 to demonstrate this.

22.10 Penalty optimization of finite element models

As a basic example of penalty minimization of finite element models consider the three-bar truss shown in Fig. 22.4. Assuming the bars have equal area A and performing an elastic analysis one obtains the single equilibrium equation required as

$$\frac{EAv}{L} + 2\frac{EA}{2L}(\cos 60^{\circ})^{2}v = \frac{EAv}{L}(1 + \tfrac{1}{4}) = Q \qquad (22.63)$$

This is associated with the vertical displacement at node 1 which we denote as v. The displacement solution is $v = 0.8QL/EA$ and the member stresses are $Ev/L = 0.8Q/A$ and $Ev/4L = 0.2Q/A$. In FSD one resets $A_1 = 0.8A$ and $A_2 = 0.2A$ if $\sigma_{\mathrm{lim}} = Q/A$ and we shall denote these areas as $\{A_0\}$.

Repeating the analysis with these areas and continuing this process one obtains a fully stressed design. Whilst this is an improved design it is not the

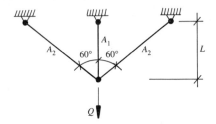

Fig. 22.4 Three-bar truss optimization problem

optimum result. The SUMT procedure introduced in Section 22.8, however, will yield the optimum solution if correctly applied; although the optimum solution is obvious our numerical procedure is unaware of this.

Beginning with $\{A_0\}$ as the element areas the augmented objective function is written as

$$F = A_1 L + 4A_2 L + \Sigma\beta \left|\frac{A_i - A_i^*}{A_i^*}\right|^2 + \beta|A^-|^2 \qquad (22.64)$$

where $A_i^* = A_i\sigma_i/\sigma_{\lim}$ is the area required to maintain $\sigma = \sigma_{\lim} = Q/A$. The first penalty term provides the 'overstress constraint'; again the squared terms are step functions which operate only when the constraint is violated.

In the second penalty term in eqn (22.64) A^- denotes a negative bar area resulting from the latest search step and clearly such results must be constrained out. The addition of other constraints involves little further complication.

The steepest descent gradient vector is given by

$$\{g_i\} \doteq \left\{\frac{\partial F}{\partial A_i}\right\} = \{L_i + 2\beta(A_i/A_i^* - 1)/A_i^* + 2\beta A_i^-\} \qquad i = 1 \to 3. \quad (22.65)$$

Using eqns (22.64) and (22.65), with $E = A = L = Q = 1$ for simplicity, the procedure is as follows:

(1) $\beta = 10$.

    (a) Using $\{A_0\} = \{0.8, 0.2, 0.2\}$ one calculates

$$F = 2.6241, \qquad \{g_1\} = \{-4.4402, 2, 2\}$$

    and using quadratic search (eqn (22.31)) with this gradient one obtains $\alpha^* = 0.0962$ as the optimum step length, giving

$$\{A\} = \{1.2271, 0.0076, 0.0076\}, \qquad F = 1.2575.$$

    (b) Using this last result for $\{A\}$ the new gradient vector is $\{g_2\}$ $= \{1, 2, 2\}$. Averaging this with $\{g_1\}$ gives

$$\{\bar{g}\} = \{-1.7201, 2, 2\}$$

    and with quadratic search in this direction (again beginning from $\{A_0\}$) one obtains $\alpha^* = 0.1167$, yielding

$$\{A\} = \{1.0007, -0.0334, -0.0334\}, \qquad F = 1.0119$$

(2) $\beta = 100$.

Using a predictor with $\{g_1\}$ calculated for the last areas obtained above and a corrector with $\{\bar{g}\} = (\{g_1\} + \{g_2\})/2$ where $\{g_2\}$ is calculated from the last predictor result we obtain

$$\{A\} = \{1.0027, -0.0014, -0.0014\}, \qquad F = 1.0029.$$

Clearly the correct solution is $\{A\} = \{1, 0, 0\}$ and $F = 1$ so that increasing the penalty factor by a factor of ten has improved the accuracy of the solution by one decimal place, as might be expected.

22.11 Concluding remarks

This chapter, besides dealing with optimization of finite element models, has other useful links with earlier chapters. Examples are the quadratic search technique (which is based on quadratic interpolation), the method of steepest descent (which provides the search direction in stress space for non-linear material problems and, of course, penalty factors and Lagrange multipliers. In the next chapter a relationship between penalty factors and Lagrange multipliers is developed and several programs using steepest descent are given.

References

1. Box, M. J. Davies, D., and Swann, W. H. (1969). *Nonlinear Optimization Techniques*. ICI Monograph No. 5, Oliver and Boyd. Edinburgh.
2. Connor, J. J. and Brebbia, C. A. (1977). *Finite Element Techniques for Fluid Flow*. Butterworths, London.
3. Michell, A. G. M. (1904). The limits of economy in frame structures. *Phil. Mag.*, Series 6, **8**. 589.
4. Whittle, P. (1971). *Optimization Under Constraints*. Wiley, London.
5. Calladine, C. R. (1973). *Engineering Plasticity*. Arnold, London.
6. Rozvany, G. I. N. (1976). *Optimal Design of Flexural Systems*. Pergamon, Oxford.
7. Reinschmidt, K. F., Cornell, C. A., and Brotchie, J. F. (1966). Iterative design and structural optimization. *J. Struct. Div. ASCE*, **92** (ST6), 281.
8. Lowe, P. G. and Melchers, R. E. (1972). On the theory of optimal constant thickness fibre-reinforced plates. *Int. J. Mech. Sci.* **14**, 311.
9. Wang, C. (1973). *Computer Methods in Advanced Structural Analysis*. Intext, New York.
10. Mohr, G. A. (1980). Dual pivot selection rule for the Simplex method. Civil Engineering Research Report, CIT, Melbourne.
11. Hadley, G. (1962). *Linear Programming*. Addison-Wesley, New Jersey.
12. Majid, K. I. (1972). *Nonlinear Structures*. Butterworths, London.
13. Argyris, J. H. (1965). Continua and discontinua. *Proceedings of the First Conference on Matrix Methods in Structural Mechanics*, Wright-Patterson AFB, Ohio.
14. Livesley, R. K. (1975). *Matrix Analysis of Structures*, (2nd edn). Pergamon, Oxford.
15. Powell, M. J. D. (1964). An efficient method of finding the minimum of a function of several variables without calculating derivatives. *Comput. J.*, **7**, 155.
16. Hooke, R. and Jeeves, T. A. (1961). Direct search solution of numerical and statistical problems. *J. Applied Computer Methods*, **8**, 212.

17. Kowalik, J. and Osborne, M. R. (1971). *Methods for Unconstrained Optimization Problems.* Elsevier, New York.
18. Rosenbrock, H. H. (1960). An automatic method for finding the greatest or least value for a function. *Comput. J.*, **3**, 175.
19. Smith, C. S. (1966). The automatic computation of maximum likelihood estimates. N.C.B. Scientific Dept. Report S.C. 846/MR/40.
20. Birkhoff, G. and Maclane, S. (1968). *A Survey of Modern Algebra.* Macmillan, New York.
21. Hestenes, M. R. and Steifel, E. L. (1952). Methods of conjugate gradients for solving linear systems. *J. Res. Nat. Bur. Standards*, Sec. B, **49**, 409.
22. Fletcher, R. and Reeves, C. M. (1964). Function minimization by conjugate gradients. *Comput. J.*, **7**, 149.
23. Khosla, P. K. and Rubin, S. G. (1981). A conjugate gradient iterative method. *Comput. Fluids*, **9**, 109.
24. Cauchy, A. L. (1847). Méthode générale pour la résolution des systemes d'équations simultanées. *Comput. Rend. Acad. Sci. Paris*, **25**, 536.
25. Booth, A. D. (1957). *Numerical Methods.* Butterworths, London.
26. Irons, B. M. and Ahmad, S. (1980). *Techniques of Finite Elements.* Ellis-Horwood, Chichester.
27. Bertsekas, D. P. (1982). *Constrained Optimization and Lagrange Multiplier Methods.* Academic Press, New York.
28. Fletcher, R. and Powell, M. J. D. (1963). A rapidly convergent descent algorithm for minimization. *Comput. J.*, **6**, 163.
29. Davidon, W. C. (1959). Variable metric method for minimization. R & D Report ANL-599, US Atomic Energy Commision. Argonne, Illinois.
30. Broyden, C. G. (1970). The convergence of a class of double rank minimization algorithm. *J. Inst. Math. Appl.*, **6**, 76.
31. Fletcher, R. (1970). A class of methods for nonlinear programming with termination and convergence properties. In *Integer and Nonlinear Programming*, (ed. J. Abadie). North-Holland, Amsterdam.
32. Goldfarb, D. (1970). A family of variable-metric methods derived by variational means. *Math. Comp.*, **24**, 23.
33. Shanno, D. F. (1970). Conditioning of quasi-Newton methods for function minimization. *Math. Comp.*, **24**, 647.
34. Roberts, S. M. and Lyvers, H. I. (1961). The gradient method in process control. *Ind. Eng. Chem.*, **53**, 877.
35. Goldfarb, D. and Lapidus, L. (1968). Conjugate gradient method for nonlinear programming problems with linear constraints. *I. & E. C. Fundamentals*, **7**, 142.
36. Leitmann, G. (1962). *Optimization Techniques with Applications to Aerospace Systems.* Academic Press, New York.
37. Carroll, C. W. (1961). The created response surface technique for optimizing nonlinear restrained systems. *Op. Res.* **9**, 169.
38. Fiacco, A. V. and McCormick, G. P. (1964). The sequential unconstrained minimization technique for nonlinear programming, a primal-dual method. *Management Sci.*, **10**, 360.
39. Box, M. J. (1965). A new method of unconstrained optimization and a comparison with other methods. *Comput. J.*, **8**, 42.

40. Gill, P. E., Murray, W., and Wright, M. H. (1982). *Practical Optimization.* Academic Press London.
41. Byrne, S. J. and Coope, I. D. (1983). Current techniques for nonlinear programing. *New Zealand Opns Res.*, **11**(1), 9.
42. Hestenes, M. R. (1969). Multiplier and gradient methods. *J. Opt. Th. & Appl.*, **4**, 303.
43. Powell, M. J. D. (1969). A method for nonlinear constraints in minimization. problems. In *Optimization*, (ed. R. Fletcher). Academic Press, London.
44. Fletcher, R. (1981). *Practical Methods of Optimization*, Vol. 2. Wiley, London.
45. Bertsekas, D. P. (1976). Multiplier methods; a survey. *Automatica*, **12**, 133.
46. Coope, I. D. and Fletcher, R. (1980). Some numerical experience with a globally convergent algorithm for nonlinearly constrained optimization. *J. Opt. Th. & Appl.*, **32**(1), 1.
47. Mohr, G. A. (1981). A predictor-corrector method for constrained nonlinear optimization. Civil Engineering Research Report, CIT. Melbourne.
48. Rosow, M. P. and Taylor, J. E. (1973). A finite element method for the optimal design of variable thickness sheets. *J. Amer. Inst. Aeron. Astron.*, **11**, 11.
49. Mohr, G. A. (1979). Elastic and plastic predictions of slab reinforcement requirements. *Trans I.E. Aust.*, **CE21**(1), 16.
50. Francavilla, A. and Ramakrishnan, C. V. (1975). Structural shape optimisation using penalty factors. *J. Struct. Mech.*, **3**, 403.
51. Mohr, G. A. (1979). Design of shell shape usng finite elements. *Comput. Struct.*, **10**, 745.
52. Norris, C. H., Wilbur, J. B., and Utku, S. (1975). *Elementary Structural Analysis*, (3rd edn). Prentice-Hall, Englewood Cliffs, NJ.
53. Sander, P. and Fleury, C. (1978). A mixed method of structural optimization. *Int. J. Num. Meth. Engng*, **13**, 385.
54. Niordson, N. I. and Pedersen, P. (1972). A review of optimal structural design. *Proceedings of the Thirteenth International Congress on Theoretical and Applied Mechanics*, Moscow.
55. Thevendran, V. (1982). Minimum weight design of a spherical shell under a concentrated load at the apex. *Int. J. Num. Meth. Engng*, **18**, 1091.
56. Tomas, J. A. (1979). Optimization of automobile components by means of the finite element method. *Proceedings of the Third Australian International Conference on Finite Element Methods*, University of NSW, Sydney.
57. Arora, J. S. and Haug, E. J. (1976). Efficient optimal design of structures by generalized steepest descent programming. *Int. J. Num. Meth. Engng*, **10**, 747.
58. Bergan, P. G. and Hanssen, L. (1976). A new approach for deriving good element stiffness matrices. In *The Mathematics of Finite Elements and Applications*, Vol. 2, (ed. J. R. Whiteman). Academic Press, London.
59. Pedersen, P. (1973). Some properties of linear strain triangles and optimal finite element models. *Int. J. Num. Meth. Engng*, **7**, 415.
60. Rachowicz, W., Oden J. T., and Demkowicz, L. (1989). Toward a universal h-p adaptive finite element strategy. Part 3: design of h-p meshes. *Comp. Meth. Appl. Mech. Engng*, **77**, 181.

23
Generalization of constraint techniques and optimization programs

We have seen the application of constraints to finite element formulations at many points in preceding chapters. Of particular note are the penalty factor, Lagrange multiplier, and basis transformation techniques introduced in Chapter 4 and frequently used subsequently. The present chapter begins by examining the relationships between these methods, giving particular attention to conversion of Lagrange multiplier formulations to obtain equivalent penalty formulations. Application of the latter to integral constraints is then discussed in relation to both variational and weighted residual formulations. Finally, the chapter closes with several programs for the optimization of finite element models of problems in both solids and fluids.

23.1 Introduction

Throughout the text we have seen much application of three important extensions of the classical finite element method.

23.1.1 Penalty factor techniques

The variational penalty factor finite element problem is to minimize the integral[1,2]

$$I(d) = \tfrac{1}{2} \int \{\varepsilon\}^{\mathrm{t}} \{\sigma\} \, \mathrm{d}V - \{d\}^{\mathrm{t}} \{q\} + \tfrac{1}{2} \beta \int [g\{d\}]^2 \, \mathrm{d}V \qquad (23.1)$$

leading to the element equations

$$k\{d\} - \{q\} + \beta (\int g^{\mathrm{t}} g \, \mathrm{d}V)\{d\} = \{0\} \qquad (23.2)$$

if $\{q\}$ is again understood to include the *interelement reactions*.

As we saw in Section 19.5, however, penalty factors can also be used to include constraints in weighted residual formulations. For a governing equation of the form $F(\phi) = 0$, subject to constraints $g(\phi) = 0$, we can write

$$\int w_i F(\phi) \, \mathrm{d}V + \beta \int w_i g(\phi) \, \mathrm{d}V = 0. \qquad (23.3)$$

Then, as we have seen in the case of unconstrained formulations, with

Galerkin weighting and judicious integration by parts eqn (23.3) will give an identical result to that of eqn (23.2) so long as $F(\phi)$ is self-adjoint.

23.1.2 Lagrange multiplier techniques

The variational Lagrange multiplier finite element problem is to minimize the expression[1,2]

$$I(d, \lambda) = \tfrac{1}{2}\int \{\varepsilon\}^{t}\{\sigma\}\,dV - \{d\}^{t}\{q\} + \{\lambda\}^{t}g\{d\} \tag{23.4}$$

where $g\{d\}$ are now discrete constraints, leading to

$$\begin{bmatrix} k & g^{t} \\ g & 0 \end{bmatrix} \begin{Bmatrix} \{d\} \\ \{\lambda\} \end{Bmatrix} = \begin{Bmatrix} \{q\} \\ \{0\} \end{Bmatrix}. \tag{23.5}$$

As we saw in Section 19.5 Lagrange multiplier formulations can also be obtained by the Galerkin method. In general we write the weighted residual Lagrange multiplier approach as

$$\int w_{i}F(\phi)\,dV + \lambda_{i}\int w_{i}g(\phi)\,dV = 0 \tag{23.6}$$

$$\int w_{i}g(\phi)\,dV = 0 \tag{23.7}$$

and eqn (23.7) must be added or the resulting matrix equations will be underdetermined.

23.1.3 Basis transformation techniques

In these we redefine the element freedoms using $\{d^{*}\} = T\{d\}$ and obtain a global stiffness matrix as[3]

$$k = T^{t}k^{*}T \tag{23.8}$$

where k is derived for the constrained local freedoms $\{d^{*}\}$. Here k^{*} may equally well have been derived using variational or weighted residual methods.

23.2 Problems with discrete constraints

In solution of the (assembled) Lagrange multiplier problem of eqn (23.5) pivoting is generally used[4] but as in eqn (19.72) this can be avoided by adding small numbers to the diagonal of the element matrix. Including forcing terms $\{q_{c}\}$ for the constraints:

$$\begin{bmatrix} k & g^{t} \\ g & -I/\beta \end{bmatrix} \begin{Bmatrix} \{d\} \\ \{\lambda\} \end{Bmatrix} = \begin{Bmatrix} \{q\} \\ \{q_{c}\} \end{Bmatrix} \tag{23.9}$$

where I is an identity matrix and β is a large number. Here I/β acts as a set of small slack variables and the numerical success of this procedure has been confirmed in the Stokes flow problem.[5,6]

To obtain an equivalent penalty formulation one uses static condensation (see eqn (6.42), that is, expanding eqn (23.9) one obtains

$$k\{d\} + g^t\{\lambda\} = \{q\} \tag{23.10}$$

$$g\{d\} - \frac{I}{\beta}\{\lambda\} = \{q_c\}. \tag{23.11}$$

Solving eqn (23.11) for $\{\lambda\}$ and substituting the result into eqn (23.10) leads to

$$[k + \beta g^t g]\{d\} = \{q\} - \beta g^t\{q_c\} \tag{23.12}$$

and this is the equivalent penalty factor formulation (cf. eqn (23.2)) but applies only to discrete constraints not requiring integration.

Taking the two-element spar problem shown in Fig. 23.1 as an example and adding the matrix $-I/\beta$ required by eqn (23.9) to the Lagrange multiplier formulation obtained in Section 4.3 gives

$$\begin{bmatrix} c & -c & 0 & 1 & 0 \\ -c & 2c & -c & 0 & 1 \\ 0 & -c & c & 0 & -1 \\ 1 & 0 & 0 & -1/\beta & 0 \\ 0 & 1 & -1 & 0 & -1/\beta \end{bmatrix} \begin{Bmatrix} u_1 \\ u_2 \\ u_3 \\ \lambda_1 \\ \lambda_2 \end{Bmatrix} = \begin{Bmatrix} 0 \\ 0 \\ P \\ 0 \\ 0 \end{Bmatrix} \tag{23.13}$$

where $c = EA/L$. Applying eqn (23.12) to this yields

$$k_p = \begin{bmatrix} c & -c & 0 \\ -c & 2c & -c \\ 0 & -c & c \end{bmatrix} + \beta \begin{bmatrix} 1 & 0 & 0 \\ 0 & 1 & -1 \end{bmatrix} \begin{bmatrix} 1 & 0 \\ 0 & 1 \\ 0 & -1 \end{bmatrix}$$

or

$$k_p = \begin{bmatrix} c + \beta & -c & 0 \\ -c & 2c + \beta & -c - \beta \\ 0 & -c - \beta & c + \beta \end{bmatrix} \tag{23.14}$$

which, because $\beta \simeq \beta c$ is a large number, is numerically equivalent to the penalty formulation of eqn (4.5).

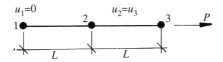

Fig. 23.1 Two-element spar problem with two discrete constraints

To deal with integrated constraints, on the other hand, the interpolation functions must be applied to the Lagrange multipliers and this situation is dealt with in the following section.

We can write a typical basis transformation in the form

$$\{d^*\} = \begin{bmatrix} I & 0 \\ g_1 & g_2 \end{bmatrix} \{d\} \tag{23.15}$$

where I operates upon the unconstrained freedoms. Partitioning k^* in a manner corresponding to eqn (23.15) and applying eqn (23.8) one obtains the global stiffness matrix as

$$k = \begin{bmatrix} I & g_1^t \\ 0 & g_2^t \end{bmatrix} \begin{bmatrix} k & k_a \\ k_a^t & k_{aa} \end{bmatrix} \begin{bmatrix} I & 0 \\ g_1 & g_2 \end{bmatrix}$$

$$= \begin{bmatrix} k + k_a g_1 + g_1^t k_a^t + g_1^t k_{aa} g_1 & k_a g_2 + g_1^t k_{aa} g_2 \\ g_2^t k_a^t + g_2^t k_{aa} g_1 & g_2^t k_a g_2 + g_2^t k_{aa} g_2 \end{bmatrix} \tag{23.16}$$

where the subscript a relates to the altered or constrained freedoms.

As β does not appear in eqn (23.16), however, we see that basis transformation is equivalent to condensing eqn (23.9) without the $-I/\beta$ entry, though this observation applies only in the case of discrete constraints.

Whilst all three methods have their attractions in different situations the relationships between them are of interest and the ease with which Lagrange multiplier formulations can be converted into more economical penalty factor or basis transformation results is of particular interest. This situation is discussed further in the following two sections.

23.3 Weighted residual formulations with integral constraints

The classical example of the appearance of integral Lagrange multiplier constraints in a finite element formulation is the Stokes flow problem of Section 19.5. If we include the small slack variables introduced in eqn (23.9) into the Galerkin weighted residual formulation of eqn (19.64) this becomes

$$\begin{bmatrix} C & 0 & -A \\ 0 & C & -B \\ -A^t & -B^t & -F \end{bmatrix} \begin{Bmatrix} \{u\} \\ \{v\} \\ \{p\} \end{Bmatrix} = \iint \begin{Bmatrix} \{f\}X \\ \{f\}Y \\ \{0\} \end{Bmatrix} dxdy$$

$$+ \int \begin{Bmatrix} \{f\}(\mu(\partial u/\partial n) - c_x p) \\ \{f\}(\mu(\partial v/\partial n) - c_y p) \\ \{0\} \end{Bmatrix} dS \tag{23.17}$$

where

$$C = \iint \mu [\{f_x\}\{f_x\}^t + \{f_y\}\{f_y\}^t]\,dxdy$$

$$A = \iint \{f_x\}\{f\}^t\,dxdy, \qquad B = \iint \{\zeta_y\}\{f\}^t\,dxdy$$

$$F = \frac{1}{\beta} \iint \{f\}\{f\}^t\,dxdy$$

$$\frac{\partial u}{\partial n} = = \frac{\partial u}{\partial x}\frac{\partial x}{\partial n} + \frac{\partial u}{\partial y}\frac{\partial y}{\partial n} = c_x\frac{\partial u}{\partial x} + c_y\frac{\partial u}{\partial y}$$

$$\{f_x\} = \left\{\frac{\partial f}{\partial x}\right\}, \qquad \{f_y\} = \left\{\frac{\partial f}{\partial y}\right\}$$

Note, however, that interpolation is now associated with the small slack variables. This occurs because the continuity constraint must strictly be written as

$$\frac{\partial u}{\partial x} + \frac{\partial v}{\partial y} - \frac{\{f\}^t\{\lambda\}}{\beta} = 0 \qquad (23.18)$$

applying interpolation to the Lagrange multipliers as the continuity constraint is to be satisfied in an integral sense. Thus at any integration point $\{f\}^t\{\lambda\}$ will yield the appropriate value of λ to be associated with the continuity constraint. Then with the application of Galerkin weighting and integration over the element one obtains the additional entry F in eqn (23.17).

As β is a large number, an entry of the form of F in eqn (23.17) is computationally equivalent to the use of an entry $-I/\beta$, the latter situation having been tested numerically,[5,6] But as we shall see the entry F in eqn (23.17) is necessary if we are to condense eqn (23.17) to obtain a completely equivalent penalty formulation.

Proceeding as in Section 19.5 the equivalent (as we shall prove below) penalty formulation is eqn (19.69):

$$\begin{bmatrix} C + \beta D_{11} & \beta D_{12} \\ \beta D_{12}^t & C + \beta D_{22} \end{bmatrix} \begin{Bmatrix} \{u\} \\ \{v\} \end{Bmatrix} = \iint \begin{Bmatrix} \{f\}X \\ \{f\}Y \end{Bmatrix} dxdy + \int \begin{Bmatrix} F_1 \\ F_2 \end{Bmatrix} dS \qquad (23.19)$$

where

$$D_{11} = \iint \{f_x\}\{f_x\}^t\,dxdy, \qquad D_{12} = \iint \{f_x\}\{f_y\}^t\,dxdy$$

$$D_{22} = \iint \{f_y\}\{f_y\}^t\,dxdy$$

$$F_1 = \{f\}\left[\mu\frac{\partial u}{\partial n} + \beta\frac{\partial u}{\partial x}c_x + \beta\frac{\partial v}{\partial y}c_x\right]$$

$$F_2 = \{f\}\left[\mu\frac{\partial v}{\partial n} + \beta\frac{\partial u}{\partial x}c_y + \beta\frac{\partial v}{\partial y}c_y\right]$$

and matrix C is as defined in association with eqn (23.17).

Observe that to obtain this result the pressure entries in the momentum equations were replaced using $p = - \beta(\partial u/\partial x + \partial v/\partial y)$. Noting that the pressures act as Lagrange multipliers in eqn (23.17) and comparing eqns (23.1) and (23.4) this step seems obvious, that is, eqn (23.1) is converted to eqn (23.4) by substituting $\{\lambda\}^t$ for $\beta g\{d\}$. (The factor of 1/2 is ignored because it can be absorbed into the large penalty factor without significantly affecting the numerical results).

This is an exceptional situation as from the form of eqn (23.17) we are able to deduce a physical significance for the Lagrange multipliers. Generally to apply penalty or Lagrange multiplier constraints in weighted residual formulations one begins from eqn (23.3) or eqns (23.6) and (23.7) and a physical interpretation of the Lagrange multipliers is not needed explicitly.

Now to show how to convert Lagrange multiplier formulations with integral constraints into penalty formulations we replace the entry I/β in eqn (23.9) by F where F is as defined in association with eqn (23.17). Then with condensation, assuming $\{q_c\} = \{0\}$, eqn (23.12) is now replaced by

$$[k + \beta(\int g^t F^{-1} g \, dV)]\{d\} = \{q\}. \tag{23.20}$$

Applying this result to eqn (23.17) the condensed left-hand side matrix becomes

$$\begin{bmatrix} C & 0 \\ 0 & C \end{bmatrix} + \beta \begin{bmatrix} A \\ B \end{bmatrix} F^{-1} [A^t \quad B^t]. \tag{23.21}$$

This yields an identical result to that in eqn (23.19) if we note that, for example, the product $\beta A F^{-1} A^t$ expands to

$$\beta \iint \{f_x\}\{f\}^t [\{f\}\{f\}^t]^{-1} \{f\}\{f_x\}^t \, dxdy = (\beta N) \iint \{f_x\}\{f_x\}^t \, dxdy \tag{23.22}$$

in view of the identity

$$\{f\}^t [\{f\}\{f\}^t]^{-1} \{f\} = N \tag{23.23}$$

for any vector of order N. The scalar constant N can thus be absorbed into the arbitrary penalty factor so that eqn (23.22) does indeed correspond to the entry D_{11} in eqn (23.19), as required. The other products in eqn (23.21) will also correspond to the entries βD_{ij} in eqn (23.19). Thus we see that the formulations of eqn (23.17) and (23.19) are completely equivalent and, more importantly, that Lagrange multiplier formulations with integral constraints are easily condensed into equivalent penalty factor formulations.

The Stokes flow formulation of eqn (19.72), in which the continuity equation was integrated by parts, can also be condensed to obtain a penalty result. In this case the outermost vectors in the product of eqn (23.23) are replaced by $\{f_x\}$ or $\{f_y\}$ but again the result is a scalar constant which is absorbed into the penalty factor and the condensed result is obtained without difficulty.

As might now be expected such condensation for the case of integral constraints can also be achieved in the context of variational formulations and this is demonstrated in the following section.

23.4 Variational formulations with integral constraints

In the foregoing section it was demonstrated that a weighted residual Lagrange multiplier formulation can be converted into an equivalent penalty formulation and such conversion is generally possible. It is also of interest to re-examine the situation with variational problems in the light of this result. To this end eqn (23.4) must be extended to

$$I(d, \lambda) = \tfrac{1}{2} \{d\}^t (\int B^t DB \, dV) \{d\} - \{d\}^t \{q\}$$

$$- \{\lambda\}^t \int \{f\} \left[g\{d\} + \frac{\{f\}^t \{\lambda\}}{2\beta} \right] dV. \qquad (23.24)$$

Here we see two extensions to the original statement

1. As the constraint is an integral one, interpolation must be applied to the (nodal) Lagrange multipliers, $\lambda = \{f\}^t \{\lambda\}$.
2. An error term is included by factoring the Lagrange multipliers and using these as small slack variables. These have arbitrary magnitude (as small as machine computation will allow), permitting the re-use of the vector $\{\lambda\}$ to form them

Minimizing eqn (23.24) with respect to $\{d\}$ and $\{\lambda\}$ and combining the results one obtains

$$\int \begin{bmatrix} B^t DB & g^t \{f\}^t \\ \{f\}g & -\{f\}^t\{f\}/\beta \end{bmatrix} dV \begin{Bmatrix} \{d\} \\ \{\lambda\} \end{Bmatrix} = \begin{Bmatrix} \{q\} \\ \{0\} \end{Bmatrix} \qquad (23.25)$$

which corresponds to eqn (23.17). With the application of condensation and employing the result of eqn (23.23) the equivalent penalty factor result, eqn (23.2), will be obtained.

Thus the introduction of small slack variables in eqn (23.24) leads to an approximate variational statement of the Lagrange multiplier problem which is equivalent to the variational penalty factor statement. Here we have arbitrarily small constraint violations. As we shall see in the following section generalization to deal with constraint violations of significant magnitude leads to a useful optimization procedure.

23.5 Optimization using a condensed Lagrange multiplier formulation

We have seen that it is possible to convert both variational and weighted residual Lagrange multiplier finite element formulations into equivalent penalty factor formulations, both in the case of discrete constraints and in that of integral constraints. To this point the constraints have been equality constraints in which negligible errors are allowed by the use of small slack variables. Now we shall examine the case of inequality constraints.

First it is useful to re-examine a familiar process, that of equilibrium iteration, using Lagrange multipliers. To this end we state the initial stiffness or modified Newton method as the problem of minimizing the integral

$$I(D, \lambda) = \tfrac{1}{2}\{D\}^t K_0\{\delta D\} - \{D\}^t\{Q\} + \{\lambda\}^t[\{Q\} - \{R\}] \quad (23.26)$$

where

$$K_0 = \sum \int B_0^t D B_0 \, dV \quad (23.27)$$

$$\{R\} = \sum \int B^* D \bar{B} \, dV \{d\} \quad (23.28)$$

are respectively the initial stiffness matrix for the system and the summed *element reactions*. In the latter B^* and $\bar{B}$ include non-linear terms in the fashion described in Chapter 12.

Minimizing eqn (23.26) with respect to $\{D\}$ gives the equilibrium equations

$$K_0\{D\} - \{Q\} - [\sum \int B^* D \bar{B} \, dV]\{\lambda\} = \{0\} \quad (23.29)$$

that is, the Lagrange multipliers are the displacement increments or corrections which provide equilibrium.

Minimizing eqn (23.26) with respect to $\{\lambda\}$ one obtains

$$\{Q\} - \{R\} = \{0\} \quad (23.30)$$

This side condition reminds us that the correct solution is obtained only when the applied loads balance the summed element reactions.

Because B^* and $\bar{B}$ themselves involve the displacements the problem must be solved iteratively. Thus, except for providing a useful formal statement of the equilibrium iteration process, the Lagrange multiplier approach leads to no new conclusions here but it does provide a simple exercise worth comparing with that to be undertaken next.

Let us now consider the problem of finite elements in which stress limits denoted by $\{\sigma_{\text{lim}}\}$ must be satisfied throughout each element (in an integral sense). To deal with this situation we generalize eqn (23.24) to

$$I(d, \lambda) = \tfrac{1}{2}\{d\}^t k\{d\} - \{d\}^t\{q\} + \{\lambda\}^t \int [D(B\{d\} + \{s\}) - \{\sigma_{\text{lim}}\}] \, dV \quad (23.31)$$

where $k = \int B^t DB \, dV$, $\{s\}$ are slack variables, and we return for convenience to a statement at element level, so that $\{q\}$ is again understood to include the interelement reactions.

If the problem is one of plane stress, for example, V is a function of the element thickness, $dV = t \, dA$. Then the problem is the continuum equivalent of that described in Section 22.2 (except that tension and compression are not considered separately) and the Lagrange multipliers, like the slack variables, play the role of strains.

Minimizing eqn (23.31) with respect to $\{s\}$ gives $\{\lambda\} = \{0\}$. This corresponds to the trivial case when there are no constraint violations. Replacing dV by $t \, dA$ and minimizing with respect to the thickness (a scalar) gives

$$\{\lambda\}^t \int [D(B\{d\} + \{s\}) - \{\sigma_{\text{lim}}\}] \, dA = 0. \tag{23.32}$$

This result makes it clear (which eqn (22.13) does not do well) that when an entry in $\{s\}$ is non-zero then the corresponding entry in $\{\lambda\}$ must be zero. We can ensure this by writing

$$\{\lambda\} + \{s\} = \{0\}. \tag{23.33}$$

Physically we interpret this condition as stating that $\{\lambda\}$ are the additional strain corrections (in addition to the $\{s\}$) required to ensure equilibrium in the non-linear problem.

Again minimizing eqn (23.31), now with respect to $\{d\}$, gives

$$\left(\int B^t DB \, dV\right)\{d\} + \left(\int B^t D^t \, dV\right)\{\lambda\} = \{q\} \tag{23.34}$$

whilst minimizing with respect to $\{\lambda\}$ gives

$$\int D(B\{d\} + \{s\}) \, dV = \int \{\sigma_{\text{lim}}\} \, dV. \tag{23.35}$$

Replacing $\{s\}$ in the latter by $-I\{\lambda\}$ (which follows from eqn (23.33)) and combining equations (23.34) and (23.35) yields

$$\int \begin{bmatrix} B^t DB & B^t D^t \\ DB & -DI \end{bmatrix} dV \begin{Bmatrix} \{d\} \\ \{\lambda\} \end{Bmatrix} = \begin{Bmatrix} \{q\} \\ \int\{\sigma_{\text{lim}}\} \, dV \end{Bmatrix}. \tag{23.36}$$

Condensing this result (by solving the second row for $\{\lambda\}$ and substituting the result into the first row) one obtains

$$\left(\int B^t DB \, dV\right)\{d\} = \{q\} - \int B^t (DB\{d\} - \{\sigma_{\text{lim}}\}) \, dV. \tag{23.37}$$

As previously seen in the case of integral constraints the integral acts over the whole of the matrix to be condensed and thus the constraint product obtained with condensation is evaluated under a single integral.

Now eqn (23.37) is usefully compared to the equilibrium iteration result of eqn (23.29), showing that the last term on the right-hand side of eqn (23.37) plays the role of element reactions.

We have now, however, the fully stressed design problem. The FSD problem is usually solved by iterating the solution of $K\{D\} = \{Q\}$ with element thicknesses adjusted so that, in truss problems, for example, $A_{new} = A_{old}(\sigma/\sigma_{lim})$ at each iteration. These adjusted areas are used to recalculate the stiffness matrices. Note that in practice a convergence factor is often required, see eqn (22.62), for example. In the present case, on the other hand, the element volume adjustments are included only on the right-hand side of eqn (23.37).

As noted in Section 22.9 FSD does not guarantee an optimal result. As we saw in Section 22.2 it is a necessary condition in problems with stress constraints but it is not a sufficient condition. This is because eqn (23.37) does not consider both tensile and compressive stress limits or vanishing members. Generalizing it to include tensile and compressive stress limits one writes

$$(\int B^t DB \, dV)\{d\} = \{q\} - \int B^t (DB\{d\} - \{\sigma_t\}) \, dV - \int B^t (DB\{d\} + \{\sigma_c\}) \, dV$$

$$(23.38)$$

where $\{\sigma_t\}$ and $\{\sigma_c\}$ are the magnitudes of the tensile and compressive stress limits respectively.

Thus after a few cycles of iteration an initially tensile element approaching zero thickness (but remaining tensile) should perhaps 'snap through' into compression to obtain the optimum solution, that is, it may impart a 'prestress' to the rest of the system, forming a useful and not intuitively obvious solution. In such cases one can at this point assume that we require $\{\sigma\} = -\{\sigma_c\}$ and use the last term in eqn (23.38) to calculate its 'reactions'. If, after further iteration, the element remains in tension this strategy is abandoned and one reverts to the tension hypothesis for this element.

If after several iterations an element remains substantially understressed it is assigned zero cross-section, so that it vanishes, and iteration continues in the modified system. An example of this situation is given in the following section.

Thus, so long as forcing terms corresponding to the constraints are available, alternative constraints can be investigated by appropriate programming. Indeed this situation closely corresponds to the situation in Section 22.10 where alternative penalty constraints are written as step functions.

When displacement constraints $g\{d\} = \{0\}$ are also present, of course, 'full stress' is no longer a necessary condition for optimality. In this case the constraint terms will again be of the form $\beta \int gg^t \, dV$ and will appear on the left-hand side. The right-hand side terms for the stress constraints must be written as step functions, that is, understress is now permissible and may be required to meet the displacement constraints.

23.6 Application of the condensed Lagrange multiplier method to optimization of finite element models

Returning to the three-bar truss problem studied in Section 22.10 the elastic solution is again given by the single equilibrium equation

$$\frac{EA_1 v}{L} + 2\left(\frac{EA_2}{2L}\right)(\cos 60°)^2 v = Q. \tag{23.39}$$

When $A_1 = A_2 = A$ this yields the displacement solution $v = 0.8QL/EA$ and the resulting member stresses are $\sigma_1 = Ev/L = 0.8Q/A$ (for the vertical member), $\sigma_2 = Ev/4L = 0.2Q/A$ (for the inclined members).

For comparative purposes we shall first obtain the FSD solution. Assuming $\sigma_{lim} = Q/A$ one can adjust the member areas so that $\sigma_i = \sigma_{lim}$, that is, $A_1 = 0.8A$ and $A_2 = 0.2A$. Repeating this process, cycling occurs: the displacement solution alternates between the values $0.8QL/EA$ and $1.176QL/EA$. To prevent this, one can use a convergence factor to limit the area changes. For example, if the area changes are halved one uses for the second iteration

$$A_1 = \tfrac{1}{2} A(1 + 0.8) = 0.9A, \quad A_2 = \tfrac{1}{2} A(1 + 0.2) = 0.6A. \tag{23.40}$$

Continuing in this fashion the solution converges after about ten iterations to $v = 0.970 \, QL/EA$, $A_1 = 0.970A$, and $A_2 = 0.243A$.

For comparison we now apply the condensed Lagrange multiplier result of the preceding section, eqn (23.37). For our three-bar truss problem the iterative procedure is written as

$$\frac{1.25EAv_{n+1}}{L} = Q - (A_1)_n L \left\{ \frac{1}{L}(\sigma_1 - \sigma_{lim}) \right\} - 2(A_2)_n (2L) \left\{ \frac{1}{4L}(\sigma_2 - \sigma_{lim}) \right\} \tag{23.41}$$

where $\sigma_1 = Ev_n/L$, $\sigma_2 = Ev_n/4L$ and the member areas are reset according to these stresses:

$$(A_1)_n = (A_1)_{n-1} \frac{\sigma_1}{\sigma_{lim}}, (A_2)_n = (A_2)_{n-1} \frac{\sigma_2}{\sigma_{lim}}. \tag{23.42}$$

After several iterations eqn (23.41) yields the FSD solution, as expected. This is not the intuitively obvious optimum result obtained in Section 22.10 using a SUMT–penalty method. At this point, however, we can observe that the Lagrange multiplier values in the vertical and inclined members are respectively

$$\lambda_1 = 0.970 - 1 = -0.030 \tag{23.43}$$

$$\lambda_2 = 0.243 - 1 = -0.757. \tag{23.44}$$

The magnitudes of these are the additional strains required to provide full stress.

Now we saw in Section 22.2 that a necessary condition for optimality is that these be zero or unity, the latter value corresponding to a vanishing member. Thus we can conclude that the inclined members must vanish and setting $A_2 = 0$ and repeating the now trivial analysis one obtains the optimum solution $A_1 = A$ and $A_2 = 0$.

Thus the solution finally assumes a 'decision' character and this is a characteristic of more general problems. In the foregoing analysis, for example, if a displacement constraint $v \leq 0.9QL/EA$ must be satisfied one observes that this has been violated after the second iteration. Such a nodal displacement constraint is then set as a boundary condition in any iteration following one in which it is violated and in such cases the FSD condition no longer applies.

In such 'mixed constraint' problems the condensed Lagrange multiplier approach does ensure a feasible solution but optimality is not guaranteed unless a sequence of increasing penalty factors is associated with the constraints. In practice, however, a large penalty factor can be associated immediately with any violated displacement constraint (as is done with the Kirchhoff constraints in thick plates and shells) and iteration will determine the optimum Lagrange multiplier values for the right-hand side stress constraints. The procedure then has much in common with the augmented Lagrangian methods described in Section 22.7 but now assumes a strong finite element character which ensures much greater economy in engineering applications.

Problems of optimal geometry, for example trusses or shells in which, within limits, the nodal elevations are variable, are not readily amenable to the approach of eqn (23.38) as the forcing terms are very difficult to construct. In such cases approximate optimality criterion models[7] or a gradient technique, such as that used in the following section, may prove more appropriate.

23.7 Simple program for truss optimization

In the following section a simple interactive BASIC program for truss optimization is given. This provides a useful computational introduction to some of the ideas discussed in Chapter 22 and the present chapter.

The program uses the following strategy.

1. Fully stressed design (FSD) is first used to obtain an improved design economically.

2. When an improved design has been obtained steepest descent (see Section 22.6.1) is used to obtain the optimum solution,[8] that is the change in the design variables $\{x_i\}$ is given by

$$\delta\{x_i\} = -\lambda\left\{\frac{\partial f}{\partial x_i}\right\} = -\lambda\{g\} \tag{23.45}$$

where λ is the optimum step length and the merit function f is the volume of material in the truss; $f = \sum A_i L_i$ where A_i, L_i are the cross-sectional areas and lengths of the truss members.

The gradient vector $\{g\}$ is first determined by steps Sx_i in each design variable (in this case the member areas A_i) in turn, using the same step factor S for each variable. Hence a re-analysis of the structure is needed to determine each entry in $\{g\}$ as

$$\{g\} = \left\{\frac{\partial f}{\partial x_i}\right\} \simeq \left\{\frac{\delta f_i}{\delta x_i}\right\} \tag{23.46}$$

where δf_i is the alteration in the merit function resulting from the perturbation δx_i for member i.

The optimum step length λ in eqn (23.45) is interactively determined by trial by the program user, by inputing the trial values (X) in line 510. When this has been determined, by noting an increase in f after several steps the value of λ which gave the lowest value of f is specified again to obtain once more the corresponding solution for $\{x_i\}$. Then a zero step length is input to commence calculation of $\{g\}$ for the next search. The procedure is terminated when no further reduction in f can be obtained.

The program deals with two load cases simultaneously by solving $\{Q\} = K\{D\}$ with two load vectors to obtain two solutions for $\{D\}$ (using the Gauss–Jordan reduction routine of the program of Section 1.11). Then in lines 1500–1710 the worst stress violation (tension or compression, load case one or two) is used to calculate the element stress ratio, T, and the elements are resized using this ratio.

In using the program the following recommendations are made

1. A step factor $S = 1$ is used for gradient vector calculation (this value is set in line 300).

2. Use only one preliminary FSD step (unless we seek only a fully stressed design).

3. In all but the first steepest descent search, use a first step length (X in line 510) of 10^{-20}. Then if no decrease in f is observed with this 'test step' the search direction is unprofitable and the current solution $\{x_i\} = \{A_i\}$ is taken as our estimate of the optimum.

Dimensioned arrays in the program are:

CO(10, 2)	nodal x and y-coordinates;
NO(10, 2)	node numbers for each element;
ES(4, 4)	element stiffness matrix;
BC(10, 2)	node numbers and b.c. flags for b.c. nodes;
D(20)	displacement solution for first load case;
D2(20)	displacement solution for second load case;
S(20, 20)	structure stiffness matrix;
Q(20)	first load case;
Q2(20)	second load case;
C(10, 2)	direction cosines for each element;
A(10)	current element cross-sectional areas;
AL(10)	element cross-sectional areas at last step;
G(10)	gradient vector of eqn (23.46).

Data input requirements for the program are

Lines	Data
1	number of nodes (NP); number of elements (NE); number of boundary conditions (NB), number of loaded nodes (NL).
1	E; tension stress limit (TLIM); compression stress limit (CLIM), taken as the same for each element.
NP	nodal coordinates (CO(I, 1), CO(I, 2)).
NE	element node numbers (NO(I, 1), NO(I, 2)); initial element cross-sectional areas (A(I)).
NL	node number (N); loads for both cases (Q(2N-1), Q(2N), Q2(2N-1), Q2(2N)).
NB	node number (BC(I, 1)); flag BC(I, 2) = 1 if $u = 0$, 2 if $v = 0$.

```
10 Dim CO(10,2),NO(10,2),ES(4,4),BC(10,2),D(20),D2(20)
20 Dim S(20,20),Q(20),C(10,2),A(10),L(10),Q2(20),AL(10),G(10)
30 Read NP,NE,NB,NL;NN=2*NP
40 Read E,TLIM,CLIM              ;Rem E,TENSION & COMPRESSION STRESS LIMITS
100 For I=1 to NP
110 Read CO(I,1),CO(I,2);Next I    ;Rem NODAL COORDINATES
120 For I=1 to NE
130 Read NO(I,1),NO(I,2),A(I);Next I;Rem ELEMENT NODE NOS AND X-AREAS
131 For I=1 to NL                          ;Rem READ NODAL LOADS
132 Read N,Q(2*N-1),Q(2*N),Q2(2*N-1),Q2(2*N);Next  ;Rem Q2( ) IS 2nd LOADING
134 For I=1 to NB                ;Rem BC(I,1)=NODE NUMBER, BC(I,2)=1 FOR
136 Read BC(I,1),BC(I,2);Next    ;Rem U=0 AND =2 FOR V=0 AT THIS NODE
140 For N=1 to NE
150 NI=NO(N,1);NJ=NO(N,2)
```

```
160 DX=CO(NJ,1)-CO(NI,1);DY=CO(NJ,2)-CO(NI,2)
180 L(N)=sqrt(DX*DX+DY*DY)           ;Rem STORE ELEMENT LENGTHS
185 C(N,1)=DX/L(N);C(N,2)=DY/L(N);Next   ;Rem STORE ELEMENT DIRECTION COSINES
187 ! "INPUT 99 TO SKIP FSD";Input FLAG  ;Rem ONE INITIAL FSD STEP RECOMMENDED
188 If FLAG=99 then Goto 300         ;Rem I.E. INPUT FLAG = 1
190 Rem
200 Gosub 1000                       ;Rem GO TO ASSEMBLY & SOLUTION ROUTINE
212 ! "NODAL DISPLACEMENTS U,V - COLUMNS 3,4 ARE FOR SECOND LOAD CASE";!
214 For I=1 to NP
215 ! %"I5",I,%"15E3",D(2*I-1),D(2*I),D2(2*I-1),D2(2*I)
216 Next I                           ;Rem OUTPUT NODAL DISPLACEMENTS
217 ! "ELEMENT STRESS RATIOS & NEW X-AREAS";!
218 V=0
219 For N=1 to NE
220 Gosub 1500                       ;Rem GO TO ELEMENT FORCE CALC. ROUTINE
221 A(N)=A(N)*T                       ;Rem FACTOR ELEMENT AREAS
222 V=V+L(N)*A(N)                     ;Rem V=TOTAL MATERIAL VOLUME
223 ! %"I5",N,%"12F3",T,A(N)
224 Next N
225 ! "VOL = ",V
228 ! "INPUT 1 TO CONT FSD";Input FLAG   ;Rem WHERE STEEPEST DESCENT TO FOLLOW
230 If FLAG<>1 then Goto 250         ;Rem RECOMMEND ONLY ONE FSD STEP
240 Goto 190
250 Rem                              ;REM SF=1 DOUBLES ELEMENT AREAS ONE BY
300 SF=1                             ;Rem ONE IN SEARCH VECTOR CALCULATION
310 B=0;Rem PENALTY NOT USED - WHEN IT IS IT SHOULD BE READ IN HERE
315 For N=1 to NE
320 AL(N)=A(N);Next                  ;Rem COPY OLD ELEMENT AREAS
325 VL=V                             ;Rem AND OLD MATERIAL VOLUME
330 ! "ELEMENT NO. & INCREMENT IN V FACTORING ITS AREA CAUSES"
340 For NK=1 to NE       ;Rem LOOP FOR SEARCH VECTOR CALCULATION **************
350 DA=SF*A(NK);A(NK)=A(NK)+DA        ;Rem INCREMENT ELEMENT AREA
360 Gosub 1000                       ;Rem GO TO ASSEMBLY AND SOLUTION
364 V=0
365 For N=1 to NE
366 Gosub 1500                       ;Rem GO TO STRESS CALC. ROUTINE
370 CV=T-1                           ;Rem CV=CONSTRAINT VIOLATION
371 If CV<0 then CV=0
375 V=V+L(N)*A(N)*T+B*CV^2           ;Rem V=MATERIAL VOLUME WHICH IS THE
376 Next N                           ;Rem THE MERIT FUNCTION
380 DV=V-VL                          ;Rem INCREMENT IN MERIT FUNCTION
382 A(NK)=A(NK)-DA                    ;Rem RETURN ELEMENT TO ORIGINAL AREA
385 ! %"I5",NK,%"12F5",DV
390 G(NK)=DV/DA                      ;Rem GRADIENT FOR THIS ELEMENT
400 Next NK           ;Rem END LOOP FOR SEARCH VECTOR CALCULATION **********
500 Rem       ;REM COMMENCE SEARCH IN GRADIENT VECTOR DIRECTION £££££££££££££££
510 ! "INPUT STEP LENGTH - ZERO ENDS SEARCH AT LAST POINT";Input X
512 If X=0 then Goto 310             ;Rem X=STEP LENGTH FOR SEARCH DIR'N
515 For NK=1 to NE
520 A(NK)=AL(NK)-G(NK)*X;Next NK      ;Rem ADJUST ELE. AREAS AS PER EQN 22.55
530 Gosub 1000                       ;Rem GO TO ASSEMBLY AND SOLUTION
535 V=0
536 For N=1 to NE
537 Gosub 1500                       ;Rem GO TO STRESS CALC. ROUTINE
538 A(N)=A(N)*T                       ;Rem FACTOR ELE. AREA BY STRESS RATIO
539 CV=T-1                           ;Rem CONSTRAINT VIOLATION
540 If CV<0 then CV=0
542 V=V+L(N)*A(N)+B*CV^2             ;Rem V=MERIT FUNCTION (TOTAL MATERIAL
544 Next N                           ;Rem VOLUME),B=PENALTY FACTOR (=0 HERE)
545 ! "V = ",V,"    ELEMENT AREAS = "
550 ! %"10F3",A(1),A(2),A(3),A(4)    ;Rem PRINT V & ELEMENT AREAS
560 Rem LATTER FORMAT DESIGNED ONLY FOR THE PRESENT TEST PROBLEM
580 Goto 500  ;Rem END OF LOOP FOR SEARCH ££££££££££££££££££££££££££££££££££££
990 End
```

```
1000 Rem                    ;Rem START ASSEMBLY & DISPLACEMENT SOLUTION ************
1002 For I=1 to NN;For J=1 to NN
1003 S(I,J)=0;Next;Next
1004 For N=1 to NE
1005 CA=C(N,1);SA=C(N,2);F=E*A(N)/L(N)
1010 ES(1,1)=F*CA*CA;ES(1,2)=F*SA*CA   ;Rem FILL TOP LH CORNER OF ESM (EQN 1.34)
1020 ES(2,1)=F*SA*CA;ES(2,2)=F*SA*SA
1030 For I=1 to 2
1040 For J=1 to 2                       ;Rem COPY TO REST OF ESM
1050 ES(I,J+2)=-ES(I,J);ES(I+2,J)=-ES(I,J);ES(I+2,J+2)=ES(I,J)
1060 Next J;Next I
1070 For I=1 to 2;For J=1 to 2
1100 For IL=1 to 2;IE=2*(I-1)+IL;NR=2*NO(N,I)-2+IL
1120 For JL=1 to 2;JE=2*(J-1)+JL;NC=2*NO(N,J)-2+JL
1130 S(NR,NC)=S(NR,NC)+ES(IE,JE)       ;Rem ADD ESM TO STRUCTURE MATRIX
1140 Next JL;Next IL;Next J;Next I
1150 Next N
1160 For I=1 to NB                      ;Rem LOOP TO ENFORCE BOUNDARY CONDITIONS
1170 NF=2*BC(I,1)-2+BC(I,2)
1180 For J=1 to NN
1190 S(NF,J)=0;S(J,NF)=0;Next J         ;Rem ZERO ROW & COLUMN
1200 S(NF,NF)=1;Q(NF)=0;Q2(NF)=0        ;Rem UNIT PIVOT & ZERO RHSs
1210 Next I
1220 For I=1 to NN                      ;Rem START GAUSS-JORDAN REDUCTION £££££
1230 D(I)=Q(I);D2(I)=Q2(I);Next
1240 For I=1 to NN
1250 X=S(I,I);D(I)=D(I)/X;D2(I)=D2(I)/X
1260 For J=I+1 to NN
1270 S(I,J)=S(I,J)/X;Next J
1280 For K=1 to NN
1290 If K=I then 1340
1300 X=S(K,I);D(K)=D(K)-X*D(I);D2(K)=D2(K)-X*D2(I)
1310 For J=I+1 to NN
1320 S(K,J)=S(K,J)-X*S(I,J);Next J
1340 Next K
1350 Next I                             ;Rem END GAUSS-JORDAN REDUCTION £££££££
1410 Return          ;Rem END ASSEMBLY & DISPLACEMENT SOLUTION **************
1500 Rem           ;REM START STRESS CALCULATION ROUTINE $$$$$$$$$$$$$$$$$$$$
1530 CA=C(N,1);SA=C(N,2)
1540 NI=NO(N,1);NJ=NO(N,2)
1550 U2=CA*D(2*NJ-1)+SA*D(2*NJ)
1560 U1=CA*D(2*NI-1)+SA*D(2*NI)
1570 T=E*(U2-U1)/L(N)                   ;Rem ELEMENT TENSION (FIRST LOAD CASE)
1580 U2=CA*D2(2*NJ-1)+SA*D2(2*NJ)
1590 U1=CA*D2(2*NI-1)+SA*D2(2*NI)
1600 If T>=0 then T=T/TLIM
1610 If T<0 then T=abs(T)/CLIM
1620 T2=E*(U2-U1)/L(N)                  ;Rem ELEMENT TENSION (SECOND LOAD CASE)
1630 If T2>=0 then T2=T2/TLIM
1640 If T2<0 then T2=abs(T2)/CLIM
1650 If T2>T then T=T2                  ;Rem TAKE LARGEST STRESS FACTOR (OF TWO)
1710 Return         ;Rem END STRESS CALCULATION ROUTINE $$$$$$$$$$$$$$$$$$$$$$
5000 Data 5,4,8,1                       ;Rem DATA FOR FOUR BAR TRUSS OF FIG. 23.2
5010 Data 30000,20,15                   ;Rem E & STRESS LIMITS
5020 Data 0,0                           ;Rem NODAL COORDINATES
5030 Data 3.6,0
5040 Data 7.2,0
5050 Data 10,0
5060 Data 3.6,-4.8
5070 Data 1,5,1.44                      ;Rem ELE. NODE NOS & X-AREAS
5080 Data 2,5,2.00
5090 Data 3,5,1.80
5100 Data 4,5,2.00
5110 Data 5,0,-10,5,-6                  ;Rem TWO LOAD CASES AT NODE 5
```

```
5120 Data 1,1                        ;Rem BOUNDARY CONDITIONS
5130 Data 1,2
5140 Data 2,1
5150 Data 2,2
5160 Data 3,1
5170 Data 3,2
5180 Data 4,1
5190 Data 4,2
```

The data in lines 5000–5190 is for the optimization of the truss of Fig. 23.2. Hence the data lines give the following information:

Line 5000   5 nodes, 4 elements, 8 boundary conditions, 1 loaded node.

Line 5010 $E = 30,000$, TLIM = 20, CLIM = 15.

Line 5020–5060   nodal x- and y-coordinates.

Line 5070–5100   element node numbers and cross-sectional areas.

Line 5110   at node 5, first load case is $X = 0$, $Y = -10$ and second load case is $X = 5$, $Y - 6$.

Line 5120–5190   each pair of lines specifies both u and v to be suppressed at nodes $1 \rightarrow 4$.

Then, using recommendations 1–3 above, the results shown in Table 23.1 are obtained.

The present FSD with steepest descent procedure gives a better result than that obtained by Wang[9] using stepwise (sequential) linear programming (SLP). Note, however, that at the end of each LP step Wang scales all the element areas by the largest stress factor encountered in any element and this cannot yield the true optimum solution. Note also that all solutions given by the present method are valid solutions because each element is factored according to its stress level at each step.

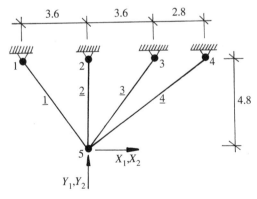

Fig. 23.2 Truss carrying two alternative loadings

Table 23.1 Optimization solutions for the problem of Fig. 23.2 using FSD with steepest descent

	A_1	A_2	A_3	A_4	f
Initial	1.440	2.000	1.800	2.000	45.040
FSD (1 step)	0.231	0.264	0.104	0.143	4.421
Search 1 ($\lambda = 0.09$)	0.296	0.189	0.139	0.085	4.200
Search 2 ($\lambda = 10^{-20}$)	0.316	0.224	0.123	0.055	4.152
($\lambda = 0.03$)	0.310	0.213	0.126	0.064	4.148
Search 3 ($\lambda = 10^{-20}$)	Test step increased f so stop				4.178
Stepwise LP solution[9]	0.364	0.261	0.121	0.084	4.837

The present procedure could easily be applied to the optimization of frame structures. In these, optimization of the sizes of the elements of the frame would follow the same process used for the foregoing truss problem, except that section limits (that is, a minimum allowable size for each element) must be specified, but inclusion of these in finite element programs is a trivial exercise.

Finally, in relation to optimum truss problems, the classical solutions of Michell (see Section 22.2) are worth remembering. These are for very special cases but for the case of line (rather than point) support Rozvany and Gollub[10] report that 'for this less restricted case solutions correspond to fields of constant strain, which give layouts consisting of a finite number of straight members' and such solutions may have more practical applications.

23.8 Program for optimal arch shapes

The following section introduces a FEMSS program for the optimization of arch shapes which uses the quadratic arch element of Section 11.1[11] and steepest descent (eqn (23.45)).

Here the merit function is the *factor weight* (FWT) which is the material volume scaled according to the stress ratios in each element, that is, we assume unit density. This is calculated in subroutine STRUB (in workspace STRESS) as

$$f = \sum_{i=1}^{NE} ARCL*W*D*SFACT \qquad (23.47)$$

where ARCL, W, and D are the element length, breadth, and depth, respectively, and SFACT is the largest value resulting from dividing the two extreme fibre stresses (TS and CS) by the tensile and compressive stress limits (TLIM and CLIM).

In the example problem studied, the penalty factor $\beta = 1$ (line 1010 of data) but shear stresses are not taken into consideration as the arch is very thin (span/thickness = 160).

As shown in Table 11.2 integration at the two cubic Gauss points gives best results[12, 13] and these are also used for the geometry calculations described in Section 11.2 (lines 65–225 in ELSUB/ELMAT). For the optimization problem of Fig. 23.3 studied here, however, straight elements are used, as movement of the central node of an element during gradient vector calculations produces a 'local arching' effect which leads to poor results. Hence only nodes 3, 5, 7, and 9 in Fig. 23.3 are specified as the shape function variables (lines 1140–1180 of data) and vertical movements of these (10 per cent increments) are used to form the gradient vector (lines 358–470 of the main or calling program).

The results for the case of a horizontally distributed load are given in Table 23.2. After completion of steepest descent searching the resulting shape is

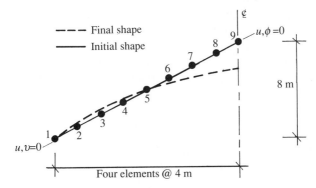

Fig. 23.3 Initial shape of arch in data lines of program listing and final shape after application of steepest descent

Table 23.2 Optimization results for an arch with a horizontally distributed load

	y_3	y_5	y_7	y_9	f
Initial	2	4	6	8	71.11
Search 1 ($\lambda = 0.04$)	3.06	4.39	5.28	7.35	26.42
Search 2 ($\lambda = 0.03$)	2.49	5.13	4.99	6.94	16.30
Search 3 ($\lambda = 0.012$)	2.54	4.80	5.39	6.67	11.94
Search 4 ($\lambda = 0.035$)	2.67	4.50	5.35	6.25	8.95
Search 5 ($\lambda = 0.005$)	2.54	4.53	5.37	6.19	8.70
Search 6 ($\lambda = 0.0003$)	2.53	4.53	5.37	6.19	8.69
Design shape	2.8	4.8	6.0	6.4	4.14
Scaled shape					
($\lambda = 2.19$)	6.13	10.51	13.14	14.02	2.95
OCM technique[14, 15]	5.6	9.3	11.6	13.5	–
Theoretical	6.06	10.40	12.99	13.86	–

closely parabolic as expected, largely eliminating the bending stresses. In this thin arch, however, the remaining bending stresses still swamp the extensional stresses.

To obtain the final optimum shape, therefore, the arch shape is taken as parabolic; this is the 'design shape' for elimination of bending. Scaling of this shape is then used (lines 344–352 of main program) to obtain the final optimum shape. A search with a gradually increasing scale factor is used to minimize the factor weight, which is now predominantly a function of extensional stresses.

The results are in reasonable agreement with the exact solution[14] and those of an optimality criterion method[14,15] given in Table 22.3. In this, constant strain shell elements are elevated according to their bending stresses, in order to reduce them, and scaling is used simultaneously to minimize the factor weight, calculated only from the extensional stresses. The present method is to be recommended, though, because it does not presume that bending should be eliminated.

Observing Fig. 23.4 the exact solution is given by the general cable theorem,[16] that is, y is proportional to the bending moment in the equivalent beam divided by the abutment thrust H, giving

$$y = -\frac{qx^2}{2H} \quad \text{where } H = \frac{qL^2}{8h}. \tag{23.48}$$

Then the thrust in the arch at any point is

$$T = H\sqrt{1 + \left(\frac{dy}{dx}\right)^2} \tag{23.49}$$

where $dy/dx = -qx/H = -8hx/L^2$. Hence the factor weight of the arch is

$$f = \int \frac{T}{WDC} \cdot dS = \int_{-L/2}^{L/2} \frac{T\sqrt{1 + (dy/dx)^2}}{WDC} \cdot dx$$

$$= \frac{qL(3L^2 + 16h^2)}{24hWDC} \tag{23.50}$$

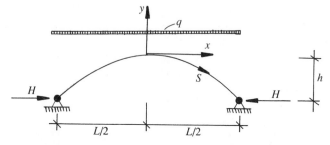

Fig. 23.4 Weightless parabolic arch with horizontally distributed load

where W, D, and C are the arch width, depth, and extensional stress limit. Minimizing f with respect to h gives $h/L = \sqrt{(3/16)} = 0.433$ or $h = 13.86$ with $L = 32$ as in Table 23.2.

An alternative numerical procedure is to base the factor weight calculation only on the extensional stresses (from the outset) and again apply steepest descent. This gives an arch of similar height to that expected but, with the triangular initial shape of Fig. 23.3, the solution tends towards a 'W' shape (with $f = 2.73$). Such a solution would involve extremely high flexural stresses and is not feasible.

Finally, for the case of a surface distributed load, the steepest descent/scaling method is applied once again, giving the results of Table 23.3. These are reasonable but not as good as those for the horizontally distributed load case, in part because the true optimum shape is not now parabolic, as assumed prior to scaling. Overall, however, the effectiveness of steepest descent techniques in arch and shell problems is demonstrated as it will be in fluids problems in the following two sections.

The following program uses the FEMSS system introduced in Section 15.7 (see also Appendix 1). Data input requirement are

Lines	Data
1	number of nodes (NP); number of elements (NE); number of b.c. nodes (NB); number of property sets (NT); half bandwidth (NBW, always 9); number of 'shape function' nodes (NSFV).
1	PENF, POW, giving the penalty factor as $\beta = \text{PENF} \times 10^{\text{POW}}$.
NT lines	property sets (PROP(N, I), $N = 1 \to \text{NT}$): E, v, W (width); D (depth); QX (horiz. UDL); QS (surface UDL); TLIM; CLIM for $I = I \to 8$.
NP lines	nodal coordinates $x = \text{CORD}(N, 1)$, $y = \text{CORD}(N, 2)$.
NE lines	node numbers for each element (NOP(N, M), $M = 1 \to 3$) Note that in line 200 each element is given property set 1 (see also Table 15.1)
NB lines	boundary node number N; non-zero flags to indicate specified freedoms (IBC(N, I), $I = 1 \to 3$); values for specified freedoms (SPEC(N, I), $I = 1 \to 3$); for example see data lines 1100, 1110.
NQ lines	node number NQ and direct loads R(1), R(2), R(3). Use a 0, 0, 0, 0 line to terminate such loads.
NSFV lines	'shape function data': SFCORD (N, 1) = node number; SFCORD(N, 2) = 1 or 2 for x or y permitted to move.

Table 23.3 Optimization results for an arch with a surface distributed load

	y_3	y_5	y_7	y_9	f
Initial	2	4	6	8	79.50
Search 1 ($\lambda = 0.04$)	3.25	4.37	5.14	7.28	32.80
Search 2 ($\lambda = 0.03$)	2.68	5.09	5.31	6.73	13.63
Design shape	2.8	4.8	6.0	6.4	4.14
Scaled shape					
($\lambda = 1.71$)	4.79	8.21	10.26	10.94	3.529
OCM technique[14,15]	4.3	7.4	9.2	10.2	–
Theory (approx.)[14]	4.27	7.32	9.15	9.76	–

In the program listing the solution routine SOLSUB/SOLVE is omitted as it is identical to that in Section 20.7, except that lines 935–960 for printing the nodal freedom solutions are omitted. Instead the factor weight (lines 342, 350, 570 of the main routine) and shape function coordinates (line 585) are of primary concern.

The data in lines 1000–1180 of the first subroutine follows the requirements listed above and the coding of the element routine closely follows the details given in Sections 11.1 and 11.2.

The present procedure can easily be applied to three-dimensional and shell problems as can that of Mohr.[15] Once again, as noted at the close of the preceding section, section limits (minimum and perhaps maximum permissible element thicknesses) will generally need to be specified. Additional details such as these are easily added to finite element optimization programs.

```
5 Rem PROGRAM FOR OPTIMUM SHAPE OF ARCHES
10 Include 'ELMAT','SOLVE','STRESS'
20 Access * from *
30 Open £8,"B:EMATS"
40 Open £9,"B:ELSTR"
50 Open £10,"B:STIFM"
60 Dim IBC(100,3),Q(300),SPEC(100,3),IMAT(100),CORD(100,2),PROP(20,8)
70 Dim NOP(100,3),CI(2,2),WF(2),R(3),SFCORD(10,2),G(10)
80 Def shared data NP,NE,NB,NT,NBW,NCN,NDF,PENF,FWT
90 Def shared data IBC( ),SPEC( ),Q( ),CI( ),WF( )
100 Def shared data IMAT( ),CORD( ),PROP( ),NOP( )
110 NDF=3;NCN=3
120 CI(1,1)=-1/sqrt(3.);CI(2,1)=1/sqrt(3.)
130 WF(1)=0.5;WF(2)=0.5
140 Read NP,NE,NB,NT,NBW,NSFV;NBW=NBW-NDF
150 Read PENF,POW;PENF=PENF*10^POW   ;Rem PENALTY FACTOR - SEE SEC. 11.1
160 For N=1 to NT;For I=1 to 8;Read PROP(N,I);Next;Next
170 For N=1 to NP;Read CORD(N,1),CORD(N,2);Next
180 For N=1 to NE;For I=1 to NCN
190 Read NOP(N,I);Next;Next
200 For N=1 to NE;IMAT(N)=1;Next      ;Rem SAME PROPERTIES FOR ALL ELES
210 For I=1 to NB
```

```
220 Read N,IBC(N,1),IBC(N,2),IBC(N,3),SPEC(N,1),SPEC(N,2),SPEC(N,3)
230 Next      ;Rem IBC ARE B.C. FLAGS AND SPEC ARE SPECIFIED BOUNDARY VALUES
240 Read NQ,R(1),R(2),R(3)            ;Rem READ NODAL LOADS
250 If NQ=0 then Goto 280
260 For K=1 to NDF;IC=(NQ-1)*NDF+K;Q(IC)=Q(IC)+R(K);Next
270 Goto 240
280 Rem
290 For N=1 to NSFV
300 Read SFCORD(N,1),SFCORD(N,2);Next;Rem SHAPE FUNCTION VARIABLES
310 For N=1 to NE                     ;Rem I.E. NODE NUMBER AND 1 OR 2
320 ELSUB N;Next                      ;Rem FOR X OR Y
330 Filepos(8)=0;SOLSUB;Filepos(8)=0
340 Filepos(9)=0;STRUB;Filepos(9)=0   ;Rem INITIAL SOLUTION
342 ! "FWT = ",FWT
344 Goto 353 ;Rem SKIP SHAPE SCALING
345 Input X                           ;Rem SCALING FACTOR
346 For N=1 to NP;CORD(N,2)=CORD(N,2)*X;Next
347 For N=1 to NE;ELSUB N;Next
348 Filepos(8)=0;SOLSUB;Filepos(8)=0  ;Rem SOLUTION FOR SCALED SHAPE
349 Filepos(9)=0;STRUB;Filepos(9)=0
350 ! "FWT = ",FWT
351 For N=1 to NP;CORD(N,2)=CORD(N,2)/X;Next
352 Goto 345
353 Rem
355 Rem
358 For NK=1 to NSFV
360 VL=FWT
370 SN=SFCORD(NK,1);SXY=SFCORD(NK,2)
380 DC=CORD(SN,SXY)*0.1               ;Rem PERTURB SF COORDINATES
390 CORD(SN,SXY)=CORD(SN,SXY)+DC
400 For N=1 to NE;ELSUB N;Next
410 Filepos(8)=0;SOLSUB;Filepos(8)=0
420 Filepos(9)=0;STRUB;Filepos(9)=0
430 V=FWT
440 DV=V-VL
450 G(NK)=DV/DC                       ;Rem CALCULATE GRADIENT
460 CORD(SN,SXY)=CORD(SN,SXY)-DC       ;Rem RESTORE COORD TO ORIG. VALUE
470 Next NK
480 For NK=1 to NSFV;! G(NK);Next      ;Rem TO GIVE AN IDEA OF STEP LENGTHS
490 Input X                           ;Rem STEP LENGTH FOR SEARCH
500 If X=0 then Goto 355              ;Rem END SEARCH WITH THIS GRADIENT
510 For NK=1 to NSFV
520 SN=SFCORD(NK,1);SXY=SFCORD(NK,2)
530 CORD(SN,SXY)=CORD(SN,SXY)-G(NK)*X;Next;Rem CHANGE SF COORDS, EQN 23.45
540 For N=1 to NE;ELSUB N;Next
550 Filepos(8)=0;SOLSUB;Filepos(8)=0
560 Filepos(9)=0;STRUB;Filepos(9)=0
570 ! "FWT = ",FWT                    ;Rem (STRESS) FACTOR WEIGHT
580 For NK=1 to NSFV;SN=SFCORD(NK,1);SXY=SFCORD(NK,2)
585 ! CORD(SN,SXY);Next               ;Rem NEW COORDS OF SHAPE FUNCTION
590 Goto 490
600 End
1000 Data 9,4,2,1,9,4                 ;Rem DATA FOR ARCH OF TABLE 22.3
1010 Data 1,0                         ;Rem USE PENALTY FACTOR = 1
1020 Data 20E6,0.2,4,0.2,-10,0,1000,1000;Rem HORIZ. DIST. LOAD CASE
1030 Data 0,0                         ;Rem INITIAL TRIANGULAR ARCH COORDS
1035 Data 2,1
1040 Data 4,2
1045 Data 6,3
1050 Data 8,4
1055 Data 10,5
1060 Data 12,6
1065 Data 14,7
1070 Data 16,8
```

```
1080 Data 1,2,3              ;Rem NODES NO. SETS FOR ELEMENTS
1090 Data 3,4,5
1095 Data 5,6,7
1097 Data 7,8,9
1100 Data 1,1,1,0,0,0,0      ;Rem SIMPLY SUPPORT ABUTMENT
1110 Data 9,1,0,1,0,0,0      ;Rem SYMMETRY AT CROWN
1120 Data 0,0,0,0            ;Rem NO POINT LOADS
1140 Data 3,2                ;Rem Y AT NODE 3 ALLOWED TO VARY
1160 Data 5,2                ;Rem Y AT NODE 5 ALLOWED TO VARY
1170 Data 7,2                ;Rem Y AT NODE 7 ALLOWED TO VARY
1180 Data 9,2                ;Rem Y AT NODE 9 ALLOWED TO VARY
```

```
 10 Rem ELMAT
 20 Def shared proc ELSUB @N
 30 Dim B(3,9),C(9,9),S(9,9),F(3),DF(3),TA(2)
 40 L=IMAT(N);E=PROP(L,1);PR=PROP(L,2);W=PROP(L,3) ;Rem QX = HORIZ UDL
 50 D=PROP(L,4);QX=PROP(L,5);QS=PROP(L,6)          ;Rem QS = SURFACE UDL
 60 I=NOP(N,1);J=NOP(N,2);K=NOP(N,3)
 65 CORD(J,2)=(CORD(I,2)+CORD(K,2))/2      ;Rem FOR STRAIGHT ELES
 70 DX=CORD(J,1)-CORD(I,1);DY=CORD(J,2)-CORD(I,2)
 75 TA(1)=DY/DX;DS=sqrt(DX*DX+DY*DY);ARCL=DS;Rem CHORD SLOPE FOR 1ST INTEG. PT
 80 Q(3*I-1)=Q(3*I-1)+DX*QX/2+DS*QS/2;Rem LOADS FOR FIRST TWO NODES
 85 Q(3*J-1)=Q(3*J-1)+DX*QX/2+DS*QS/2
 90 DX=CORD(K,1)-CORD(J,1);DY=CORD(K,2)-CORD(J,2)
 95 TA(2)=DY/DX;DS=sqrt(DX*DX+DY*DY);ARCL=ARCL+DS;Rem SLOPE FOR 2ND INTEG PT
 97 Rem ARCL = APPROX ELE ARC LENGTH (SUM OF TWO CHORDS BETWEEN NODES)
100 Q(3*J-1)=Q(3*J-1)+DX*QX/2+DS*QS/2;Rem LOADS FOR 2ND AND 3RD NODES
110 Q(3*K-1)=Q(3*K-1)+DX*QX/2+DS*QS/2
160 Goto 225                      ;Rem FOR STRAIGHT ELES
170 AA=sqrt((CORD(J,1)-CORD(I,1))^2+(CORD(J,2)-CORD(I,2))^2)
180 BB=sqrt((CORD(K,1)-CORD(J,1))^2+(CORD(K,2)-CORD(J,2))^2)
190 CC=sqrt((CORD(K,1)-CORD(I,1))^2+(CORD(K,2)-CORD(I,2))^2)
200 SP=(AA+BB+CC)/2
210 AREA=sqrt(SP*(SP-AA)*(SP-BB)*(SP-CC))
220 CURV=4*AREA/(AA*BB*CC) ;Rem ELEMENT CURVATURE (EQN 11.14): NOT USED NOW
225 CURV=0                        ;Rem FOR STRAIGHT ELES
230 For II=1 to 2;X=CI(II,1) ;Rem £££££ COMMENCE ELEMENT INTEGRATION LOOP
240 For I=1 to 3;For J=1 to 9;B(I,J)=0;Next;Next
250 F(1)=(X*X-X)/2;F(2)=1-X*X;F(3)=(X*X+X)/2 ;Rem AS PER EQN 11.5
260 DF(1)=(2*X-1)/ARCL;DF(2)=-4*X/ARCL;DF(3)=(2*X+1)/ARCL ;Rem AS PER EQN 11.6
270 For J=1 to 3
280 JU=3*J-2;JV=3*J-1;JT=3*J
290 CA=1/sqrt(1+TA(II)*TA(II));SA=sqrt(1-CA*CA) ;Rem ELE DIRECTION COSINES
300 B(1,JU)=CA*DF(J)-SA*F(J)*CURV
310 B(1,JV)=SA*DF(J)+CA*F(J)*CURV
320 B(2,JU)=CA*DF(J)*CURV          ;Rem FILL MATRIX OF EQN 11.4
330 B(2,JV)=SA*DF(J)*CURV
340 B(2,JT)=-DF(J)
350 B(3,JU)=-SA*DF(J)
360 B(3,JV)=CA*DF(J)
370 B(3,JT)=-F(J);Next
380 For J=1 to 9
390 C(3*II-2,J)=E*W*D*B(1,J)        ;Rem ELEMENT STRESS MATRIX
400 C(3*II-1,J)=E*W*D^3*B(2,J)/12
410 C(3*II,J)=5*E*W*D*B(3,J)/(12*(1+PR));Next
420 SMUL=WF(II)*ARCL
430 For I=1 to 9;For J=1 to 9
440 S(I,J)=S(I,J)+PENF*B(3,I)*C(3*II,J)*SMUL
450 For K=1 to 2;KA=3*II-3+K        ;Rem ESM AS PER EQN 11.7
460 S(I,J)=S(I,J)+B(K,I)*C(KA,J)*SMUL
470 Next;Next;Next
475 Next II                ;Rem £££££ END ELEMENT INTEGRATION LOOP
```

```
480 For I=1 to 9;For J=1 to 9
490 Write £8,S(I,J);Next;Next      ;Rem FILE ELE STIFFNESS MATRICES
500 For I=1 to 6;For J=1 to 9
510 Write £9,ARCL,C(I,J);Next;Next ;Rem FILE ELE STRESS MATRICES
520 Return;Proc end
530 End


10 Rem STRESS
20 Def shared proc STRUB
30 Dim F(9),B(9,9),R(9)
35 FWT=0
40 For N=1 to NE                   ;Rem LOOP ON ELEMENTS
50 For I=1 to 6;For J=1 to 9
60 Read £9,ARCL,B(I,J);Next;Next   ;Rem READ ELE LENGTH & STRESS MATRIX
70 For I=1 to NCN;M=NOP(N,I)
80 K=(I-1)*NDF
90 For J=1 to NDF;IJ=J+K
100 R(IJ)=DIS(J,M);Next J          ;Rem ELEMENT DISPLACEMENT VECTOR
110 Next I
120 For I=1 to 6;F(I)=0;For J=1 to 9
130 F(I)=F(I)+B(I,J)*R(J);Next;Next ;Rem ELE GAUSS POINT STRESSES
150 L=IMAT(N);W=PROP(L,3);D=PROP(L,4)
160 TLIM=PROP(L,7);CLIM=PROP(L,8)  ;Rem TENSION & COMP. STRESS LIMITS
170 ES=(F(1)+F(4))/(2*W*D)         ;Rem AVERAGE EXTENSIONAL STRESS
180 BS=3*(F(2)+F(5))/(W*D*D)       ;Rem AVERAGE BENDING STRESS
190 TS=ES+BS;CS=ES-BS              ;Rem EXTREME FIBRE STRESSES
200 SFACT=abs(TS/TLIM)
205 TEMP=abs(CS/CLIM)
210 If TEMP>SFACT then SFACT=TEMP  ;Rem MAXIMUM STRESS FACTOR
215 Rem SFACT=ABS(ES)/CLIM         ;REM USE THIS FOR SHAPE SCALING
220 FWT=ARCL*W*D*SFACT+FWT         ;Rem 'FACTOR WEIGHT'
225 ! N,F(1),F(2),F(4),F(5)        ;Rem GAUSS POINT EXT. & BENDING STRESSES
230 Next N
240 Return;Proc end
250 End
```

Note: Solution routine SOLSUB is as in Section 20.7 except that the loops in lines 80 and 250 are redimensioned to 9 and lines 935–960 are omitted.

Lines 65, 160, and 225 of the element routine (ELSUB, in workspace ELMAT) restrict to 'straight' elements, a restriction found necessary in optimization applications with steepest descent. For simple analysis applications these lines can be removed when curved elements are desired, although with practical numbers of elements satisfactory results are obtained when these are straight.

23.9 Program for optimization of fluid flows

In the following section a FEMSS program for the optimization of fluid flow is given. This is a modification of the uvp program of Section 20.7 in which steepest descent is used to modify part of the boundary to optimize a fluid flow. Note, however, that the program of Section 20.7 is coded for straight sided elements, whereas in the present program the elements are isopara-metric.

As an example, consider the problem of Fig. 23.5. This is similar to that of Fig. 20.1(b) except that Fig. 23.5 shows the whole domain, not a quadrant. In Fig. 23.5 vertical movement of nodes 9 and 10 is permitted, 20 per cent increments in these being used to calculate the gradient vector in line 460 of the program's main or calling routine. This vertical movement is used to maximize the exit velocity, u_{out}, at node 11 (note that a positive sign is used in eqn (23.45) for maximization).

With isoparametric elements, such movements, if not excessive, present no difficulty although in the present example node 11 is not kept equidistant between nodes 10 and 12 as it should be; nor is node 9 between nodes 8 and 10 though the distortion here is much less. Nevertheless the problem provides a useful qualitative example of optimization of fluid flows.

Note that $p = 1$ at inlet (specified as a negative value as required in eqn (19.72)) and zero at node 10 to force the flow. We also have $u = 1$ at inlet and this must be specified at node 2. Here this is also specified at node 1 to increase the outlet velocity.

At node 9 no boundary conditions are specified. Strictly, however, the normal velocity at such a point should be zero. This requires a skew boundary condition which, for a moving node, is a complicated exercise omitted in the present approximate approach.

As is generally advisable the midside node pressures are suppressed (lines 390–490 of the element routine) to avoid the chequerboard syndrome problems discussed in Section 19.5.

Since they are based on eqn (19.72) the present elements are designed for viscous flow but inviscid flows can be dealt with by using a small artificial viscosity.[17] In the present example, $\mu = 0.1$ (in the third data line) and, in view of the 'penalty factor' α of eqn (19.73) this is the smallest permissible value with the eight-digit computation of MEGABASIC. Thus with the 10^{-5} factor

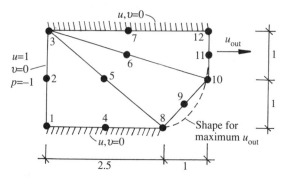

Fig. 23.5 Mesh for optimization of an exit flow (nozzle) problem

for α and $\mu = 0.1$ we have a 'shift' of 10^{-6}, leaving the minimum of two digits to work with. For smaller values of μ, therefore, a larger factor for α must be used. (As we discovered in Section 23.2 α is really a small slack variable, that is, an inverse penalty factor).

Table 23.4 shows the results for the problem of Fig. 23.5. Only a single search is used to obtain an approximate solution, sufficient results with this search being shown to indicate the simple search procedure used, culminating in selection of $\lambda = 0.7$ as giving the optimum shape of the outlet shown (by a dashed line) in Fig. 23.5.

There is no exact solution known for the present problem but the approximate result is in accordance with expectations. For example, one can compare with the exponential shapes of the 'cones' of various sound-producing instruments[18] (here, of course, the direction of 'flow' is reversed).

The following program listing uses the FEMSS system introduced in Section 15.7. Data input requirements are:

Lines	Data
1	number of nodes (NP); number of elements (NE); number of b.c. nodes (NB); number of property sets (NT); half bandwidth (NBW); number of 'shape function' nodes (NSFV).
1	factor for α in eqn (19.73). As noted above, use a larger value than 10^{-5} when viscosities less than 0.1 are to be used.
NT lines	property sets (PROP(N, I), N = 1 → NT); μ, element thickness (TH); density (DENS, not used); body forces in x, y-directions (FX, FY) for I = 1 → 5.
NP lines	nodal coordinates $x = $ CORD(N, 1), $y = $ CORD(N, 2).
NE lines	node numbers for each element (NOP(N, M), M = 1 → 6). Note that in line 320 each element is given property set 1 (see also Table 15.1).
NB lines	boundary condition node number N, non-zero flags to indicate specified freedoms (IBC(N, I), I = 1 → 3); values for specified freedoms (SPEC(N, I), I = 1 → 3); for example, see data lines 1180–1250.
NQ lines	node number NQ and direct loads R(1), R(2), R(3). Use a 0, 0, 0, 0 line to terminate such loads (or if there are no such loads).
NSFV lines	'shape function' data: SFCORD(N, 1) = node number; SFCORD (N, 2) = 1 or 2 for x- or y-coordinate permitted to change.

Table 23.4 Steepest descent maximization of outlet velocity in Fig. 23.5

	u_{11}	y_9	y_{10}
Initial	1.0737	0.5	1.0
Search 1 ($\lambda = 0.2$)	1.1535	0.4143	1.1175
($\lambda = 0.4$)	1.2066	0.3286	1.2351
($\lambda = 0.6$)	1.2306	0.2429	1.3526
($\lambda = 0.7$)	1.2307	0.2001	1.4114
($\lambda = 0.8$)	1.2223	0.1572	1.4702

The data in lines 1000–1280 of the first (main) subroutine follows the requirements listed above and is for the problem of Fig. 23.5. Particular note should be made of the boundary condition data (lines 1180–1250) and the comments on this earlier in this section.

The coding of the element routine ELSUB (in workspace ELMAT) closely follows the details of eqn (19.72) and Section 7.3 (that is, element is coded isoparametrically and can have curved sides, unlike that in Section 20.7).

The solution routine SOLSUB (in workspace SOLVE) is omitted from the following listing as it is identical to that in Section 20.7.

Finally, noting the comments on the use of a small artificial viscosity earlier in the section another interesting (analysis only) problem for the present program is that of Fig. 20.1(b). As the viscosity used diminishes, the outlet velocity solution should approach the approximate values obtained in Section 20.1.

```
5 Rem PROGRAM FOR FLOW OPTIMIZATION
10 Include 'ELMAT','SOLVE'
20 Access * from *
30 Open £8,"B:EMATS"
50 Open £10,"B:STIFM"
60 Dim FNE(100),IBC(100,3),Q(300),SPEC(100,3),IMAT(100),CORD(100,2),PROP(20,5)
70 Dim NOP(100,6),CI(6,2),WF(6),R(5),SFCORD(10,2),G(10)
80 Def shared data NP,NE,NB,NT,NBW,NCN,NDF,PENF,FNE( )
90 Def shared data IBC( ),SPEC( ),Q( ),CI( ),WF( )
100 Def shared data IMAT( ),CORD( ),PROP( ),NOP( )
110 NDF=3;NCN=6
120 CI(1,1)=0.5;CI(1,2)=0.5;CI(2,1)=0;CI(2,2)=0.5;CI(3,1)=0.5;CI(3,2)=0
130 WF(1)=1/6;WF(2)=1/6;WF(3)=1/6        ;Rem INTEGRATION AT MIDSIDE NODES
150 Read NP,NE,NB,NT,NBW,NSFV;NBW=NBW-NDF
155 Read PENF,POW;PENF=PENF*10^POW       ;Rem PENALTY FACTOR - SEE EQN 19.73
160 For N=1 to NT;For I=1 to 5;Read PROP(N,I);Next;Next
170 For N=1 to NP;Read CORD(N,1),CORD(N,2);Next
180 For N=1 to NE;For I=1 to 6
190 Read NOP(N,I);Next;Next
320 For N=1 to NE;IMAT(N)=1;Next
325 For I=1 to NB
330 Read N,IBC(N,1),IBC(N,2),IBC(N,3),SPEC(N,1),SPEC(N,2),SPEC(N,3)
335 Next     ;Rem IBC ARE B.C. FLAGS AND SPEC ARE SPECIFIED BOUNDARY VALUES
340 Read NQ,R(1),R(2),R(3)               ;Rem READ NODAL FORCING TERMS
350 If NQ=0 then Goto 380
```

```
360 For K=1 to NDF;IC=(NQ-1)*NDF+K;Q(IC)=Q(IC)+R(K);Next
370 Goto 340
380 Rem
385 For N=1 to NSFV
387 Read SFCORD(N,1),SFCORD(N,2);Next        ;Rem SHAPE FUNCTION VARIABLES
390 For N=1 to NE                            ;Rem I.E. NODE NUMBER AND 1 OR 2
400 ELSUB N;Next                             ;Rem FOR X OR Y
410 Filepos(8)=0
415 SOLSUB;Filepos(8)=0                      ;Rem INITIAL SOLUTION
420 Rem
430 For NK=1 to NSFV                         ;Rem FORM GRADIENT VECTOR
440 VL=DIS(1,11)                             ;Rem MERIT FUNC. FOR THIS PROBLEM
450 SN=SFCORD(NK,1);SXY=SFCORD(NK,2)
460 DC=CORD(SN,SXY)*0.2
470 CORD(SN,SXY)=CORD(SN,SXY)+DC             ;Rem PERTURB SF COORDINATES
480 For N=1 to NE;ELSUB N;Next
490 Filepos(8)=0;SOLSUB;Filepos(8)=0
500 V=DIS(1,11)
510 DV=V-VL
520 G(NK)=DV/DC                              ;Rem CALCULATE GRADIENT
525 CORD(SN,SXY)=CORD(SN,SXY)-DC             ;Rem RESTORE COORD TO ORIG. VALUE
530 Next NK
540 ! G(1),G(2)                              ;Rem GRADIENTS (THIS PROBLEM ONLY)
550 Input X                                  ;Rem STEP LENGTH FOR SEARCH
560 If X=0 then Goto 420                      ;Rem END SEARCH WITH THIS GRADIENT
570 For NK=1 to NSFV
580 SN=SFCORD(NK,1);SXY=SFCORD(NK,2)
590 CORD(SN,SXY)=CORD(SN,SXY)+G(NK)*X;Next   ;Rem CHANGE SF COORDS, EQN 23.45
600 For N=1 to NE;ELSUB N;Next               ;Rem USING + FOR MAXIMIZATION
610 Filepos(8)=0;SOLSUB;Filepos(8)=0
620 ! DIS(1,11),CORD(9,2),CORD(10,2)         ;Rem RESULTS (THIS PROBLEM ONLY)
630 Goto 550
640 End


1000 Data 12,3,8,1,30,2                      ;Rem MESH OF FIG. 23.5
1010 Data 1,-5                               ;Rem PENALTY FACTOR OF EQN 19.73
1020 Data 0.1,1,0,0,0                        ;Rem VISCOSITY = 0.1, CANNOT BE LESS
1030 Data 0,0                                ;Rem WITH 8 FIG. COMPUTATION UNLESS
1040 Data 0,1                                ;Rem SMALLER PENALTY FACTOR USED
1050 Data 0,2
1060 Data 1.25,0
1070 Data 1.25,1
1080 Data 1.75,1.5
1090 Data 1.75,2
1100 Data 2.5,0
1110 Data 3,0.5
1120 Data 3.5,1
1130 Data 3.5,1.5
1140 Data 3.5,2
1150 Data 1,8,3,4,5,2
1160 Data 3,8,10,5,9,6
1170 Data 3,10,12,6,11,7
1180 Data 1,1,1,1,1,0,-1
1190 Data 2,1,1,1,1,0,-1                      ;Rem INLET VELOCITY = 1
1200 Data 3,1,1,1,0,0,-1
1210 Data 4,1,1,0,0,0,0                       ;Rem U,V = 0 (RIGID BOUNDARY)
1220 Data 7,1,1,0,0,0,0
1230 Data 8,1,1,0,0,0,0                       ;Rem DITTO
1240 Data 10,0,0,1,0,0,0                      ;Rem P=0 TO FORCE FLOW
1250 Data 12,1,1,0,0,0,0
1260 Data 0,0,0,0
1270 Data 9,2                                 ;Rem SHAPE FUNCTION COORD DATA
1280 Data 10,2
```

```
10 Rem ELMAT                              ;REM AS PER SEC. 20.7 BUT IPE
20 Def shared proc ELSUB @N               ;Rem CODING OF SEC. 15.8 USED
30 Dim S(18,18),F(6),TEMP(2,6),XY(6,2),DL(2,6),TJ(2,2),T(18,18),TT(18,18)
40 L=IMAT(N);VISC=PROP(L,1);TH=PROP(L,2);DENS=PROP(L,3)
50 FX=PROP(L,4);FY=PROP(L,5)              ;Rem COLLECT ELEMENT PROPERTIES
60 For M=1 to 6;K=NOP(N,M);XY(M,1)=CORD(K,1);XY(M,2)=CORD(K,2);Next
70 A=0
80 For II=1 to 3      ;Rem START INTEGRATION LOOP ££££££££££££££££££££££££££££
90 F1=4*CI(II,1);F2=4*CI(II,2)
100 DL(1,1)=F1-1;DL(1,2)=0;DL(1,3)=F1+F2-3;Rem MATRIX OF EQN 7.15
110 DL(1,4)=F2;DL(1,5)=-F2;DL(1,6)=4-2*F1-F2
120 DL(2,1)=0;DL(2,2)=F2-1;DL(2,3)=F1+F2-3
130 DL(2,4)=F1;DL(2,5)=4-F1-2*F2;DL(2,6)=-F1
140 For I=1 to 2;For J=1 to 2;TJ(I,J)=0;For K=1 to 6
150 TJ(I,J)=TJ(I,J)+DL(I,K)*XY(K,J)        ;Rem JACOBIAN, SEE EQN 7.16
160 Next;Next;Next
170 DJ=TJ(1,1)*TJ(2,2)-TJ(1,2)*TJ(2,1)
180 DD=TJ(1,1);TJ(1,1)=TJ(2,2)/DJ;TJ(2,2)=DD/DJ ;Rem INVERT JACOBIAN
190 TJ(1,2)=-TJ(1,2)/DJ;TJ(2,1)=-TJ(2,1)/DJ
200 C1=CI(II,1);C2=CI(II,2);CC=C1+C2
210 F(1)=2*C1*C1-C1;F(2)=2*C2*C2-C2;F(3)=1-3*CC+2*CC*CC
220 F(4)=4*C1*C2;F(5)=4*C2*(1-CC);F(6)=4*C1*(1-CC)
230 For I=1 to 2;For J=1 to 6;TEMP(I,J)=0;For K=1 to 2
240 TEMP(I,J)=TEMP(I,J)+TJ(I,K)*DL(K,J);Next;Next;Next ;Rem AS PER EQN 7.17
250 VOL=TH*WF(II)*abs(DJ);A=A+VOL
260 For I=1 to 6;For J=1 to 6
270 G=VOL*VISC*(TEMP(1,I)*TEMP(1,J)+TEMP(2,I)*TEMP(2,J))
280 S(3*I-2,3*J-2)=S(3*I-2,3*J-2)+G        ;Rem FILL ELEMENT MATRIX AS PER
290 S(3*I-1,3*J-1)=S(3*I-1,3*J-1)+G        ;Rem EQN 19.72
300 S(3*I-2,3*J)=S(3*I-2,3*J)-VOL*F(I)*TEMP(1,J)
310 S(3*I-1,3*J)=S(3*I-1,3*J)-VOL*F(I)*TEMP(2,J)
320 S(3*I,3*J-2)=S(3*I,3*J-2)-VOL*TEMP(1,I)*F(J)
330 S(3*I,3*J-1)=S(3*I,3*J-1)-VOL*TEMP(2,I)*F(J)
340 SUM=SUM+G*G
350 Next;Next
360 For K=1 to 6
370 S(3*K,3*K)=S(3*K,3*K)+sqrt(SUM)*PENF/36;Next  ;Rem ADD FACTOR OF EQN 19.73
372 For I=1 to 6;NF=NOP(N,I)*NDF-2
374 Q(NF)=Q(NF)+VOL*FX
376 Q(NF+1)=Q(NF+1)+VOL*FY;Next
380 Next II           ;Rem END INTEGRATION LOOP ££££££££££££££££££££££££££££££
385 Rem GOTO 495
390 For I=1 to 18;T(I,I)=1;Next
400 T(12,3)=0.5;T(12,6)=0.5;T(12,12)=0
410 T(15,6)=0.5;T(15,9)=0.5;T(15,15)=0
420 T(18,9)=0.5;T(18,3)=0.5;T(18,18)=0     ;Rem SUPPRESSION OF MIDSIDE NODE
430 For I=1 to 18;For J=1 to 18            ;Rem PRESSURES AS PER EQN 4.29
440 TT(I,J)=0;For K=1 to 18
450 TT(I,J)=TT(I,J)+S(I,K)*T(K,J);Next;Next;Next
460 For I=1 to 18;For J=1 to 18
470 S(I,J)=0;For K=1 to 18
480 S(I,J)=S(I,J)+T(K,I)*TT(K,J);Next;Next;Next
490 S(12,12)=1;S(15,15)=1;S(18,18)=1
495 Rem
500 For I=1 to 18;For J=1 to 18
510 Write £8,S(I,J);Next;Next              ;Rem FILE THE ELEMENT MATRICES
520 ! N,A                                 ;Rem DIAGNOSTIC WRITE ONLY
530 Return;Proc end
600 End
```

The versatility of such programs is worth a final remark, although some improvements could be made, such as the addition of coding to keep midside nodes in elements with moving nodes approximately central; cf. line 65 of ELSUB in Section 23.8. We have already noted possible application to inviscid flows using a small artificial viscosity. It should also be noted that with the use of a large viscosity the present program can also be used to model slow creeping plastic flows such as those in hot metal extrusion processes.[19]

The problem studied approximately in the present section has no analytical solution. In the following section, however, a simple optimization program for one-dimensional lubrication problems is given and, in this case, a very simple analytical solution is available against which to test it.

23.10 Program for optimization of one-dimensional lubrication

As a final offering the present section gives a simple program for the optimization of one-dimensional lubrication problems. This uses the Gauss–Jordan reduction routine introduced in Section 1.11 and the element formulation of Section 19.8, explicit details for the one-dimensional case being given in Section 20.3.

Here the 'shape function' variables are the nodal thicknesses (CORD(I, 2)), and the merit function can be either the overall friction factor F/P or the total pressure P on the bearing, that is, the hydrodynamic 'lift'. These are given by[20]

$$P = \int p \, dx = \sum \int_0^{L_e} \{ f \}^t \{ p \} \, dx = \sum \bar{p} L_e \qquad (23.51)$$

$$F = \int \tau \, dx = \int \mu \frac{\partial u}{\partial y} \, dx = \sum \frac{\mu U L_e}{\bar{h}} \qquad (23.52)$$

where $\bar{p}$ and $\bar{h}$ are the average pressure and thickness for an element. These and their ratio are calculated in lines 740–810 of the program.

As an example consider the problem of Fig. 23.6. The program yields the solution for the nodal pressures obtained in eqn (20.18) and a friction factor of 2.420, in good agreement with the exact solution which is $F/P = 2.431H/L$.[21]

As a simple optimization exercise we permit only $k = h_2/h_1$ to vary, as analytical solutions are available for this case.[21] This is done by specifying the thickness at node 5 (h_2) as the only 'shape function' variable and linearly interpolating the values at nodes 2, 3, 4 from those of h_2 and h_1 (the latter being fixed at 0.5). Note that two or three extra lines need to be added to the program listing given for this case.

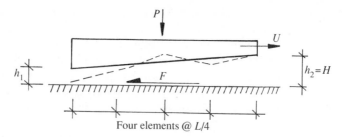

Fig. 23.6 Slider bearing problem. Dashed line is a result for min. F/P (all nodes permitted to move)

Then if we choose the friction factor ($F/P = \text{FFACT}$) as the merit function and seek to minimize this, steepest descent (with $\lambda = 1.34$) yields a minimum value of $F/P = 2.1875$ when $h_5 = 1.4255$ so that $k = 2.8510$.

The exact solution for the friction factor is [21]

$$\frac{F}{P} = \frac{1}{6L} \frac{h_1(k-1)(4\ln k - 6k^*)}{\ln k - 2k^*} \tag{23.53}$$

where $k^* = (k-1)/(k+1)$ for which the minimum value is $F/P = 2.3111$ when $k = 2.533$. Considering the logarithmic nature of eqn (2.53) the numerical solution with only four linear elements is reasonable.

Alternatively we can choose the 'lift' P as the merit function. Seeking a minimum we find after search that when $h_5 = 0.5$ (that is, $k = 1$), P is zero, agreeing with the exact solution,[21] and that if search is continued further P becomes negative.

If instead we seek to maximize P (when the sign in line 260, that is eqn (23.45), must be reversed) we obtain with $\lambda = 1.205$ a maximum value of $P = 0.5719$ when $h_5 = 1.0284$ (that is, $k = 2.0568$) which, with so few and simple elements, compares well with the analytical solution $P_{\max} \simeq 0.64$ when $k = 2.2$.[21]

Finally, again seeking to minimize F/P and now allowing all nodes to move, search tends towards the shape shown by a dashed line in Fig. 23.6; the thickness at node 1 is approximately zero. The reason for this is seen by putting $L_e = L/4$, $\mu = U = 1$ in eqns (23.51) and (23.52) and taking the ratio, giving

$$\frac{F}{P} = \frac{\sum 1/\bar{h}}{\sum \bar{p}}. \tag{23.54}$$

Such a shape, with the bearing 'closed', gives much increased pressures in the denominator of eqn (23.54), with F and $P \simeq 2$ (that is, $F/P \simeq 1$), the increase in F resulting mainly from the small average element thicknesses.

The 'local widening' at the middle of the bearing also decreases F/P, as expected. For example, an 'inclusion' in the bearing will obviously reduce its efficiency. This is easily verified by a single analysis with node thicknesses 0.5, 0.625, 0.875, 0.875, 1.0, yielding $F/P = 2.2356$, whereas with node thicknesses 0.5, 0.625, 0.625, 0.875, 1.0 we obtain $F/P = 2.7127$ compared to the value of 2.420 obtained with the original data of lines 1010–1050, and such investigations provide another useful application of the present program.

The following program listing uses the linear element of Section 20.3. The element k matrix is given by eqn (20.15) (lines 530–560) and the element load vector is given by eqn (20.16) (lines 570–580). assuming that μ and U are unity. The solution routine is the miniscule Gauss–Jordan reduction routine used in Section 1.11, but as in Section 20.6 this is modified to deal with non-zero boundary conditions (lines 610 and 625).

Data requirements for the program are

Lines	Data
1	number of nodes (NP); number of elements (NE); number of b.c. nodes (NB); number of 'shape function' nodes (NSFV).
NP lines	nodal x-coordinates (CORD(I, 1)) and thicknesses h (CORD(I, 2)).
NE lines	node numbers for each element, NOP(I, 1), NOP(I, 2).
NB lines	boundary condition node number IBC(I, 1) and value of pressure specified at this node IBC(I, 2).
NSFV lines	'shape function' data: SFCORD(I, 1) = node number; SFCORD(I, 2) = 1 or 2 for x or thickness h to vary.

The data in lines 1000–1150 is that for the problem of Fig. 23.6 with $L, H = 1$ (μ, $U = 1$ is already assumed in the coding for k and $\{q\}$) and the last described case studied, that is, the minimization of F/P with all node thicknesses variable.

With the insertion of the lines

265 CORD(1, 2) = 0.5; CORD(2, 2) = 0.375 + CORD(5, 2)/4
266 CORD(3, 2) = 0.25 + CORD(5, 2)/2; CORD (4, 2) = 0.125
        + 3*CORD(5, 2)/4

with lines 1000 and 1115 changing to
    1000 Data 5, 4, 2, 1
    1115 Data 5, 2
the previous, more successful, cases can be studied, that is those in which node 5 is moved in the search, nodes 2–4 proportionately following that movement to maintain the linear wedge shape of Fig. 23.6.

```
10 Rem PROGRAM FOR OPTIMIZATION OF 1-D LUBRICATION PROBLEM
20 Dim CORD(10,2),NOP(10,2),Q(10),SSM(10,10),IBC(10,2),SFCORD(10,2),G(10)
30 Read NP,NE,NB,NSFV              ;Rem NSFV = NO. SHAPE FUNCTION VARIABLES
40 For I=1 to NP;Read CORD(I,1),CORD(I,2);Next ;Rem X & T FOR EACH ELE.
50 For I=1 to NE;Read NOP(I,1),NOP(I,2);Next   ;Rem NODE NOS FOR EACH ELE.
60 For I=1 to NB;Read IBC(I,1),IBC(I,2);Next   ;Rem NODE NO. & SPEC. PRESSURE
70 Gosub 500                                ;Rem INITIAL SOLUTION
80 For N=1 to NP;! Q(N);Next                 ;Rem PRINT INITIAL SOLUTION
90 ! "FFACT = ",FFACT
100 For I=1 to NSFV;Read SFCORD(I,1),SFCORD(I,2);Next
110 Rem          ;REM ABOVE 1ST NO. IS NODE, 2ND IS 1 OR 2 TO VARY X OR T
120 For NK=1 to NSFV               ;Rem LOOP TO CALCULATE GRADIENT VECTOR
130 VL=FFACT                       ;Rem LAST VALUE OF MERIT FUNCTION
140 SN=SFCORD(NK,1);SXY=SFCORD(NK,2)
150 DC=CORD(SN,SXY)*0.2            ;Rem CHANGE IN SHAPE FUNCTION
160 CORD(SN,SXY)=CORD(SN,SXY)+DC
170 Gosub 500
180 V=FFACT;DV=V-VL
190 G(NK)=DV/DC                    ;Rem GRADIENT FOR THIS VARIABLE
200 CORD(SN,SXY)=CORD(SN,SXY)-DC;Next NK
210 For NK=1 to NSFV;! G(NK);Next
220 Input X                        ;Rem STEP LENGTH FOR SEARCH
230 If X=0 then 110                ;Rem ENDS SEARCH WITH THIS GRADIENT
240 For NK=1 to NSFV
250 SN=SFCORD(NK,1);SXY=SFCORD(NK,2)
260 CORD(SN,SXY)=CORD(SN,SXY)-G(NK)*X;Next ;Rem CHANGE SF, EQN 23.45
270 Gosub 500;! "FFACT = ",FFACT
280 For NK=1 to NSFV;SN=SFCORD(NK,1);SXY=SFCORD(NK,2)
290 ! CORD(SN,SXY);Next                     ;Rem PRINT CURRENT RESULTS
300 Goto 220
310 End
500 Rem £££££££££ SOLUTION ROUTINE
505 For I=1 to NP;Q(I)=0           ;Rem INITIALIZATION
506 For J=1 to NP;SSM(I,J)=0;Next;Next
510 For N=1 to NE
520 I=NOP(N,1);J=NOP(N,2)
530 T=(CORD(I,2)+CORD(J,2))/2      ;Rem AVERAGE ELEMENT THICKNESS
540 L=CORD(J,1)-CORD(I,1);X=T^3/(6*L);Rem ELEMENT LENGTH
550 SSM(I,I)=SSM(I,I)+X;SSM(I,J)=SSM(I,J)-X
560 SSM(J,I)=SSM(J,I)-X;SSM(J,J)=SSM(J,J)+X ;Rem DEPLOY ELE. MATRIX
570 X=(CORD(J,2)-CORD(I,2))/2      ;Rem dT/dx FOR ELEMENT
580 Q(I)=Q(I)+X;Q(J)=Q(J)+X;Next N    ;Rem ADD ELE. 'LOADS' TO Q( )
590 For N=1 to NB
600 K=IBC(N,1);X=IBC(N,2)
610 For I=1 to NP;Q(I)=Q(I)-X*SSM(I,K)
620 SSM(K,I)=0;SSM(I,K)=0;Next I    ;Rem SET BOUNDARY CONDITIONS
625 SSM(K,K)=1;Q(K)=X;Next N
630 For I=1 to NP                  ;Rem COMMENCE GAUSS JORDAN REDUCTION
640 X=SSM(I,I);Q(I)=Q(I)/X         ;Rem X = PIVOT
650 For J=I+1 to NP                ;Rem REDUCE ONLY TO RIGHT OF PIVOT
660 SSM(I,J)=SSM(I,J)/X;Next J      ;Rem FOR EQN SOLUTION (NOT INVERSION)
670 For K=1 to NP
680 If K=I then 720
690 X=SSM(K,I);Q(K)=Q(K)-X*Q(I)    ;Rem X = 'ROW MULTIPLIER'
700 For J=I+1 to NP
710 SSM(K,J)=SSM(K,J)-X*SSM(I,J);Next J
720 Next K
730 Next I
740 F=0;P=0
750 For N=1 to NE
760 I=NOP(N,1);J=NOP(N,2)
770 L=CORD(J,1)-CORD(I,1);T=(CORD(I,2)+CORD(J,2))/2
780 AVP=(Q(I)+Q(J))/2
790 P=P+AVP*L;F=F+L/T                       ;Rem ELE. CONTRIBUTIONS TO P & F
800 Next
```

```
810 FFACT=F/P                    ;Rem FRICTION FACTOR (THE MERIT FUNCTION)
820 Return
1000 Data 5,4,2,5
1010 Data 0,0.5                  ;Rem X & T FOR EACH NODE
1020 Data 0.25,0.625
1030 Data 0.5,0.75
1040 Data 0.75,0.875
1050 Data 1,1
1060 Data 1,2                    ;Rem NODE NOS FOR EACH ELEMENT
1070 Data 2,3
1080 Data 3,4
1090 Data 4,5
1100 Data 1,0                    ;Rem NODE NO. & SPECIFIED PRESSURE
1110 Data 5,0
1115 Data 1,2                    ;Rem NODE NO. & 2 INDICATES T VARIES
1120 Data 2,2                    ;Rem HERE ALL NODE THICKNESSES VARIABLE
1130 Data 3,2
1140 Data 4,2
1150 Data 5,2
```

The program provides another useful example of the application of gradient methods to the optimization of finite element models and also, through eqns (23.51) and (23.52), a very simple example of the types of calculation required in aerodynamical problems.[17]

23.11 Conclusions

In a very broad ranging text we have dealt with simple skeletal problems, the basic theories of solid and fluid mechanics and of optimization, techniques for formulating finite elements, and programming techniques for solving finite element problems.

From the beginnings of the 1950s FEM has come a long way and in many respects we could view the method as fully developed. For example, from the wide range of linear through to quintic elements derived in the text we can select elements most appropriate in terms of simplicity and acceptable accuracy for the problem we wish to solve. For simple one-dimensional problems, for example, linear elements should in most cases prove satisfactory. For plane stress or thick plate and shell problems, on the other hand, a quadratic basis is, practically speaking, the minimum choice, whilst for thin plate or shell problems a cubic basis (which may be transformed to a quadratic one as in Section 8,10) is the practical minimum.

Having made our choice of element and the appropriate numerical integration (see Section 15.10 and 15.11), and chosen a mesh with due consideration given to the complexity of the displacement, stress, velocity, and pressure field likely to be encountered (see Section 1.1) we should also obtain solutions with, for example, one or more coarse trial meshes and examine the convergence (and extrapolateability) of the solution (see Section 3.9). Such examination, of course, is particularly important when singularities are present and it can be used to detect and determine the nature of these (see Section 15.9). Hence, whilst the question of accuracy remains a subject of continuing research,[22,23] the foregoing recommendation remains a necessary one.

In the way of applications there is still, however, much to be done. In the mechanics of solids large displacement problems (particularly those involving the large curvature correction introduced in Chapter 12) require further research. In the mechanics of fluids compressible flows and transonic flows remain something of a challenge (see Sections 19.9 and 19.10).

Finally, irrespective of the class of FEM problem, optimization of finite element models presents the greatest new challenge for the method, one for which it seems ideally suited because of the approximate physical images of mathematical problems which finite elements are able to form. For this purpose we can use

(1) Stepwise linear programming (SLP);[9,24]

(2) 'random search methods' (with exploratory changes in each design variable in turn), such as Box's 'complex' method,[25]

(3) first-order gradient methods such as steepest descent[8,26]

(4) second-order gradient methods such as Davidon's method[27] (of which the so-called 'DFP' and 'BFGS' methods are minor variations);

(5) simple optimality criterion methods.[7,28,29]

Stepwise or sequential linear programming is rather unwieldy, LP being designed for a very restricted type of problem. Random search methods, for example the 'simplex method' (see Section 22.5), require too many steps to be practical in finite element applications.

Gradient methods, on the other hand, are 'sensible' search methods, search being, theoretically at least, in the 'best' direction, that of steepest descent until close to the optimum. Second-order gradient methods improve this direction near the optimum but accurate evaluation of the Hessian matrix is very laborious when there are more than a few variables, whereas steepest descent methods can be much improved by the predictor–corrector modification introduced in Section 22.8.

A sensible recommendation for finite element applications, therefore, might indeed be the sort of combination of optimality criteria (such as FSD) and steepest descent (or the 'residual' method of Section 23.5) used in the programs of Sections 23.7–23.10.

These programs are, of course, interactive, a considerable advantage in optimization with FEM,[30] and can be extended to deal with an even wider range of problems. The first two, for example, being easily extended to deal with frame and shell problems.

In the latter instance a large number of elements with thirty df or more may be involved and calculation of $\{g\}$ on an element by element basis may be impractical. This difficulty can be overcome by using a polynomial shape function $z = f(x, y)$ and using the relatively few coefficients of this as the design variables. Indeed the linear interpolation used when only h_2/h_1 is varied in Section 23.10 provides a very simple example of such a process, one which has also been used in three dimensional problems.[31]

When we also seek to optimize the thickness of the shell, another shape function $t = f(x, y)$ and, of course, section limits will be needed, as they will be in most structural problems.

There are, of course, many other possible applications of FEM and optimization, for example to dynamical problems with frequency constraints[32, 33] as well as to more complex fluids problems that the simple exploratory exercises of Sections 23.9 and 23.10. The latter is a relatively unexplored area of considerable interest.

A much more demanding type of problem, however, is that of optimization of fluid flows coupled with optimization of an associated 'structure' leaving much scope for future work. Less demanding are the types of problems discussed in Appendix 6 and application of optimization to finite element models of these should prove a subject of considerable interest in the future.

References

1. Mohr, G. A. and Caffin, D. A. (1985). Penalty factors, Lagrange multipliers and basis transformation in the finite element method. *Trans I. E. Aust.*, **CE27** (3), 174.
2. Reddy, J. N. (1982). On penalty function methods in the finite element analysis of flow problems. *Int. J. Num. Meth. Fluids*, **2**, 151.
3. Mohr, G. A. (1981). Finite element formulation by nested interpolations: application to cubic elements. *Comput. Struct.*, **14**, 211.
4. Thomas, G. R. and Gallagher, R. H. (1976). A triangular element based on generalized potential energy concepts. In *Finite Elements for Thin Shells and Curved Members*, (ed. D. G. Ashwell and R. H. Gallagher). Wiley, London.
5. Mohr, G. A. (1984). Finite element analysis of viscous flows. Proceedings of the International Conference on Computational Techniques and Applications, University of Sydney, (ed. J. Noye and C. Fletcher). North-Holland, Amsterdam.
6. Mohr, G. A. (1984). Finite element analysis of viscous fluid flow. *Comput. Fluids*, **12**, 217.
7. Sander, P. and Fleury, C. (1978). A mixed method of structural optimization. *Int. J. Num. Meth. Engng*, **13**, 385.
8. Gellatly R. A. and Gallagher, R. H. (1966). A procedure for automated minimum weight structural design, I: Theoretical basis. *Aero. Quart.*, **17**, 216.
9. Wang, C. K. (1973). *Computer Methods in Advanced Structural Analysis*. International Textbook Co., New York.

10. Rozvany, G. I. N. and Gollub, W. (1990). Michell layouts for various combinations of line supports—I. *Int. J. Mech. Sci.*, **32**, 1021.
11. Mohr, G. A. and Garner, R. (1983). Reduced integration and penalty factors in an arch element. *Int. J. Struct.*, **3**(1), 9.
12. Irons, B. M. and Ahmad, S. (1980). *Techniques of Finite Elements*. Wiley, Chichester.
13. Barlow, J. (1989). More on optimal stress points—reduced integration, element distortions and error estimation. *Int. J. Num. Meth. Engng*, **28**, 1487.
14. Mohr, G. A. (1976). *Analysis and Design of Plate and Shell Structures using Finite Elements*. Ph.D. thesis, University of Cambridge.
15. Mohr, G. A. (1979). Design of shell shape using finite elements. *Comput. Struct.* **10**, 745.
16. Norris, C. H., Wilbur, J. B., and Utku, S. (1975). *Elementary Structural Analysis*, (3rd edn). Prentice-Hall, Englewood Cliffs, NJ.
17. Argyris, J. H., St Doltsnis, I., and Friz, H. (1989). Hermes space shuttle: exploration of reentry aerodynamics. *Comp. Meth. Appl. Mech. Engng*, **73**, 1.
18. Strutt, J. W. (3rd Baron Rayleigh) (1945). *The Theory of Sound*, (2nd edn). Dover, New York.
19. Crandall, S. H. (1956). *Engineering Analysis*. McGraw-Hill, New York.
20. Mohr, G. A. and Broome, N. D. (1982). A finite element lubrication model. Fourth Australian International Conference on Finite Element Methods in Engineering, University of Melbourne.
21. Yuan, S. W. (1967). *Foundations of Fluid Mechanics*. Prentice-Hall, Englewood Cliffs, NJ.
22. Zienkiewicz, O. C. and Morgan, K. (1983). *Finite Elements and Approximation*. Wiley, New York.
23. Rachowicz, W., Oden, J. T., and Demkowicz, L. (1989). Toward a universal h-p adaptive finite element strategy. Part 3: design of h-p meshes. *Comp. Meth. Appl. Mech. Engng*, **77**, 181.
24. John, K. V., Ramakrishnan, C. V., and Sharma, K. G. (1987). Minimum weight design of trusses using improved move limit method of sequential linear programming. *Comput. Struct.*, **27**, 583.
25. Tomas, J. A. (1979). Optimization of automobile components by means of the finite element method. *Proceedings of the Third Australian International Conference on Finite Element Methods*. University of NSW, Sydney.
26. Thevendran, V. and Thambiratnam, D. P. (1988). Minimum weight design of conical concrete water tanks. *Comput. Struct.*, **29**, 699.
27. Thevendran, V. (1982). Minimum weight design of a spherical shell under a concentrated load at the apex. *Int. J. Num. Meth. Engng*, **18**, 1091.
28. Fleury, C. (1980). An efficient optimality criteria approach to the minimum weight design of elastic structures. *Comput. Struct.*, **11**, 163.
29. Allwood, R. J. and Chung, Y. S. (1985). An optimality criteria method applied to the design of continuous beams of varying depth with stress, deflection and size constraints. *Comput. Struct.* **20**, 947
30. Sikiotis, E., Saouma, V., Long, M., and Rogger, W. (1990). Finite element based optimization of complex structures on a Cray-MP supercomputer. *Comput. Struct.*, **36**, 901.

31. Yao, T. and Choi, K. K. (1989). 3-D shape optimal design and automatic finite element regridding. *Int. J. Num. Meth. Engng*, **28**, 369.
32. Kam, T. Y. and Chang, R. R. (1989). Optimal design of laminated composite plates with dynamic and static considerations. *Comput. Struct.*, **32**, 387.
33. Sadek, E. A. (1989). An optimality criterion method for dynamic optimization of structures. *Int. J. Num. Meth. Engng*, **28**, 579.

Appendices

Appendix 1
Programming for FEMSS
using MEGABASIC

In the text a number of BASIC programs are included to introduce the reader to practical applications of FEM. These are MEGABASIC,[1] a language designed for microcomputers which has many attractive features.

1. 1 Mb of core can be addressed (minimum of 128 kb needed).

2. Up to sixty-four separate programs can be stored (in concatenated form if desired) and accessed in core simultaneously; any of these can be linked to a MAIN program and run as subroutines of this.

3. Loops, functions, and so on, can be nested without limit.

4. Arrays may have variable dimensions and consume up to 65 kb.

5. Code is semi-compiled so that execution is up to about ten times faster than interpretive BASICS and programs are more compact than compiled BASICS.

6. An efficient line editor is built in.

7. A background mode facility is also built in.

8. The handling of subroutines and common data is neater than in FORTRAN.

9. Format statements are neater than in FORTRAN or BASIC.

10. Using the DEF SHARED PROC command users can write their own computer language easily, that is, form their own commands.

11. It is compatible with most operating systems.

Hence MEGABASIC proves useful in forming the FEMSS system introduced in Section 15.7. Figure A1.1 illustrates the simple program structure used in this system in which only four workspaces are used, these being read from files of the same name on disc A. A simple introduction to the FEMSS system is given by the following program in which hypothetical stiffness and

Workspaces
File name   Proc. name   Data files on disc B

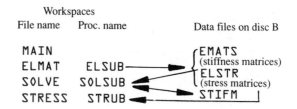

Fig. A1.1 Basic structure of FEMSS

stress matrices (with entries equal to the element number) are created and
filed to disc in the usual way (files EMATS, ELSTR on disc B).

```
10 Rem MAIN
20 Rem USE CREATE "B:EMATS", CREATE "B:STIFM", CREATE "B:ELSTR" TO FORM
30 Rem EMPTY FILES ON DISC B IF THIS HAS NOT BEEN DONE BEFORE FOR THIS DISC
40 Include 'ELMAT','SOLVE','STRESS'
50 Access * from *                  ;Rem ACCESS EVERYTHING FROM EVERYTHING
60 Dim NOP(100,2),Q(100)            ;Rem ELE. NODE NOS., LOAD MATRIX
70 Open £8,"B:EMATS"                ;Rem STORES ELE STIFFNESS MATRICES"
80 Open £9,"B:ELSTR"                ;Rem STORES ELE STRESS MATRICES
90 Open £10,"B:STIFM"               ;Rem STORES REDUCED STRUCTURE MATRIX
100 Def shared data NP,NE,NDF,NCN,NOP( ),Q( );Rem SHARED WITH THE SUBROUTINES
110 Read NP,NE,NDF,NCN              ;Rem £ NODES, ELES, D.F./NODE, NODES/ELE.
120 For N=1 to NE;For I=1 to NCN;Read NOP(N,I);Next;Next
130 For N=1 to NE;ELSUB N;Next      ;Rem CALL ELSUB FOR EACH ELEMENT
140 Filepos(8)=0;Filepos(9)=0       ;Rem 'REWIND' THESE FILES
150 SOLSUB                          ;Rem CALL SOLUTION ROUTINE
160 STRUB                           ;Rem CALL STRESS CALCULATION ROUTINE
170 End
180 Data 3,2,1,2                     ;Rem NP = 3, NE = 2, NDF = 1, NCN = 2
190 Data 1,2                         ;Rem NODE NUMBERS FOR FIRST ELEMENT
200 Data 2,3                         ;Rem NODE NUMBERS FOR SECOND ELEMENT
```

```
10 Rem ELMAT
20 Def shared proc ELSUB @N         ;Rem DECLARES A SUBROUTINE, N=ARGUMENT
30 Dim S(2,2),C(1,2)                ;Rem ELE STIFFNESS & STRESS MATRICES
40 For J=1 to 2
50 For I=1 to 2;S(I,J)=N;Next       ;Rem FORM HYPOTHETICAL ELE MATRICES
60 For I=1 to 1;C(I,J)=N;Next
70 Next
80 For I=1 to 2;For J=1 to 2
90 Write £8,S(I,J);Next;Next        ;Rem FILE ELE STIFFNESS MATRICES
100 For I=1 to 1;For J=1 to 2
110 Write £9,C(I,J);Next;Next       ;Rem FILE ELE STRESS MATRICES
120 Return;Proc end
130 End
```

```
10 Rem SOLVE
20 Def shared proc SOLSUB
30 Dim SK(100,100),ESM(2,2),D(100)  ;Rem STRUCTURE MATRIX, 'ESM', DISPL VECTOR
40 Def shared data D( )             ;Rem SHARE WITH STRESS SUBROUTINE
50 For N=1 to NE
60 For I=1 to 2;For J=1 to 2
70 Read £8,ESM(I,J);Next;Next;Next N;Rem READ ESM FROM FILE £8
80 ! "LAST ELEMENT STIFFNESS MATRIX"
90 For I=1 to 2;For J=1 to 2
100 ! %"10E2",ESM(I,J);Next;Next    ;Rem LAST ESM TO SCREEN
```

```
110 NN=NP*NDF
120 For I=1 to NN;For J=1 to NN
130 SK(I,J)=10*I+J;Write £10,SK(I,J);Rem FILE HYPOTHETICAL STRUCTURE MATRIX
140 Next;Next
150 Filepos(10)=filepos(10)-5*NN     ;Rem BACKSPACE FILED STRUCTURE MATRIX -N.B.
160 For J=1 to NN                     ;Rem - MEGABASIC USES 5 BYTES/NUMBER
170 Read £10,D(J);Next                ;Rem LAST ROW OF SK INTO D( )
180 !;! "SECOND LAST ROW OF STRUCTURE MATRIX"
190 For J=1 to NN;! %"I5",D(J);Next  ;Rem LAST ROW OF SK TO SCREEN
200 Return;Proc end
210 End


10 Rem STRESS
20 Def shared proc STRUB
30 Dim F(1,2)                         ;Rem SPACE FOR ELE STRESS MATRICES
40 For N=1 to NE
50 For I=1 to 1;For J=1 to 2
60 Read £9,F(I,J);Next;Next           ;Rem READ ELE STRESS MATRICES
70 Next
80 !;! "LAST ELEMENT STRESS MATRIX"
90 For I=1 to 1;For J=1 to 2
100 ! %"10F2",F(I,J);Next;Next        ;Rem LAST ELE STRESS MATRIX TO SCREEN
110 Return;Proc end
120 End
```

In subroutine SOLSUB a hypothetical structure stiffness matrix, with entries corresponding to the locations in the matrix, is created and filed as STIFM on disc B. Usually this is recalled for back substitution to obtain the displacement solution for the structure but here we simply read back and print the last row of the structure matrix.

Finally subroutine STRUB is called and this reads back the element stress matrices. Usually these are multiplied by the element nodal displacements to calculate the element stresses but here we simply print the last stress matrix to indicate successful access to this file.

The reader may notice some new features, such as the DEF SHARED PROC command, in the foregoing demonstration program. It may be helpful, therefore, to have an extremely brief look at some elementary MEGABASIC commands.

Workspace operations

USE MATS   creates an empty workspace MATS or, if this already exists, makes it the current workspace.

USE ⟨carriage return⟩ followed by repeated use of the space bars moves forwards through the workspaces (repeated use of the backspace key moves backwards through them). Carriage return stops in the current workspace.

SHOW   lists the current workspaces.

INCLUDE 'ELMAT', 'SOLVE', 'STRESS loads these files from disc A into workspaces of the same name (if these are not already present) and runs these when the current workspace program is RUN.

ACCESS ⟨package list⟩ [FROM⟨package list⟩]
                 For example, ACCESS 'SUB1', 'SUB2' FROM 'MAIN' allows MAIN to access workspaces SUB1 and SUB2. When the optional FROM clause is omitted the workspace containing the ACCESS command is permitted access to SUB1 and SUB2. * means all workspaces and this abbreviation is used in line 50 of the first routine of the above program.

SHOW ACCESS shows the access permitted (by prior use of the ACCESS command) between other workspaces and the current one.

SAVE PROG saves the program in the current workspace on disc A as PROG. pgm (.pgm is the default file type).

RUN　　　　　　　runs the program in the current workspace.

CREATE "B: MATS" creates file MATS on disc B (for disc A, omit B:).

LOAD B: MAIN loads MAIN.pgm from disc B into the current workspace which will take the name MAIN if not already named.

OPEN #8, "B: MATS" opens the file MATS on disc B for access using channel number 8 (this may be 0–31 but 0–7 are reserved for actual devices, for examples 0 for console and 1 for printer). Note that the printer used for the programs listed in the present text replaces # by £

WRITE #8,25,12,1988 writes the numbers 25,12,1988 to file #8.

FILEPOS (8) = 0 rewinds the file to the start. This may be used to permit random access, as in line 150 of subroutine SOLSUB in the above program.

READ #8,X,Y,Z reads the next three entries of this file.

CLOSE　　　　　closes all files. This is automatically done at end of program execution.

DIR　　　　　　lists files on disc A; use DIR 2 for disc B.

DESTROY "B:MATS" erases file MATS on disc B.

Program creation and alteration

ENTER 300,20 provides automatic line numbering in steps of 20 starting with line number 300.

REN 100,10,20,80 renumbers line 20 to 80 commencing with line number 100 and using steps of 10

CHANGE 1,100, "OLD", "NEW" replaces the string OLD by NEW up to line 100. (To go to the end of the file replace 100 by $).

DEL 1,100 deletes lines 1 to 100.

Line editing

ED	commences line editing of the program in the current workspace.
ED 100	commences editing at line 100.
CTRL-C	(or any command) terminates editing, otherwise editing is automatically terminated when the last line has been edited.
CTRL-A	copies the current character in the current line.
CTRL-Z	deletes the current character in the current line.
CTRL-Y	commences an insertion in the current line (CTRL-Y again ends the insertion).

Input/output statements

INPUT/PRINT and READ/WRITE.   As noted above these are associated with a channel number; if this is omitted the default number 0 (the console) is used. Note that the abbreviation ! may be used for PRINT.

LI list the program in the current workspace on the screen.

LI   #1   lists the program in the current workspace on the printer.

! % "I5", N, % "15E6", X, % "C10F3", Y prints N in integer format, X in exponential format, and Y in fixed point format with commas (for example, 1,510.560).

A$ = "15E6"; ! % A$, X, Y, Z   prints X, Y, Z in exponential format.

Subroutines

Subroutines may be created within a single program using the usual GOSUB and RETURN commands.

The DEF SHARED PROC command illustrated in the foregoing demonstration program is preferable as it allows separate subroutine files to be combined (using INCLUDE . . .) and data sharing is properly controlled (using DEF SHARED DATA). Such subroutines are then terminated using RETURN; PROC END.

The DEF FUNC command (DEF SHARED FUNC is to be shared with other subroutines) can be used when only a single result from a numeric expression (for example, $F = X*X + X$) is to be returned. When more than one line is used RETURN; FUNC END is used to terminate.

The text gives a number of useful programs. These, and some others available in the literature, are worth collecting together.

1. Direct current (d.c) networks. These are a useful introduction to some FEM concepts and extension to a.c. networks provides an interesting exercise.[2]

2. 'Arbitrary' networks. For example those for project management are discussed in Appendix 6.

3. Truss analysis. See Section 1.11.

4. Frame analysis. Except for dealing with distributed loads these are a straightforward extension of the truss program.[2]

5. Constraint techniques. It is somewhat artificial but the program of Section 4.5 provides an introduction to these.

6. Interpolation functions. The Gauss–Jordan reduction routine of Mohr and Milner[2] is useful for forming, for example, the higher-order interpolation matrices of Chapter 5 because it uses pivotal condensation.

7. Plane stress and strain. The linear strain triangle program of Section 15.8 provides a useful introduction to this type of problem and to programming of isoparametric elements.

8. Thin plate bending. Yang's program[3] provides an introduction to this type of problem using rectangular bicubic elements but the program of Section 15.8 (using triangular elements derived using basis transformation) is much more useful.

9. Shells. The program of Section 15.7 provides a useful introduction to general shell analysis using curved elements. Alternatively a very useful facet element is obtained by combining the elements of Section 8.10 and 10.6 using eqn (10.16). It should also be noted that axisymmetric shell problems, following the formulation of Section 11.3, are simpler than those of frame analysis.

10. Non-linear material and large displacement problems. The frame program of Mohr and Milner[2], for example, is very easily extended to deal with large displacements by the simple expedient of 'load stepping' or, more desirably, by the use of numerical integration to form the tangent stiffness matrix (eqn (12.89); see also eqn (5.137)) and, more importantly, calculate residual loads following eqn (12.87).[4]

11. Eigenvalue analysis. The program of Section 13.11 is mathematically quite sophisticated but very economical. The compact determinant search program of Williams and Howson[5] is worth noting by those interested in skeletal structural analysis.

12. Analysis of vibration'. The program of Mohr and Milner[2] provides a limited introduction to the use of time stepping techniques and those of Hughes and Zienkiewicz[6,7] are useful for more advanced applications, though somewhat cumbersome.

13. Heat flow. The program of Heubner[8] provides a useful introduction.

14. Potential flow. The program of Section 20.6 provides an easy passage into this class of problem.

15. Viscous flow. The program of Section 20.7 may be used for simple viscous flow problems that is, those of Stokes flow.

16. Boundary element method. Brebbia[9] gives simple programs using constant linear elements for the analysis of pseudoharmonic problems which the user should have little difficulty mounting on a microcomputer.

17. Optimization. The program of Wang[10] provides an introduction. However, it uses sequential linear programming, which is somewhat unwidely, and the approach used in the programs of Chapter 23 is much more attractive.

All the programs referred to in the literature above, like those of the present text, can readily be mounted on a microcomputer.

In the present text a 640 kb machine with 8088(2) chips and an 8087(2) numeric co-processor running at 4.7/10 MHz was used. This gave quite acceptable run-times, except perhaps for the 30 df shell element which required some 3 minutes of computation time for one element.

Only dual floppy discs were used (as this configuration is currently more widely available to students), the first for program storage and the second for temporary matrix files. It was found that transfer times to and from these matrix files were not a significant part of total computation time but a hard disc is preferable from the reliability point of view.

With 80386 and 80486 machines now available running at up to 60 MHz, speed and not storage capacity is the only remaining problem and implementation of FEM on PCs is becoming increasingly practical. Moreover, innovations such as parallel processing[11] can further increase solution speed. For the purposes of teaching introductory comes, however, the FEMSS system described in this appendix exemplifies the use of microcomputers for didactic purposes.[12]

References

1. Cochran, C. (1984). *Megabasic, version 4.1.* The American Planning Corp., Alexandria, Va.
2. Mohr, G. A. and Milner, H. R. (1986). *A Microcomputer Introduction to the Finite Element Method.* Pitman, Melbourne.
3. Yang, T. Y. (1986). *Finite Element Structural Analysis.* Prentice-Hall, Englewood Cliffs, NJ.
4. Mohr, G. A. and Milner, H. R. (1981). Finite element analysis of large displacements in flexural systems. *Comput. Struct.*, **13**, 553.
5. Williams, F. W. and Howson, W. D. (1977). Compact calculation of natural frequencies and buckling loads for plane frames. *Int. J. Num. Meth. Engng*, **11**, 1967.
6. Hughes, T. J. R. (1987). *The Finite Element Method – Linear Static and Dynamic Finite Element Analysis.* Prentice-Hall, Englewood Cliffs, NJ.
7. Zienkiewicz, O. C. (1977). *The Finite Element Method,* (3rd edn). McGraw-Hill, London.
8. Heubner, K. H. (1975). *The Finite Element Method for Engineers.* Wiley., New York.
9. Brebbia, C. A. (1984). *The Boundary Element Method for Engineers.* Pentech Press, Plymouth.
10. Wang, C. K. (1973). *Computer Methods in Advanced Structural Analysis.* International Textbook Co., New York.
11. Farhat, C. and Wilson, E. (1988). A parallel active equation solver. *Comput. Struct.*, **28**, 289.
12. Vasko, T. and Dicheva, D. (1987). Computers in education – an international overview. *Scientific World*, **31**(2), 11.

Appendix 2
Basic equations of the displacement method

Displacement interpolation

$$u = \{f\}^t \{d\}. \tag{A2.1}$$

Strain interpolation

$$\{\varepsilon\} = B\{d\}. \tag{A2.2}$$

where B is a differential function of $\{f\}$, for example, for plane stress and strain

$$
\begin{array}{cc}
u \text{ columns} & v \text{ columns}
\end{array}
$$
$$
B = \begin{bmatrix}
\{\partial f/\partial x\}^t & \{0\}^t \\
\{0\}^t & \{\partial f/\partial y\}^t \\
\{\partial f/\partial y\}^t & \{\partial f/\partial x\}^t
\end{bmatrix}. \tag{A2.3}
$$

Element equilibrium equations

$$(k + k_b)\{d\} = \{q_r\} + \{q_t\} + \{q_b\} + \{q_0\} = \{q\} \tag{A2.4}$$

where

$$k = \int B^t D B \, dV \tag{A2.5}$$

$$k_b = \mu \int F F^t \, dS \tag{A2.6}$$

$$\{q_t\} = \int F \{T\} \, dS \tag{A2.7}$$

$$\{q_b\} = \int F \{X\} \, dV \tag{A2.8}$$

$$\{q_0\} = -\int B^t D \{\varepsilon_0\} \, dV \tag{A2.9}$$

and $\{q_r\}$ are the interelement reactions which are assumed to be approximately self-cancelling between elements, to not be developed at free boundaries, and to be absorbed at restrained boundaries.

System equations

$$K\{D\} = \{Q\} \tag{A2.10}$$

where

$$K = \sum k + \sum k_b, \quad \{d\} = c\{D\}, \quad \{Q\} = \sum\{q\}. \qquad \text{(A2.11)}$$

Element stresses

$$\{\sigma\} = D(B\{d\} - \{\varepsilon_0\}). \qquad \text{(A2.12)}$$

Appendix 3
Basic equations for isoparametric elements for plane stress and strain

Isoparametric mapping of the derivatives (in a triangle)

$$\begin{bmatrix} \partial\{f\}^{\iota}/\partial x \\ \partial\{f\}^{\iota}/\partial y \end{bmatrix} = J^{-1}\begin{bmatrix} \partial\{f\}^{\iota}/\partial L_1 \\ \partial\{f\}^{\iota}/\partial L_2 \end{bmatrix} = J^{-1}S \qquad (A3.1)$$

where

$$J = S[\{x\} \ \{y\}].$$

Note that in evaluating $\partial\{f\}^{\iota}/\partial L_1$ and $\partial\{f\}^{\iota}/\partial L_2$ the area coordinate L_3 is eliminated from each entry in $\{f\}$ using the identity $L_3 = 1 - L_1 - L_2$ before forming the expressions for the local derivatives.

Strain interpolation matrix

$$B = \begin{array}{cc} u \text{ columns} & v \text{ columns} \\ \begin{bmatrix} \partial\{f\}^{\iota}/\partial x & \{0\}^{\iota} \\ \{0\}^{\iota} & \partial\{f\}^{\iota}/\partial y \\ \partial\{f\}^{\iota}/\partial y & \partial\{f\}^{\iota}/\partial x \end{bmatrix} \end{array} . \qquad (A3.3)$$

Element stiffness matrix

$$k = \sum_{i=1}^{n} B^{\iota}DB(t|J|_{\text{abs}}\omega_i) \qquad (A3.4)$$

where the summation is over n numerical integration points with associated weights ω_i.

Element stresses
These are the same as eqn (A2.12) at any desired point in the element. Most commonly the stress sampling points are the most heavily weighted integration points but in some cases nodal values, after averaging between elements, are acceptable.

Appendix 4
Natural equations for thin plates

Natural slopes (see Section 5.8)

$$\phi_a = L_{21} \frac{\partial w}{\partial s_a} = \frac{\partial w}{\partial L_2} - \frac{\partial w}{\partial L_1} \tag{A4.1}$$

$$= x_{21} \phi_x + y_{21} \phi_y. \tag{A4.2}$$

The first result is used in forming the inverse interpolation matrix A^{-1} and the second result is used to transform from Cartesian to natural slopes in a matrix T.

Natural curvatures (see Section 5.10)

$$\chi_a = \frac{\partial^2 w}{\partial L_1^2} + \frac{\partial^2 w}{\partial L_2^2} - \frac{2\partial^2 w}{\partial L_1 \partial L_2} \tag{A4.3}$$

$$= L_{21}^2 (c_{ax}^2 \chi_x + c_{ay}^2 \chi_y + 2 c_{ax} c_{ay} \chi_{xy}) \tag{A4.4}$$

The first result is used in the formation of curvature interpolation matrix B and the second result is used to transform from Cartesian to natural curvatures in matrix T in elements with curvature freedoms.

Element stiffness matrix

$$k = T^t A_i^t k^* A_s T \tag{A4.5}$$

where A_s is the truncated interpolation matrix, that is, with the rigid body rows removed. Correspondingly matrix B has the rigid body columns removed.

$$k^* = \int S^t DS \, dV \quad \text{(numerical integration in practice)} \tag{A4.6}$$

$$S = HB \tag{A4.7}$$

$$H = \begin{bmatrix} c_{ax}^2 & c_{ay}^2 & \sqrt{2 c_{ax} c_{ay}} \\ c_{bx}^2 & c_{by}^2 & \sqrt{2 c_{bx} c_{by}} \\ c_{cx}^2 & c_{cy}^2 & \sqrt{2 c_{cx} c_{cy}} \end{bmatrix}^{-1} \tag{A4.8}$$

$$B = \begin{bmatrix} (\partial^2\{M\}^{\mathrm{t}}/\partial L_1^2 + \partial^2\{M\}^{\mathrm{t}}/\partial L_2^2 - 2\partial^2\{M\}^{\mathrm{t}}/\partial L_1\partial L_2)/L_{21}^2 \\ (\partial^2\{M\}^{\mathrm{t}}/\partial L_2^2 + \partial^2\{M\}^{\mathrm{t}}/\partial L_3^2 - 2\partial^2\{M\}^{\mathrm{t}}/\partial L_2\partial L_3)/L_{32}^2 \\ (\partial^2\{M\}^{\mathrm{t}}/\partial L_3^2 + \partial^2\{M\}^{\mathrm{t}}/\partial L_1^2 - 2\partial^2\{M\}^{\mathrm{t}}/\partial L_3\partial L_1)/L_{13}^2 \end{bmatrix} \quad (A4.9)$$

where, as noted in Section 8.4, when the transformation of eqn (A4.8) is used corresponding adjustments to the twist modulus in matrix D must be made: it must be doubled when k^* is calculated and multiplied by $\sqrt{2}$ when the plate moments are calculated.

Appendix 5
Some elementary calculus

In this appendix we summarize a principal result of Chapter 17, namely the application of perturbations[†] to the functional or differential equations governing the physical problem. The purpose of this is to pose the problem in some alternative and perhaps more convenient form. Such perturbation techniques form perhaps the most fundamental and powerful tool of calculus as new results can be obtained without appeal to special theorems.

A5.1 Minimization of functionals using perturbation

As an example consider the functional of eqn (3.1):

$$\Pi_e = \tfrac{1}{2}\int \{\varepsilon\}^t\{\sigma\}\,dV - \int\{\bar{u}\}^t\{T\}\,dS. \qquad (A5.1)$$

Allowing a perturbation $\{\delta d\}$ and associated variations $\{\delta\varepsilon\} = B\{\delta d\}$ and $\{\delta\bar{u}\} = F^t\{\delta d\}$ and noting that $\{\sigma\} = DB\{d\}$ we obtain the result of eqn (3.12);

$$\{\delta d\}^t\int B^t DB\,dV\{d\} = \{\delta d\}^t\int F\{T\}\,dS + \{\delta d\}^t\{q_r\} \qquad (A5.2)$$

if we include the interelement reactions $[q_r]$. Then deleting the perturbation $\{\delta d\}$ as a common factor we obtain the same result as that obtained by the theorem of minimum potential energy (eqn (3.8)). In the context of structural mechanics the foregoing procedure is often referred to as the method of virtual work.

A5.2 Formation of functionals from a differential equation

If we take for example the differential equation for the shape of a beam (the elastica), $EId^4/dx^4 = q$, and allow a change in shape given by $\delta v = g(x)$ then

---

[†] Not to be confused with the *fluxions* or rates of Newtonian calculus. In these the perturbation has an associated change in dimension by which we generally divide and take the limit to calculate a rate of change in some quantity.

the associated virtual work is

$$\delta F = \int \left(\frac{EId^4v}{dx^4} - q \right) \delta v \, dx. \tag{A5.3}$$

Integrating by parts twice using the standard expression $\int uv' = uv| - \int vu'$ we obtain eqn (17.11):

$$\delta F = EI[\delta vv_{xxx} - \delta v_x v_{xx}]_0^L - \int_0^L (EI\delta v_{xx}v_{xx} - \delta vq) \, dx. \tag{A5.4}$$

The boundary terms in eqn (A5.4) are the interelement reactions. Omitting these and taking the arbitrary perturbation $\delta v = v$ we obtain

$$F(v) = \int [\tfrac{1}{2}EI(v_{xx})^2 - qv] \, dx \tag{A5.5}$$

which is the required functional. Note that the factor $1/2$ must be introduced because $\delta(\tfrac{1}{2}v_{xx}^2) = \tfrac{1}{2}(\delta v_{xx}v_{xx} + v_{xx}\delta v_{xx})$ which follows from the product rule of differentiation (as does the standard expression for integration by parts noted above). This point is also important in non-linear problems (see, for example, eqn (12.86)).

A5.3 Formation of a differential equation from a functional

Taking for example the functional of eqn (A5.5) and allowing perturbation δv and integrating by parts twice we obtain

$$\delta F(v) = EI(\delta v_x v_{xx} - \delta vv_{xxx})| + \int [EI\delta vv_{xxxx} - q\delta v] \, dx. \tag{A5.6}$$

Once again omitting the boundary terms (see Section 17.3) and deleting the common factor δv we obtain the required differential equation.

Note that the boundary terms cannot always be neglected. In the finite element method we can always neglect them as interelement reactions which, at least in the limit of mesh refinement, should be self-cancelling. But in elements at the boundary, the boundary terms correspond to boundary reactions in the structural mechanics context or boundary fluxes in the fluid mechanics context. In the former case such reactions are only developed when the boundary is fixed when the corresponding equation is omitted from the solution. In fluid mechanics problems, however, there may be boundary fluxes or flows (see, for example, eqn (19.74)) which must be taken into account. Integration by parts, therefore, performs a very useful function in revealing these boundary terms.

A5.4 Direct attack upon the differential equation

Suppose that we wish to form equations that satisfy the differential equation $L(\{\bar{u}\}) = 0$ in some way. In the finite element method we substitute interpolations $\{\bar{u}\} = F^{t}\{d\}$ which, because they are based on finite polynomials rather than transcendental functions, introduce a residual error $R = L(F^{t}\{d\})$. Multiplying R by weighting functions w_i at n points, integrating over the element volume, and equating the result to zero we obtain

$$\int w_i R \, dV = \int w_i L(F^{t}\{d\}) = 0, \qquad i = 1 \rightarrow n. \tag{A5.7}$$

Choosing as perturbations $w_i = \{\delta\bar{u}\} = F^{t}\{\delta d\}$ we obtain

$$\{\delta d\}^{t}(\int F F^{*t} \, dV)\{d\} = \{\delta d\}^{t}\{0\} \tag{A5.8}$$

where F^* is formed by differentiation in accordance with the terms in the differential equation $L(\bar{u}) = 0$ and the common factor $\{\delta d\}^{t}$ is deleted.

As noted in Section 17.1 this is, of course, completely equivalent to the Galerkin method in which the weighting functions are chosen as the interpolation functions. Beginning with a perturbation argument, however, makes it a good deal clearer why we might do this, namely that there is little reason to use approximating functions (interpolation functions in FEM) for the perturbations different from those used for the variables which we have perturbed. More importantly, use of the same interpolations for the perturbation will yield the most accurate solution in the least squares sense.

Equations (A5.8) could be used to form a solution but, generally, integration by parts is applied to reduce the order of derivatives in F^*, and increase the order in F from zero. Our motivation for reducing the order of the derivatives is not merely that of simplicity but that of increased accuracy. If the differential equation has even-ordered derivatives only, for example, we can obtain a solution with a least squares character. This will obviously minimize the truncation error in the solution resulting from the necessary use of approximating functions, as distinct from infinite series which must, if we use enough terms, be able to express the exact solution. When we use an infinite series approach, such as the Fourier series method, we are ultimately limited to numerical expression of the solution to some limited accuracy. In the finite element method, therefore, we commence the approximation process at the outset by choosing finite polynomials as the basis of our interpolation functions and increase the accuracy of the solution by using more elements.

A5.5 Integration by parts in two dimensions (from first principles)

Consider as an example the two-dimensional expression

$$\iint v \nabla^2 u \, dx \, dy = \iint v \left(\frac{\partial^2 u}{\partial x^2} + \frac{\partial^2 u}{\partial y^2} \right) dx \, dy. \tag{A5.9}$$

To reduce the order of the derivatives we integrate $\partial^2 u/\partial x^2$ with respect to x and $\partial^2 u/\partial y^2$ with respect to y to obtain

$$\iint v \frac{\partial^2 u}{\partial x^2}\, dx\, dy = \int v \frac{\partial u}{\partial x}\, dy \bigg| - \iint \frac{\partial v}{\partial x} \frac{\partial u}{\partial x}\, dx\, dy \qquad \text{(A5.10)}$$

$$\iint v \frac{\partial^2 u}{\partial y^2}\, dx\, dy = \int v \frac{\partial u}{\partial y}\, dx \bigg| - \iint \frac{\partial v}{\partial y} \frac{\partial u}{\partial y}\, dx\, dy. \qquad \text{(A5.11)}$$

Note that the direction cosines of the outwardly directed normal at the boundary are $c_x = \partial x/\partial n$, $c_y = \partial y/\partial n$ and hence that $c_x = \partial y/\partial S$, $c_y = \partial x/\partial S$. As noted in Section 17.4 we chose $c_y = \partial x/\partial S$ to ensure that $\int dS$ around the boundary yields a positive result. This considerably simplifies the calculations, avoiding the need to consider the parity of the boundary terms in detail. Then

$$dx = c_x\, dn = c_y\, dS. \qquad dy = c_y\, dn = c_x\, dS. \qquad \text{(A5.12)}$$

Hence the boundary terms of eqns (A5.10) and (A5.11) may be combined to yield

$$\int v \frac{\partial u}{\partial x}\, dy + \int v \frac{\partial u}{\partial y}\, dx = \int v \frac{\partial u}{\partial x} c_x\, dS + \int v \frac{\partial u}{\partial y} c_y\, dS = \int v \frac{\partial u}{\partial n}\, dS \qquad \text{(A5.13)}$$

if we note that

$$\frac{\partial u}{\partial n} = \frac{\partial u}{\partial x} \frac{\partial x}{\partial n} + \frac{\partial u}{\partial y} \frac{\partial y}{\partial n} = \frac{c_x \partial u}{\partial x} + \frac{c_y \partial u}{\partial y}. \qquad \text{(A5.14)}$$

Thus our original expression $\iint v \nabla^2 u\, dx\, dy$ reduces to

$$- \iint (v_x u_x - v_y u_y)\, dx\, dy + \int v u_n\, dS. \qquad \text{(A5.15)}$$

In finite element practice we satisfy this differential equation in an integral sense over a subdomain or element and hence introduce the interpolation $u = \{f\}^t\{u\}$. If we choose v to be the perturbation $\delta u = \{f\}^t\{\delta u\}$, eqn (A5.15) yields

$$\int \{f\} \frac{\partial u}{\partial n}\, dS = \iint (\{f_x\}\{f_x\}^t + \{f_y\}\{f_y\}^t)\, dx\, dy\, \{u\} \qquad \text{(A5.16)}$$

which is of the customary form $\{q\} = k\{u\}$ with k expressible in the form $\iint B^t DB\, dV$. Here D is simply a unit matrix but when the interior integral is associated with an anisotropic permeability, for example, it will not be.

If all direction cosines are assumed to be positive then as noted above, the calculations are greatly simplified and, more importantly, we can proceed from first principles. This in turn allows us to more clearly identify the form of the boundary integrals. The interested reader can easily prove that, proceeding instead with $c_y = - \partial x / \partial S$ and considering the sign changes that occur in all quadrants, the same result is obtained.[1]

Reference

1. Zienkiewicz, O. C. (1977). *The Finite Element Method* (3rd edn). McGraw-Hill, London.

Appendix 6
Other problems with
a finite element character

There are many other problems, besides those of computational mechanics, that possess a finite element character. Important examples occur in the fields of economics and management, including

(1) product distribution problems;

(2) economic distribution problems;

(3) project scheduling problems.

In product distribution problems the nodes are the manufacture and distribution points whilst the elements are the routes between these nodes. The 'stiffness' of these routes or elements is the unit cost of moving an item between its pair of nodes. The analyst seeks to minimize the total cost of distributing items through the network to meet the demand at each point of sale.[1] Such problems, like those of traffic flow, have a relationship to those of fluid flow in pipe networks[2,3] and, like the latter, can be solved in matrix form.

In economic distribution problems capital is distributed to meet various needs to best advantage overall. An example is Project LINK which was commenced in 1968 to relate the major macroeconomic models being used in each of the main regions of the world.[4] To this end such quantities as a world trade matrix are formed. In the simplest cases the problem is linear and is again comparable to the network problems of Chapter 1.

In project sheduling problems the nodes are points in time at which one or several 'elements' of the complete project finish and others begin. With each element we associate a duration required for its completion. The 'stiffness' of each element is then the increment in resource commitment which results in a unit decrement in its duration. Such a concept is used in the flowline scheduling method where elements are assigned a 'modulus' K and the problem can be posed in algebraic form using these moduli.[5]

As an example of this class of problem we introduce the critical path method (CPM) in its original form in which the elements are 'rigid', that is,

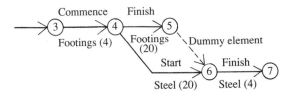

Fig. A6.1 Arrow notation flow chart

their durations are fixed, so that matrix solution is not necessary. Figure A6.1 shows a segment of a flow chart for a construction project using arrow notation, as distinct from precedence notation[5] which is more conductive to posing the problem in matrix form. The numbered circles are events (we shall call them nodes) at which activities (or elements) of the project end and others begin. Then each element is defined by two nodes and an estimated duration (shown in parentheses).

Here both the footing and steel elements have been split into two parts to form a sensible schedule. The dummy element enforces the proper sequence of work and in this case will have zero duration, that is, steel finishing can commence as soon as the footings are finished. Dummy elements can have non-zero duration, for example to allow curing of concrete, or may have negative duration to allow overlap of elements.

In such networks a critical path, which is the longest connected sequence of elements, exists and this is the duration of the complete project. Elements lying on this path cannot be delayed without delaying project completion but non-critical elements can be delayed a certain amount called their *float*.

For each element, therefore, we can define an earliest start (ES), earliest finish (EF), latest start (LS), and latest finish (LF) time. The ES and EF times are the earliest at which an element can commence and finish without affecting preceding elements and the LS and LF times are the latest at which an element can finish without delaying completion of the whole project. These four times are related by the simple results

$$EF = ES + D, \qquad LF = LS + D \qquad (A6.1)$$

where D is the duration of the element.

We analyse the project by constructing a flowchart and carrying out a critical path analysis to find the critical path and its elements. As an example consider the simple flowchart shown in Fig. A6.2. The critical elements are represented by double lines which in such a simple case can be identified using the tabulation of Table A6.1. Here the alternative path times to each

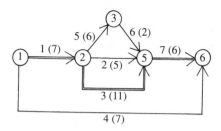

Fig. A6.2 Flow chart for a hypothetical project. Element numbers are marked on each arrow and their durations follow in parentheses

node are calculated and the longest of these is the 'latest time' T for that node. This is then the ES time for the following elements and their EF times are given by adding their durations to this ES time using eqns A6.1.

The operations of Table A6.1 are called the *forward pass* and we now follow this with a *backward pass*. First we attach the time 24 days to node 6, this being the project time. This is the LF time for elements meeting at this node and their LS times are calculated by subtracting their durations.

Continuing this backward pass operation throughout the flowchart the LF and LS times for each element are obtained. Then the elements for which ES = LS and EF = LF are *critical* and lie on the critical path. Elements which are not critical possess an amount of float F, which is calculated as

$$F = LS - ES = LF - EF. \tag{A6.2}$$

This is the amount of time by which such elements may be delayed without affecting project completion.

For larger problems, of course, a computer program such as the simple BASIC program which follows, is helpful.

Table A6.1 Tabular determination of T for each node

Node	Path times	Latest times
1	0	0
2	7	7
3	13	13
5	18, 12, 15	18
6	21, 10, 24, 7	24

```
5 Rem CRITICAL PATH METHOD PROGRAM
6 Rem
10 Dim N(10,2),D(10),C(10),W(10),T(10),P(3,10)
20 Dim S1(10),F1(10),S2(10),F2(10)
30 A$="        ";Read Z,E          ;Rem INPUT NO. NODES AND ELEMENTS
40 For I=1 to 3;For J=1 to 10
50 P(I,J)=0;Next;Next              ;Rem INTITIALIZE PRECEDENCE MATRIX
60 For I=1 to E
70 Read N(I,1),N(I,2),D(I),C(I),W(I);Rem INPUT ELEMENT DATA -N.B. ELEMENTS
80 Next                           ;Rem MUST BE IN NUMERICAL ORDER
90 Q=0;Q2=0                       ;Rem INITIALIZE TOTAL PROJECT COSTS
100 For K=1 to Z;T(K)=0;Next      ;Rem INITIALIZE LF TIMES FOR NODES
110 For K=1 to E;J=N(K,2);I=1
120 For M=1 to 3;If P(M,J)=0 then Goto 140
130 I=I+1;Next M                  ;Rem COLLECT PRECEEDING ELEMENT
140 P(I,J)=K;Next K               ;Rem NUMBERS FOR EACH NODE
150 For K=1 to E                  ;Rem COMMENCE FORWARD PASS **********
160 I=N(K,1);J=N(K,2);S1(K)=T(I)  ;Rem S1(K)=ES TIME FOR ELEMENT
170 F1(K)=T(I)+D(K)               ;Rem F1(K)=EF TIME FOR ELEMENT
180 If F1(K)>T(J) then T(J)=F1(K) ;Rem INCREASE NODE LF TIME IF NECESSARY
190 Q=Q+D(K)*C(K);Q2=Q2+D(K)*C(K)*W(K)
200 Next                          ;Rem END FORWARD PASS       **********
210 ! "JOBTIME = ",T(Z);! "JOBCOST = ",Q;! "PESSIMISTIC COST VARN = ",Q2
220 For I=1 to E;F2(I)=1000;Next   ;Rem INTITIALIZE LF TIMES
230 For K=1 to E                  ;Rem COMMENCE BACKWARD PASS XXXXXXXXX
240 R=E-K+1;I=N(R,1);J=N(R,2)     ;Rem MOVE BACWARDS THROUGH ELEMENTS
250 If J=Z then F2(R)=T(Z)        ;Rem LF TIME FOR LAST ELEMENT
260 S2(R)=F2(R)-D(R)              ;Rem LS = LF - DURATION
270 For M=1 to 3;S=P(M,I);If S=0 then Goto 290
280 If F2(S)>S2(R) then F2(S)=S2(R) ;Rem LF OF PREC ELES = LS OF THIS ELEMENT
290 Next
300 If S1(R)=S2(R) then F=F+D(R)*W(R);Rem INCREMENT PROJECT DELAY ESTIMATE
310 Next                          ;Rem END BACKWARD PASS      XXXXXXXXX
320 ! "PESSIMISTIC JOB DELAY = ",F;! A$,"ES",A$,"EF",A$,"LS",A$,"LF"
330 For K=1 to E;! %"10I",S1(K),F1(K),S2(K),F2(K);Next
340 End
400 Data 6,7                      ;Rem DATA FOR PROBLEM OF FIG A6.2
410 Data 1,2,7,10,0.1
420 Data 2,5,5,10,0.1
430 Data 2,5,11,10,0.1
440 Data 1,6,7,10,0.1
450 Data 2,3,6,10,0.1
460 Data 3,5,2,10,0.1
470 Data 5,6,6,10,0.1
```

The data input requirements are simply

Lines	Data
1	number of nodes (Z); number of elements (E).
E lines	first node number ($N(I,1)$); second node number ($N(I,2)$); duration ($D(I)$); cost ($C(I)$); and 'wastage' ($W(I)$).

The costs estimated for each activity are unit costs (that is, costs per unit time) and the 'wastages' are the estimated fractional increases in duration (and hence cost) of each element. Thus, the *pessimistic* estimate of element duration is given by adding this increase.

The data listed is for the problem of Fig. A6.2 and the program output is simply the ES, EF, LS, and LF times for each element and the optimistic and pessimistic estimates of total project time and cost. Note that in the problem of Fig. A6.2 there is no node 4. This omission is deliberate and allows for the insertion of 'dummy elements' or extra elements.

The program can easily be extended to provide *bar charts* for the project and these are usually fitted to a calendar. Extension of such programs to deal with resource scheduling of trades, equipment, materials, and so on, requires a large amount of coding and input of the resource data for such programs can itself be laborious. Resource sheduling can, however, easily be carried out manually with reference to a bar chart.

Such a program can easily be modified to allow a 'stiffness' and time limits (min. and max.) and the initial solution is with the average of these. Then steepest descent can be used to obtain an optimum solution for the scheduling problem.

References

1. Enrick, N. L. (1965). *Management Operations Research*. Holt, Rinehart and Winston, New York.
2. Salter, R. J. (1976). *Highway Traffic Analysis and Design*, (2nd edn). Macmillan, London.
3. Mohr, G. A. (1983). A bilinear rule for macroscopic traffic flow. *Australian Road Research*, 13(1), 38.
4. Klein, L. R., Pauly, P., and Voisin, P. (1982). The world economy – a global model. *Perspectives in Computing*, 2(2), 4.
5. Mohr, W. E. (1982). *Project Management and Control*, (3rd edn). Melbourne University Press.

Glossary of Symbols

$\{a\}$	coefficients of dimensionless interpolation polynomial
a, b	local coordinates in quadrilateral element
a, b, c	natural coordinates (that is, parallel to the sides) in a triangular element
A	element area, dimensionless interpolation matrix
b, d	breadth and depth of beam element (Chapter 1)
B	strain interpolation matrix
c	element damping matrix, particle damping parameter (Chapter 1), Chezy coefficient (Chapter 19), speed of sound (Chapter 19)
$\{c\}$	coefficients of Cartesian interpolation, cost vector (Chapter 22)
c, s	direction cosines of line element
c_x, c_y	direction cosines of normal to element boundary
c_{ax}, c_{ay}	direction cosines of side 12 of triangular element
c_v, c_p	specific heats at constant volume and constant pressure
C	system damping matrix (also K_D), B^t (Chapters 1, 3), capacitance (Chapter 1), Cartesian interpolation matrix (Chapters 3, 5)
C_0, C_1	zeroth and first-order displacement continuity
d	dimensionless coordinates ($0 \rightarrow 1$) in line element
$\{d\}$	nodal displacements for element
$\{d_N\}$	dimensionless natural freedoms for element
D	modulus matrix, diffusion coefficient in Fick's law (Chapter 18)
$\{D\}$	system displacement vector
D_N	natural modulus matrix for triangular or tetrahedral element
D_T	tangent modulus matrix
D_{XY}	twist modulus of plate or shell ($= Gt^3/12$ in isotropic case)
e_a, e_b, e_c	'eccentricities' of nodes in triangular element

$\{e\}$ or $\{e_i\}$	eigenvector
E	Young's modulus
f	yield function for plastic flow, Coriolis parameter (Chapter 19)
$\{f\}$	vector of interpolation functions
$\{f_x\}$, $\{f_{xx}\}$	first and second partial derivatives of $\{f\}$ with respect to x
$\{f_a\}$	first derivative of $\{f\}$ with respect to the coordinate S_a on side 12 of triangular element
F	interpolation matrix defined in eqn (2.10), gradient vector for plastic flow (Chapter 12)
F_t, F_c	tensile and compressive strengths of a material
F_y	yield strength of a material
F_{u1}, F_{v1}	Cartesian force components at node 1 of an element
$\{g\}$	steepest descent gradient vector
G	shear modulus ($= E/2(1 + v)$ in isotropic case)
$G(x, s)$	Green's function
h	characteristic length of element
H	Hessian transformation matrix, hardening parameter for plastic flow (Chapter 12), vector iteration matrix ($= M^{-1}K$ or $K^{-1}M$) for eigenvalue determination (Chapter 13), static pressure head (Chapters 16, 18)
I	unit or identity matrix
$\hat{\mathbf{i}}, \hat{\mathbf{j}}, \hat{\mathbf{k}}$	unit vectors parallel to Cartesian axes
j	'jot' or $\sqrt{-1}$
J	Jacobian matrix
J_1, J_2, J_3	first, second, and third stress invariants
k	element stiffness matix
k_D	element damping matrix
k_0, k_G, k_L	initial, geometric, and large displacement stiffness matrices for an element
k_x, k_y	Cartesian permeability coefficients for seepage
K	system stiffness matrix, bulk modulus (Chapter 16)
K_T	tangent stiffness matrix
K_Δ	upper triangular matrix after forward pass of Gauss reduction
L	length of line element, inductance (Chapter 1)
L_a, L_b	side lengths of rectangular element
L_a, L_b, L_c	side lengths of triangular element
L_1, L_2, L_3	area coordinates in triangular element
m	element mass matrix, particle mass (Chapter 13)
m_x, m_y, m_{xy}	unit moments in a plate or shell
M	system mass matrix, Mach number (Chapters 19, 20)

M_i	moment at node i of beam element
M_b	bending couple in a beam element
M_T	torsional moment in beam
M_∞	free stream Mach number
$\{M\}$	vector of modes in interpolation polynomial
n	normal coordinate
N	element stress matrix
N_x, N_y, N_{xy}	unit extensional forces in a shell
p	static pressure head, distributed load intensity on beam (Chapter 1)
P	axial force in line element, plan length of arch element (Chapter 11)
P_r	Prandtl number
q	flow rate in pipe
q_x, q_y	Cartesian flow rates
$\{q\}$	nodal load vector for element
$\{q_0\}, \{q_b\}, \{q_t\} \{q_r\}$	initial load, body force, surface traction, and interelement reaction load vectors for element
Q_x, Q_y	shear force resultants in thick plate or shell
$\{Q\}$	system load vector
$\{Q_n\}$	direct nodal loads to system
$\{Q(t)\}$	vector of time-dependent loads on system
r	radial coordinate, radius vector (Chapter 11)
R	residual error, electrical resistance (Chapter 1), universal gas constant (Chapters 16, 20)
R_x, R_y, R_{xy}	curvatures of shell shape
R_a, R_b, R_c	natural curvatures of shape of triangular shell element
s	dimensionless coordinate $(-1 \rightarrow 1)$ in line element or along the side of an element
s_a, s_b, s_c	coordinates along side of a triangular element
s_x, s_y	curvilinear coordinates on surface of a shell
S	strain interpolation matrix based on Cartesian coordinates, finite set (Chapter 4)
t	element thickness, time coordinate (Chapters 13, 14, 18, 19)
T	basis transformation matrix, temperature (Chapters 16, 18, 19)
$\{T\}$	vector of surface tractions
u, v	Cartesian velocity components (Chapters 16, 19, 20)
u, v, w	Cartesian displacement components
U	strain energy
U_e	element strain energy
U, V	Cartesian velocities of bearing surface (Chapter 19)

V	element volume, electrical voltage (Chapter 1)
V_x, V_y	unit shear forces
w_i	weighting factor for residual error
W_p	plastic work
W_e, W_i	external and internal work in an element
x, y, z	Cartesian coordinates
$\{x\}, \{y\}$	design and slack variables in linear programming
X, Y, Z	Cartesian body force intensities
$\{X\}$	vector of Cartesian body force intensities
Y	complex admittance
Z	complex impedance
α	step length in search techniques, slope of line element (Chapter 1), coefficient of linear thermal expansion (Chapter 3)
α_x, α_y	coefficients of anisotropic thermal expansion
α, β, γ	vertex angles of triangular element, natural extensional freedoms (Chapter 10)
β	penalty factor, shear parameter (Chapter 1)
γ	specific heat ratio (Chapter 19)
γ_x, γ_y	Cartesian transverse shearing strains in plate or shell
δ	time step length
Δ	area of triangular element
$\varepsilon_x, \varepsilon_y, \varepsilon_{xy}$	Cartesian strains (extensional), eddy viscosities (Chapter 19)
$\varepsilon_r, \varepsilon_z, \varepsilon_{rz}, \varepsilon_\phi$	strains in axisymmetric solid
ε_{ij}	strain tensor
$\{\varepsilon_0\}$	vector of initial strains
$\varepsilon_e, \varepsilon_p$	elastic and plastic strain components
η	similarity parameter for transient flow
θ	temperature rise, angle of twist per unit length (Chapters 16, 17)
$\kappa_x, \kappa_y, \kappa_z$	Cartesian coefficients of thermal conductivity
$\kappa_a, \kappa_b, \kappa_c$	artificial natural curvatures in plate or shell (Chapter 8)
λ	Lagrange multiplier or eigenvalue
Λ_e	element kinetic energy
μ	dynamic viscosity, boundary stiffness parameter (Chapter 3)
ν	Poisson's ratio, kinematic viscosity (Chapters 16, 19, 20)
Π	total potential energy
Π_e	element potential energy
ρ	mass density
$\sigma_x, \sigma_y, \sigma_{xy}$	Cartesian stresses (extensional)

σ_r, σ_c	radial and circumferential stresses
σ_{ij}	stress tensor
ϕ	potential flow function, St Venant's warping function (Chapters 16 and 17), magnetic potential (Chapter 18)
ϕ_x, ϕ_y, ϕ_z	Cartesian slope freedoms
ϕ_a, ϕ_b, ϕ_c	dimensionless natural slopes in triangular plate or shell element
Φ	viscous dissipation function
$\chi_x, \chi_y, \chi_{xy}$	Cartesian curvatures of deformation in plate or shell
χ_a, χ_b, χ_c	dimensionless natural curvatures in triangular plate or shell element
ψ	stream flow function, Prandtl's stress function (Chapters 16, 18), electrical potential (Chapter 18)
ψ_a, ψ_b, ψ_c	dimensioned natural slopes in triangular plate or shell element
ω	angular frequency of vibration, vorticity (Chapter 19)
ω_i	integration point weighting factor
Ω	shear parameter (Chapter 1)
Ω_i	integration point weights in interval $-1 \to 1$ in line element
$(\ )_a, (\ )_b, (\ )_c$	natural quantities referred to sides of triangular element
$(\ )_e$	element quantity
$(\ )_i$	nodal quantity, iteration number
$(\ )_0$	initial quantities
$O(\ )$	order of truncation error
$d(\ )$	infinitesimal increment
$\delta(\ )$	finite increment
δ_{ij}	Kronecker delta
$\delta(x, s)$	Dirac delta function
Σ	summation
Π	chain product
$\subset$	Boolean subset
$(\ )_{abs}$	absolute (unsigned) value
$\text{sgn}(\ )$	sign of $(\ )$
$\dim(\ )$	dimension of a linear space
$\text{Min}(\ )$	minimum value of $(\ )$
$\text{Max}(\ )$	maximum value of $(\ )$
$\|\ \|$	step function (Chapters 4, 22), modulus (Chapter 21)
$\|\ \|^F$	heavyside step function
$\int dx, \int dS, \int dV$	line, surface, and volume integrals
$I(\ )$	integral of a function
$F(\ )$	functional

$\dfrac{d(\ )}{dx}$ ordinary derivative

$\dfrac{\partial(\ )}{\partial x}$ partial derivative

$\dfrac{D(\ )}{Dt}$ total derivative

$$\nabla(\ ) = \mathrm{grad}(\ ) = \frac{\hat{\mathbf{i}}\partial(\ )}{\partial x} + \frac{\hat{\mathbf{j}}\partial(\ )}{\partial y} + \frac{\hat{\mathbf{k}}\partial(\ )}{\partial z}$$

$$\nabla\cdot(\{a, b, c\}) = \mathrm{div}(\{a, b, c\}) = \frac{\partial a}{\partial x} + \frac{\partial b}{\partial y} + \frac{\partial c}{\partial z}$$

$$\nabla\times(\{a, b.\ c\}) = \hat{\mathbf{i}}\left(\frac{\partial c}{\partial y}\cdot - \frac{\partial b}{\partial z}\right) + \hat{\mathbf{j}}\left(\frac{\partial a}{\partial z} - \frac{\partial c}{\partial x}\right) + \hat{\mathbf{k}}\left(\frac{\partial b}{\partial x} - \frac{\partial a}{\partial y}\right) \text{ (curl)}$$

Matrix notation:

Matrices are denoted by capital letters to simplify notation and this should be clear in context, scalar quantities within matrix expressions frequently being enclosed to identify them as such (for example eqn (3.32)). Where the entries in a matrix are to be indicated, however, square brackets are used.

$\{\ \}$	denotes a vector
$\mid\ \mid$	determinant of a matrix
$\langle\ \rangle$	diagonal matrix with diagonal entries enclosed
$(\ )^{-1}$	inverse of a matrix
$(\ )^{t}$	transpose of a matrix
$\{0\}$	null vector
I	unit or identity matrix
O	null matrix

Author index

Subject index